ISNM 65:
International Series of Numerical Mathematics
Internationale Schriftenreihe zur Numerischen Mathematik
Série internationale d'Analyse numérique
Vol. 65

Birkhäuser Verlag
Basel · Boston · Stuttgart

Anniversary Volume on Approximation Theory and Functional Analysis

Edited by

P. L. Butzer, Aachen
R. L. Stens, Aachen
B. Sz.-Nagy, Szeged

1984

Birkhäuser Verlag
Basel · Boston · Stuttgart

Proceedings of the Conference Held at the Mathematical Research Institute at Oberwolfach, Black Forest, July 30 – August 6, 1983

Editors

Prof. P. L. Butzer
Rhein.-Westf. Techn. Hochschule Aachen
Lehrstuhl A für Mathematik
Templergraben 55
D–5100 Aachen (FRG)

Priv.-Doz. Dr. R. L. Stens
Rhein.-Westf. Techn. Hochschule Aachen
Lehrstuhl A für Mathematik
Templergraben 55
D–5100 Aachen (FRG)

Prof. B. Sz.-Nagy
József Attila University
Aradi vértanúk tere 1
H–6720 Szeged (Hungary)

Library of Congress Cataloging in Publication Data

Anniversary volume on approximation theory and functional analysis.
(International series of numerical mathematics ; v. 65)
Lectures given at the International Conference on Approximation Theory and Functional Analysis.
1. Approximation theory -- Congresses. 2. Functional analysis -- Congresses. I. Butzer, Paul Leo, 1928– II. Stens, R. L., 1948– . III. Szokefalvi-Nagy, Béla, 1913– . IV. International Conference on Approximation Theory and Functional Analysis (1983: Oberwolfach Mathematical Research Institute)
V. Mathematisches Forschungsinstitut Oberwolfach.
VI. Series.
QA221.A538 1984 511'.4 84–18451
ISBN 3–7643–1574–1

CIP-Kurztitelaufnahme der Deutschen Bibliothek

Anniversary volume on approximation theory and functional analysis : proceedings of the conference held at the Math. Research Inst. at Oberwolfach, Black Forest, July 30 – August 6, 1983 / ed. by P. L. Butzer ... –
Basel ; Boston ; Stuttgart : Birkhäuser, 1984.
(International series of numerical mathematics ; Vol. 65)
ISBN 3–7643–1574–1

NE: Butzer, Paul L. [Hrsg.]; Mathematisches Forschungsinstitut ⟨Oberwolfach⟩; GT

Printed in Germany
ISBN 3-7643-1574-1

Lyubomir Iliev
Born on April 20, 1913 in Veliko Tarnovo

Ralph Phillips
Born on June 23, 1913 in Los Angeles

Béla Szőkefalvi-Nagy
Born on July 29, 1913 in Kolozsvár

Adriaan Cornelis Zaanen
Born on June 14, 1913 in Rotterdam

Previous Conferences on Approximation Theory, Functional Analysis and Related Topics

held since 1963 at the Oberwolfach Mathematical Research Institute and conducted from Aachen, in collaboration with a colleague from abroad since 1968

1. On Approximation Theory, August 4–10, 1963. (Eds. of the Proceedings: P.L. Butzer, J. Korevaar; ISNM, vol. 5, Birkhäuser Verlag, Basel 1964 (second edition 1972), XVI + 261 pages)

2. Arbeitstagung über Approximationstheorie, March 1–8, 1964. (Without Proceedings)

3. Harmonische Analysis und Integraltransformationen, August 2–10, 1965. (Without Proceedings)

4. Abstract Spaces and Approximation, July 18–27, 1968. (Eds.: P.L. Butzer, B. Sz.-Nagy; ISNM, vol. 10, Birkhäuser Verlag, Basel 1969, 423 pages)

5. Linear Operators and Approximation I, August 14–22, 1971. (Eds.: P.L. Butzer, J.-P. Kahane, B. Sz.-Nagy; ISNM, vol. 20, Birkhäuser Verlag, Basel 1972, 506 pages)

6. Linear Operators and Approximation II, March 30 – April 6, 1974. (Eds.: P.L. Butzer, B. Sz.-Nagy; ISNM, vol. 25, Birkhäuser Verlag, Basel 1974, XX + 588 pages)

7. Linear Spaces and Approximation, August 20–27, 1977. (Eds.: P.L. Butzer, B. Sz.-Nagy; ISNM, vol. 40, Birkhäuser Verlag, Basel 1978, 685 pages)

8. Functional Analysis and Approximation, August 9–16, 1980. (Eds.: P.L. Butzer, B. Sz.-Nagy, E. Görlich; ISNM, vol. 60, Birkhäuser Verlag, Basel 1981, 482 pages)

Anniversary Commemorations

This volume forms a record of the lectures given at the international Conference on Approximation Theory and Functional Analysis held at the Oberwolfach Mathematical Research Institute from 30. July to 6. August 1983. It commemorates at the same time several anniversaries:

(i) The 70th anniversary in 1983 of the birth of four of our many Oberwolfach participants during the past years, namely Professors L. Iliev (Sofia), R. Philipps (Stanford), B. Szőkefalvi-Nagy (Szeged) and A.C. Zaanen (Leiden). Consequently the articles of this volume are dedicated to these four distinguished mathematicians as a token of friendship and esteem.

(ii) The 20th anniversary of the Oberwolfach conference "On Approximation Theory" held August 4–10, 1963. This inaugurated a series of conferences conducted from Aachen, in collaboration with Béla Sz.-Nagy since 1968, and in 1968 also with J.-P. Kahane (Paris). It was the first to deal with approximation *theory* in the Western world.[1])

The conference in 1963 was also the first of any of the Oberwolfach Conferences (since the founding of the Institute in 1944) to be published in an accessible volume of proceedings, making the results – lectures, problems, remarks, etc. – available to the wider mathematical community. The proceedings of 1963 were edited by P.L. Butzer and J. Korevaar (Amsterdam), printed at Professor Barner's chair in Freiburg i. Br., and published by Birkhäuser, Basel.

At this point we extend our sincere thanks to Professor Martin Barner, Director of the Oberwolfach Mathematical Research Institute, for the outstanding cooperation during these twenty years. He and his team at Freiburg and Oberwolfach, with Mr. H.G. Förstendorf as administrative official, have not only helped with their expert planning and ad-

1) A conference "On Numerical Approximation" had taken place in Madison, Wisc. on April 21–23, 1958 (Proc. edited by R.E. Langer; Univ. of Wisconsin Press, Madison 1959, x + 462 pp.). There also were two conferences in the USSR with associated proceedings, namely: "Investigations in Modern Problems of the Constructive Theory of Funct." (Russ.) (Proc. First Conf. Constructive Theory Funct.; Leningrad Oct. 1959; V.I. Smirnov, ed.), Gos. Izdat. Fiz.-Mat. Lit., Moscow 1961, 368 pp, and "Studies of Contemporary Problems of Constructive Theory of Functions" (Russ.) (Proc. Second All – Union Conf. Constructive Theory Funct.; Baku Oct. 8–13, 1962; I.I. Ibragimov, ed.) Izdat. Akad. Nauk Azerbaĭdžan SSR, Baku 1965, 638 pp.

vice but have always provided the best possible facilities and surroundings to conduct such conferences. There is no doubt that their interest and concern contributed substantially towards making the conferences rewarding scientific experiences as well as memorable social events for successive international groups of participants.

P.L. Butzer
Aachen

Preface

These Proceedings include 42 of the 49 invited conference papers, three papers submitted subsequently, and a report devoted to new and unsolved problems based on two special problem sessions and as augmented by later communications from the participants. In addition, there are four short accounts that emphasize the personality of the scholars to whom the proceedings are dedicated. Due to the large number of contributors, the length of the papers had to be restricted.

This volume is again devoted to recent significant results obtained in approximation theory, harmonic analysis, functional analysis, and operator theory. The papers solicited include in addition survey articles that not only describe fundamental advances in their subfields, but many also emphasize basic interconnections between the various research areas. They tend to reflect the range of interests of the organizers and of their immediate colleagues and collaborators.

The papers have been grouped according to subject matter into ten chapters. Chapter I, on operator theory, is devoted to certain classes of operators such as contraction, hyponormal, and accretive operators, as well as to suboperators and semigroups of operators. Chapter II, on functional analysis, contains papers on function spaces, algebras, ideals, and generalized functions. Chapter III, on abstract approximation, is concerned with the comparison of approximation processes, the gliding hump method, certain interpolation spaces, and n-widths. Whereas Chapter IV deals with approximation of functions of one or two real variables by linear approximation processes or with best approximation of those functions by polynomials, also on disjoint intervals, Chapter V is mainly devoted to the approximation of functions of a complex variable, including the Shannon sampling theorem. More or less classical interpolation problems are handled in Chapter VI, and orthogonal polynomials and functions, abstract harmonic analysis and their applications are dealt with in Chapter VII. Chapter VIII includes papers on ordinary and partial differential equations, and on difference equations, in particular concerning the approximation of solutions and, finally, Chapter IX contains papers on probability theory and statistics, and a result in integral geometry. The volume closes with Chapter X on new and unsolved problems.

The number of mathematicians, in East and West, who are actively engaged in approximation theory and related topics has very much increased since 1963. This fact and our policy to include a number of mathematicians who have not attended our previous conferences, as well as the limited capacity of the Oberwolfach Institute, explain why we were unfortunately unable to reinvite all those who had participated in some of the foregoing conferences. We could not even invite all of our past and present collaborators nor students, not to mention the many other active workers elsewhere.

The editors' sincere appreciation is due to all the participants who attended the conference, many of whom travelled from other continents for the week of the conference; further, to Dieter Pontzen (now Siemens, Munich) for his competent handling of a good part of the general editorial work; to Dietmar Schulz for compiling the new and unsolved problems; to Rolf J. Nessel for valuable advice during the preparation of the conference; to the coworkers and research assistants from Aachen for their help in organizing the conference; to the secretaries of the Aachen Lehrstuhl A für Mathematik for retyping many of the papers and for their aid in preparing this volume; and last but not least to Carl Einsele of Birkhäuser Verlag for his cooperation over the past twenty years.

February 1984

P.L. Butzer	R.L. Stens	B. Sz.-Nagy
Aachen	Aachen	Szeged

Contents

VII. Orthogonal Functions and Harmonic Analysis

VIII. Differential and Difference Equations

IX. Probability Theory and Miscellaneous Topics

X. New and Unsolved Problems

Zur Tagung

Diese Tagung setzte die Reihe der alle zwei bis drei Jahre von Aachen und Szeged aus organisierten Konferenzen über Approximationstheorie, Funktionalanalysis und verwandte Gebiete fort. Die Leitung hatten diesmal P.L. Butzer, R.L. Stens (beide Aachen) und B. Szőkefalvi-Nagy (Szeged).

Es nahmen insgesamt 61 Mathematiker aus 15 Ländern (Belgien, Bulgarien, Großbritannien, Indien, Kanada, Niederlande, Österreich, Polen, Rumänien, Schweden, Spanien, Süd-Afrika, Ungarn, USA und Deutschland) teil, von denen viele zum erstenmal überhaupt in Oberwolfach waren. Wie schon bei den vorangegangenen Tagungen dieser Reihe hatten die Veranstalter besonderen Wert darauf gelegt, nicht nur bekannte Namen aus den verschiedenen Fachgebieten, sondern auch eine große Anzahl junger Wissenschaftler einzuladen. Das Oberwolfacher Institut bietet bekanntlich sehr gute Gelgenheiten zu Gesprächen und Diskussionen zwischen Jung und Alt, wie es an anderen Tagungsorten nur selten möglich ist. So war die Tagung denn auch durch eine besonders kollegiale, ja sogar herzliche Atmosphäre gekennzeichnet. Sowohl in den Vorträgen als auch in den Diskussionen und Gesprächen am Rande gab jeder sein Bestes, um diese Jubiläumstagung zu einem Erfolg werden zu lassen.

Das umfangreiche wissenschaftliche Programm (49 Vorträge) konnte nur dadurch bewältigt werden, daß man traditionsgemäß schon am Sonntag morgen begann und erst am späten Freitag abend endete. Umso erstaunlicher war es, daß das Interesse selbst bei den letzten Vorträgen noch außergewöhnlich hoch war. Es wurde ein breites Spektrum von Themen aus verschiedenen Gebieten der Approximationstheorie und der Funktionalanalysis behandelt. Das offizielle Programm wurde abgerundet durch zwei Sitzungen, in denen neue und ungelöste Probleme vorgestellt wurden. Professor F.L. Gilfeather, Lincoln, befand sich schon in Deutschland auf dem Weg nach Oberwolfach, als er seine Teilnahme absagen mußte, um ein Amt bei der National Science Foundation in Washington zu übernehmen. Sein geplanter Vortrag wurde ins Tagungsbuch aufgenommen.

Neben den mathematischen Vorträgen wurde in einer Feierstunde über das Leben der Professoren L. Iliev, R. Phillips, B. Szőkefalvi-Nagy und A.C. Zaanen berichtet, die an mehreren der bisherigen Tagungen dieser Reihe teilgenommen haben und in diesem Jahr ihr 70. Lebensjahr vollendeten. Wegen einer gerade überstandenen schweren Operation konnte Professor Iliev leider nicht persönlich teilnehmen.

Zum Schluß ein Wort des Dankes an das gastgebende Institut. Die hervorragende Betreuung durch das gesamte Personal trug wesentlich zum Gelingen der Tagung bei.

Tagungsleiter: P.L. Butzer R.L. Stens B. Sz.-Nagy

List of Participants

C. Bennett, Dept. of Mathematics and Statistics, University of South Carolina, Columbia, SC 29208, USA

H. Berens, Mathematisches Institut, Universität Erlangen-Nürnberg, Bismarckstraße 1 1/2, D-8520 Erlangen, Fed. Rep. Germany

P.L. Butzer, Lehrstuhl A für Mathematik, Rheinisch-Westfälische Technische Hochschule Aachen, Templergraben 55, D-5100 Aachen, Fed. Rep. Germany

W.C. Connett, Dept. of Mathematical Sciences, University of Missouri, 8001 Natural Bridge Road, St. Louis, MO 63121, USA

R.A. DeVore, Dept. of Mathematics and Statistics, University of South Carolina, Columbia, SC 29208, USA

W. Dickmeis, Lehrstuhl A für Mathematik, Rheinisch-Westfälische Technische Hochschule Aachen, Templergraben 55, D-5100 Aachen, Fed. Rep. Germany

W. Engels, Lehrstuhl A für Mathematik, Rheinisch-Westfälische Technische Hochschule Aachen, Templergraben 55, D-5100 Aachen, Fed. Rep. Germany

D. Gaşpar, Dept. of Mathematics, University of Timişoara, Bul. V. Pârvan 4, 1900 Timişoara, Roumania

M. v. Golitschek, Institut für Angewandte Mathematik und Statistik, Universität Würzburg, Am Hubland, D-8700 Würzburg, Fed. Rep. Germany

J.J. Grobler, Potchefstroom University for CHE, Potchefstroom 2520, South Africa

K. Gustafson, Dept. of Mathematics, University of Colorado, Boulder, CO 80309, USA

M. de Guzmán, Mathemáticas, Universidad Autonoma de Madrid, Madrid 34, Spain

P.R. Halmos, Dept. of Mathematics, Indiana University, Swain Hall East, Bloomington, IN 47405, USA

W.K. Hayman, Dept. of Mathematics, Imperial College of Science and Technology, Huxley Building, Queen's Gate, London SW7 2BZ, Great Britain

E. Hewitt, Dept. of Mathematics GN-50, University of Washington, Seattle, WA 98195, USA

C.B. Huijsmans, Subfaculteit der Wiskunde en Informatica, Rijksuniversiteit te Leiden, Wassenaarseweg 80, Postbus 9512, 2300 RA Leiden, The Netherlands

K.G. Ivanov, Institute of Mathematics with Computer Center, Bulgarian Academy of Sciences, P.O. Box 373, Sofia 1090, Bulgaria
J.W. Jerome, Dept. of Mathematics, Northwestern University, Lunt Hall, Evanston, IL 60201, USA
L. Kérchy, Bolyai Institute, József Attila University, Aradi vértanúk tere 1, H-6720 Szeged, Hungary
T.H. Koornwinder, Centrum voor Wiskunde en Informatica, Postbus 4079, 1009 AB Amsterdam, The Netherlands
J. Korevaar, Mathematisch Instituut, Universiteit van Amsterdam, Roetersstraat 15, 1018 WB Amsterdam, The Netherlands
L. Leindler, Bolyai Institute, József Attila University, Aradi vértanúk tere 1, H-6720 Szeged, Hungary
G. Lumer, Faculté des Sciences, Université de l'Etat, Avenue Maistriau 15, B-7000 Mons, Belgium
C. Markett, Lehrstuhl A für Mathematik, Rheinisch-Westfälische Technische Hochschule Aachen, Templergraben 55, D-5100 Aachen, Fed. Rep. Germany
P. Masani, Dept. of Mathematics and Statistics, University of Pittsburgh, Pittsburgh, PA 15260, USA
M.W. Müller, Lehrstuhl für Mathematik VIII, Universität Dortmund, Postfach 500500, D-4600 Dortmund 50, Fed. Rep. Germany
D.H. Mugler, Dept. of Mathematics, University of Santa Clara, Santa Clara, CA 95053, USA
J. Musielak, Institute of Mathematics, A. Mickiewicz University, Matejki 48/49, 60-830 Poznań, Poland
R.J. Nessel, Lehrstuhl A für Mathematik, Rheinisch-Westfälische Technische Hochschule Aachen, Templergraben 55, D-5100 Aachen, Fed. Rep. Germany
P. Nevai, Dept. of Mathematics, Ohio State University, Columbus, OH 43210, USA
J. Peetre, Dept. of Mathematics, Lund Institute of Technology, Box 725, S-22007 Lund, Sweden
F. Peherstorfer, Institut für Mathematik, Johannes Kepler Universität, A-4040 Linz, Austria
R. Phillips, Dept. of Mathematics, Stanford University, Stanford, CA 94305, USA
D. Pontzen, Bahnhofsplatz 3, D-8011 Baldham, Fed. Rep. Germany

C.R. Putnam, Dept. of Mathematics, Purdue University, West Lafayette, IN 47907, USA

P. Révész, Mathematical Institute, Hungarian Academy of Sciences, Reáltanoda u. 13–15, H-1053 Budapest V, Hungary

S.D. Riemenschneider, Dept. of Mathematics, University of Alberta, Edmonton, Alberta T6G 2G1, Canada

S. Ries, Lehrstuhl A für Mathematik, Rheinisch-Westfälische Technische Hochschule Aachen, Templergraben 55, D-5100 Aachen, Fed. Rep. Germany

J. Rovnyak, Dept. of Mathematics, University of Virginia, Mathematics-Astronomy Building, Charlottesville, VA 22903-3199, USA

P.O. Runck, Institut für Mathematik, Johannes Kepler Universität, A-4040 Linz, Austria

D.C. Russell, Dept. of Mathematics, York University, 4700 Keele Street, Downsview, Ontario M3J 1P3, Canada

R.B. Saxena, Dept. of Mathematics and Astronomy, Lucknow University, C-268 Niralanagar, Lucknow 226007, India

W. Schempp, Lehrstuhl für Mathematik I, Universität Siegen, Hölderlinstraße 3, D-5900 Siegen, Fed. Rep. Germany

F. Schipp, Dept. of Numerical Analysis and Computer Sciences, Eötvös Loránt University, Múzeum körút 6–8, H-1088 Budapest VIII, Hungary

G. Schmeisser, Mathematisches Institut, Universität Erlangen-Nürnberg, Bismarckstraße 1 1/2, D-8520 Erlangen, Fed. Rep. Germany

D. Schulz, Lehrstuhl A für Mathematik, Rheinisch-Westfälische Technische Hochschule Aachen, Templergraben 55, D-5100 Aachen, Fed. Rep. Germany

Bl. Sendov, Institute of Mathematics with Computer Center, Bulgarian Academy of Sciences, P.O. Box 373, Sofia 1090, Bulgaria

H.S. Shapiro, Dept. of Mathematics, Royal Institute of Technology, S-10044 Stockholm 70, Sweden

R.C. Sharpley, Dept. of Mathematics and Statistics, University of South Carolina, Columbia, SC 29208, USA

W. Splettstößer, Lehrstuhl A für Mathematik, Rheinisch-Westfälische Technische Hochschule Aachen, Templergraben 55, D-5100 Aachen, Fed. Rep. Germany

E.L. Stark, Lehrstuhl A für Mathematik, Rheinisch-Westfälische Technische Hochschule Aachen, Templergraben 55, D-5100 Aachen, Fed. Rep. Germany

R.L. Stens, Lehrstuhl A für Mathematik, Rheinisch-Westfälische Technische Hochschule Aachen, Templergraben 55, D-5100 Aachen, Fed. Rep. Germany

J. Szabados, Mathematical Institute, Hungarian Academy of Sciences, Reáltanoda u. 13–15, H-1053 Budapest V, Hungary

B. Sz.-Nagy, Bolyai Institute, József Attila University, Aradi vértanúk tere 1, H-6720 Szeged, Hungary

K. Tandori, Bolyai Institute, József Attila University, Aradi vértanúk tere 1, H-6720 Szeged, Hungary

V. Totik, Bolyai Institute, József Attila University, Aradi vértanúk tere 1, H-6720 Szeged, Hungary

R.S. Varga, Institute for Computational Mathematics, Kent State University, Kent, OH 44242, USA

P. Vértesi, Mathematical Institute, Hungarian Academy of Sciences, Reáltanoda u. 13–15, H-1053 Budapest V, Hungary

E. van Wickeren, Lehrstuhl A für Mathematik, Rheinisch-Westfälische Technische Hochschule Aachen, Templergraben 55, D-5100 Aachen, Fed. Rep. Germany

A.C. Zaanen, Subfaculteit der Wiskunde en Informatica, Rijksuniversiteit te Leiden, Wassenaarseweg 80, Postbus 9512, 2300 RA Leiden, The Netherlands

L. Zsidó, Mathematisches Institut A, Universität Stuttgart, Pfaffenwaldring 57, D-7000 Stuttgart 80, Fed. Rep. Germany

Program of the Sessions

Sunday, July 31
10.00 Opening
Morning session. Chairman: B. Sz.-Nagy
10.15 P.R. Halmos: Subdiscrete convergence and infranormal matrices
11.15 M. de Guzmán: New methods for maximal convolution operators
First afternoon session. Chairman: E. Hewitt
15.00 J. Peetre: Invariant Banach spaces connected with the holomorphic discrete series
15.50 L. Leindler: Strong approximation
16.30 J. Musielak: Some embedding theorems for modular classes
Second afternoon session. Chairman: R. Phillips
17.10 C.R. Putnam: Positive commuting perturbations of selfadjoint operators and hyponormality
17.50 R.L. Stens: The cardinal interpolation series
Evening session. Chairman: W.C. Connett
19.45 P. Nevai: Orthogonal polynomials and their derivatives

Monday, August 1
First morning session. Chairman: H.S. Shapiro
8.45 W.K. Hayman: The best harmonic approximation to a continuous function in a closed Jordan domain
9.40 G. Schmeisser: Reconstruction of entire harmonic functions from given values
10.20 K. Gustafson: Graph theory in the approximation theory of fluid dynamics
Second morning session. Chairman: J. Peetre
11.10 C. Bennett: Bounded mean oscillation and bounded lower oscillation
11.50 R.C. Sharpley: The K-functional for H^1 and H^∞
First afternoon session. Chairman: K. Gustafson
15.45 L. Kérchy: Subspace lattices connected with C_{11}-contractions
16.25 L. Zsidó: On generation of one-parameter operator groups
Second afternoon session. Chairman: P. Masani
17.10 K. Tandori: Über die Mittel von orthogonalen Funktionen
17.50 F. Pehersdorfer: Extremalpolynome in der L^1- und L^2-Norm auf zwei disjunkten Intervallen
Evening session. Chairman: Bl. Sendov
19.45 J. Rovnyak: Some extremal problems with constraints.

Tuesday, August 2

First morning session. Chairman: G. Lumer

8.45 J.W. Jerome: Fixed point and implicit function theorems and their applications

9.40 D.H. Mugler: Green's functions for the finite difference heat, Laplace, and wave equations

Second morning session. Chairman: J. Korevaar

10.30 H. Berens: On maximal extensions of accretive operators in the plane

11.10 W. Dickmeis: On condensation of singularities on a set of full measure

11.50 S.D. Riemenschneider: n-widths of some smoothness spaces

First afternoon session. Chairman: P.O. Runck

15.15 Bl. Sendov: Averaged moduli of smoothness

16.05 J. Szabados: Polynomial approximation on disjoint intervals

Second afternoon session. Chairman: R.A. DeVore

16.55 V. Totik: The necessity of a new kind of modulus of smoothness

17.50 K.G. Ivanov: Approximation of functions of two variables by algebraic polynomials

Evening session. Chairman: P.L. Butzer

19.30 Birthday lectures by R. Phillips and G. Lumer, P.R. Halmos, J. Korevaar, Bl. Sendov. Party

Wednesday, August 3

First morning session. Chairman: H. Berens

8.45 M. v. Golitschek: Shortest path algorithms for the approximation by nomographic functions

9.40 C. Markett: Product formulas for Bessel, Whittaker, and Jacobi functions via the solution of an associated Cauchy problem

Second morning session. Chairman: M. de Guzmán

10.30 W.C. Connett: Harmonic analysis for eigenfunction expansions: The existence of a positive convolution and some of its consequences

11.10 J. Korevaar: $\sqrt{\delta}$

11.20 T.H. Koornwinder: On a new class of generalized functions introduced by J.J. Lodder

Thursday, August 4

First morning session. Chairman: A.C. Zaanen

8.45 R. Phillips: The spectrum of the Laplacian for domains in the hyperbolic space

9.35 P. Masani: Fourier integration with respect to a vector measure

Second morning session. Chairman: W.K. Hayman

10.30 R.A. DeVore: Differentiation in $\mathbb{R}^n$

11.10 R.B. Saxena: Uniform convergence of some poised problems of Hermite-Birkhoff interpolation

11.50 D.C. Russell: Spline interpolation of power-dominated data

First afternoon session. Chairman: R.S. Varga

15.30 G. Lumer: An exponential representation of Hille-Yosida type for evolution operators

16.20 C.B. Huijsmans: Ideals in C(X)

Second afternoon session. Chairman: L. Leindler

17.00 P. Vértesi: Recent results on the divergence of Lagrange interpolation

17.40 J.J. Grobler: Spectral properties of positive operators

Evening session. Chairman: H.S. Shapiro

19.45 First problem session

Friday, August 5

First morning session. Chairman: R.J. Nessel

8.45 E. Hewitt: A class of positive trigonometric sums

9.40 R.S. Varga: On the Bernstein conjecture for $|x|$

Second morning session. Chairman: J. Rovnyak

10.35 W. Schempp: Functional analytic aspects of radar detection

11.20 Second problem session

First afternoon session. Chairman: J. Musielak

15.30 R.J. Nessel: Some negative results in connection with Marchaud-type inequalities

16.25 P. Révész: Estimation of the regression function via orthogonal expansion

Second afternoon session. Chairman: J. Szabados

17.10 F. Schipp: Martingales, bases, Hardy spaces and a.e. convergence

17.50 D. Gaşpar: Über die Wold-Zerlegung isometrischer Halbgruppen

Evening session. Chairman: J.W. Jerome

19.30 H.S. Shapiro: Exact quadrature formulas for analytic and harmonic functions

Closing address

International Series of
Numerical Mathematics, Vol. 65

TO ACADEMICIAN L. ILIEV ON THE OCCASION OF HIS 70th ANNIVERSARY

Bl. Sendov
Center of Mathematics and Mechanics
Bulgarian Academy of Sciences
Sofia, Bulgaria

Academician Lyubomir Iliev enters his seventies with unfailing creative spirit, with an enormous experience as an organizer of science and education of national and worldwide importance. His road as a scientist and still more as an organizer of science is most closely related to the remarkable advance of Bulgarian mathematics and the birth and development of computer technology in our country during the last three decades. The happy fusion of mathematical talent and organizational and political insight allowed Acad. Iliev to establish himself as the most authoritative and most popular leader of the Bulgarian mathematical community.

Academician Lyubomir Iliev was born on April 20th, 1913 in the town of Veliko Tarnovo. He has never broken the ties with his native town even though he left barely 19 years old after having accomplished his secondary education. The vivid interest of Acad. Iliev towards the history of Bulgaria, his native town having once been its capital, is one of the components of his continued contact. And what is more - he is a passionate investigator of the ancient history and the peoples which have inhabited our land. Few people know that he is a zealous numismatist and has devoted his studies to coins minted on the Balkan peninsula. This historic and archaelogical hobby has also contributed to the ties with his native town Veliko Tarnovo.

Acad. Iliev received his university degree in mathematics at the Faculty of Physics and Mathematics in the University of Sofia "Kliment Ohridsky" in 1936. After that he specialized for two years with Acad. N. Obreshkov, heading at that time the Chair of Algebra. In 1938 he obtained a doctor's degree. Acad. Iliev spent the academic year of 1940 in Germany under the tuition of the eminent mathematicians Prof. O. Perron and Prof. C. Caratheodory at the University of Munich.

His mathematical creative work is based on the solid qualification of a professional mathematician in a classical domain. For details see [1]. Speaking in general, he is a specialist in the field of complex analysis. He has obtained most of his results in the field of analytic inextensibility and overconvergence of power and other series which have found place in his first monograph. The problem for distribution of the zeros of polynomials and entire functions having a definite integral representation was also very fruitful for him. This problem had focussed the attention of the leading Bulgarian mathematicians of that time who were Prof. Iliev's teachers. His results in the field of univalent functions are widely known.

In the last few years Acad. Iliev succeeded in getting interesting results concerning spline functions. In this research he uses a logic for their generation that is implied by the numerous previous investigations he achieved for various systems of polynomials and entire functions.

The mathematical creative work of Acad. Iliev is marked by its own specific style. His contributions are remarkable for the elegance of the results and conciseness of the proofs. In many cases the proof is a clever trick, a combination of various well-known propositions which, taken together, yield numerous new results. Many of his works are not lengthy but they are rich in results and new ideas. Due to this they have been often cited and have drawn the attention of many Bulgarian and foreign mathematicians. These studies were the inspiration of a long series of further investigations.

From 1938 till 1941 Acad. Iliev was a high school teacher in Sofia. He often recalls these years on various occasions. The young mathematician possessing a doctor's degree in mathematics had to sit for an additional exam in order to occupy a teacher's position in Sofia. Probably these three years of teaching of his priori inclination to education were the reason for the continued attention of Acad. Iliev to the problems of secondary school and mathematical education in general. He is the author of a number of high school textbooks in mathematics and books of solved problems. He is one of the pioneers and ardent advocates of the system for training teachers in mathematics that would have a profound knowledge of elementary mathematics at the University of Sofia. He is also one of the authors of the first textbooks towards this aim. The Chair of Educational Methods in Teaching Mathematics was founded at the Mathematical Faculty of the University at his urgent request and under his auspices. The considerable experience of Acad. Iliev on the problems of secondary education in mathematics is still brought to a wide audience thanks to his papers and reports on this subject.

In 1941 Acad. Iliev was nominated an Assistant Professor at the University of Sofia "Kliment Ohridsky", and he has never lost his ties with this university ever since. At that time there were two or three assistant professors in mathematics. The young assistant professor Iliev had seminars with students in calculus, analytic geometry, algebra and others.
These were years loaded with intensive teaching and very active mathematical performance. These were years of radical changes in university education and organization.

In 1947 Acad. Iliev was nominated an Associate Professor at the Chair of Analysis, and in 1952 he became a regular Professor and Head of this Chair. His great contribution to the development and advance of university education in our country is closely related to his leading role at this Chair, later named Chair of Complex Analysis, which he has been heading for more than 30 years now, as well as to his activity as a Vice-Dean of the Faculty of Mathematics and Physics (1950-1951) and Vice-Rektor of the University of Sofia (1951-1960).

The Chair of Analysis will remain in the history of the University of Sofia "Kl. Ohridsky" because it took the leading role in organizing the specializations in computational mathematics, probability theory and mathematical statistics, as well as many others. This Chair was the first in the University to be connected with the first computer centre in Bulgaria, set up in April 1961.

We owe much credit to Acad. Iliev for his participation in setting up the electronic computer technology in our country. These activities are related to his position as a Deputy Director of the Mathematical Institute with Computer Centre at the Bulgarian Academy of Sciences (1961-1963) and further as a Director of this Institute from 1963 on. His position as a Scientific Secretary General of the Bulgarian Academy of Sciences (1961-1968) as well as his mandate as Vice-President of the Academy contributed to his efforts in this respect.

Acad. Iliev engaged every bit of his organizational talent in setting up the first computer centre in the country and in the construction of the first Bulgarian computer. It was necessary to train new specialists in new fields and at the same time apply their knowledge in creating the novel and unknown technology. Only the people who were most actively involved in this really epical period in the history of Bulgarian computational mathematics can appreciate what Acad. Iliev meant to this exploit. Acad. Iliev employed all resources of international cc-operation in upholding this aim. His activity as a Bulgarian representative in IFIP (International Federation of Information Processing) was entirely put in service of this aim, as well as his participation as a member of the IFIP Council and Vice-President of this most authoritative international Federation on the problems,of computerization.

In his other activities as a member of the Board of the Bulgarian Academy of Sciences Acad. Iliev is at the service of international co-operation. In the field of mathematics he has contributed a lot as a member of the Board of the International Mathematical Centre "S. Banach", the Balkan Mathematical Union, as well as IIASA.

Acad. Iliev meets his happy 70th anniversary as a most respected teacher of many generations of Bulgarian mathematicians and as an honoured architect of the Bulgarian mathematical school.

[1] Mathematical Structures, Computational Mathematics, Mathematical Modelling. Publ. House of the Bulgarian Academy of Sciences, Sofia 1975, pp. 17-27.

International Series of
Numerical Mathematics, Vol. 65

RECOLLECTIONS

Ralph Phillips
Department of Mathematics
Stanford University
Stanford, California

I am somewhat amazed by the fact that I am now seventy years old and am expected to take my place among the senior citizens of the mathematical community. The fact is that I do not feel like an elder citizen and am still somewhat shy in the company of my betters. Nevertheless I have been a mathematician for over forty years and some recollections of my early experiences may therefore be of interest.

My studies at Los Angeles High, in the city of Los Angeles where I was born on June 23, 1913, were not remarkable. I took as much mathematics as I could, not because I was especially interested in mathematics but rather because it was easy for me. I was in fact more interested in football than mathematics, which at that time I considered to be an unmanly occupation. I entered U.C.L.A. in 1931. U.C.L.A. had undergone a metamorphosis from teachers college to university only two years before and most of the faculty were holdovers from that earlier state. This made little difference to me since I had no clear idea of where I was going. In fact I changed my major from business to engineering and finally to a double major in mathematics and physics.

After staying on an extra year at U.C.L.A., I enrolled in the graduate physics program at the University of Michigan in 1936, attaching myself to Samuel Goudsmit, who tried his best to make a physicist out of me. He soon realized that this was hopeless and arranged for me to get a fellowship in the mathematics department. Thus it was that I did not pursue a mathematical career in earnest until the third year of my graduate studies.

The next two years were very enjoyable. I found myself at last in a comfortable environment with new vistas opened up to me in courses by T.H. Hildebrandt, R.L. Wilder and Y. Rainich.I was especially attracted to the youthful field of functional analysis and wrote my thesis under Hildebrandt on vector valued integrals, then the subject of recent papers by S. Bochner, N. Dunford, B.J. Pettis and I. Gelfand. I tried to put all of these integrals into a single setting dependent only on the topology imposed on the space.

After getting my Ph.D. along with a postdoctoral fellowship from the University of Michigan in 1939, I was able to spend the academic year 1939-1940 at the Institute for Advanced Study in Princeton. At that time the senior professors at the Institute gave lecture courses in Fine Hall at Princeton University and I attended those given by Herman Weyl and John von Neumann. Also at the Institute that year was a very fine crop of young mathematicians: W. Ambrose, H. Dowker, P. Erdös, P. Halmos, W. Strodt and H. Wallman, all of whom have made their mark on mathematics. At the time we were content to organize a seminar on almost periodic functions. For me the high point of the year was my collaboration with S. Bochner, who was very kind to me both professionally and socially. The one embarassment for me while I was at the Institute was learning from Garrett Birkhoff that he had worked out the main idea of my thesis in a paper published back in 1935.

Late in the spring I visited J.D. Tamarkin at Brown University. He was a wonderful warm person and an excellent mathematician, one of the few classical mathematicians who then appreciated the potential of functional analysis. Paranthetically let me say that I believe that potential is best realized when functional analysis is used in combination with other branches of mathematics, rather than as a study in itself. Tamarkin suggested that I look at a problem of Marcinkiewicz, which I did during the following summer. As a result he recommended me to Gabor Szegö for a position at Stanford University later that summer. Tamarkin told me at the time and Szegö later confirmed that I was not given an appointment at Stanford because I was Jewish. Although Hildebrandt had warned me, when I was a student, that there was an unofficial quota imposed on the number of Jews in American universities, this was my first encounter with antisemitism in the mathematical community.

The following year I was a visiting instructor at the University of Washington, replacing A. Taub who was spending the year at the Institute. As I recall, we were required to teach twelve hours per week. At the time I was young and foolish enough to teach an additional graduate course in topology at the request of some of the students. It is not by coincidence that one of the students in the class is now my wife. I was one of three instructors hired that year at the University of Washington. The other two were retained, but it was clear at the time that I was by no means the poorest of the lot. Again I attribute this to anti-semitism.

The following summer, while I was visiting at the University of Michigan, I received a telegram from Ted Martin asking me if I would accept an instructorship at M.I.T. In those days such a telegram was tantamount to an offer and I telegraphed back an affirmative reply. However the offer was not forthcoming. An explanation was given to me a few months later by Ted Martin while I was an instructor at Harvard. On the basis of Bochner's recommendation, M.I.T. had sent me the telegram. In the mean time a letter from Hildebrandt had arrived indicating that I was Jewish. This put the chairman of the mathematics department at M.I.T., H.B. Phillips, in an "impossible" position. To round out the picture let me note that Hildebrandt was genuinely fond of me and was merely doing what was expected of him at the time.

During the 1941-42 academic year while I was at Harvard, Leon Alaoglu, Ted Martin, Charles Rickart, Marshall Stone and I had a seminar on functional analysis. The Gelfand paper on normed rings was not well known in the U.S.A. at the time and I volunteered to talk on it. After the first presentation, Garrett Birkhoff was told of the elegance of the theory and he asked me to give a repeat performance for him and after that he asked for a third presentation for his father.

Shortly after the United States entered the war in December of 1941, I joined the Radiation Laboratory at M.I.T. and began working on radar anti-aircraft devices in the division headed by I.A. Getting. After a few months I was appointed group leader of the theoretical section. We were mathematical advisors for the division. We helped with the design of the servomechanism controls for radar tracking and the design of a companion anti-aircraft gun director. The group had some very talented people in it among whom were Hugh Dowker, Leonard Eisenbud (physicist), Paul Samuelson (economist), Abe Seidenberg, Tony Svoboda, Clifford Truesdell and Witold Hurewicz. Many of the ideas developed in the course of our work are to be found in the book on servomechanisms published in the Rad. Lab. Series, vol. 25.

By 1946 anti-semitism in academia had essentially disappeared and for the first time I was offered a position with the possibility of tenure at New York University. There I met Peter and Anneli Lax, Bernie Friedman, Louis Nirenberg and Joe Keller. As I recall we all attended a class on partial differential equations given by Max Shiffman. Even then N.Y.U., under Courant's direction and with mathematicians such as K. Friedrichs and Fritz John, was first class in analysis. It was here that I obtained my initial exposure to rigorous applied mathematics; unfortunately it took many such exposures to really take.

Primarily because I did not like living in New York City, I accepted a position at the University of Southern California in Los Angeles for the following year, even though it was a comedown both professionally and financially.

It took me several years to get back into the swing of things mathematically after my four year term at the Rad. Lab. I began by studying Einar Hille's monograph on Functional Analysis and Semigroup Theory. I pestered Hille with a continuous flow of letters which he answered with great care and encouragement. Later on I sent him preprints of several papers on semigroups of operators and in 1952 he asked me to collaborate with him on the second edition of his book. It took us three years to finish the job. Although the new edition is one-and-a half times the size of the original and contains several new developments of mine, it is in spirit very much like the original edition.

At about this same time I helped to organize a joint Cal Tech, U.C.L.A., U.S.C. bimonthly seminar in analysis. After a leisurely dinner, the group would meet in my study, a former artist's studio behind my house, which was about equidistant from the three schools. The sessions were well attended and quite lively, usually lasting two or three hours. Among the participants were R. Arens, R. Blattner, H.F. Bohnenblust, E. Coddington, H.O. Cordes, P. Curtis, C. Deprima, H. Dye, A. Erdeli, R. Finn, S. Karlin, W. Luxemburg, J. Mc Gregor, I. Singer, F. Spitzer and C. Wilcox.

My collaboration with Peter Lax dates back to 1958 when I visited the Courant Institute for a semester. We have worked very well together - so well in fact that it is often impossible to say who first suggested the seminal ideas and who did the refining. I have also enjoyed collaborating with Henry Dye, Gunther Lumer, Leonard Sarason, Michael Crandell,and more recently with Peter Sarnak. Aside from the fun of working with someone else, it seems to me that the end product of our collaboration is much better than either of us would have produced by himself.

To bring this chronicle up to date, let me note that I taught at U.C.L.A. from 1958 to 1960 and have been at Stanford University ever since. As my intent here is to recall my early experiences, I shall say no more about these years although they constitute the major portion of my career.

Finally let me list my Ph.D. students all of whom have taught me a lot and helped to sustain my spirit. They are, in chronological order: A.V. Balkrishnan, H. Potter Kerfoot, Allen Sims, John D. Brooks, Dale Thoe, Norman Shenk, Georg E. Schmidt, Gerd Grubb, Stanly Steinberg, Michael Reed, James Ralston, J. Thomas Beale, Andrew Majda, Chong-Kiu Chan, Daniel Bondy, John N. Palmer, Peter Trudinger, Alex Woo and Bettina Wiskott.

In closing let me say that the satisfaction of a mathematician comes for the most part from within, since most of us write for a very small audience. In this and other respects it is a lonely profession. In spite of this I have found the life of a mathematician very fulfilling and warmly recommend it to the younger generation.

International Series of
Numerical Mathematics, Vol. 65

BÉLA SZŐKEFALVI-NAGY

P.R. Halmos
Department of Mathematics
Indiana University
Bloomington, Indiana 47405

When Paul Butzer asked me to take advantage of this meeting and tell you something about the celebrant, our friend Béla Szőkefalvi-Nagy, I was pleased and at the same time embarrassed. I have known and admired Nagy for almost a quarter of a century, but I didn't know as much about his life as I should have liked to know. All I could think of at first was a report consisting of two sentences — "He is a great mathematician. He is a nice guy." — but surely those sentences are so well known that they need not be said again. To solve the problem, I did some historical research (that helped some), and I made use of some friendly spies (that helped a lot). My most effective spies were Ferenc Márta, a chemist colleague and a long-time friend of Nagy, and Ciprian Foiaş. Let me tell you what I learned.

The first question every functional analyst who arrives in Hungary asks is "Where is Szeged?". Szeged seems to be the Göttingen of Hungary, and the person who was chiefly instrumental in making it that was Nagy. Even so it wasn't always clear that Nagy would become one of _us_. His first two papers were published in the Zeitschrift für Physik — it looked as if he might become one of _them_. But all is well that ends well, and Nagy became the mathematician we know, a functional analyst and, in particular, an operator theorist par excellence.

After his university years he went on to continue his studies in Leipzig (where he was in touch with van der Waerden, E. Hopf, Koebe, and Heisenberg) and in Paris (where his contacts were Favard, Hadamard, and Denjoy). His professional career began at the Teacher Training School in Szeged, where he was the only mathematics professor, and, with the help of only one assistant, he taught everything: analysis, algebra, and even descriptive geometry.

His first bona fide mathematical paper appeared in the prestigious Mathematische Annalen (in 1936), and that was the beginning of an impressive deluge: by now he has published over 150 articles and some of the most important and highly respected books in operator theory. The first book was first written in French, for the Actualité series, but a war came along and interfered; it had to be re-written in German, and it became famous as the Ergebnisse report "Spektraldarstellung linearer Transformationen des Hilbertschen Raumes". It appeared in 1942. The other major books should at least be mentioned. Riesz-Nagy (Functional Analysis) appeared in 1952 in French, and then kept reappearing in other languages; the text on real functions and orthogonal expansions appeared in 1954 (first in Hungarian and ten years later in English); and Nagy-Foiaş (Harmonic analysis of operators on Hilbert space) appeared in 1967 (first in French and later in English and in Russian). Foiaş told me, by the way, thay when he, Foiaş, proposed to Nagy that they write their joint book, the first answer was no. In 1953 Nagy wrote the first of what turned out to be an infinite sequence of papers (Sur les contractions de l'espace de Hilbert); II and III appeared in 1957 and 1958. The others were written in collaboration with Foiaş (the last one, so far, was XII, in 1966), and Foiaş's proposal was made sometime early in the 1960's. No, said Nagy then, we are not ready for that. Then VII appeared in 1964, and, as it turned out, there was a slip in it; a non-linear operator was treated as if it were linear, and repair was needed. Then Nagy said yes — all right, let's write the book, and the rest is history.

He travelled widely, and, for instance, paid extended visits to Columbia University and to Indiana University. He was invited to address not one but three different International Congresses, but he wasn't able to come to the first of them. The 1958 Congress in Edinburgh was too near in time to the troubles of 1956, and, as a result, Nagy was not allowed to leave Hungary. It was my pleasure and honor temporarily to be Nagy — he sent me his manuscript, and I presented it. By the time Stockholm (1962) and Nice (1970) came along, Nagy himself was Nagy.

How and when did it all start? You could say that it started when Nagy was born on the 29th of July 1913, in Kolozsvár (now Cluj), but that wouldn't give full credit to his enviable family background. Not only was his mother a teacher (of natural science), but his father, Gyula Szőkefalvi-Nagy, was a professor of mathematics and a contributing member of the profession, first in Kolozsvár and later in Szeged. Some of the credit belongs to his wife Jolán, a history teacher and a talented singer whom he married in 1941, and with whom he produced a number of children, a perfect number in the technical sense of the word. They have, to be more precise, six children, born in arithmetic progression, with initial term 1942 and constant difference equal to two years. They are all intellectual professionals (a musician, three physicists, a librarian, and an economist), and between them they have presented our friend with (the last time I counted) a total of ten grandchildren.

What Riesz was to him, he in turn was to several young people. You will perhaps not be surprised to learn that he has had a perfect number of students. He has always kept his eyes open looking for young students to encourage and to help, looking not only in Szeged but in all of Hungary.

The young people under his wing view him with respect, you could almost say awe. They are scared because they know he is severe, but they know that he is fair. He makes use of their attitude to spur them on. He values accomplishment; he demands perfection, or at least a good try for it. His judgments as an examiner do not depend on mood or other personal variables. To pass, you must know the subject; bluffing will do you no good. He treats his juniors (even people like Durszt and Foiaş) the way a father would, or an uncle. When Korányi and Pukánszky left, he felt as if he had lost two sons.

Some of his research was done in collaboration; the number of his collaborators is as perfect as the number of his children. He thinks it is important, in collaboration, that each member of the team try to do more than the other. Collaboration with him is a strenuous and highly rewarding game of ping-pong; you tell him something, and you can be sure that he will bounce the ball back with something new added in increased elegance or greater depth.

He is conscientious about every aspect of the profession. Example: he himself typed the original version of the Riesz-Nagy book — both parts. Example: when a new concept arises, he gives a lot of serious thought to naming it properly. Is "quasi-similar" really the right phrase? — he will think about that for days. About substance, as opposed to language, he thinks even longer and harder — he always wants to be sure that he understands analysis from what might be called the pictorial point of view; he sees operator theory as a part of the synthetic (Euclidean) geometry of Hilbert space.

There was a time, believe it or not, when Nagy didn't know about Hilbert space. When he was a high school student, he was sufficiently talented and interested in languages that it looked possible that he would become a linguist. He is still very fond of languages and knows many of them. He knows Rumanian, for instance, but he is modest about it. When he and Foiaş worked together they spoke French — that is till about 1970. Then, after they had both spent some time in Bloomington, their most frequent working language became English.

He can and does work, in the sense of real, honest mathematical work, almost everywhere — on an airplane, or at home, or on his strenuous daily walk, or while swimming in the Tisza (where he prefers the rough, dangerous part to the easy public beaches) — everywhere, that is, except at his office in the university. There he does the administrative and bureaucratic and correspondence chores that we all need to do, and he does them in an organized, no nonsense manner. He is efficient (do you realize how unusual that is? — in the Hungarian language there isn't even a word for the concept!).

He is reliable and punctual. He is on time for his lectures and for his appointments. When you agree to meet him at 1 o'clock, you had better be there at 1 o'clock — he will be. He doesn't like carelessness, sloppiness, or, in language, confusion and ambiguity. He appreciates good work, craftsmanship, and he compliments it wherever it occurs.

I would love to see his reprint collection at home — so far I have only heard about it. He files reprints systematically, of course, his own in one file, and those of others alphabetically in another, except for the ones of special current interest to him, which go into a third. He is not an indiscriminate reprint collector, however. His collection is highly selective — it contains only the ones he considers best. I would like to know how many and which reprints of... never mind, maybe I don't really want to know.

His main hobbies are music, and long walks in natural, fresh-air surroundings, where he can hear the songs of birds. He was active and successful in a project of improving the environment of Szeged and establishing there a part of the national park system, a bird refuge.

He is not a social butterfly — he likes people, he likes company, but he has arranged his life so that he spends most of it with his work and his family. He especially values the help of Jolán, his wife — she protects him from others and helps him to live in a way that is both pleasant and productive.

There is more to a life, especially to a life like that of Béla Szőkefalvi-Nagy, more than can be said in a few minutes or in a few pages. Perhaps my original, first approximation is the best summary after all. When all is said and done, the final report is still that he is an outstanding mathematician and a warm human being.

International Series of
Numerical Mathematics, Vol. 65

ADRIAAN CORNELIS ZAANEN

J. Korevaar
Mathematisch Instituut
Universiteit van Amsterdam
Amsterdam

Aad Zaanen was born June 14, 1913 in Rotterdam, where his father was a building contractor (Bau-Unternehmer); there were four other children. Earlier Zaanens lived in the country South East of Rotterdam, a low-lying area of dairy farms that my ancestors also come from. That is perhaps one reason why I have always felt close to Zaanen on a personal level, although I am not so close to him as a mathematician.

Zaanen indeed has all the best qualities of the good farmer, and I use the word "Bauer" here without any disrespect. He is industrious, methodical, thorough, reliable, well-balanced, not perturbed by a bit of heavy weather, even-tempered, a person of sound judgment, wisdom and kindness: a good person to go to when you have problems. He showed that good judgment and wisdom also in choosing a partner in life who would work hard alongside of him. Would you say then that he is shrewd? No, he is much too kind to be called that. Above all, he has shown the steady work habits and good organization of the farmer without which he could not have achieved his impressive production.

Zaanen went to secondary school in Rotterdam, to the no-nonsense five-year school called H.B.S., a type that does not exist anymore. Those schools provided excellent education in mathematics, the sciences and modern languages. Many prospective students of science or engineering went to the H.B.S., although in those days only a small percentage of the pupils continued on to higher education. Zaanen's mathematics teacher was Van der Corput, a brother of the well-known mathematician. He had an engineering degree and was a good teacher, but not at all research-minded (as some high-school teachers were). In fact, it was the director of the school, a physics Ph.D. from Leiden, who stimulated Aad to go to Leiden to study mathematics and physics. That was in 1930.

The decision to specialize in mathematics came very soon. Not that physics was bad at Leiden. On the contrary, the physicists were world famous. In theoretical physics there had been or were Lorentz, Ehrenfest, Kramers, and there was outstanding experimental low temperature physics: for many years the laboratory of Kamerlingh Onnes in Leiden provided the coldest spot on earth! So the physics was great, but the physicists of around 1930 did not teach well (it was still that way when I arrived at Leiden ten years later). The mathematics program for the first two years was much better organized.

Don't assume now that the mathematics program at Leiden was modern or even up to date. Before world war II, mathematical education in the Netherlands was on the whole rather backward. In the period 1900-1940 only a few mathematics professors were really active in research. Of course there was Brouwer in Amsterdam, who did very fundamental work in topology and foundations. Van der Corput in Groningen made important contributions to analytic number theory and asymptotics. One might also mention the applied mathematicians Biezeno (applied mechanics) and Tinbergen (econometrics). Finally, there were some classical-type geometers, notably Schouten. Those few leaders had a large number of good students, but practically none of them obtained a professorship until after the war. A notable exception was Van der Waerden, who went to Germany. Turning to analysis: the important developments in France (integration and related developments) and Germany (integral equations around Hilbert) at the beginning of the century received only sporadic attention in the Netherlands before 1940. Other small countries did much better: Hungary produced Fejér, the brothers Riesz and others, Poland had its famous school of functional analysis, in Sweden there were Carleman and others. In Switzerland, Plancherel was active since his Göttingen Habilitation of 1909; his work on singular integral equations led very early to the L^2 theory of Fourier transforms.

Let us go back now to Leiden of 1930. The mathematics professors were Droste for analysis and Van der Woude for geometry. Droste was no research mathematician but he was a solid teacher, Van der Woude was old-fashioned but full of enthusiasm. More important, there was a stimulating new teacher, the young lecturer Kloosterman, who had spent several years abroad and was already a number theorist of some renown. Kloosterman was responsible for first and second year analysis. It was he who offered topics courses on subjects that were not yet part of the regular university program, such as group theory and linear operators in Hilbert space. (I briefly saw the same scene at Leiden ten years later, before the University was closed down by the occupation forces.)

In 1938, Zaanen obtained the Ph.D. degree in mathematics with Droste as his advisor. He showed that series of eigenfunctions of certain Sturm-Liouville problems behave very much like Fourier series. The thesis shows the influence of Zygmund's book which had appeared in 1935. The other Polish "blue books" would get his attention later. After the degree, Zaanen became a secondary school teacher. There were no university positions for young people then. Even high-school positions were hard to get: it was the height of the depression and one was lucky to get a temporary part time job. Zaanen's first position was in Rotterdam and it turned out to be very important for his later life: this is where he met his future wife, Ada Jacoba van der Woude (no relation to the mathematician). She was a pupil at the school, although not of him. They were married a few years later, during the war, and from what I have seen, they have been very good for each other.

During nine years of secondary school teaching, Zaanen struggled away from his largely old-fashioned university education by studying Marshall Stone's book on Hilbert space operators and Banach's book on linear operators. From those books one could learn about theory and how to apply it to concrete cases. At this time, Zaanen became interested in integral equations with symmetrizable kernels and symmetrizable operators, on which he soon started to publish.

This is a good moment to say something about Zaanen's university career. In 1947 he became the mathematics professor at the new University of Indonesia in Bandung, which was the successor to the pre-war Institute of Technology. It was rather enterprising to go there with a young family, while the Indonesian war of independence against the Dutch was still in progress and the Dutch were only partly in control of the country. However, Mrs. Zaanen whole-heartedly supported the decision to go. The Zaanens spent three busy but happy years there. When they left, Indonesia had become independent.

Zaanen then became a professor at the Delft Institute of Technology. His six years there were very important for his work as I will indicate below. In 1956 he moved to his Alma Mater, where he succeeded his Ph.D. advisor and remained until his retirement in 1982. His 26 years at Leiden coincided with turbulent times for the Dutch universities: rapid expansion, student und junior staff uprisings around 1970 followed by so-called democratization and finally, the onset of a period of reductions.

In Indonesia, Zaanen had continued to work on integral equations, now in the setting of L^p spaces and the somewhat more general Orlicz spaces. The results can be found in his remarkable book "Linear Analysis" of 1953. It is a big book, perhaps typical of a self-taught person: the first part is like a textbook, and then there is a treatise on recent developments, including original contributions. At Delft, Zaanen had the good fortune to meet up with Luxemburg, who became his assistant, Ph.D. student and long-term collaborator. The thesis problem led directly to the consideration of more general spaces than Orlicz spaces which Zaanen and Luxemburg called Banach function spaces and for which they obtained many nice results. Similar function spaces were also considered elsewhere at that time, notably in Canada by Halperin, Ellis and Lorentz.

After he got his degree, Luxemburg moved to Canada and subsequently to the U.S., more precisely, to the California Institute of Technology. However, the fruitful collaboration continued, thanks also to some visiting professorships of Zaanen at "Cal. Tech.". At the hands of Luxemburg and Zaanen, the Banach function spaces led to still more general spaces, the lattice or Riesz spaces, where the partial ordering plays a fundamental role. The axioms for such spaces go back to Freudenthal and Kantorovich; other contributors include F. Riesz (naturally), G. Birkhoff and Nakano. Kantorovich, incidentally, later received a Nobel prize, but that was for his work in economics. The joint results of Luxemburg and Zaanen on Riesz spaces were published in a series of papers in the Proceedings of the Netherlands Academy of Sciences and in a first book on Riesz spaces (1971; a second volume by Zaanen has appeared this year).

After Luxemburg left for overseas, Zaanen completed a nice book on integration, of which there exist two editions. Then came along a dozen Ph.D. students at Leiden, most of them in the general area of Riesz spaces and the corresponding operators. I should tell you at this point that Dutch Ph.D.

theses are fairly substantial pieces of work, corresponding to 2-5 published papers. During these busy years Zaanen also found time to contribute generously to various administrative chores. He turned out to be an ideal administrator in difficult times, thanks to his patience and his skill to listen and mediate. That he found time for all these activities is certainly a tribute to Mrs. Zaanen, who shielded him from household chores and took the principal burden of educating four lively boys. Before I finish, I like to illustrate her devotion to him with a little anecdote. During one of their visits to the U.S., the Zaanens drove across the country. Asked about her impressions, she said: "It is a big country, but rather dull: roads, roads, roads, and lots of dead animals on the roads." And he said: "It is a beautiful country, and there is so much variety!" You see, she had been doing all the driving, and he could look around. Actually, they like to travel together: at this time, they have just returned from a stay at one of the universities in South Africa.

When A.C. Zaanen retired, his former students organized an interesting symposium in his honor. The Proceedings have been published under the appropriate title "From A to Z", Math. Centre Tracts 149 (1982). This was a well-deserved tribute to the founder of a notable Dutch school of functional analysis. We are very proud of him in the Netherlands. I am sure that you will join me now in wishing him and Mrs. Zaanen many more good years together!

I Operator Theory

International Series of
Numerical Mathematics, Vol. 65

SUBNORMAL SUBOPERATORS AND THE SUBDISCRETE TOPOLOGY

P.R. Halmos*
Department of Mathematics
Indiana University
Bloomington, Indiana 47405

Dedicated to the memory of Errett Bishop

A _suboperator_ is a bounded linear transformation from a subspace of a Hilbert space into the whole space. The main purpose of this work is to raise some questions about the extension properties of suboperators (e.g., which are subpositive?), and, in particular, to pose the problem of characterizing _subnormal_ suboperators. The _subdiscrete topology_ of operators on a Hilbert space is the topology of discrete convergence (pointwise ultimate equality). The problem of finding a simple proof of Bishop's theorem (the set of subnormal operators is strongly closed) is reduced to the problem of finding a direct proof of the following assertion: if every finite restriction of an operator T is nearly subnormal, then every finite restriction of T is subnormal.

1. Introduction

The purpose of this study is not to answer questions but to ask them. The subject is a part of the theory of operators on Hilbert space, and the questions arose in the course of an attempt to find a simple proof of a theorem of Bishop's. The theorem is a topological characterization of an algebraic concept. The attempt suggested a generalization of the algebraic concept and an enlargement of the pertinent topology. The questions concern properties of the generalization and of the enlargement, and relations between the new concepts and the old ones. The questions seem to be interesting and challenging, and they are certainly elementary; they make sense in the finite-dimensional case and are unanswered even there.

2. Suboperators

In its most general use the word "operator" has come to mean a (usually linear) transformation with some regularity properties, between vector spaces

*Supported in part by a grant from the National Science Foundation.

with some analytic structure. According to its most special definition, which is the one to be used in this paper, an operator is a bounded linear transformation from a Hilbert space into itself; that is the definition that has the richest algebraic implications. The generalization that is usually considered the mildest keeps the Hilbert space structure and keeps the boundedness assumption but allows the domain and the range to be in two different spaces. Even in this mild generalization the algebraic loss is large: the identity operator has disappeared, operators can no longer be multiplied, and it becomes impossible to compare an operator with its adjoint and hence, for instance, impossible to consider the important class of Hermitian operators. There is, however, an even milder generalization, namely the one in which the domain space and the target space are not completely unrelated, but the former is a subspace of the latter. To avoid any possible terminological misunderstanding, I shall use a new word for the objects to be considered; if H is a Hilbert space and H_0 is a (closed) subspace of H, I define a *suboperator* to be a bounded linear transformation from H_0 into H.

The set of all suboperators defined on various subspaces of H and mapping them into H is larger than the set of all operators on H, but, for some purposes, it is not too large. It can, for instance, be naturally endowed with several algebraic structures (such as a partial multiplication and partial comparability with the adjoint). The only one to be studied here is the order structure defined by extension, i.e., the possibility of comparing a suboperator with an operator (a "total" operator). Sample questions: which suboperators are sub-Hermitian, subunitary, subpositive, subprojective, subnormal? The definitions are surely guessable. Thus, for instance, a suboperator is sub-Hermitian if it has a Hermitian extension (to H). The remaining definitions can be formulated exactly the same way; they require the existence of extensions that are unitary, or positive operators, or projections, or normal operators.

The first problem for each of these classes is characterization. The point is that the definitions are existential; to apply them as tests is difficult. The desideratum is an "intrinsic" characterization expressed in terms of the behavior of the given suboperator on its domain. For some of the sample questions above the characterization turns out to be trivial, and for others it is not yet solved.

3. Matrices and adjoints

If Q $(H_0 \to H)$ is a suboperator, then, for every h in H_0, the image Qh is uniquely representable in the form $f+g$, with f in H_0 and g in $H_0^\perp$. Write $f = Ah$ and $g = Bh$, and note that A is an operator on the Hilbert space H_0, whereas B is a bounded linear transformation from H_0 into $H_0^\perp$. In analogy with the customary way of representing operators by matrices, the suboperator Q is represented by the "column matrix" $\binom{A}{B}$; this representation is often helpful in calculations.

If T is an operator on H, then the decomposition $H = H_0 \oplus H_0^\perp$ induces a matrix representation of T, $T = \begin{pmatrix} A & X \\ B & Y \end{pmatrix}$, where $A\colon H_0 \to H_0$ and $B\colon H_0 \to H_0^\perp$, as before, whereas $X\colon H_0^\perp \to H_0$ and $Y\colon H_0^\perp \to H_0^\perp$. To say that an operator T is an extension of a suboperator Q with domain H_0 is exactly the same as to say that the column matrix $\binom{A}{B}$ of Q is equal to the first column of the matrix of T.

The concept of adjoint is defined for every bounded linear transformation between Hilbert spaces, and it is defined in particular for suboperators. The defining identity for the adjoint Q^* of a suboperator Q is, as usual, $(f, Q^*g) = (Qf, g)$, valid whenever $f \in H_0$ and $g \in H$. If $Q = \binom{A}{B}$, then $Q^* = (A^* \ B^*)$, where the "row matrix" $(A^* \ B^*)$ maps a vector $f+g$ in H (f in H_0, g in $H_0^\perp$) onto $A^*f + B^*g$ in H_0. In computationally usable symbolism: $(A^* \ B^*)\binom{f}{g} = A^*f + B^*g$; that is, $(A^* \ B^*)$ acts on $\binom{f}{g}$ via the usual row-by-column multiplication.

The familiar formal manipulations with matrices are easy to justify for suboperators. If, for instance, $Q = \binom{A}{B}$ is a suboperator and $T = \begin{pmatrix} C & X \\ D & Y \end{pmatrix}$ is an operator (both matrix representations determined by the same decomposition $H = H_0 \oplus H_0^\perp$), then the product TQ is always defined and is represented by the matrix $\begin{pmatrix} C & X \\ D & Y \end{pmatrix}\binom{A}{B} = \binom{CA + XB}{DA + YB}$. Similarly, the products Q^*Q and QQ^* are always defined and are represented by $(A^* \ B^*)\binom{A}{B} = A^*A + B^*B$ and $\binom{A}{B}(A^* \ B^*) = \begin{pmatrix} AA^* & AB^* \\ BA^* & BB^* \end{pmatrix}$.

4. Sub-Hermitian and subunitary

To say that a suboperator $Q = \binom{A}{B}$ is sub-Hermitian means that X and Y can be found so that $\begin{pmatrix} A & X \\ B & Y \end{pmatrix}$ is Hermitian. If that is so, then, clearly, $A = A^*$; in other words, a necessary condition that Q be sub-Hermitian is that its compression to H_0 be Hermitian. (Recall the definition of that compression in terms of the projection P_0 on H with range H_0 : it is the restriction to H_0 of the operator P_0QP_0 .) If, conversely, A is Hermitian, then Q is sub-Hermitian; to find a Hermitian extension, put $X = B^*$ and let Y be an arbitrary Hermitian operator on $H_0^\perp$. Conclusion: $\binom{A}{B}$ is sub-Hermitian if and only if A is Hermitian.

To say that a suboperator Q is subisometric means that it has an extension to an operator T that is an isometry. If that is so, then $\|Qf\| = \|Tf\| = \|f\|$ for every f in H_0 ; in other words, a necessary condition that Q be subisometric is that Q map H_0 isometrically into H . Since T maps $H_0^\perp$ isometrically into $(QH_0)^\perp$, it is also necessary that $\dim H_0^\perp \leqq \dim(QH_0)^\perp$, i.e., that co-dim $H_0 \leqq$ co-rank Q . (If $\dim H_0 < \infty$, the latter condition is no condition at all — it holds for all Q .) If, conversely, Q is isometric and the co-dimension of the domain of Q is less than or equal to the co-rank of Q , then Q is subisometric; just define T to be Q on H_0 and to be an arbitrary isometry from $(\text{dom } Q)^\perp$ into $(\text{ran } Q)^\perp$. (Note that since Q is isometric, the range of Q is closed.)

The theory of subunitary suboperators goes the same way: just replace the dimension inequalities in the preceding paragraph by equations. The result is that Q is subunitary if and only if Q is isometric and the co-dimension of the domain of Q is equal to the co-rank of Q .

5. Subpositive

To say that a suboperator $Q = \binom{A}{B}$ is subpositive means that X and Y can be found so that $\begin{pmatrix} A & X \\ B & Y \end{pmatrix}$ is positive. If that is so, then, clearly, A and Y are positive, $X = B^*$, and, moreover,

$$\left(\begin{pmatrix} A & B^* \\ B & Y \end{pmatrix}\begin{pmatrix} \alpha f \\ \beta g \end{pmatrix}, \begin{pmatrix} \alpha f \\ \beta g \end{pmatrix}\right) = \left(\begin{pmatrix} \alpha Af + \beta B^*g \\ \alpha Bf + \beta Yg \end{pmatrix}, \begin{pmatrix} \alpha f \\ \beta g \end{pmatrix}\right) =$$

$$= (Af, f)|\alpha|^2 + (g, Bf)\bar{\alpha}\beta + (Bf, g)\alpha\bar{\beta} + (Yg, g)|\beta|^2 \geqq 0$$

whenever $f \in H_0$, $g \in H_0^\perp$, and α and β are complex scalars. The latter condition is equivalent to the positiveness of the 2×2 matrix

$$\begin{pmatrix} (Af, f) & (g, Bf) \\ (Bf, g) & (Yg, g) \end{pmatrix} .$$

That in turn implies that the determinant is positive, i.e., that

$|(Bf, g)|^2 \leqq (Af, f) \cdot (Yg, g)$. Various conditions expressible in terms of A and B only follow from this inequality. One of them is $\ker A \subset \ker B$, another form of which is $\ker \sqrt{A} \subset \ker B$. (If $Af = 0$ for some f in H_0 , then $(Bf, g) = 0$ for all g in $H_0^{\perp}$, and therefore $Bf = 0$. Recall that $\ker \sqrt{A} = \ker A$ for all positive operators A .) These kernel inclusions imply that $\overline{\operatorname{ran} B^*} \subset \overline{\operatorname{ran} A}$ and $\overline{\operatorname{ran} B^*} \subset \overline{\operatorname{ran} \sqrt{A}}$ (recall that A is Hermitian). What is somewhat less obvious is that the closures can be removed from the last relation; that is the main thrust of the proposition that follows. (Caution: it is not necessarily true that $\operatorname{ran} \sqrt{A} = \operatorname{ran} A$.)

PROPOSITION 1. _If A and Y are positive operators on Hilbert spaces H_0 and H_1 respectively, and if B is a bounded linear transformation from H_0 into H_1 such that $|(Bf, g)|^2 \leq (Af, f) \cdot (Yg, g)$ whenever $f \in H_0$ and $g \in H_1$, then $\sqrt{A}$ is a right factor of B ; that is, there exists a bounded linear transformation C from H_0 into H_1 such that $B = C\sqrt{A}$._

PROOF. Define a mapping C from $\operatorname{ran} \sqrt{A}$ ($\subset H_0$) into H_1 by $C\sqrt{A}f = Bf$. Since the assumed inequality implies that $\ker \sqrt{A} \subset \ker B$, it follows that the proposed definition of C is unambiguous; a routine elementary argument proves that C is linear. If $f' = \sqrt{A}f$, then $\|Cf'\|^2 = \|C\sqrt{A}f\|^2 = \|Bf\|^2 = \sup\{|(Bf, g)|^2 : \|g\| \leqq 1\} \leqq (Af, f) \cdot \sup\{(Yg, g) : \|g\| \leqq 1\} = \|\sqrt{A}f\|^2 \cdot \|Y\| = \|Y\| \cdot \|f'\|^2$, so that C is bounded and therefore continuous. Extend C to a bounded linear transformation on H_0 (by first extending it to the closure of $\operatorname{ran} \sqrt{A}$ and then defining it to be 0 on the orthogonal complement of that closure); the result satisfies the stated conclusion.

The converse of Proposition 1 is true also.

PROPOSITION 2. _If A is a positive operator on a Hilbert space H_0 , and if B and C are bounded linear transformations from H_0 to a Hilbert space H_1 , such that $B = C\sqrt{A}$, then the 2×2 matrix_

$$\begin{pmatrix} (Af, f) & (g, Bf) \\ (Bf, g) & (CC^*g, g) \end{pmatrix}$$

is positive whenever $f \in H_0$ and $g \in H_1$.

PROOF. The diagonal entries of the matrix are positive by assumption; all that needs to be checked is that the determinant is positive. To prove that, note that if $f \in H_0$ and $g \in H_1$, then $|(Bf, g)|^2 = |(C\sqrt{A}f, g)|^2 = |(\sqrt{A}f, C^*g)|^2 \leq \|\sqrt{A}f\|^2 \cdot \|C^*g\|^2 = (Af, f)\cdot(CC^*g, g)$.

COROLLARY 1. <u>A necessary and sufficient condition that</u> $\sqrt{A}$ <u>be a right factor of</u> B <u>is that there exist a positive operator</u> Y <u>such that</u> $|(Bf, g)|^2 \leq (Af, f)\cdot(Yg, g)$.

Corollary 1 is a very slight sharpening of some known range inclusion and factorization theorems; see [5, #59].

COROLLARY 2. <u>A suboperator</u> $\binom{A}{B}$ <u>is subpositive if and only if</u> $A \geq 0$ <u>and</u> $\operatorname{ran} B^* \subset \operatorname{ran}\sqrt{A}$.

The characterization of subpositive suboperators is the subtlest result so far, but the characterization of subprojections lies slightly deeper still. The result is easy to state, but the proof is mildly involved.

PROPOSITION 3. <u>A suboperator</u> $\binom{A}{B}$ <u>is a subprojection if and only if</u> A <u>is Hermitian and</u> $A - A^2 = B^*B$.

PROOF. The necessity of the condition $A = A^*$ is trivial, and the necessity of the other condition is scarcely less so. Indeed, if $T = \begin{pmatrix} A & X \\ B & Y \end{pmatrix}$ is a projection, then X must be B^*, and the idempotence of T, i.e., the relation

$$\begin{pmatrix} A & B^* \\ B & Y \end{pmatrix} = \begin{pmatrix} A^2 + B^*B & AB^* + B^*Y \\ BA + YB & BB^* + Y^2 \end{pmatrix},$$

implies the asserted equation.

Sufficiency takes more of an argument. Observe first that $A^2 \geq 0$ (because A is Hermitian), whence it follows that A $(= A^2 + B^*B)$ is positive. The equation implies also that $A^2 \leq A$, whence it follows that $A \leq 1$; for a quick proof use the spectral theorem.

It is convenient to consider the polar decomposition $B = VP$, where P is a positive operator on H_0 and V is a partial isometry from H_0 into $H_0^\perp$ such that $\ker V = \ker P$; see [5, #134]. Since $P^2 = B^*B = A - A^2$, so that A commutes with P^2, it follows that A commutes with P .

The problem is to find a suitable definition of Y. The idempotence equation shows that if a Y exists, it must be such that $YB = B(1 - A)$. It is natural, therefore, to try to define Y on $\operatorname{ran} B$ by writing

$YBf = B(1 - A)f$ for all f in H_0, but the question of unambiguity must be faced. All is well: if $Bf = 0$, then $P^2f = B^*Bf = 0$, whence $Pf = 0$, and therefore $B(1 - A)f = VP(1 - A)f = V(1 - A)Pf = 0$.

Is the Y that the definition produces bounded? Yes; the argument goes as follows: $\|YBf\|^2 = \|B(1 - A)f\|^2 = (B(1 - A)f, B(1 - A)f) = (B^*B(1 - A)f, (1 - A)f) = (P^2(1 - A)f, (1 - A)f) = \|P(1 - A)f\|^2 = \|(1 - A)Pf\|^2 \leq \|1 - A\|^2(P^2f, f) = \|1 - A\|^2 \|Bf\|^2$. Consequence: Y can be extended to a bounded operator on the closure of $\operatorname{ran} B$ (a subspace of $H_0^\perp$).

No matter how Y is defined on $\operatorname{ran}^\perp B$ $(= \ker B^*)$, the matrix $\begin{pmatrix} A & B^* \\ B & Y \end{pmatrix}$ will satisfy three of the four equations that characterize idempotence; but it still remains to define Y on $\ker B^*$ suitably so as to satisfy the fourth equation, $Y - Y^2 = BB^*$. The simplest definition works: put $Yg = 0$ whenever $g \in \ker B^*$. It follows that if $g \in \ker B^*$, then $(Y - Y^2)g = 0$ and, of course, $BB^*g = 0$; the only thing that remains to be checked is that if $g \in \operatorname{ran} B$, then too $(Y - Y^2)g = BB^*g$. For that purpose assume $g = Bf$ and compute: $(1 - Y)Yg = (1 - Y)YBf = (1 - Y)B(1 - A)f = B(1 - A)f - YB(1 - A)f = B(1 - A)f - B(1 - A)^2f = B(1 - A)(1 - (1 - A))f = B(A - A^2)f = BB^*Bf = BB^*g$—q.e.d.

Final step: Y is Hermitian. Reason: if $g \in \ker B^*$, then $(Y(Bf + g), Bf + g) = (YBf, Bf) = (B(1 - A)f, Bf) = (P^2(1 - A)f, f) = ((1 - A)Pf, Pf)$, which is real, and in fact positive.

6. Subnormal suboperators

Extension questions about suboperators can be expressed as extension questions about submatrices. Given a suboperator $Q: H_0 \to H$, choose an orthonormal basis for H_0 and extend it to an orthonormal basis for H. To simplify the language, order the resulting basis so that the vectors in H_0 come first. It follows that the matrix of every extension of Q will begin the same way as every other — Q determines the beginning — and every extension problem becomes the problem of filling in the rest so as to obtain a matrix of specified type.

Here are some trivial but illuminating examples, based on the preceding results. Question: what can the first column of a Hermitian matrix be? Answer: any vector whose first coordinate is real. Question: what can the first column of a unitary matrix be? Answer: any unit vector. Question: what can the first column of a positive matrix be? Answer: any vector with strictly positive first coordinate, or the zero vector. Question: what can the first column of a projection be? Answer: any vector $\langle\alpha_1, \alpha_2, \alpha_3, \ldots\rangle$ with α_1 real and such that $\alpha_1 - \alpha_1^2 = \sum_{j=2}^{\infty} |\alpha_j|^2$. The same questions can be asked about the first <u>two</u> columns, and the characterizations discussed above supply the answers.

The hardest and most important questions about suboperators, so far in their short history, concern the subnormal ones. The first natural question is the only easy one: what can the first column of a normal matrix be? Answer: any vector at all. More explicitly: every square summable sequence of complex numbers can be the first column of a normal matrix. Proof: (1) multiply by a suitable complex number α of modulus 1 so as to make the first coordinate real; (2) extend the result to a Hermitian matrix; (3) divide by α.

The result of this easy reasoning can be stated as follows: the set N of normal operators is <u>transitive</u>. The usual geometric definition of (simple) transitivity (for a class C of linear transformations) can be expressed this way: every linear transformation from a 1-dimensional subspace to the whole space has an extension that belongs to C . The transitivity of N can therefore be expressed this way: every suboperator with a 1-dimensional domain is subnormal.

Is the set of normal operators <u>doubly</u> transitive? (To obtain the definition of double transitivity from that of simple transitivity, replace 1 by 2 .) Equivalent formulations: can any two columns occur as the beginning of a normal matrix?; or, is every suboperator with a 2-dimensional domain subnormal? The answer is no. Counterexample:

$$Q = \begin{pmatrix} 0 & 0 \\ 1 & 0 \\ 0 & 0 \\ 0 & 0 \end{pmatrix} .$$

Proof: each column of a normal matrix has the same length (norm) as the corresponding row. It follows that if the indicated 4×2 matrix constituted the beginning of a 4×4 normal matrix N , then the second row of N would have to be 0 , but, since it begins with 1 , that is impossible.

[An alternative approach goes like this. Since $Qe_1 = e_2$ and $Qe_2 = 0$, where e_1 and e_2 are the first two basis vectors, any normal extension N would have a point of nilpotence: $Ne_1 \neq 0$, $N^2 e_1 = 0$. Assertion: if a normal operator N is "locally nilpotent" — e.g., $N^2 f = 0$ for some f —

then it is "locally zero" — i.e., then $Nf = 0$. Proof: easy by a direct argument, and even easier by use of the spectral theorem.]

Since not every suboperator with a 2-dimensional domain is subnormal, it makes sense to ask which ones are.

PROBLEM 1. <u>Is there an intrinsic characterization of subnormality for a suboperator</u> $\binom{A}{B}$ <u>with domain</u> H_0 <u>in terms of the geometric behavior of</u> A <u>and</u> B <u>on the subspace</u> H_0 ?

The answer is not known; even the usually easy low-dimensional cases are shrouded in mystery.

A usable necessary condition for subnormality is near the surface. To say that a suboperator $Q = \binom{A}{B}$ is subnormal means that X and Y can be found so that $N = \begin{pmatrix} A & X \\ B & Y \end{pmatrix}$ is normal. The normality of N, i.e., the equality of N^*N and NN^*, implies that $A^*A + B^*B = AA^* + XX^*$, and hence that $A^*A - AA^* + B^*B$ is positive. This positivity condition is important enough to deserve a name; the history of the "total" version of the subject strongly suggests that the name be <u>hyponormal</u>. It is possible that for sufficiently small matrices hyponormality is sufficient for subnormality, but except in the completely trivial case of matrices of size 1 even that is not known.

PROBLEM 2. <u>If</u> A <u>and</u> B <u>are</u> 2×2 <u>matrices such that the suboperator</u> $\binom{A}{B}$ <u>is hyponormal, does it follow that</u> $\binom{A}{B}$ <u>is subnormal?</u>

As soon as "sufficiently small" grows beyond 2, the necessary condition ceases to have any chance of being sufficient; here is an example discovered by Eric Nordgren. If

$$A = \begin{pmatrix} 0 & 0 & 1 \\ 1 & 0 & 0 \\ 0 & 1 & 0 \end{pmatrix} \quad \text{and} \quad B = \begin{pmatrix} 1 & 0 & 0 \\ 0 & 0 & 0 \\ 0 & 0 & 0 \end{pmatrix},$$

then $\binom{A}{B}$ is hyponormal but not subnormal. Note that A in the example is normal (in fact unitary); this observation makes it obvious that $\binom{A}{B}$ is hyponormal. To prove that $\binom{A}{B}$ is not subnormal, note that in

$$Q = \begin{pmatrix} 0 & 0 & 1 \\ 1 & 0 & 0 \\ 0 & 1 & 0 \\ 1 & 0 & 0 \\ 0 & 0 & 0 \\ 0 & 0 & 0 \end{pmatrix}$$

the third column has length 1 , and that, therefore, the third row of any candidate for a normal extension N must also have length 1 . Consequence: the third row of N must be

$$0\ 1\ 0\ 0\ 0\ 0\ .$$

It follows that the third row of N is orthogonal to every other row, and hence, by normality again, the third column of N must be orthogonal to every other column. In particular, therefore, the first row of N must be

$$0\ 0\ 1\ 0\ 0\ 0\ ,$$

which has length 1 . That, however, is impossible: the length of the first column of N is $\sqrt{2}$.

A few fragmentary results are available. Thus, for example, $\binom{A}{A^*}$ is always subnormal (a possible normal extension is $\begin{pmatrix} A & A^* \\ A^* & A \end{pmatrix}$), but $\binom{A}{A}$ can fail to be (if $A = \begin{pmatrix} 0 & 1 \\ 0 & 0 \end{pmatrix}$, then the first column of $\binom{A}{A}$ is 0 but the first row is not). The subnormality of $\binom{A}{B}$ is not a trivial problem even when A is normal. If, however, A is normal and B commutes with A , then $\binom{A}{B}$ is always subnormal (a possible normal extension is $\begin{pmatrix} A & B^* \\ B & A \end{pmatrix}$). If A is normal and the spectrum of A consists of not more than two numbers (an extremely special case), and, in particular, if A is a normal matrix of size 2 , then $\binom{A}{B}$ is always subnormal; the proof can be made to depend on the fact that some affine transform $\alpha A + \beta$ is Hermitian.

A case of some importance is the one in which the underlying Hilbert space H is infinite-dimensional but the domains of the subnormal suboperators under consideration are finite-dimensional subspaces of H . The relation between possible normal extensions of finite size and those of infinite size is unknown.

PROBLEM 3. Does every finite restriction of a normal operator have a finite normal extension? (Unabbreviated: if N is a normal operator on H and H_0 is a finite-dimensional subspace of H , can the restriction $N|H_0$ be extended to a normal operator defined on a finite-dimensional space H_1 with $H_0 \subset H_1 \subset H$?)

7. Subnormal operators

An important and by now classical special case of subnormality occurs when $B = 0$, i.e., when Q maps H_0 into H_0, or, in other words, when H_0 is invariant under Q (and hence H_0 is invariant under any possible normal extension of Q). If $\binom{A}{0}$ is subnormal, the operator A itself is called subnormal. When subnormal operators were discovered, they looked like a promising generalization of the normal ones, and, indeed, the promise has been amply fulfilled. The theory became an industry with many exciting connections to classical and modern analysis.

Each of the rich stock of examples of normal operators lives on an infinite-dimensional Hilbert space; an easy trace argument shows that on finite-dimensional spaces the only subnormal operators are the normal ones. One reason why the theory of suboperators is of interest is that it provides a context for the study of subnormality in spaces that can be of either finite or infinite dimension.

The characterization problem for subnormal suboperators is a generalization of the characterization problem for subnormal operators. Several solutions of the special problem have been offered. An early result [2, 4] is that an operator A on H_0 is subnormal if and only if for every finite set $\{f_0, f_1, \ldots, f_n\}$ of vectors in H_0 the matrix $\langle (A^i f_j, A^j f_i)\rangle$ is positive. As a typical application, consider the proof that the set S of subnormal operators is norm closed. If $\{A_n\}$ is a sequence (or, for that matter, a net) of subnormal operators converging in the norm to A, then $(A_n^i f_j, A_n^j f_i) \to (A^i f_j, A^j f_i)$ for each i and j (because each of the mappings $A \mapsto (A^i f_j, A^j f_i)$ is norm continuous); it follows that the limit matrix $\langle (A^i f_j, A^j f_i)\rangle$ is positive, and hence that A is subnormal.

Bishop's topological characterization of subnormality [1] is chronologically the second one; it is still one of the deepest results of the theory. It asserts that an operator is subnormal if and only if it is the pointwise limit of a net of normal operators. Equivalently, in the customary technical language, S is the strong closure of N.

The "only if" part of Bishop's characterization theorem is the easier part to prove. Suppose, indeed, that A is subnormal; the assertion is that every strong neighborhood of A contains a normal operator. A typical basic strong neighborhood of A is a set of the form $\{T: \|Af - Tf\| < \varepsilon\|f\|$ for all f in $M\}$, where M is a finite-dimensional subspace of H_0 and ε is a positive number. Since A is subnormal, there exists a normal operator N on a

Hilbert space H that includes H_0 such that N agrees with A on H_0; there is no loss of generality in assuming that H_0 and H have the same dimension. Let U be an isomorphism between H_0 and H (i.e., an isometry from H_0 onto H) that agrees with the identity on both M and NM. If $T = U^*NU$, then T is a normal operator on H_0; if $f \in M$, then $Tf = U^*NUf = U^*Nf = Nf = Af$, and that completes the proof.

Three comments. (1) The proof deals with the prescribed positive number ε the most efficient way possible; it works for every ε simultaneously, because, in fact, it works for $\varepsilon = 0$. This is very important in what follows. (2) A small modification of the proof can be used to show that every subnormal operator is the strong limit of a sequence of normal operators. (3) If it were known a priori that every operator in the strong closure of N is the strong limit of a sequence, the "if" part of Bishop's theorem would also have an easy proof. Indeed: since strongly convergent sequences are bounded, and multiplication is strongly continuous on bounded sets, the proof that showed that S is norm closed would then show that S is strongly closed also. Possible unboundedness is the source of the difficulty of Bishop's theorem; it is what caused some attempted proofs in the literature to go wrong.

Bishop's original proof gets around the difficulty. Given a strongly convergent set of normal operators, he proves that their spectral measures converge in an appropriate sense to a positive operator measure, and then uses Naimark's theorem about restrictions of spectral measures to produce the desired normal extension in the limit. A more recent proof by Conway and Hadwin [3] has an ingenious new idea: it shows that a strongly convergent net of normal operators can be truncated (and thus be made bounded) so as to remain strongly convergent to the same limit. Bishop's theorem is simple — it is easy to state and to understand; it is a pity that all proofs contain surprising and seemingly ad hoc technical and conceptual complications.

Is there a theorem like Bishop's for suboperators? A possible partial statement would be that the set of subnormal suboperators — or perhaps the set of those with a prescribed fixed domain — is closed with respect to some appropriate topology. If H is finite-dimensional, then the distinctions among the standard topologies disappear — and even then the statement turns out to be false. A simple example, due to José Barría, looks like this. If

$$T_\varepsilon = \left(\begin{array}{ccc|ccc} 0 & 0 & 1 & 1 & 0 & 0 \\ 1 & 0 & 0 & 0 & 0 & \varepsilon \\ 0 & 1 & 0 & 0 & \varepsilon & 0 \\ \hline 1 & 0 & 0 & 0 & 0 & -1/\varepsilon \\ 0 & \varepsilon & 0 & 0 & -1 & 0 \\ 0 & 0 & \varepsilon & -1/\varepsilon & 0 & 0 \end{array}\right)$$

for each positive number ε, then the matrix T_ε is normal (by brutal verification). Consequence: if Q_ε is the 6×3 matrix consisting of the first three columns of T_ε, then Q_ε is subnormal. Since Q_ε converges as $\varepsilon \to 0$ to the non-subnormal Nordgren matrix studied above, the conclusion follows: the set of 6×3 subnormal matrices is not a closed set.

PROBLEM 4. What is the closure of the set of subnormal suboperators?

The question is intentionally vague. Does it refer to spaces of finite dimension or infinite, does it ask about a fixed domain or all domains, and does it mean strong closure or norm? It doesn't matter: the answer is unknown in all but the degenerate trivial cases.

8. Subdiscrete topology

The strong operator topology can be described as follows: topologize H by the ordinary (metric) topology of vectors, and form the Cartesian product H^H of " H copies" of H topologized by the product topology; the strong operator topology is exactly the relativization of that topology to the subset $\mathcal{B}(H)$ of H^H consisting of all (bounded) operators. The proof of the easy part of Bishop's theorem suggests a related but different procedure. Topologize H by the discrete topology, and then form the topological product H^H; the relativization of the topology so obtained to $\mathcal{B}(H)$ will be called the subdiscrete topology. (A similar topologization of suboperators can be defined by focusing attention on H^{H_0}, where H_0 is a subspace of H.)

The strong topology is sometimes described as the topology of pointwise convergence; a net $\{A_n\}$ of operators converges to A strongly if and only if, for each f, the images $A_n f$ get arbitrarily near to Af when n is sufficiently large. In similar terms the subdiscrete topology could be described as the topology of ultimate equality; a net $\{A_n\}$ converges to A subdiscretely if and only if, for each f, the images $A_n f$ become equal to

Af when n is sufficiently large. Rudimentary forms of the concept have occurred in mathematics before. Thus, for example, most teachers who want to present a pointwise convergent sequence of functions that is not uniformly convergent construct, in fact, a sequence that is at each point ultimately equal to its limit. A more official version occurs in the theory of formal power series; see [6, Chapter VII].

If M is a finite-dimensional subspace of H, the set $\{T: Tf = 0$ for all f in $M\}$ is a subdiscrete neighborhood of 0; the definition of the subdiscrete topology implies that the collection of all sets of that form is a base of the subdiscrete topology at 0. From this in turn it follows that an operator S belongs to the subdiscrete closure of a set C of operators if and only if for every finite-dimensional subspace M there exists an operator T in C such that the restrictions $S|M$ and $T|M$ are equal. In compressed but suggestive language: S belongs to the subdiscrete closure of C if and only if every finite restriction of S has an extension in C.

It is obvious from the definition that the subdiscrete topology is larger than the strong topology (more open sets), and, consequently, that it is a Hausdorff topology. What the "$\varepsilon = 0$" proof of the easy part of Bishop's theorem shows is that S is included not only in the strong closure but even in the subdiscrete closure of N — a stronger statement. Perhaps this observation, which was the original motivation for considering the subdiscrete topology at all, together with the possible intrinsic interest of that topology, and its relation to extensions (mentioned in the preceding paragraph), will be accepted as sufficient justification for the brief study that follows.

The vector operations on operators are subdiscretely continuous; the verifications are routine. A more useful statement is that operator multiplication is subdiscretely continuous; the proof goes as follows. If $S_n \to S$ and $T_n \to T$ subdiscretely (nets), then for each of f there exists an index n_0 such that, for $n \geqq n_0$, both the equations $S_n Tf = STf$ and $T_n f = Tf$ are true. It follows that if $n \geqq n_0$, then $S_n T_n f = S_n Tf = STf$ — q.e.d.

The hard part of Bishop's theorem says that S is strongly closed. The corresponding subdiscrete assertion is weaker; it is therefore not surprising that it has an easier and more accessible direct proof.

PROPOSITION 4. _The set_ S _of subnormal operators is subdiscretely closed_.

PROOF. If $S_n \to S$ (net, subdiscrete convergence), then the subdiscrete continuity of multiplication implies that $(S_n^i f_j, S_n^j f_i) \to (S^i f_j, S^j f_i)$ for each i and j.

COROLLARY 3. <u>The subdiscrete closure of</u> N <u>is</u> S.

The subdiscrete topology differs from the strong topology in many striking ways. Thus, for instance, both the set of nilpotent operators of index 2 ($T^2 = 0$) and the set of idempotent operators ($T^2 = T$) are strongly dense [5, #111]; since, however, multiplication is subdiscretely continuous, in the subdiscrete topology both those sets are closed. Another difference between the two topologies is in the following pathological behavior of the subdiscrete topology.

PROPOSITION 5. <u>Every countable set is subdiscretely closed</u>.

PROOF. If T is in the subdiscrete closure of $\{T_1, T_2, T_3, \ldots\}$, then every finite restriction of T has an extension among the T_n's, and, in particular, corresponding to every vector f there exists an index n such that $Tf = T_n f$. Otherwise expressed: $\bigcup_{n=1}^{\infty} \ker(T - T_n) = H$. By the Baire category theorem, the only way a countable union of subspaces can be H, is for one of the terms to be H.

The subdiscrete topology has many properties in common with the discrete topology; are they the same? If the underlying Hilbert space H is finite-dimensional, they are; in that case $\{T\colon Tf = 0 \text{ for all } f \text{ in } H\}$ is a singleton neighborhood of 0. If, however, the dimension of H is infinite, then there exist sets that are not subdiscretely closed; according to Corollary 3, N is one of them. Could the subdiscrete topology in the infinite-dimensional case be at least metrizable? The answer is no.

PROPOSITION 6. <u>The subdiscrete topology for operators on an infinite-dimensional Hilbert space</u> H <u>does not have a countable base at</u> 0 <u>(i.e., it does not satisfy Hausdorff's first axiom of countability)</u>.

PROOF. Suppose that $\{U_1, U_2, U_3, \ldots\}$ is a countable set of basic subdiscrete neighborhoods of 0, $U_n = \{T\colon Tf = 0 \text{ for all } f \text{ in } M_n\}$, where M_n is a finite-dimensional subspace. Since H is not finite-dimensional, there exists a vector f_0 that does not belong to $\bigcup_n M_n$. If $U = \{T: Tf_0 = 0\}$, then U is a subdiscrete open set containing 0, but no U_n is included in U. Reason: for each n there exists an operator T such that $Tf = 0$ for all f in M_n and $Tf_0 \neq 0$.

Three more minor comments about the subdiscrete topology.

(1) The set of all operators of finite rank is subdiscretely dense. Reason: if T is an arbitrary operator and M is an arbitrary subspace of finite dimension, then there exists an operator T_M of finite rank that agrees with T on M; just define T_M to be 0 on $M^{\perp}$.

(2) In the infinite-dimensional case, the subdiscrete topology is neither smaller nor larger than the norm topology. Reason: the set of normal operators is norm closed but not subdiscretely closed, and the same thing is true of the set of compact operators; on the other hand, the set of all non-zero scalars is subdiscretely closed but not norm closed.

(3) In the infinite-dimensional case, the mapping $T \longmapsto T^*$ is not subdiscretely continuous. To prove that, let S be a subnormal operator that is not normal. By Corollary 3 there exists a net $\{T_n\}$ of normal operators such that $T_n \to S$ subdiscretely. If it were true that $T_n^* \to S^*$ subdiscretely, then (by Corollary 3) S^* would be subnormal; since the only operators S such that both S and S^* are subnormal are the normal ones, that cannot happen. Since $T_n - S \to 0$ subdiscretely, it follows that adjunction is subdiscretely discontinuous at 0, and hence everywhere. (This argument is much simpler than my original one; it is due to Donald Hadwin.)

9. Epilogue

Is there any hope that the concepts introduced above will lead to a simple proof of Bishop's theorem? One possible approach they suggest is the search for a direct proof of the following assertion: if every finite restriction of an operator T is nearly subnormal, then every finite restriction of T is subnormal. If the set of subnormal suboperators (on a fixed finite-dimensional domain) were closed, the result would be trivial — but that set is not closed. The assertion itself is true: it is a consequence of Bishop's theorem.

Does subnormality have something to do with compactness? Some traces of compactness arguments are perceptible in Bishop's proof, and the Conway-Hadwin proof uses a Montel theorem for operator-valued analytic functions — which looks like compactness again. The fact that the set S of subnormal operators is subdiscretely closed can be expressed this way: if every finite restriction of an operator A has a normal extension, then A has a normal extension — and that has a curious echo of compactness in it.

REFERENCES

[1] Bishop, E., Spectral theory for operators on a Banach space, Trans. Amer. Math. Soc. 86 (1957), 414-445.

[2] Bram, J., Subnormal operators, Duke Math. J. 22 (1955), 75-94.

[3] Conway, J.B. - Hadwin, D.W., Strong limits of normal operators, Glasgow Math. J. 24 (1983), 93-96.

[4] Halmos, P.R., Normal dilations and extensions of operators, Summa Brasil. Math. 2 (1950), 125-134.

[5] Halmos, P.R., A Hilbert space problem book, Second edition, Springer-Verlag, New York, 1982.

[6] Zariski, O. - Samuel, P., Commutative algebra, Volume II. Springer-Verlag, New York, 1976.

International Series of
Numerical Mathematics, Vol. 65

SPECTRAL PROPERTIES OF POSITIVE OPERATORS

Jacobus J. Grobler
Department of Mathematics and Applied Mathematics
Potchefstroom University for Christian Higher Education
Potchefstroom

In answer to a problem posed by A.C. Zaanen we prove the spectral radius theorem of Ando-Krieger without using any representation methods. The theorem states that a positive and band irreducible abstract kernel operator on a Dedekind complete Banach lattice has a strictly positive spectral radius. As a result the abstract version of the theorems of Jentzsch and Frobenius can be derived without resorting to representation theory.

1. Introduction

We assume the reader to be familiar with the theory of Banach lattices and for general notions which we do not define we refer to [5] and [7]. In order to state our problem we fix some notation. Let E be a Banach lattice and let $L(E)$ denote the set of bounded linear operators on E. If E is Dedekind complete we denote by $L^b(E)$ the Riesz space of order bounded linear operators on E and E'_{00} the set of order continuous linear functionals on E. By $(E'_{00}\otimes E)^{dd}$ we denote the band generated in $L^b(E)$ by the finite rank operators of the form $\sum_{i=1}^n x'_i\otimes x_i$, with $x'_i \in E'_{00}$, $x_i \in E$ and $n \in \mathbb{N}$. An element $0 < u \in E$ is called an *order unit* (respectively: *weak order unit*, *topological unit*) if the ideal (respectively: band, closed ideal) generated by u is the whole space E. An operator $T \in L(E)$ is called *irreducible* (respectively: *band irreducible*) if E contains no non-trivial T-invariant closed ideals (respectively: bands). We shall call T *strongly irreducible* if Tu is a weak order unit in E for every $0 < u \in E$.

At the 1982 Oberwolfach meeting on Riesz spaces and order bounded operators, A.C. Zaanen posed the problem of finding a representation-free proof of the following fact: If $0 < T \in (E'_{00}\otimes E)^{dd}$ is strongly band irreducible, then

T^2 is a weak order unit in $(E'_{00}\otimes E)^{dd}$. In a program to prove generalizations of the classical eigenvalue theorems of O. Perron-G. Frobenius and R. Jentzsch in the setting of Banach lattices without representation methods, this was the only missing link. Though we do not solve this particular problem, we show nevertheless that it is possible to complete the program successfully. The essential step is to prove the spectral radius theorem of Ando-Krieger in the manner indicated. Our exposition is based on a result of H.H. Schaefer the existing proof of which uses the Kakutani representation of AM-spaces by spaces of continuous functions. In section 2 we present a new proof which avoids this. In section 3 we prove the main results.

2. Schaefer's theorem

If $0 < u \in F$, F a Banach lattice, we denote by F_u the principal ideal generated in F by u. The space F_u normed by the gauge p_u of the order interval $[-u,u]$ is an AM-space with unit u.

THEOREM 2.1. *Let F be a closed ideal in a Dedekind complete AM-space M with unit e, and let r(T) denote the spectral radius of the operator $0 < T \in L(F)$. If T is irreducible then $r(T) > 0$ holds.*

PROOF. For $\lambda > r(T)$, let $S := \sum_{n=1}^{\infty}\lambda^{-n}T^n$. It then follows from the inequality $TS \leqslant \lambda S$ that for every fixed $x > 0$ in F the ideal generated by $u = Sx$ is T-invariant. Hence, u is a topological unit in F. Secondly, since the spectrum of $S = T(\lambda I-T)^{-1}$ consists of all numbers $z(\lambda-z)^{-1}$ for which z belongs to the spectrum of T, $r(T) > 0$ holds if and only if $r(S)>0$ holds. To show that the latter relation is true, we prove the existence of an element $e_0 \in F$ and a number $\varepsilon>0$ such that $Se_0 \geqslant \varepsilon\, e_0 > 0$.

Let $x > 0$ be an arbitrary element of F. Let $n \in \mathbb{N}$ be such that $(x - \frac{1}{n}e)^+ > 0$ and let e_0 be the component of e in the band generated by $(x-\frac{1}{n}e)^+$ in M. Since $\frac{1}{n}e_0 \leqslant x \in F$ we have $e_0 \in F$. Let F_0 be the ideal generated by e_0 in M with norm the gauge p_{e_0} of the interval $[-e_0,e_0]$. For any $x \in F_0$ one has $p_{e_0}(x) = p_e(x)$ because $|x| \leqslant \lambda e$ if and only if $|x| \leqslant \lambda e_0$. Hence (F_0,p_{e_0}) is a closed linear subspace of M and since $e_0 \in F$, also of F.

As observed above, the element $u = Se_0$ is a topological unit of F. If we denote by P_o the band projection of M onto the band generated by e_o, we see from the inequality $u \leq ke$ which holds for some $k \in \mathbb{N}$ that $P_o u := u_o$ belongs to F_o. Furthermore, u_o is a topological unit for (F_o, p_{e_o}) (If $z \in F_o$ there exists a sequence (z_n) in F_u such that $z_n \to z$ in the p_e-norm. The elements $P_o z_n$ belong to the ideal generated by $P_o u = u_o$ and $p_{e_o}(P_o z_n - z) = p_e(P_o(z_n - z)) \leq p_e(z_n - z) \to 0$ as $n \to \infty$. The ideal generated by u_o is therefore dense in (F_o, p_{e_o}).). But, (F_o, p_{e_o}) is an AM-space with unit and hence u_o is an order unit for (E_o, p_{e_o}) (Since the ideal generated by u_o is dense in F_o, it must contain some element w belonging to the set $e_o + \frac{1}{2}[-e_o, e_o]$. Such a w satisfies $\frac{1}{2}e_o \leq w \leq ku_o$, $k \in \mathbb{N}$, showing that u_o is an order unit for F_o). Hence, there exists an $\varepsilon > 0$ such that

$$\varepsilon e_o \leq u_o = P_o u = P_o S e_o \leq S e_o,$$

and our proof is complete.

REMARK. The preceding theorem holds true without assuming the space to be Dedekind complete. For our purposes the theorem in the above form will suffice. (See also [6] and [7]).

3. The Spectral Radius Theorem

We assume henceforth that E is a Dedekind complete Banach lattice. For any band $B \subset E$ we denote its disjoint complement by B^d and the band projection onto B by P_B. It is not difficult to see that an operator $0 < T \in L(E)$ is band irreducible if and only if $P_{B^d} T P_B > 0$ for every band $\{0\} \neq B \neq E$. Moreover, using the operator S as in the proof of Theorem 2.1 one readily sees that if $E'_{oo} \neq \{0\}$ and if T is order continuous and band irreducible, then both E and E'_{oo} have weak order units ([8] Theorem 136.5). If e and e' are weak order units for E and E'_{oo} respectively, it follows from the inequality $(e' \wedge g') \otimes (e \wedge h) \leq (e' \otimes e) \wedge (g' \otimes h)$ that every element $g' \otimes h \in E'_{oo} \otimes E$ is in the band generated by $e' \otimes e$ in $L^b(E)$. This implies that $e' \otimes e$ is a weak order unit for $(E'_{oo} \otimes E)^{dd}$. (See also [8] Lemma 136.6.)

LEMMA 3.1. _Let_ E _be_ _a_ _Dedekind_ _complete_ _Banach_ _lattice_ _and_ _let_

$0 < T \in (E'_{00} \otimes E)^{dd}$ _be band irreducible. If_ e' _and_ e _are weak order units for_ E'_{00} _and for_ E _respectively then_ $S = (e' \otimes e) \wedge T$ _is band irreducible._

PROOF. Let $\{0\} \neq B \neq E$ be a band in E. Then $n(P_{B^d}[(e' \otimes e) \wedge T]P_B) \geq P_{B^d}[n(e' \otimes e) \wedge T]P_B \uparrow P_{B^d} T P_B > 0$ and it follows that $P_{B^d}[(e' \otimes e) \wedge T]P_B > 0$. By our remark above $S = (e' \otimes e) \wedge T$ is band irreducible.

Before proving our next lemma, we mention a general fact from spectral theory. Let $T : X \to X$ be a linear operator on the Banach space X belonging to $L(X)$. Suppose furthermore that $T = T_2 T_1$ where $T_1 : X \to Y$, $T_2 = Y : \to X$ with Y also a Banach space and all the operators continuous and linear. If $S := T_1 T_2 \in L(Y)$ then $r(T) = r(S)$. This follows from the spectral radius formula:

$$r(T) = \lim \|(T_2 T_1)^n\|_X^{1/n} \leq \lim \|T_2\|^{1/n} \|(T_1 T_2)^{n-1}\|_Y^{1/n} \|T_1\|^{1/n} =$$

$$= \lim \{\|(T_1 T_2)^{n-1}\|_Y^{1/n-1}\}^{1-1/n} = r(T_1 T_2) = r(S).$$

By symmetry $r(S) \leq r(T)$ also holds.

LEMMA 3.2. _Let_ E _be a Dedekind complete Banach lattice and let_ $0 < T \leq e' \otimes e \in E'_{00} \otimes E$. _If_ T _is band irreducible, then_ $r(T) > 0$ _holds._

PROOF. Let $f := \|e'\| e$ and let U be the unit ball in E. It follows from $0 \leq T \leq e' \otimes e$ that $T(U) \subset [-f, f]$. The ideal E_f normed by p_f is an AM-space with unit which we shall denote by M. The operator T can be factored as $T = i.T_0$ where $i : M \to E$ is the embedding map and $T_0 \in L(E, M)$ is defined by $T_0 x = Tx$ for all $x \in E$. Both i and T_0 are continuous. Let F be the closed ideal generated in M by $T_0(E) = T(E)$. Then T factors through F in the same way. By the remark above the operators $T = i.T_0 : E \to E$ and $S := T_0.i : F \to F$ have the same spectral radius and our proof will be complete if we can show that $S : F \to F$ is irreducible (Theorem 2.1). If $\{0\} \neq I$ is a closed ideal in F such that $SI = (T_0.i)I \subset I$, then $i(I)$ is a T-invariant ideal in E and since T is band irreducible and order continuous, $i(I)$ is order dense in E. From $0 < T \leq e' \otimes e$ it follows that T_0 maps order convergent nets in E onto norm convergent nets in F and as a consequence we find that $T_0(E) \subset I$. Since F is the closed ideal generated by $T_0(E)$, we have $I = F$ and our proof is complete.

THEOREM 3.3 *Let E be a Dedekind complete Banach lattice. If $T > 0$ is band irreducible and if $T \in (E'_{00} \otimes E)^{dd}$, then $r(T) > 0$ holds.*

PROOF. Our assumption that $0 < T \in (E'_{00} \otimes E)^{dd}$ implies that $E'_{00} \neq \{0\}$ and by a remark above there exist weak order units e' and e for E'_{00} and E respectively. Let $S := (e' \otimes e) \wedge T$, then S is band irreducible by Lemma 3.1 and $r(S) > 0$ by Lemma 3.2. Hence, $r(T) \geq r(S) > 0$ holds.

Theorem 3.3 was proved by T. Ando [1] for compact irreducible operators. The compactness condition was removed by H.J. Krieger [4] and the present author proved in [2] the result in the form stated above. The method used in this section was used by the present author in [3] to show that the representation theory for kernel operators could be avoided in proving 3.3. For the sake of completeness we reproduced the proofs.

To complete the program mentioned in the introduction we observe that the following two theorems can now be proved without representation. For the proofs we refer to [8].

THEOREM 3.4. (*Generalized Jentzsch theorem*). *Let E be a Dedekind complete Banach lattice such that E'_{00} separates the points of E and let T be a positive operator in E such that T is compact and band irreducible and T belongs to $(E'_{00} \otimes E)^{dd}$. Then the spectral radius r(T) of T is an eigenvalue of T of multiplicity one, i.e. the corresponding eigenspace is of dimension one. Furthermore, each non-zero element in the eigenspace is a weak order unit in E.*

THEOREM 3.5. (*Generalized Frobenius theorem*). *Let E and T satisfy the same conditions as in Theorem 3.4. If $\lambda_1, \lambda_2, \ldots, \lambda_k$ are the eigenvalues of T satisfying $|\lambda_i| = r(T)$ $(i = 1,2,\ldots,k)$ then these numbers are exactly the k roots of the equation $\lambda^k - r(T)^k = 0$. Furthermore, the spectrum $\sigma(T)$ of T is invariant under the rotation of the complex plane by the angle $2\pi k^{-1}$ (multiplicities included). This implies in particular that all eigenvalues $\lambda_1, \ldots, \lambda_k$ satisfying $|\lambda_i| = r(T)$ $(i = 1,2,\ldots,k)$ are of multiplicity one.*

COROLLARY 3.6. *Let E and T satisfy the same conditions as in the preceding theorems. If T is moreover strongly band irreducible, then any eigenvalue λ of T different from r(T) satisfies $|\lambda| < r(T)$.*

PROOF. Let $\lambda \neq r(T)$ be an eigenvalue of T satisfying $|\lambda| = r(T)$ and let x be an eigenvector pertaining to λ. By a wellknown argument (see [8] Theorem 137.3) it follows that $|x|$ is an eigenvector pertaining to $r(T) = r$ and since $\lambda \neq r$ the vectors x and $|x|$ are linearly independent. From the general Frobenius theorem there exists a number $k \in \mathbb{N}$ such that $\lambda^k = r^k$. The operator T^k satisfies all the conditions of the generalized Jentzsch theorem and $r(T^k) = r^k$. The eigenspace pertaining to r^k has therefore dimension one contradicting the fact that both x and $|x|$ belong to this eigenspace. Hence, all eigenvalues λ satisfying $|\lambda| = r(T)$ must be equal to $r(T)$.

Corollary 3.6 is usually presented as a part of the theorem of Jetzsch.

REFERENCES

[1] Ando, T., Positive operators in semi-ordered linear spaces. J. Fac. Sci. Hokkaido Univ., Ser. I. 13 (1957), 214-228.

[2] Grobler, J.J., On the spectral radius of irreducible and weakly irreducible operators in Banach lattices. Quaest. Math. 2 (1978), 495-506.

[3] Grobler, J.J., A short proof of the Ando-Krieger theorem. Math. Z. 174 (1980), 61-62.

[4] Krieger, H.J., Beiträge zur Theorie positiver Operatoren. Schriftenreihe der Institute für Math.,Reihe A, Heft 6. Berlin: Akademie Verlag (1969).

[5] Luxemburg, W.A.J.-Zaanen, A.C., Riesz Spaces I. North-Holland Publishing Company, Amsterdam/London 1971.

[6] Schaefer, H.H., Topologische Nilpotenz irreduzibler Operatoren. Math. Z. 117 (1970), 135-140.

[7] Schaefer, H.H., Banach Lattices and Positive Operators. Springer-Verlag, Berlin/Heidelberg/New York 1974.

[8] Zaanen, A.C., Riesz Spaces II. North-Holland Publishing Company, Amsterdam/London 1983.

International Series of
Numerical Mathematics, Vol. 65

POSITIVE COMMUTING PERTURBATIONS OF SELFADJOINT OPERATORS AND HYPONORMALITY

C. R. Putnam[1]
Department of Mathematics
Purdue University
West Lafayette

Let P be a selfadjoint operator on a separable, infinite dimensional Hilbert space. Then there exists a completely hyponormal operator T having a polar factorization T = UP, U unitary, and satisfying the condition that T*T and TT* commute, if and only if $P \geqq 0$ and $\sigma(P)$ contains at least two points, 0 is not in $\sigma_p(P)$, and, whenever $\sigma_p(P)$ is not empty, neither $\sup \sigma_p(P)$ nor $\inf \sigma_p(P)$ belongs to $\sigma_p(P)$ with a finite multiplicity.

1. Introduction

Only bounded operators on a separable, infinite dimensional Hilbert space H will be considered. Let T be completely hyponormal, so that

$$T^*T - TT^* = D \geqq 0,$$

where T has no nontrivial reducing subspace on which it is normal. In addition, let T have the polar factorization

(2) $$T = UP,\ P = (T^*T)^{\frac{1}{2}} \text{ and } U \text{ is unitary},$$

so that

(3) $$P^2 - UP^2U^* = D \geqq 0, \quad \text{where} \quad T^*T = P^2 \quad \text{and} \quad TT^* = UP^2U^*.$$

(Such a factorization (2) exists if and only if 0 is not an eigenvalue of T^*; see [2].) Let $\sigma(P)$ and $\sigma_p(P)$ denote, respectively, the spectrum and point spectrum of P.

The following was proved in [1]:

(*) <u>Let</u> P <u>be a selfadjoint operator on</u> H. <u>Then there exists a completely hyponormal operator</u> T <u>satisfying</u> (2) <u>for some unitary</u> U <u>if and only if</u>

---

[1]This work was supported by a National Science Foundation Research Grant.

(A) $P \geqq 0$ _and_ $\sigma(P)$ _contains at least two points_;

(B) $0 \notin \sigma_p(P)$;

(C) _neither_ $\max \sigma(P)$ _nor_ $\min \sigma(P)$ _is in_ $\sigma_p(P)$ _with a finite multiplicity_.

There will be proved the following analogous result:

THEOREM. _Let_ P _be a selfadjoint operator on_ H. _Then there exists a completely hyponormal operator_ T _satisfying_ (2) _for some unitary_ U _and, in addition, the condition_

(4) $$T^*T \text{ and } TT^* \text{ commute},$$

if and only if both (A) _and_ (B) _above are satisfied and also_

(C)' _whenever_ $\sigma_p(P)$ _is not empty, neither_ $\sup \sigma_p(P)$ _nor_ $\inf \sigma_p(P)$ _is in_ $\sigma_p(P)$ _with a finite multiplicity_.

2. Proof of the Theorem.

Actually, the "if" part of the Theorem was implicitly proved in [2]. In fact, the conditions (A), (B) and (C)' were shown in [2] to be sufficient conditions in order that there exist a completely hyponormal T satisfying (2). A perusal of the argument given shows, however, that the operator T constructed also satisfies (4). In fact, in the construction of [2], both operators P^2 and UP^2U^* (cf. (2) and (3)) were either (unitarily equivalent, via a common unitary operator, to) direct sums of multiplication operators on $L^2(0,1)$, or direct sums of diagonal operators on corresponding ℓ^2 spaces, or combinations of these two cases. Thus, in order to complete the proof of the Theorem it is enough to show that the conditions (A), (B) and (C)' are also necessary for the validity of both (2) and (4). The necessity of both (A) and (B) already follows from (*) but, in any case, is easily proved, so that only the necessity of (C)' need be shown.

Assume then (2) and (4) and, as may be supposed, that $\sigma_p(P)$ is not empty. Let $a = \inf \sigma_p(P^2)$ and $b = \sup \sigma_p(P^2)$ and define

(5) $$E_a = \{x: P^2x = ax\} \quad \text{and} \quad E_b = \{x: P^2x = bx\} .$$

It will be shown that each subspace E_a and E_b is either $\{0\}$ or has dimension ∞. (It is to be noted that the conditions (A), (B) and (C)' hold if and only if analogous conditions hold with P replaced by P^2.) Suppose then that (C)' fails to hold, so that either

(6a) $$0 < \dim E_a < \infty$$

or

(6b) $$0 < \dim E_b < \infty \quad .$$

First, suppose (6a). In view of (3), condition (4) is equivalent to

(7) $$P^2D = DP^2 \ ,$$

and so E_a is invariant under both P^2 and D. Further,

(8) $$E_a \subset N(D),$$

where $N(D)$ is the null space of D. For, otherwise, it follows from (6a) that there exists some $y \neq 0$ in E_a for which $Dy = cy$ with $c > 0$. By (3), $UP^2U^*y = (P^2 - D)y = (a - c)y$, so that $a - c$ is in $\sigma_p(UP^2U^*)$ and hence also in $\sigma_p(P^2)$, in contradiction with the definition $a = \inf \sigma_p(P^2)$.

Consequently, for each x in E_a, $UP^2U^*x = (P^2 - D)x = ax$ and hence $P^2U^*x = aU^*x$. Thus, E_a is invariant under U^* and, by (6a), U^* must have an eigenvector in E_a. Clearly, the span of such an eigenvector is a reducing subspace of T on which T is normal, in contradiction with the complete hyponormality hypothesis. Thus (6a) is untenable.

Next, suppose (6b). By (3), $U^*P^2U - P^2 = U^*DU \equiv D_1$. Further, (3) and (7) imply that $UP^2U^*D = DUP^2U^*$ and hence also that $P^2D_1 = D_1P^2$. An argument similar to that of above then shows that, corresponding to (8), $E_b \subset N(D_1)$. Hence, for each x in E_b, $U^*P^2Ux = (P^2 + D_1)x = P^2x = bx$, and so $P^2Ux = bUx$, so that E_b is invariant under U. Thus, by (6b), U would have an eigenvector in E_b whose span is a normal reducing subspace of T, again a contradiction. Thus, neither (6a) nor (6b) can hold, and (C)' is established. This completes the proof.

REFERENCES

[1] Clancey, K.F. - Putnam. C.R., Nonnegative perturbations of selfadjoint operators. J. Funct. Anal. 50 (1983), 306 - 316.

[2] Putnam, C.R., Absolute values of completely hyponormal operators. J. Operator Theory 8 (1982), 319 - 326.

International Series of
Numerical Mathematics, Vol. 65

THE RESOLVENT OF GENERALIZED SELF - ADJOINT OPERATORS AND A RELATED GROWTH PROBLEM

Lâszlo Zsidõ
Mathematisches Institut A
Universität Stuttgart

A theorem of J.R.Partington on the resolvent of hermitian operators ([11], Th.5) is extended to a wider class of operators and a related problem on the growth of subharmonic functions is discussed.

1. The Resolvent of m - Self - Adjoint Operators.

Let X be a complex Banach space and X' its dual. The Hahn-Banach theorem ensures that for each $x \in X$

$$J(x) = \{\phi \in X' ; \|\phi\| = \|x\| ; \phi(x) = \|x\|^2\} \neq \emptyset .$$

(a good reference for the task of the "duality mapping" J in the functional analysis is [2]). For a closed linear operator T: $X \supset \mathfrak{D}_T \to X$ we denote the spectrum of T by $\sigma(T)$.

Now T is called hermitian, if for every $x \in \mathfrak{D}_T$ there exists some $\phi \in J(x)$ with

$$\phi(T x) \in R$$

and $\sigma(T)$ contains neither $\{\zeta \in C; \operatorname{Im}\zeta > 0\}$, nor $\{\zeta \in C; \operatorname{Im}\zeta < 0\}$. Using [9], Lemma 3.1 it es easy to see that if T is hermitian then

$$\sigma(T) \subset R \quad \text{and} \quad \|(\lambda - T)^{-1}\| \leq \frac{1}{|\operatorname{Im}\lambda|} \quad \text{for} \quad \lambda \in C \setminus R .$$

On the other hand, if the above condition is fulfiled then by the proof of [13], Th.2

$$\phi(T x) \in R \text{ for each } x \in \mathfrak{D}_T \text{ and each } \phi \in J(x) ,$$

so T is hermitian.

Recently J.R.Partington has proved.

THEOREM 1.1 ([11], Th.5). *If* T *is a hermitian operator in a complex Banach space, then*

$$\|(\lambda - T)^{-1}\| \leqq (1+\pi^2)^{\frac{1}{2}} d(\lambda, \sigma(T))^{-1} \quad \text{for} \quad \lambda \in C \setminus \sigma(T)\,,$$

where $d(\lambda, \sigma(T))$ *denotes the distance of* λ *to* $\sigma(T)$.

Accordingly to [14] and [4], Ch.5,§ 4, we call a closed linear operator $T\colon X \supset \mathfrak{D}_T \to X$ m - *self - adjoint* for some integer $m \geqq 0$, if $\sigma(T) \subset R$ and there is some constant $c > 0$ with

$$\|(\lambda - T)^{-1}\| \leqq \frac{c}{|\operatorname{Im}\lambda|^{m+1}} \quad \text{for} \quad \lambda \in C \setminus R\,.$$

In this case we denote the least constant c, for which the above condition holds, by c_T.

It is natural to ask, whether resolvent estimates like in the above theorem hold for m - self - adjoint operators. We note that there are 0 - self - adjoint operators in Hilbert spaces, which are not similar to hermitian operators ([10], § 4, Section 1), so even the case of 0 - self - adjoint operators can not be reduced to the above theorem.

The following result in the same time extends and sharpenes Th.1.1:

THEOREM 1.2. *If* T *is an* m - *self - adjoint operator in a complex Banach space, then*

$$\|(\lambda - T)^{-1}\| \leqq (2\sqrt{e})^{m+1} c_T d(\lambda, \sigma(T))^{-m-1} \quad \text{for} \quad \lambda \in C \setminus \sigma(T)\,.$$

We remark, that for a closed linear operator T in a complex Banach space the function

$$C \setminus \sigma(T) \ni \lambda \to \ln \|(\lambda - T)^{-1}\|$$

is subharmonic (accordingly to [12], 4.3 or [8], 2.1). Thus, taking in account that

$$\mathbb{C} \setminus \sigma(T) \ni \lambda \to \frac{1}{m+1} (\ln \|(\lambda - T)^{-1}\| - \ln c_T)$$

is subharmonic, Th.1.2 is a direct consequence of

THEOREM 1.3. *Let* $\Omega \subset \mathbb{C}$ *be an open set and* $u : \Omega \to [-\infty, +\infty)$ *a subharmonic function such that*

$$u(\lambda) \leq \ln|\operatorname{Im} \lambda|^{-1} \quad \text{for} \quad \lambda \in \Omega \setminus \mathbb{R} .$$

Then

$$u(\lambda) \leq \ln(2\sqrt{e}\, d(\lambda, \partial\Omega)^{-1}) \quad \text{for} \quad \lambda \in \Omega .$$

PROOF. Let $\lambda \in \Omega$ and $0 < r < d(\lambda, \partial\Omega)$ be arbitrary. Since the closed disc $B_r(\lambda) = \{\zeta \in \Omega; |\zeta - \lambda| \leq r\}$ is entirely contained in Ω, we have by the subharmonicity of u

$$\pi r^2 u(\lambda) \leq \int_{B_r(\lambda)} u(\xi + i\eta)\, d\xi\, d\eta .$$

But $u(\xi + i\eta) \leq \ln|\eta|^{-1}$, so we get

$$\pi r^2 u(\lambda) \leq \int_{B_r(\lambda)} \ln|\eta|^{-1} d\xi\, d\eta =$$

$$= \int_{x^2+y^2 \leq r^2} \ln|y + \operatorname{Im} \lambda|^{-1} dx\, dy =$$

$$= \int_{-r}^{r} \left(\int_{-\sqrt{r^2-x^2}}^{\sqrt{r^2-x^2}} \ln|y + \operatorname{Im} \lambda|^{-1} dy \right) dx .$$

Now we remark, that for every $a > 0$ and $b \in \mathbb{R}$

$$\int_{-a}^{a} \ln|y + b|^{-1} dy \leq \int_{-a}^{a} \ln|y|^{-1} dy .$$

Indeed, if $|b| \leq a$ then

$$\int_{-a}^{a} \ln|y|^{-1} dy - \int_{-a}^{a} \ln|y + b|^{-1} dy =$$

$$\int_{a-|b|}^{a} \ln|y|^{-1} dy - \int_{a}^{a+|b|} \ln|y|^{-1} dy \geq 0$$

and if $|b| > a$ then

$$\int_{-a}^{a} \ln|y|^{-1}dy - \int_{-a}^{a} \ln|y|^{-1}dy =$$

$$= \left(\int_{0}^{a} \ln|y|^{-1}dy - \int_{a}^{2a} \ln|y|^{-1} dy\right) +$$

$$+\left(\int_{0}^{|b|-a} \ln|y|^{-1} - \int_{2a}^{|b|+a} \ln|y|^{-1}dy\right) \geqq 0 .$$

By the above remark we have

$$\pi r^2 u(\lambda) \leqq \int_{-r}^{r} \left(\int_{-\sqrt{r^2-x^2}}^{\sqrt{r^2-x^2}} \ln|y|^{-1}dy\right) dx =$$

$$= \int_{x^2+y^2 \quad r^2} \ln|y|^{-1} dx\, dy =$$

$$= \int_{0}^{r} \left(\int_{0}^{2\pi} \ln|\rho \sin\Theta|^{-1} d\Theta\right) \rho\, d\rho =$$

$$= \int_{0}^{r} (2\pi \ln \rho^{-1} + 2\pi \ln 2)\rho\, d\rho =$$

$$= \pi r^2\left(\frac{1}{2} + \ln r^{-1}\right) + \pi r^2 \ln 2 =$$

$$= \pi r^2 \ln (2\sqrt{e}\, r^{-1}) ,$$

$$u(\lambda) \leqq \ln(2\sqrt{e}\, r^{-1}) .$$

Since $0 < r < d(\lambda,\partial\Omega)$ is arbitrary, we conclude that

$$u(\lambda) \leqq \ln (2\sqrt{e}\, d(\lambda, \partial\Omega)^{-1}) .$$

2. A general Growth Estimate for Subharmonic Functions.

We consider here the following problem:

Let $f: [0,+\infty) \to [0,+\infty)$ be an increasing function, $\Omega \subset C$ an open set and $u: \Omega \to [-\infty,+\infty)$ a subharmonic function such that

$$u(\lambda) \leqq f(|\mathrm{Im}\ \lambda|^{-1}) \quad \text{for} \quad \lambda \in \Omega \setminus R .$$

Then ask for a reasonable upper estimate for $u(\lambda)$ only in terms of f and $d(\lambda, \partial\Omega)$.

We note that Th. 1.3 gives a solution of the above problem in the case $f(t) = \ln^+ t = \max\{0, \ln t\}$.

By a classical result of N.Sjöberg, N.Levinson and A.Beurling, if $\int_1^{+\infty} \frac{\ln f(t)}{t^2}\, dt < +\infty$ then u is bounded above on $\{\lambda \in \Omega;$ $d(\lambda, \partial\Omega) \geqq \varepsilon\}$ by a constant depending only on f and $\varepsilon > 0$ (for recent treatments of this result we send to [15],[5], Th.3, 1 , Lemma II, [7],[6],[3], Th. 1.12,[16], Propositions 4.1 and 4.4). So the problem consists in giving an appropriate estimate for this constant. This is essentially due in [3] from which we extract the next theorem. We note that there is no loss of generality if we assume that f is strictly increasing and continuous, because else f can be replaced by the equivalent growth modulus

$$t \to \frac{t}{1+t} + \int_t^{t+1} f(s)\, ds .$$

THEOREM 2.1. _Let_ $f: [0,+\infty) \to [0,+\infty)$ _be a continuous strictly increasing function such that_ $f(1) = 1$ _and_ $\int_1^{+\infty} \frac{\ln f(t)}{t^2}\, dt < +\infty$. _Let us define_ $1 = \alpha_1 < \alpha_1 < \alpha_2 < \dots$ _by_

$$f(\alpha_j) = e^{j-1} \quad \text{for} \quad e^{j-1} < \lim_{t\to+\infty} f(t) ,$$

$$\alpha_j = +\infty \quad \text{else.}$$

Then

$$\sum_{j=1}^{\infty} \frac{1}{\alpha_j} < +\infty$$

and for every open set $\Omega \subset C$ _and every subharmonic function_ u: $\Omega \to [-\infty,+\infty)$ _with_

$$u(\lambda) \leqq f(|\mathrm{Im}\,\lambda|^{-1}) \quad \text{for} \quad \lambda \in \Omega \setminus R$$

we have

$$u(\lambda) \leqq e^k \quad \text{for} \quad \lambda \in \Omega,\ d(\lambda, \partial\Omega) \geqq \frac{4(e+1)}{\pi} \sum_{j=k}^{\infty} \frac{1}{\alpha_j} .$$

PROOF. (accordingly to [3]). We have for each $n \geqq 2$

$$\sum_{j=1}^{n} \frac{1}{\alpha_j} = 1 + \sum_{j=2}^{n} (j-1)\left(\frac{1}{\alpha_j} - \frac{1}{\alpha_{j+1}}\right) + (n-1)\frac{1}{\alpha_{n+1}} \leqq$$

$$\leq 1 + \sum_{j=2}^{n} \int_{\alpha_j}^{\alpha_{j+1}} \frac{j-1}{t^2}\, dt + \int_{\alpha_{n+1}}^{+\infty} \frac{n}{t^2}\, dt \leq$$

$$\leq 1 + \int_{\alpha_2}^{+\infty} \frac{\ln f(t)}{t^2}\, dt ,$$

so

$$\sum_{j=1}^{\infty} \frac{1}{\alpha_j} \leq 1 + \int_{\alpha_2}^{+\infty} \frac{\ln f(t)}{t^2}\, dt < +\infty .$$

Now let u be as in the statement and let

$$\lambda \in \Omega,\ d(\lambda, \partial\Omega) \geq \frac{4(e+1)}{\pi} \sum_{j=k}^{\infty} \frac{1}{\alpha_j}$$

be arbitrary. We show that the assumption $u(\lambda) > e^k$ contradicts the upper boundedness of the upper semi-continuous function u on the compact subsets of Ω.

We claim that if $u(\lambda) > e^k$ then there exists $\lambda_1 \in \Omega$ with

$$|\lambda - \lambda_1| < \frac{4(e+1)}{\pi} \frac{1}{\alpha_k} \quad \text{and} \quad u(\lambda_1) > e^{k+1} .$$

Indeed, denoting

$$D_1 = \{z \in C;\ |z - \lambda| < \frac{4(e+1)}{\pi} \frac{1}{\alpha_k},\ |\mathrm{Im}\, z| < \frac{1}{\alpha_k}\}$$

$$D_2 = \{z \in C;\ |z - \lambda| < \frac{4(e+1)}{\pi} \frac{1}{\alpha_k},\ |\mathrm{Im}\, z| \geq \frac{1}{\alpha_k}$$

we should else have by the subharmonicity of u

$$\pi \left(\frac{4(e+1)}{\pi} \frac{1}{\alpha_k}\right)^2 e^k < \pi \left(\frac{4(e+1)}{\pi} \frac{1}{\alpha_k}\right)^2 u(\lambda) \leq$$

$$\leq \int_{D_1} u(\xi + i\eta)\, d\xi\, d\eta + \int_{D_2} u(\xi + i\eta)\, d\xi\, d\eta \leq$$

$$\leq \text{area}\,(D_1)\, e^{k+1} + \left[\pi \left(\frac{4(e+1)}{\pi} \frac{1}{\alpha_k}\right)^2 - \text{area}\,(D_1)\right] e^{k-1} ,$$

$$\pi \left(\frac{4(e+1)}{\pi} \frac{1}{\alpha_k}\right)^2 (e^k - e^{k-1}) < \text{area}\,(D_1)(e^{k+1} - e^{k-1}) \leq$$

$$\leq \left(2 \frac{4(e+1)}{\pi} \frac{1}{\alpha_k}\right) \left(2 \frac{1}{\alpha_k}\right)(e^{k+1} - e^{k-1}) =$$

$$= \pi \left(\frac{4(e+1)}{\alpha_k}\right)^2 (e^k - e^{k-1}) ,$$

which is absurd.

Repeating the above reasoning, we deduce the existence of $\lambda_2, \lambda_3, \ldots \in \Omega$ such that for each $j \geq 1$

$$|\lambda_j - \lambda_{j+1}| < \frac{4(e+1)}{\pi} \frac{1}{\alpha_{k+1}} \quad \text{and} \quad u(\lambda_{j+1}) > e^{k+j+1} .$$

Consequently u is not upper bounded on the compact set

$$\{\lim_{j\to\infty} \lambda_j, \lambda_1, \lambda_2, \ldots\} \subset \{z \in C;\ |\lambda - z| < \frac{4(e+1)}{\pi} \sum_{j=k}^{\infty} \frac{1}{\alpha_k}\} \subset \Omega$$

and the proof is complete.

Let f and $\alpha_1, \alpha_2, \ldots$ be as in Th.2.1. We denote

$$c_f = \sup_{\substack{k \geq 1 \\ \alpha_k < +\infty}} \alpha_k \sum_{j=k}^{B} \frac{1}{\alpha_j} \in [1, +\infty] .$$

Then we have:

COROLLARY 2.2. <u>Let</u> f: $[0,+\infty) \to [0,+\infty)$ <u>be a continuous strictly increasing function such that</u> $f(1) = 1$, $\int_1^{+\infty} \frac{\ln f(t)}{t^2} dt < +\infty$ <u>and</u> $c_f < +\infty$. <u>Then for every open set</u> $\Omega \subset C$ <u>and every subharmonic function</u> $u: \Omega \to [-\infty, +\infty)$ <u>with</u>

$$u(\lambda) \leq f(|\mathrm{Im}\,\lambda|^{-1}) \quad \underline{\text{for}} \quad \lambda \in \Omega \setminus R$$

<u>we have</u>

$$u(\lambda) \leq \max\{e, e^2 f(\frac{4(e+1)}{\pi} c_f\, d(\lambda, \partial\Omega)^{-1})\} \quad \underline{\text{for}} \quad \lambda \in \Omega$$

PROOF. Let $\lambda \in \Omega$ be arbitrary. In view of Th.2.1, if

$$d(\lambda, \partial\Omega) \geq \frac{4(e+1)}{\pi} \sum_{j=1}^{\infty} \frac{1}{\alpha_j}$$

then

$$u(\lambda) \leq e$$

and if

$$\frac{4(e+1)}{\pi} \sum_{j=k}^{\infty} \frac{1}{\alpha_j} > d(\lambda, \partial\Omega) \geq \frac{4(e+1)}{\pi} \sum_{j=k+1}^{\infty} \frac{1}{\alpha_j}$$

for some $k \geq 1$ with $\alpha_k < +\infty$, then

$$u(\lambda) \leq e^{k+1} = e^2 f(\alpha_k) \leq$$

$$\leqq e^2 f\left(\frac{4(e+1)}{\pi}\alpha_k \sum_{j=k}^{\infty}\frac{1}{\alpha_j}\, d(\lambda,\partial\Omega)^{-1}\right) \leqq$$
$$\leqq e^2 f\left(\frac{4(e+1)}{\pi} c_f\, d(\lambda,\partial\Omega)^{-1}\right).$$

For example, if $f(t) = t^{\alpha}$ for some $\alpha > 0$ then

$$c_f = \frac{e^{1/\alpha}}{e^{1/\alpha}-1} .$$

Finally we remark that, using Th.2.1 it is easy to see that if f and $\alpha_1,\alpha_2,\ldots$ are as in Th.2.1 and, additionally

$$(*) \qquad \int_1^{+\infty} \frac{(\ln f(t))^2}{t^2}\, dt \to +\infty$$

or, equivalently,

$$\sum_{j=1}^{\infty} \frac{j}{\alpha_j} < +\infty$$

then there exists an increasing function $g\colon [0,+\infty) \to [0,+\infty)$ with $\int_1^{+\infty} \frac{\ln g(t)}{t^2}\, dt < +\infty$ such that, for every subharmonic function $u\colon \Omega \to [-\infty,+\infty)$ satisfying the estimate.

$$u(\lambda) \leqq f(|\mathrm{Im}\,\lambda|^{-1}) \quad \text{for} \quad \lambda \in \Omega \setminus R ,$$

we have

$$u(\lambda) \leqq g(d(\lambda,\partial\Omega)^{-1} \quad \text{for} \quad \lambda \in \Omega .$$

If $c_f < +\infty$ then $(*)$ is satisfied:

$$\sum_{j=1}^{\infty} \frac{1}{\alpha_j} = \sum_{k=1}^{\infty} \sum_{j=k}^{\infty} \frac{1}{\alpha_j} \leqq \sum_{k=1}^{\infty} c_f \frac{1}{\alpha_k} < +\infty .$$

However, for example, $f(t) = e^{t^{\varepsilon}}$ with some $0 < \varepsilon < \frac{1}{2}$ satisfies $(*)$ in spite of $c_f = +\infty$.

PROBLEM. Is $(*)$ necessary to the conclusion of the above remark ?

3. The Resolvent of Generalized Self - Adjoint Operators.

Let X be a complex Banach space. Accordingly to [3], § 6 we call a closed linear operator $T\colon X \supset \mathfrak{D}_T \to X$ generalized self - adjoint, if $\sigma(T) \subset R$ and there is some increasing function F:

$[0,+\infty) \to [0,+\infty)$ with $\int_1^{+\infty} \frac{\ln^+ \ln^+ F(t)}{t^2}\, dt < +\infty$ such that

$$\|(\lambda - T)^{-1}\| \leq F\left(\frac{1}{|\operatorname{Im} \lambda|}\right) \text{ for } \lambda \in \mathbb{C} \setminus \mathbb{R} .$$

An F satisfying the above inequality is called growth function for T.

From Cor. 2.2 follows immediately:

THEOREM 3.1. *Let T be a generalized self - adjoint operator in a complex Banach space and $F = c\, e^f$ a growth function for T, where $c > 0$ and $f: [0,+\infty) \to [0,+\infty)$ is a continuous strictly increasing function with $f(1) = 1$, $\int_1^{+\infty} \frac{\ln f(t)}{t^2}\, dt < +\infty$ and $c_f < +\infty$. Then we have for all $\lambda \in \mathbb{C} \setminus \sigma(T)$*

$$\|(\lambda - T)^{-1}\| \leqq \max\{c\, e^e,\ c^{1-e^2} F\left(\frac{4(e+1)}{\pi}\, c_f\, d(\lambda, \sigma(T))^{-1}\right)^{e^2}\} .$$

We note that the growth functions which occur in Th.3.1 are essentially the increasing functions $F: [0,+\infty) \to [0,+\infty)$ with $\int_1^{+\infty} \frac{\ln^+ \ln^+ F(t)}{t^2}\, dt < +\infty$ satisfying

$$\sup_{t \geqq 1} t \int_t^{+\infty} \frac{\ln^+ \ln^+ F(s)}{s^2}\, ds < +\infty .$$

Each F as above satisfies

$$(**) \qquad \int_1^{+\infty} \frac{(\ln^+ \ln^+ F(t))^2}{t^2}\, dt < +\infty,$$

but the converse implication do not hold. By the last remark in Section 2, if an increasing function $F: [0,+\infty) \to [0,+\infty)$ satisfies (**) then there exists an increasing function $G: [0,+\infty) \to [0,+\infty)$ with $\int_1^{+\infty} \frac{\ln^+ \ln^+ G(t)}{t^2}\, dt < +\infty$ such that, for every generalized self - adjoint operator T with growth function F,

$$\|(\lambda - T)^{-1}\| \leqq G(d(\lambda,\sigma(T))^{-1}) \text{ for } \lambda \in \mathbb{C} \setminus \sigma(T) .$$

Now resolvent functions are particular analytic functions and the resolvent functions of generalized self - adjoint operators T are still more particular: If T is densely defined then its resolvent function is by [3], Th.6.9 the Cauchy transform of a multiplicative operator valued ultradistribution on R with support $\sigma(T)$. So, independently from the solution of the Problem in Section 2, it is meaningful to ask:

PROBLEM. _Can (**) be replaced in the above statement by_

$$\int_1^{+\infty} \frac{\ln^+ \ln^+ F(t)}{t^2}\, dt < +\infty \;?$$

REFERENCES

[1] Beurling, A., _Analytic continuation across a linear boundary_. Acta Math. 128 (1972), 153 - 182.

[2] Ciorănescu, I., _Duality Mappings in the Non - Linear Functional Analysis_ (in Rumanian). Editura Academici, Bucureşti 1974; English review, MR 52 4038.

[3] Ciorănescu, I. - Zsidó, L., _ω - Ultradistributions and their application to operator theory_. Spectral Theory (Banach Center Pullications, Vol.8), Polish Scientific Publishers, Warsaw 1982, 77 - 220.

[4] Colojoară, I. - Foiaş, C., _Theory of Generalized Spectral Operators_. Gordon ánd Breach, New York / London / Paris 1968.

[5] Domar, Y., _On the existence of a largest subharmonic minorant of a given function_. Arkiv för Mat. 3 (1958), 429 - 440.

[6] Dyn'kin, E.M., _Functions with given estimate for ∂f / ∂z̄ and N. Levinson's theorem_ (in Russian). Mat.Sbornik 89(131) (1972), 182 - 190; English transl. in Math. USSR Sbornik 18 (1972), 181 - 189.

[7] Gurariĭ, V.P., _On Levinson's theorem concerning families of analytic functions_ (in Russian). Zap. Naucn. Sem. Leningrad. Otdel. Mat. Inst. Steklov. (LOMI) 19 (1970), 215 - 220; English transl. in Sem. Math. Inst. Leningrad 19 (1972), 124 - 127.

[8] Hayman, W.K. - Kennedy, P.B., _Subharmonic Functions_, Vol. 1. Academic Press, London / New York / San Francisco 1976.

[9] Lumer, G. - Phillips, R.S., _Dissipative Operators in a Banach Space_. Pacific J. Math. 11 (1961), 679 - 698.

[10] Markus, A.S., _Some criteria for the completeness of a system_

of root vectors of a linear operator in a Banach space (in Russian). Math. Sbornik 70(112) (1966), 526 - 561; English transl. in Amer. Math. Soc. Transl. (2) 85 (1969), 51 - 91.

[11] Partington, J.R., The resolvent of a Hermitian operator on a Banach space. J. London Math. Soc. (2) 27 (1983), 507 - 512.

[12] Radó, T., Subharmonic Functions. Springer - Verlag, Berlin 1937.

[13] Stampfli, J.G. - Williams, J.P., Growth conditions and the numerical range in a Banach algebra. Tôhoku Math. J. 20 (1968), 417 - 424.

[14] Tillmann, H.G., Eine Erweiterung des Funktionalkalküls für lineare Operatoren. Math. Ann. 151 (1963), 424 - 430.

[15] Wolf, F., On majorants of subharmonic and analytic functions. Bull. Amer. Math. Soc. 48 (1942), 925 - 932.

[16] Zsidó, L., Invariant subspaces of compact perturbations of linear operators in Banach spaces. J. Operator Theory 1 (1979), 225 - 260.

International Series of
Numerical Mathematics, Vol. 65

SUBSPACE LATTICES CONNECTED WITH C_{11}-CONTRACTIONS

László Kérchy
Bolyai Institute
József Attila University
Szeged

In this paper the investigation of invariant subspace lattices of C_{11}-contractions, begun in [8] and [9], is continued. Special emphasis is put on the behaviour of hyperinvariant subspaces under quasi-similarity. Moreover, a negative answer is given to a problem of Sz.-Nagy and Foiaş.

1. Introduction

The operators considered are always linear, bounded, acting on complex, separable Hilbert spaces. Moreover, terminology and notation of [10] is freely used.

Let $T \in L(H)$ be a C_{11}-contraction on the Hilbert space H, and $L \in \mathrm{Lat}\, T$ an invariant subspace of T. We know by [10, Theorem II.4.1] that there exists a unique decomposition $L = M \oplus N$ of L into the orthogonal sum of the subspaces M and N such that the matrix of $T|L$ with respect to this decomposition is of the form $T|L = \begin{bmatrix} T_{11} & T_{12} \\ 0 & T_{22} \end{bmatrix}$, where $T_{11} \in C_{11}$ and $T_{22} \in C_{\cdot 0}$. We call the subspace M the C_{11}-part of the subspace L (according to T), and we denote it by $L^{(1)}$ (more precisely by $L_T^{(1)}$).

It is easy to show that $L^{(1)}$ is the largest one of the invariant subspaces M included in L such that $T|M \in C_{11}$. From this fact we can conclude that the mapping $\varphi : \mathrm{Lat}\, T \to \mathrm{Lat}\, T$, $\varphi : L \to L^{(1)}$ is isotone, that is, $L \subset M$ implies $L^{(1)} \subset M^{(1)}$. On the other hand, φ is obviously idempotent and inclusive, that is, $(L^{(1)})^{(1)} = L^{(1)}$ and $L^{(1)} \subset L$ for every $L \in \mathrm{Lat}\, T$. Hence, using a simple argument taken from lattice theory (cf. [4,p. 111]), we can prove that the set of the fixed points of φ, which coincides with the range of φ, is closed under spanning. This set will be denoted by $\mathrm{Lat}_1 T(:= \{L \in \mathrm{Lat}\, T : L = L^{(1)}\})$, and will be called as the C_{11}-invariant subspace lattice of T. The set $\mathrm{Lat}_1 T$ is actually a complete lattice under set-inclusion as partial ordering. The least upper bound of a subset $A \subset \mathrm{Lat}_1 T$ is the span of the subspaces belonging to $A : \vee\{L : L \in A\}$, while the greatest lower bound of A is the C_{11}-intersection of these subspaces: $\cap^{(1)}\{L : L \in A\} := (\cap\{L : L \in A\})^{(1)}$.

It was proved in [8] that $\mathrm{Lat}_1 T = \{(\mathrm{ran}\, A)^- : A \in \{T\}'\}$, where $\{T\}' := \{S \in L(H) : ST = TS\}$ denotes the commutant of T. It follows that $\mathrm{Lat}_1 T \subset \mathrm{Lat}'' T := \mathrm{Lat}\{T\}''$, where $\{T\}'' = \{\{T\}'\}'$ is the bicommutant of T. We remark that if U is a unitary operator, then $\mathrm{Lat}_1 U = \mathrm{Lat}'' U = \{\mathrm{ran}\, P : P \in \{U\}'$ is an (orthogonal) projection$\}$.

We define the C_{11}-hyperinvariant subspace lattice of T as the C_{11}-part of its hyperinvariant subspace lattice, that is, $\mathrm{Hyplat}_1 T := \mathrm{Hyplat}\, T \cap \mathrm{Lat}_1 T$, where $\mathrm{Hyplat}\, T := \mathrm{Lat}\{T\}'$.

PROPOSITION 1. <u>For every</u> C_{11}-<u>contraction</u> T $\mathrm{Hyplat}_1 T$ <u>is a closed sublattice of</u> $\mathrm{Lat}_1 T$, <u>that is, for every subset</u> $A \subset \mathrm{Hyplat}_1 T$ <u>the subspaces</u> $\vee A$ <u>and</u> $\cap^{(1)} A$ <u>also belong to</u> $\mathrm{Hyplat}_1 T$.

PROOF. It is clearly enough to prove that $L^{(1)} \in \mathrm{Hyplat}\, T$ for every subspace $L \in \mathrm{Hyplat}\, T$. Let $S \in \{T\}'$ be an arbitrary operator. Then $L \in \mathrm{Lat}\, S$, and $S|L$ commutes with $T|L$. Let us consider the matrices of the operators $T|L$ and $S|L$ in the decomposition $L = L^{(1)} \oplus M$:

$$T|L = \begin{bmatrix} T_{11} & T_{12} \\ 0 & T_{22} \end{bmatrix}, \qquad S|L = \begin{bmatrix} S_{11} & S_{12} \\ S_{21} & S_{22} \end{bmatrix}.$$

In virtue of the definition of $L^{(1)}$ we know that $T_{11} \in C_{11}$ and $T_{22} \in C_{\cdot 0}$. On account of the commuting relation we obtain that $T_{22} S_{21} = S_{21} T_{11}$, and so $S_{21}^* T_{22}^{*n} = T_{11}^{*n} S_{21}^*$ for every natural number n. Hence, for every vector $h \in M$, we have $\lim_{n\to\infty} \| T_{11}^{*n} S_{21}^* h \| = \lim_{n\to\infty} \| S_{21}^* T_{22}^{*n} h \| = 0$, since $T_{22}^* \in C_{0\cdot}$. But now $T_{11}^* \in C_{1\cdot}$ implies that $S_{21}^* h = 0$, and we conclude that $S_{21} = 0$, that is $L^{(1)} \in \mathrm{Lat}\, S$.

2. Hyperinvariant Subspaces and Quasi-similarity

Now we intend to study the C_{11}-hyperinvariant subspace lattice of T. If U is a unitary operator, then its adjoint U* belongs to {U}', and so $\mathrm{Hyplat}_1 U = \mathrm{Hyplat}\, U$. Moreover, as it was proved by R.G. Douglas and C. Pearcy in [6], Hyplat U consists of the spectral subspaces of U.

Let us now assume that $T \in L(H)$ is a completely non-unitary (c.n.u.) C_{11}-contraction. Let $\Theta_T(e^{it})$ denote its characteristic function, and $\Delta_T(e^{it}) := [I - \Theta_T(e^{it})^* \Theta_T(e^{it})]^{1/2}$. Let us consider the system B(C) of the Borel measurable subsets of the unit circle $C = \{z \in \mathbb{C} : |z| = 1\}$. We say that the Borel sets $\alpha, \beta \in B(C)$ are T-equivalent, if rank $\Delta_T(e^{it}) = 0$ a.e. on the symmetric difference $\alpha \Delta \beta$ of α and β. It is immediate that this is in fact an equivalence relation on B(C); let $B_T(C)$ denote the set of equivalence classes. If $\alpha, \beta \in B(C)$ and rank $\Delta_T(e^{it}) = 0$ a.e. on the set $\alpha \setminus \beta$, then we say that the equivalence class $\hat{\alpha}$ corresponding to α is less than or equal to the equivalence class $\hat{\beta}$ corresponding to $\beta : \hat{\alpha} \leqslant \hat{\beta}$. It can be easily shown that $B_T(C)$ is a countably complete and distributive lattice under this (well-defined) relation.

B. Sz.-Nagy and C. Foiaş defined in [10], via the regular factorization of the characteristic function of T, an injective, order-preserving mapping q_T from the lattice $B_T(C)$ into the lattice $\mathrm{Hyplat}_1 T$ (cf. [10, Theorem VII.5.2]). Later R. I. Teodorescu has proved (cf. [12]) that q_T is surjective. Therefore, q_T is a lattice-isomorphism between the lattices $B_T(C)$ and $\mathrm{Hyplat}_1 T$. Taking into account that for an arbitrary C_{11}-contraction T $\mathrm{Hyplat}_1 T$ is the direct sum of the C_{11}-hyperinvariant subspace lattice of its singular unitary part and the one of its orthogonal complement (cf. [8, Lemmas 1 and 2]), we obtain the following

PROPOSITION 2. *For all arbitrary C_{11}-contractions T, $\mathrm{Hyplat}_1 T$ is a countably distributive, complete lattice.*

Now we shall prove a uniqueness property, "rigidity" of the C_{11}-hyperinvariant subspaces.

THEOREM 3. *If* $T \in L(H)$ *is a* C_{11}*-contraction*, $H_1, H_2 \in \mathrm{Hyplat}_1 T$, *and* $T|H_1$ *is quasi-similar to* $T|H_2$, *then* $H_1 = H_2$.

PROOF. On account of [8, Lemmas 1 and 2] we may assume that T is a c.n.u. contraction. (We recall that quasi-similar unitary operators are even unitarily equivalent, cf. [5].)

By Teodorescu's result for $j = 1,2$ there exists an $\alpha_j \in B(C)$ such that $H_j = H_{\alpha_j}$, where $H_{\alpha_j} := q_T(\hat{\alpha}_j)$. Let $\Theta_j(e^{it})$ denote the characteristic function of $T|H_{\alpha_j}$, and $\Delta_j(e^{it}) := [I - \Theta_j(e^{it})^* \Theta_j(e^{it})]^{1/2}$. From the definition of H_{α_j} in [10] we know that $\mathrm{rank}\, \Delta_j(e^{it}) = h_{\alpha_j}(e^{it})\, \mathrm{rank}\, \Delta_T(e^{it})$ a.e., where h_{α_j} denotes the characteristic function of α_j and $0 \cdot \infty$ is defined to be 0. But by [9, Corollary 1] $T|H_{\alpha_1}$ is quasi-similar to $T|H_{\alpha_2}$ if and only if $\mathrm{rank}\, \Delta_1(e^{it}) = \mathrm{rank}\, \Delta_2(e^{it})$ a.e.. Hence, we have $h_{\alpha_1}(e^{it}) \cdot \mathrm{rank}\, \Delta_T(e^{it}) = h_{\alpha_2}(e^{it}) \cdot \mathrm{rank}\, \Delta_T(e^{it})$ a.e., and so α_1 and α_2 are T-equivalent. Therefore, we conclude $H_1 = H_{\alpha_1} = H_{\alpha_2} = H_2$.

The following theorem shows that quasi-similarity implies an isomorphism between the C_{11}-hyperinvariant subspace lattices.

THEOREM 4. *If* T_1 *and* T_2 *are quasi-similar* C_{11}*-contractions, then there exists a unique lattice-isomorphism* $q : \mathrm{Hyplat}_1 T_1 \to \mathrm{Hyplat}_1 T_2$ *such that* $T_1|L$ *is quasi-similar to* $T_2|q(L)$ *for every* $L \in \mathrm{Hyplat}_1 T_1$.

PROOF. The uniqueness immediately follows by the preceding theorem.

In virtue of [8, Lemmas 1 and 2] we may assume again that T_1, T_2 are c.n.u. contractions. Since they are quasi-similar, it follows that $\mathrm{rank}\, \Delta_{T_1}(e^{it}) = \mathrm{rank}\, \Delta_{T_2}(e^{it})$ a.e.. Hence we have that $B_{T_1}(C) = B_{T_2}(C)$, and we obtain with an argument similar to the one in the proof of Theorem 3 that $q = q_{T_2} \circ q_{T_1}^{-1}$ will be an appropriate mapping.

We shall denote the mapping q by q_{T_1,T_2}. Now we examine how the isomorphism q_{T_1,T_2} can be implemented by quasi-affinities intertwining T_1 and T_2. Let $X \in I(T_1,T_2)$ be an arbitrary quasi-affinity, where $I(T_1,T_2)$ denotes the set of intertwining operators $S : ST_1 = T_2S$. It can be easily proved (cf.

[8, Lemma 7]) that for any subspace $L \in \mathrm{Hyplat}_1 T_1$ the subspace $\vee\{BXL: B \in \{T_2\}'\}$ belongs to $\mathrm{Hyplat}_1 T_2$. Let q_X denote the mapping $q_X : \mathrm{Hyplat}_1 T_1 \to \mathrm{Hyplat}_1 T_2$, $q_X : L \to \vee\{BXL : B \in \{T_2\}'\}$.

THEOREM 5. _If_ T_1 _and_ T_2 _are quasi-similar_ C_{11}_-contractions, then for any quasi-affinity_ $X \in I(T_1, T_2)$ q_X _coincides with_ q_{T_1,T_2}.

PROOF. Let $Y \in I(T_2, T_1)$ be another quasi-affinity, and let $L \in \mathrm{Hyplat}_1 T_1$ be an arbitrary subspace. Since for every $A \in \{T_1\}'$ and $B \in \{T_2\}'$ we have $AYBX \in \{T_1\}'$, it follows that $q_Y(q_X(L)) \subset L$. Hence, we obtain that $T_1|L \overset{i}{<} T_2|q_X(L) \overset{i}{<} T_1|q_Y(q_X(L)) \overset{i}{<} T_1|L$. (We recall that an operator S_1 can be injected into an operator S_2, $S_1 \overset{i}{<} S_2$, if $I(S_1, S_2)$ contains an injection.) Now, applying [8, Lemma 6], we infer that $T_1|L$ is quasi-similar to $T_2|q_X(L)$, and so by Theorem 3 we conclude that $q_{T_1,T_2}(L) = q_X(L)$.

It is easy to see (cf. [10, Proposition II.5.1] and [7, Proposition 4.1]) that if $\mathrm{Hyplat}_1 T_2$ is of a special form, then q_{T_1,T_2} has a simpler implementation.

PROPOSITION 6. _If_ T_1 _and_ T_2 _are quasi-similar_ C_{11}_-contractions and_ $\mathrm{Hyplat}_1 T_2 = \{(\mathrm{ran}\ B)^- : B \in \{T_2\}''\}$, _then for every quasi-affinity_ $X \in I(T_1, T_2)$ _and for every subspace_ $L \in \mathrm{Hyplat}_1 T_1$ _we have_ $(XL)^- \in \mathrm{Hyplat}_1 T_2$. _Moreover, the mapping_ $q_X^* : \mathrm{Hyplat}_1 T_1 \to \mathrm{Hyplat}_1 T_2$, $q_X^* : L \to (XL)^-$ _coincides with_ q_{T_1,T_2}.

The assumption $\mathrm{Hyplat}_1 T_2 = \{(\mathrm{ran}\ B)^- : B \in \{T_2\}''\}$ of this proposition is obviously fulfilled if $T_2 = U$ is a unitary operator. We are going to show that it is true also when T_2 has the following property:

(P) every injection $X \in \{T_2\}'$ is a quasi-affinity.

(For a detailed study of this property see [8], [9] and [2].)

THEOREM 7. _If the_ C_{11}_-contraction_ $T \in L(H)$ _has the property_ (P), _then_ $\mathrm{Hyplat}_1 T = \{(\mathrm{ran}\ A)^- : A \in \{T\}''\}$.

PROOF. We know by [8, Proposition 5] that there exists a basic system $\{H_n\}_n$ consisting of subspaces of $\mathrm{Hyplat}_1 T$ such that $T_n = T|H_n$ is similar to a unitary operator for every n. (We recall that $\{H_n\}_n$ is a basic system if $H = H_n \dot{+} (\vee_{k \neq n} H_k)$, for every n, and $\cap_n (\vee_{k \geq n} H_k) = \{0\}$, cf. [1].)

Let $M \in \mathrm{Hyplat}_1 T$ be an arbitrary subspace. We infer by Proposition 2 that $M = M \cap^{(1)} (\vee_n H_n) = \vee_n (M \cap^{(1)} H_n)$. For any operator $S_n \in \{T_n\}'$, the operator $S_n P_n$ belongs to $\{T\}'$, where P_n denotes the projection onto the subspace H_n corresponding to the decomposition $H = H_n \dot{+} (\vee_{k \neq n} H_k)$. Hence the subspace $M_n = M \cap^{(1)} H_n$ belongs to $\mathrm{Hyplat}_1 T_n$. But T_n is similar to a unitary operator and so there exists an operator $A_n \in \{T_n\}''$ such that $(\mathrm{ran}\ A_n)^- = M_n$.

Let $\{c_n\}_n$ be a sequence of positive numbers such that $\sum_n c_n \|A_n\| \|P_n\| < \infty$, and let us define the operator $A \in L(H)$ in the following way: $Ah := \sum_n c_n A_n P_n h$, where $h \in H$. Then it is easy to verify that $A \in \{T\}''$. On the other hand, we have $(\mathrm{ran}\ A)^- = (A(\vee_n H_n))^- = \vee_n (AH_n)^- = \vee_n M_n = M$, which completes the proof.

3. Examples

Let us assume again that $T \in L(H)$ is a c.n.u. C_{11}-contraction. As it was proved in Section 2, for every sequence $\{\alpha_n\}_n$ of Borel sets in $B(C)$, we have

$$H_\alpha = \bigcap_n^{(1)} H_{\alpha_n} \qquad \text{and} \qquad H_{\alpha'} = \bigvee_n H_{\alpha_n} ,$$

where $\alpha = \cap_n \alpha_n$ and $\alpha' = \cup_n \alpha_n$. (Here H_β, $\beta \in B(C)$, denotes the C_{11}-hyperinvariant subspace corresponding to $\hat{\beta} \in B_T(C) : H_\beta = q_T(\hat{\beta})$.)

B. Sz.-Nagy and C. Foiaş have proved (cf. [10, Theorem VII.6.2.]) that if the characteristic function of T has a scalar multiple, then even the relation

$$H_\alpha = \bigcap_n H_{\alpha_n}$$

holds. Moreover, they raised the question, whether this relation is true for every c.n.u. C_{11}-contraction (cf. [10, p. 322]). We shall show that $\mathrm{Hyplat}_1 T$ is generally not a sublattice of Hyplat T, even in the case when T has a cyclic vector. Therefore, the answer to the question posed by Sz.-Nagy and Foiaş is n e g a t i v e .

THEOREM 8. *There exists a* C_{11}*-contraction* T *such that* T *has a cyclic vector* *and* Hyplat$_1$T *is not a sublattice of* Hyplat T. (*Consequently*, Lat$_1$T *is not a sublattice of* Lat T.)

We shall prove this theorem using the technique elaborated in [3]. The following lemma is an immediate generalization of [3, Lemma 3.2].

LEMMA 9. *Let* $N \geq 2$ *be a natural number or* $N = \omega$ *the first infinite ordinal. Let us assume that, for every number* $n < N$, $T_n \in L(H_n)$ *is an injective contraction such that* ran $T_n \neq H_n$. *Then for every* $n < N$ *there exists a non-zero vector* $f_n \in H_n$ *such that the operator* T *acting on the space* $\mathbb{C} \oplus (\oplus_{n<N} H_n)$ *and defined by the matrix*

$$T = \begin{bmatrix} 0 & 0 & 0 & \cdots \\ f_1 & T_1 & 0 & \cdots \\ f_2 & 0 & T_2 & \cdots \\ \cdot & & & \\ \cdot & & & \\ \cdot & & & \end{bmatrix}$$

will be an injective contraction.

Here f_n denotes the linear transformation $f_n : \mathbb{C} \to H_n$, $f_n : \lambda \to \lambda f_n$ also. We note that the adjoint of the contraction $T \in L(\mathbb{C} \oplus (\oplus_{n<N} H_n))$ constructed in this lemma is the operator $T^* \in L(\mathbb{C} \oplus (\oplus_{n<N} H_n))$ of the form

$$T^* = \begin{bmatrix} 0 & f_1^* & f_2^* & \cdots \\ 0 & T_1^* & 0 & \cdots \\ 0 & 0 & T_2^* & \cdots \\ \cdot & & & \\ \cdot & & & \\ \cdot & & & \end{bmatrix},$$

where f_n^* denotes the adjoint of the linear transformation f_n, that is, $f_n^* : H_n \to \mathbb{C}$, $f_n^* : h_n \to \langle h_n, f_n \rangle$.

PROOF OF THEOREM 8. For every Borel set $\alpha \in B(C)$ let M_α denote the operator of multiplication by e^{it} on the space $L^2(\alpha)$. (We consider the Lebesgue measure on α.)

Let $\alpha_1, \alpha_2, \alpha_3 \in B(C)$ be pairwise disjoint sets of positive measure. In virtue of [3, Proposition 3.1] for $j = 1,2,3$ there exists a non-invertible C_{11}-contraction $T_j \in L(H_j)$ such that T_j is quasi-similar to M_{α_j}. Applying Lemma 9 we infer that there exist non-zero vectors $f_j \in H_j$ ($j = 1,2,3$) such that the operators

$$\begin{bmatrix} 0 & 0 \\ & \\ f_1 & T_1 \end{bmatrix} \in L(\mathbb{C} \oplus H_1) \quad , \quad \begin{bmatrix} 0 & 0 & 0 \\ f_2 & T_2^* & 0 \\ f_3 & 0 & T_3^* \end{bmatrix} \in L(\mathbb{C} \oplus H_2 \oplus H_3)$$

will be injective contractions. Now let us define the operator $T \in L(\mathbb{C} \oplus H_1 \oplus H_2 \oplus H_3)$ by the matrix

$$T = \begin{bmatrix} 0 & 0 & f_2^* & f_3^* \\ f_1 & T_1 & 0 & 0 \\ 0 & 0 & T_2 & 0 \\ 0 & 0 & 0 & T_3 \end{bmatrix} .$$

It is easy to verify that T is a C_{11}-contraction. On the other hand, on account of [3, Theorem 1.7] and [10, Proposition II.3.4] it follows that T is quasi-similar to $M_{\alpha_1} \oplus M_{\alpha_2} \oplus M_{\alpha_3}$, which is unitarily equivalent to the unitary operator M_α, where $\alpha = \alpha_1 \cup \alpha_2 \cup \alpha_3$. Therefore, T has a cyclic vector. But this implies by [11] that $\{T\}' = \{T\}''$ and so $\text{Hyplat}_1 T = \text{Lat}_1 T$. Since the subspaces $\mathbb{C} \oplus H_1 \oplus H_2$ and $\mathbb{C} \oplus H_1 \oplus H_3$ are obviously C_{11}-invariant for T, we obtain that they belong to $\text{Hyplat}_1 T$ also. But their intersection $(\mathbb{C} \oplus H_1 \oplus H_2) \cap (\mathbb{C} \oplus H_1 \oplus H_3) = \mathbb{C} \oplus H_1$ is not in $\text{Hyplat}_1 T$, and so we conclude that $\text{Hyplat}_1 T$ is not a sublattice of Hyplat T.

REMARK 10. It can be achieved that the contraction T constructed in the proof of Theorem 8 be c.n.u. and that the set $\mathbb{C} \setminus (\alpha_1 \cup \alpha_2 \cup \alpha_3)$ be of positive measure. Then, considering the regular factorization of the characteristic function Θ_T of T corresponding to the invariant subspace $\mathbb{C} \oplus H_1$ we obtain a negative answer to the question raised in [10, p. 322]. (Cf. [10, Proposition VI.3.5 and Proposition VII.2.1].)

Namely, there exists a contractive analytic function $\Theta(\lambda)$, outer from both sides, such that $\Theta(e^{it})$ is isometric on a set of positive Lebesgue measure and $\Theta(\lambda)$ has a non-outer regular divisor. (This is the case even if we assume that rank $[I-\Theta(e^{it})^*\Theta(e^{it})]\leq 1$ a.e., cf.[9, Corollary 1].)

REMARK 11. It can be asked now, whether the conjecture of Sz.-Nagy and Foiaş remains true if we assume that the sequence $\{\alpha_n\}_n$ of Borel sets is monotone decreasing. The following example shows that the answer is negative for this question also.

Let $\{\alpha_n\}_{n=1}^{\infty}$ be a sequence of pairwise disjoint Borel sets of positive measure, and for every n let $T_n \in L(H_n)$ be a non-invertible C_{11}-contraction, quasi-similar to M_{α_n}. We infer by Lemma 9 that there exists a sequence $\{f_n\}_{n=1}^{\infty}$ of non-zero vectors such that $f_n \in H_n$ for every n and that the operators

$$\begin{bmatrix} 0 & 0 \\ f_1 & T_1 \end{bmatrix} \in L(\mathbb{C}\oplus H_1) \quad , \quad \begin{bmatrix} 0 & 0 & 0 & 0 & \cdots \\ f_2 & T_2^* & 0 & 0 & \cdots \\ f_3 & 0 & T_3^* & 0 & \cdots \\ \cdot & & & & \\ \cdot & & & & \\ \cdot & & & & \end{bmatrix} \in L(\mathbb{C}\oplus(\bigoplus_{n=2}^{\infty} H_n))$$

are injective contractions. Let us define now the operator $T \in L(\mathbb{C}\oplus(\oplus_{n=1}^{\infty} H_n))$ by the matrix

$$T = \begin{bmatrix} 0 & 0 & f_2^* & f_3^* & \cdots \\ f_1 & T_1 & 0 & 0 & \cdots \\ 0 & 0 & T_2 & 0 & \cdots \\ 0 & 0 & 0 & T_3 & \cdots \\ \cdot & & & & \\ \cdot & & & & \\ \cdot & & & & \end{bmatrix} .$$

It can be easily seen that T is a C_{11}-contraction, and T is quasi-similar to the operator M_α, where $\alpha = \cup_{n=1}^{\infty} \alpha_n$. Hence, T has a cyclic vector and $\mathrm{Hyplat}_1 T = \mathrm{Lat}_1 T$.

Let N_2 denote the set of natural numbers, greater than one. To every non-empty subset I of N_2 there corresponds a C_{11}-hyperinvariant subspace $H_I = \mathbb{C}\oplus H_1 \oplus (\oplus_{n\in I} H_n)$ of T. Let us consider now a decreasing sequence $\{I_n\}_{n=1}^{\infty}$ of non-empty subsets of N_2 such that $\cap_{n=1}^{\infty} I_n = \phi$. Then $\{H_{I_n}\}_{n=1}^{\infty}$ will be a decreasing sequence consisting of C_{11}-hyperinvariant subspaces of T such that

$$\bigcap_{n=1}^{\infty} H_{I_n} = \mathbb{C} \oplus H_1 \notin \mathrm{Hyplat}_1 T.$$

REFERENCES

[1] Apostol, C., Operators quasi-similar to a normal operator. Proc. Amer. Math. Soc. 53 (1975), 104-106.

[2] Bercovici, H., C_0-Fredholm operators. II. Acta Sci. Math. (Szeged) 42 (1980), 3-42.

[3] Bercovici, H. - Kérchy, L., Quasi-similarity and properties of the commutant of C_{11}-contractions. Acta Sci. Math. (Szeged) 45 (1983), 67-74.

[4] Birkhoff, G., Lattice Theory. Amer. Math. Soc. Coll. Publ. Volume XXV.

[5] Douglas, R.G., On the operator equation S*XT = X and related topics. Acta Sci. Math. (Szeged) 30 (1960), 19-32.

[6] Douglas, R.G., - Pearcy, C., On a Topology for Invariant Subspaces. J. Funct. Anal. 2 (1968), 323-341.

[7] Fialkow, L., A note on quasisimilarity. II. Pac. J. Math. 70 (1977), 151-162.

[8] Kérchy, L., On invariant subspace lattices of C_{11}-contractions. Acta. Sci. Math. (Szeged) 43 (1981), 281-293.

[9] Kérchy, L., On the commutant of C_{11}-contractions. Acta. Sci. Math. (Szeged) 43 (1981), 15-26.

[10] Sz.-Nagy, B. - Foiaş, C., Harmonic Analysis of Operators on Hilbert Space, North Holland - Akademiai Kiadó, Amsterdam - Budapest 1970.

[11] Sz.-Nagy, B. - Foiaş, C., Vecteurs cycliques et commutativité des commutants. Acta Sci. Math. (Szeged) 32 (1971), 177-183.

[12] Teodorescu, R.I., Factorisations regulierès et sousespaces hyperinvariants. Acta Sci. Math. (Szeged) 40 (1978), 389-396.

International Series of
Numerical Mathematics, Vol. 65

ON THE WOLD DECOMPOSITION OF ISOMETRIC SEMIGROUPS

Dumitru Gaşpar and Nicolae Suciu
Department of Mathematics
University of Timişoara
Timişoara

The study begun in [3] is continued. Namely the modified unilateral translation part, which appears in the Wold-type decomposition given in [3], is characterized. The ultraevanescent part, i.e., the orthocomplement of both (unilateral and modified unilateral) translation parts, is also considered. Finally, conditions on self-duality of isometric semigroups in terms of unilateral, modified unilateral translation and ultraevanescent semigroups are discussed.

1. Introduction

Some basic definitions from prediction theory and dilation theory of isometric semigroups on Hilbert spaces are needed. Let G be an abelian group. We write 1 for the unit element of G. A *unitary representation* of G on the Hilbert space H, is a map $g \to U_g$ of G in the algebra B(H) of all linear bounded operators on H, such that $U_1 = I$, $U_{g^{-1}} = U_g^*$ and $U_{g_1 g_2} = U_{g_1} U_{g_2}$ for any $g, g_1, g_2 \in G$. Obviously, U_g is a unitary operator on H for any $g \in G$. We denote our map briefly $\{U_g\}_{g \in G}$ and we shall call it also a *unitary group* of operators on H. Let now S be a subsemigroup of G. A map $s \to T_s$ of S into B(H) will be called an *isometric* (*unitary*) *semigroup*, if its values T_s are isometric (unitary) operators on H and, for any $s, \sigma \in S$, $T_{s\sigma} = T_s T_\sigma$. Obviously, we have $T_1 = I$.

If the closed subspace M reduces the isometric (unitary) semigroup $\{T_s\}_{s \in S}$ (i.e., $T_s M \subset M$ and $T_s^* M \subset M$ for all $s \in S$), then it is clear that $\{T_s | M\}_{s \in S}$ is an isometric (unitary) semigroup too. Of great interest for the isometric semigroups which arise in prediction theory, is the problem of finding the orthogonal decomposition $\bigoplus_\alpha H_\alpha$ for H such that all H_α reduce $\{T_s\}_{s \in S}$

and $\{T_s^\alpha\}_{s \in S}$ are special isometric semigroups (throughout such situations $T_s^\alpha = T_s|H_\alpha$, for every index α), which admit a "good" description. Such decompositions are given in [7] and [3] under the following general hypothesis on S:

(1) $$S \cap S^{-1} = \{1\} \text{ and } SS^{-1} = G.$$

For particular cases on G and S such decompositions are given earlier (see [10], [4], [6]).

A complete description is admitted, under hypothesis (1), by a unitary semigroup, since it can be uniquely extended to a unitary group for which spectral representation holds (see [8]). Two other "special" isometric semigroups are defined with the aid of the bilateral translation-group $\{b_g\}_{g \in G}$ on the space $l^2(G,X)$ of X - valued square summable functions $(h_g)_{g \in G}$ by

$$b_s(h_g)_{g \in G} = (b_{s^{-1}g})_{g \in G} \qquad (s \in G),$$

X being a Hilbert space, which is called the d e f e c t s p a c e of $\{b_s\}_{s \in S}$. For $E \subset G$, we denote $l^2(E,X)$ the set of those elements of $l^2(G,X)$ which are supported on E. An isometric semigroup $\{T_s\}_{s \in S}$ on H is called a u n i l a t e r a l (resp., m o d i f i e d u n i l a t e r a l) t r a n s l a t i o n, if there is a Hilbert space X (resp. Y) such that $\{T_s\}_{s \in S}$ is unitarily equivalent to the restriction of $\{b_s\}_{s \in S}$ to $l^2(S,X)$ (resp., to $l^2(G \setminus S^{-1},Y)$).

Let now $\{T_s\}_{s \in S}$ be a fixed isometric semigroup on H such that S satisfies (1).

In [7] Ion Suciu has proved that there is a unique decomposition of H under the form

(2) $$H = H_u \oplus H_t \oplus H_r,$$

such that H_u, H_t, H_r reduce $\{T_s\}_{s \in S}$, $\{T_s^u\}_{s \in S}$ is unitary, $\{T_s^t\}_{s \in S}$ is a unilateral translation with the defect space $R = H \ominus \bigvee_{\sigma^{-1}s \notin S^{-1}} T_\sigma^* T_s H$, and $\{T_s\}_{s \in S}$ is r i g h t e v a n e s c e n t, i.e., H_r does not contain a closed subspace which reduces $\{T_s^r\}_{s \in S}$ to a unitary semigroup, or to a unilateral translation.

We shall call the space R the r i g h t d e f e c t s p a c e of the given semigroup.

On the other hand, in [3] we have proved that there is a unique orthogonal decomposition of H under the form

(3) $$H = H_u \oplus H_m \oplus H_1,$$

such that H_u, H_m, H_1 reduce $\{T_s\}_{s \in S}$, $\{T_s^u\}_{s \in S}$ is unitary, $\{T_s^m\}_{s \in S}$ is a modified unilateral translation with the defect space L, and $\{T_s^1\}_{s \in S}$ is *left evanescent*, i.e., H_1 does not contain a closed subspace which reduces $\{T_s^1\}_{s \in S}$ to a unitary, or to a modified unilateral translation. The defect space L is given in terms of the space $K(\supset H)$ of the *minimal unitary extension* $\{U_g\}_{g \in G}$ of $\{T_s\}_{s \in S}$ (i.e., $\{U_g\}_{g \in G}$ is a unitary group on K such that, for any $s \in S$, $U_s H \subset H$, $U_s|H = T_s$ and $\bigvee_{g \in G} U_g H = K$; see [5]). Namely, $L = H^\perp \ominus P_{H^\perp} \bigvee_{g \notin S} U_g H^\perp$, where $H^\perp = K \ominus H$. L will be called the *left defect space* of $\{T_s\}_{s \in S}$.

Let us recall (see [3]) that the defect subspaces R and L are wandering for $\{U_g\}_{g \in G}$ and

(4) $$\bigoplus_{s \in S} U_s R = H_t \ , \qquad \bigoplus_{g \notin S^{-1}} U_g L = H_m \ .$$

In what follows we shall study the modified unilateral part of $\{T_s\}_{s \in S}$, and the restriction of $\{T_s\}_{s \in S}$ to $H_r \cap H_1$, i.e., the ultraevanescent part of $\{T_s\}_{s \in S}$. We also consider the dual semigroup in the sense of [1] , and we give some sufficient conditions for the unitary equivalence of the given semigroup to its dual.

2. The results

2.1. Modified unilateral translation part. We characterize the closed subspaces of H which reduce the isometric semigroup $\{T_s\}_{s \in S}$ to a modified unilateral translation. For this we need firstly

LEMMA 1. *Let* **T** *be the von-Neumann algebra generated by* $\{T_s\}_{s \in S}$ *on* H, *and* **U** *be the von-Neumann algebra generated by the minimal unitary extension* $\{U_g\}_{g \in G}$ *on* K. *Then the map* $U \to U|H$ *defines a strong (weak) continuous* *-*isomorphism of the commutant* **U'** *of* **U** *onto the commutant* **T'** *of* **T**.

PROOF. If $U \in$ **U'**, then U commutes with P_H (the orthoprojector of K on H) and with U_g $(g \in G)$. This implies that $\rho(U) = U|H$ commutes with T_s and T_s^* $(s \in S)$, therefore $\rho(U) \in$ **T'**. It is now easy to verify that ρ is linear, multiplicative and involutive, in the sense that $\rho(U)^* = \rho(U^*)$, $U \in \mathbf{U}'$. For the surjectivity, let $T \in \mathbf{T}'$. Since $\{U_s\}_{s \in S}$ is a minimal extension on K for $\{T_s\}_{s \in S}$ in the sense of Douglas ([2]), it results from Theorem 2 of [2] that there is a uniquely determined operator $U \in B(K)$ such that $UU_s = U_sU$ $(s \in S)$, $UH \subset H$, $\rho(U) = T$ and $||U|| = ||T||$. Now if $g \in G$, then by (1) it is of the form $\sigma^{-1}s$ with σ, $s \in S$ and because of $U_g = U^*_{\sigma}U_s$ it results that U commutes with U_g. Moreover U commutes with P_H. Indeed, for each $g_i = \sigma_i^{-1}s_i \in G$ $(\sigma_i, s_i \in S)$ and $h_i \in H$ $(i = 1,2,\ldots,n)$ we have

$$P_HU(\sum_i U_{g_i}h_i) = P_H(\sum_i U^*_{\sigma_i}UU_{s_i}h_i) = \sum_i(P_HU^*_{\sigma_i}|H)UU_{s_i}h_i =$$

$$= \sum_i T^*_{\sigma_i}UU_{s_i}h_i = U(\sum_i T^*_{\sigma_i}U_{s_i}h_i) = UP_H(\sum_i U_{g_i}h_i).$$

Now by the minimality condition on K, it results that U commutes with P_H, also finally $U \in \mathbf{U}'$. This means that ρ is a *- isomorphism of $\mathbf{U}'$ onto $\mathbf{T}'$. The strong (weak-) continuity can be obtained as in [2].

THEOREM 2. <u>Let</u> M <u>be a reducing subspace for</u> $\{T_s\}_{s \in S}$. <u>The following are equivalent</u>:

(i) $\{T_s|M\}_{s \in S}$ <u>is a modified unilateral translation;</u>

(ii) <u>there is a wandering subspace</u> D <u>of</u> $K \ominus H$ <u>for</u> $\{U_g\}_{g \in G}$ <u>(i.e.,</u> $U_gD \perp U_{g'}D$, $g \neq g'$), <u>such that</u> $M = \bigoplus_{g \notin S} U^*_gD$;

(iii) <u>there is a closed subspace</u> L_o <u>of</u> L, <u>such that</u> $M = \bigoplus_{g \notin S} U^*_gL_o$;

(iv) $M \subset H_m$, <u>where</u> H_m <u>is the subspace which appears in the decomposition</u> (3).

PROOF. The implication (i)$\Longrightarrow$(iv) is the Theorem 3 of [3], and (iii)$\Longrightarrow$(ii) is trivial. The proof of (ii)$\Longrightarrow$(i) coincides with a corresponding part of the proof of Theorem 6 of [3]. For (iv)$\Longrightarrow$(iii) let $M \subset H_m$ be a reducing subspace for $\{T_s\}_{s \in S}$. Now if by K_m we denote the space of the minimal unitary extension of $\{T^m_s\}_{s \in S}$ (i.e., $K_m = \bigvee_{g \in G} U_gH_m$) and by L_m the left defect space of $\{T^m_s\}_{s \in S}$, then from Theorem 6 and Corollary 7 of [3] we have $H_m = \bigoplus_{g \notin S} U^*_gL_m$. Since by Lemma 4 of [3], $K_m \ominus H_m$ reduces the dual semigroup of $\{T^m_s\}_{s \in S}$ and since L_m is the right defect space for the dual semigroup $\{T^{m\#}_s\}_{s \in S}$, from Proposition 1(i) of [3] it results that $L_m \subset L$. Let now P_m (resp. P^m_M) be the orthoprojector of K_m (resp. H_m) onto H_m (resp. M). Since M reduces $\{T^m_s\}_{s \in S}$, P^m_M commutes with T^m_s and with T^{m*}_s, $s \in S$. Then by Lemma 1, there is an orthoprojector $P \in B(K_m)$ which commutes with P_m and $U_g|K_m$, for each $g \in G$, such that $P|H_m = P^m_M$. This implies easily that P commutes with the orthoprojector Q_m of K_m onto L_m. Thus we have

$$M = P^m_M H_m = PH_m = \bigoplus_{g \notin S} U^*_g P Q_m K_m = \bigoplus_{g \notin S} U^*_g Q_m P K_m = \bigoplus_{g \notin S} U^*_g L'_m,$$

where $L'_m = Q_m P K_m \subset L_m \subset L$ is a closed subspace of L. This finishes the proof.

COROLLARY 3. _Let $\{T_s\}_{s \in S}$ be an isometric completely non - unitary (i.e., without unitary part) semigroup on H, and M a reducing subspace for $\{T_s\}_{s \in S}$. Then $\{T_s|M\}_{s \in S}$ is left evanescent if and only if $M \subset H_1$ (see decomposition (3))._

PROOF. Let us suppose that M reduces $\{T_s\}_{s \in S}$ to a left evanescent semigroup. Then $M = (M \cap H_m) \oplus (M \cap H_1)$ holds, and since $M \cap H_m$ reduces $\{T_s\}_{s \in S}$, we obtain from Theorem 2 that $\{T_s|M \cap H_m\}_{s \in S}$ is a modified unilateral translation. But $M \cap H_m$ also reduces $\{T_s|M\}_{s \in S}$, therefore $M \cap H_m = \{0\}$. This means $M = M \cap H_1 \subset H_1$. Conversely, suppose $M \subset H_1$. Since $\{T_s\}_{s \in S}$ is completely non-unitary and M reduces $\{T_s\}_{s \in S}$, $\{T_s|M\}_{s \in S}$ is completely non-unitary too. Now if N ($\subset M$) reduces $\{T_s|M\}_{s \in S}$ to a modified unilateral translation, then N reduces $\{T_s\}_{s \in S}$ too, and by Theorem 2 we have $N \subset H_m$. This gives $N = \{0\}$, which proves that $\{T_s|M\}_{s \in S}$ is left evanescent.

More information can be obtained if the left defect space L is one-dimensional.

PROPOSITION 4. _An isometric semigroup $\{T_s\}_{s \in S}$ is a modified unilateral translation with one - dimensional left defect space if and only if $\{T_s\}_{s \in S}$ is not left evanescent and has not non-trivial reducing subspaces._

PROOF. Suppose $\{T_s\}_{s \in S}$ is a modified unilateral translation on H and dim L = = 1. Then through the Corollary 7 of [3] we have $H = H_m$ and therefore $\{T_s\}_{s \in S}$

is not left evanescent. Then, since L does not contain non-trivial closed subspaces, by Theorem 2(iii) we obtain that $\{T_s\}_{s \in S}$ has not non-trivial reducing subspaces. Conversely, let us suppose that $\{T_s\}_{s \in S}$ is not left evanescent and it has not non-trivial reducing subspaces. Let $k_0 \in L$, $k_0 \neq 0$ and L_0 be the closed subspace of L generated by $\{k_0\}$. It is easy to check (as in the proof of Theorem 6 of [3]) that the subspace $M = \bigoplus_{g \notin S} U_g^* L_0$ reduces $\{T_s\}_{s \in S}$. Then, by hypothesis, we have $H = M$ and from Theorem 2 it results that $\{T_s\}_{s \in S}$ is a modified unilateral translation. Finally, as in the proof of Theorem 3 in [3] we obtain that the left defect space of $\{T_s\}_{s \in S}$ is isomorphic to L_0, i.e., it is one-dimensional.

2.2. Ultraevanescent part. We shall say that the isometric semigroup $\{T_s\}_{s \in S}$ on H is u l t r a e v a n e s c e n t, if it is simultaneously left and right evanescent. By combining the decompositions (2) and (3) we obtain a new Wold-type decomposition. We formulate it in the following

THEOREM 5. *There is a unique decomposition of* H *under the form*

$$H = H_u \oplus H_e \oplus H_c, \tag{5}$$

such that H_u, H_e, H_c *reduce* $\{T_s\}_{s \in S}$, $\{T_s^u\}_{s \in S}$ *is unitary,* $\{T_s^e\}_{s \in S}$ *is ultraevanescent,* $\{T_s^c\}_{s \in S}$ *is completely non-unitary, and* H_c *does not contain nontrivial subspaces which reduce* $\{T_s^c\}_{s \in S}$ *to a ultraevanescent semigroup.*

PROOF. It is enough to decompose the completely non-unitary part. For simplicity we suppose $H_u = \{0\}$. We set $H_e = H_r \cap H_l$ and $H_c = H_t \vee H_m$. It is easy to see that H_e, H_c reduce $\{T_s\}_{s \in S}$. By a Remark from [7] p.103 and by Corollary 3, it results that $\{T_s^e\}_{s \in S}$ is ultraevanescent. Now if $M \subset H_c$ reduces $\{T_s^c\}_{s \in S}$ to a ultraevanescent semigroup, then since $\{T_s|M\}_{s \in S}$ is right evanescent, we have $M \perp H_t$, hence $M \subset H_r$. Analogously, by Corollary 3 it results $M \subset H_l$. It follows $M \subset H_e$ and therefore $M = \{0\}$. Now let $H = H_e' \oplus H_c'$ be another decomposition, with the same properties. Then as before we obtain $H_e' \subset H_r \cap H_l = H_e$, therefore $H_e = H_e' \oplus H_1$. H_1 reduces $\{T_s\}_{s \in S}$ and by Corollary 3 and the mentioned Remark of [7] p.103, the semigroup $\{T_s|H_1\}_{s \in S}$ is ultraevanescent. But $H_1 \perp H_e'$, hence $H_1 \subset H_c'$ and therefore $H_1 = \{0\}$. This means that the two decompositions coincide, which finishes the proof.

REMARK. The subspace H_c from Theorem 5 has the form

$$H_c = H_t \vee H_m = H_t \oplus (H_m \cap H_r) = (H_t \cap H_l) \oplus H_m .$$

This means that the semigroup $\{T_s^c\}_{s \in S}$ is a direct sum of a unilateral translation with a modified unilateral translation. But, generally speaking, this decomposition is not uniquely determined.

This latter decomposition is uniquely determined if G is a dense subgroup of **R** and $S = G_+$ (the non-negative elements of G). Indeed, as we have proved in [3](Example B), in this case, the spaces H_t and H_m are mutually orthogonal. Hence, by Theorem 5, we have, in this particular case,

$$H = H_u \oplus H_t \oplus H_m \oplus H_e ,$$

which means that the isometric semigroup $\{T_s\}_{s \in G_+}$ consists of four summands: a unitary part, a unilateral translation part, a modified unilateral translation part and an ultraevanescent part.

Consider again a general group G, and the subsemigroup S verifying (1). Let us consider the discrete topology on G and denote $\hat{G}$ the (compact) character group of G. We shall denote by E the spectral measure on $\hat{G}$ of the minimal unitary extension $\{U_g\}_{g \in G}$ of the isometric semigroup $\{T_s\}_{s \in S}$. We are now in position to prove the following Sz.-Nagy—Foiaş - type result (see [9] chap.II).

THEOREM 6. <u>For every</u> $h \in H_t \vee H_m$, $h \neq 0$, <u>the elementary measure</u> $(E(.)h,h)$ <u>is mutually absolutely continuous with respect to the Haar (normalized) measure</u> m <u>on</u> $\hat{G}$.

PROOF. Let us recall (see for instance [8]) that E satisfies

$$\int_{\hat{G}} \langle g,x\rangle dE(x) = U_g \qquad (g \in G).$$

Since R and L are wandering subspaces for $\{U_g\}_{g \in G}$, for every $k \in R \cup L$ we have

$$\int_{\hat{G}} \langle g,x\rangle d(E(x)k,k) = (U_g k,k) = \delta_g^1 \|k\|^2 = \|k\|^2 \int_{\hat{G}} \langle g,x\rangle dm(x)$$

where $\delta_g^1 = 0$ if $g \neq 1$, and 1 for $g = 1$.

Now applying the identity Theorem for the Fourier transform on $\hat{G}$, we obtain $\|E(\Delta)k\|^2 = m(\Delta)\|k\|^2$ for every Borel set Δ of $\hat{G}$. It is now easy to see that, because of (4), and since $U_gE = EU_g$ $(g \in G)$, $m(\Delta) = 0$ implies $(E(\Delta)h,h) = 0$ for every $h \in H_t \vee H_m$, $h \neq 0$. This finishes the proof.

COROLLARY 7. <u>If the spectral measure</u> E <u>is singular with respect to the Haar measure on</u> $\hat{G}$, <u>then the given isometric semigroup is ultraevanescent.</u>

This result was obtained in the special case mentioned above by P.S.Muhly (see Proposition 6.7 of [6]).

<u>2.3. Selfdual semigroups.</u> The completely non-unitary isometric semigroup $\{T_s\}_{s \in S}$ will be called s e l f d u a l, if it is unitarily equivalent to its dual $\{T_s^{\#}\}_{s \in S}$, which is defined (see [3]) by $T_s^{\#} = U_s^{*}|H^{\perp}$, $s \in S$. Recall that we have characterized the selfdual semigroups in [3]. We give now two sufficient conditions for the selfduality of an isometric semigroup.

THEOREM 8. <u>Suppose the completely non-unitary isometric semigroup</u> $\{T_s\}_{s \in S}$ <u>has not an ultraevanescent part. We denote</u> $R_1(L_r)$ <u>the right (left) defect space for</u> $\{T_s^{1}\}_{s \in S}$ $(\{T_s^{r}\}_{s \in S})$. <u>If</u> $\dim R = \dim L_r$, <u>or</u> $\dim L = \dim R_1$, <u>then</u> $\{T_s\}_{s \in S}$ <u>is selfdual.</u>

PROOF. Since, by Theorem 5, $H_r \cap H_1 = \{0\}$, we obtain from (4) (see also the relations (6) from [3])

$$H = \bigoplus_{s \in S} U_s R \oplus \bigoplus_{g \notin S} U_g^{*} L_r \quad \text{and} \quad H^{\perp} = \bigoplus_{f \notin S} U_f R \oplus \bigoplus_{s \in S} U_s^{*} L_r .$$

If $\dim R = \dim L_r$, then there is a unitary operator U from R to L_r. Define the unitary operator W from H to $H^{\perp}$ by

$$W\Big(\sum_{s \in S} U_s r_s \oplus \sum_{g \notin S} U_g^{*} k_g\Big) = \sum_{g \notin S} U_g U^{*} k_g \oplus \sum_{s \in S} U_s^{*} U r_s ,$$

where finitely many of $r_s \in R$ and $k_g \in L_r$ differ from zero. By using Proposition 2 of [3] we obtain, for every $\sigma \in S$,

$$T_\sigma^{\#} W\Big(\sum_{s \in S} U_s r_s \oplus \sum_{g \notin S} U_g^{*} k_g\Big) = T_\sigma^{\#}\Big(\sum_{g \notin S} U_g U^{*} k_g \oplus \sum_{s \in S} U_s^{*} U r_s\Big) =$$

$$= \sum_{g \notin S} U_{\sigma^{-1} g} U^{*} k_g \oplus \sum_{s \in S} U_{s\sigma} U r_s = W\Big(\sum_{s \in S} U_{\sigma s} r_s \oplus \sum_{g \notin S} U_{\sigma^{-1} g}^{*} k_g\Big) =$$

$$= W\, T_\sigma\Big(\sum_{s \in S} U_s r_s \oplus \sum_{g \notin S} U_g^{*} k_g\Big).$$

This leads to $T^{\#}_{\sigma}W = WT_{\sigma}$, $\sigma \in S$. The case dim L = dim R_1 will be treated similarly. The proof is finished.

COROLLARY 9. *If G is a dense subgroup in* **R** *and* dim L=dim R, *then* $\{T^t_s \oplus T^m_s\}_{s \in G_+}$ *is selfdual.*

THEOREM 10. *Let* $\{T_s\}_{s \in S}$ *be a completely non-unitary isometric semigroup without ultraevanescent part. If* $H_t = H_m$ *and* dim R = dim L, *then* $\{T_s\}_{s \in S}$ *is selfdual.*

PROOF. From hypothesis and Theorem 5 it follows that $H_r = H_1 = \{0\}$ and hence we obtain

$$H = \bigoplus_{s \in S} U_s R \quad \text{and} \quad H = \bigoplus_{\sigma \in S} U^*_{\sigma} L.$$

Now using an argument similar to that of the preceding proof, we obtain the theorem.

Remark that - via Example A of [3] -this Theorem generalizes a result of J.B.Conway ([1]), which says that any pure isometry is selfdual.

REFERENCES

[1] Conway, J.B., *The dual of a subnormal operator.* J.Operator Theory 5, 2(1981), 195-211.

[2] Douglas, R.G., *On extending commutative semigroups of isometries.* Bull. London Math.Soc. 1(1969), 157-159.

[3] Gaşpar, D., - Suciu, N., *On the structure of isometric semigroups.* Proceedings of the 8th Conference of Operator Theory, Timişoara and Herculane, June 6-16, 1983 (in print).

[4] Helson, H. - Lowdenslager, D., *Prediction theory and Fourier series in several variables II.* Acta Math. 106(1961), 175-213.

[5] Ito, T., *On the commutative family of subnormal operators.* J.Faculty Sci. Hokkaido Univ., Ser.I Math. 14(1958), 1-15.

[6] Muhly, P.S., *A structure theory for isometric representations of a class of semigroups.* J.Reine Angew.Math. 225(1972), 135-154.

[7] Suciu, I., On the semi-groups of isometries. Studia Math.(PAN) 30(1968), 101-110.

[8] Suciu, I., Function Algebras. Ed.Academiei R.S.R., Bucuresti, 1973.

[9] Sz.-Nagy, B. - Foiaş, C., Harmonic Analysis of Operators on Hilbert space. North-Holland Publishing Company, Amsterdam/London, 1974.

[10] Wold, H., A study in the analysis of stationary time series. Uppsala, Stockholm, 1938.

International Series of
Numerical Mathematics, Vol. 65

ON MAXIMALLY ACCRETIVE OPERATORS IN THE PLANE

H. Berens and L. Hetzelt
Mathematisches Institut
Universität Erlangen-Nürnberg
Erlangen

For the real plane endowed with a strictly convex and smooth norm the class of maximally accretive operators coincides with the class of m-accretive operators exactly when the norm generates an inner product.

1. Introduction

By Zorn's lemma any accretive operator has a maximally accretive extension. In this respect G.J. Minty [5] proved in 1962 that if H is a Hilbert space the accretive (monotone) operator $A \subset H \times H$ is maximally accretive (monotone) exactly when there exists a $\lambda \in \mathbb{R}^+$ such that (and consequently for all $\lambda \in \mathbb{R}^+$) $I + \lambda A$ defines a surjection of A onto H. In this case A is said to be m-accretive. See the following section for the relevant definitions and notations. In contrast to Minty's result, M.G. Crandall and T.H. Liggett [3] showed in 1971 that for $\ell_p(2)$, $1 \leq p \leq \infty$, the class of m-accretive operators coincides with the class of maximally accretive operators exactly when $p = 1,2$, or ∞.

To see that for $1 < p < \infty$ the only ℓ_p-space for which coincidence occurs is $\ell_2(2)$ the authors considered the following operator A defined by $((0,0),(0,0))$, $((1,0),(0,-1))$, and $((0,1),(1,0))$ - A is just the clockwise rotation of the plane by 90^0 restricted to $(0,0),(1,0)$, and $(0,1)$. They pointed out that for $p \neq 2$ no maximally accretive extension of A is defined on the open triangle $\{(\xi,\eta) \in \ell_p(2) : \xi,\eta > 0, \xi + \eta < 1, \xi < \eta\}$ if $2 < p < \infty$, respectively on $\{(\xi,\eta) \in \ell_p(2) : \xi,\eta > 0, \xi + \eta < 1, \xi > \eta\}$ if $1 < p < 2$, in con-

The paper was completed while the first named author was visiting the University of Alberta at Edmonton. His stay was supported by the NSERC CANADA under Grant A 7687.

trast to the fact that the closure of the domain of an m-accretive operator on $\ell_p(2)$ is convex, see Lemma 1 below. Let us mention that A as well as -A are accretive operators on $\ell_p(2)$ - in $\ell_2(2)$ this means that A is atone in the sense of E.H. Zarantonello [6].

Motivated by the example we took a closer look at the extension problem in the plane and were able to prove the following

THEOREM. _Let X be a real, 2-dimensional, normed vector space with a strictly convex and smooth norm_ $|\cdot|$. _If every accretive operator in_ $X \times X$ _has an m-accretive extension then the norm generates an inner product._

In the following section we do the necessary preliminary work, while the theorem is proved in Section 3. Section 4, finally, contains further remarks and comments.

2. Definitions, Notations, and Preliminaries

Let X be a finite-dimensional real normed vector space with norm $|\cdot|$. It's elements are denoted by x,y,z.. . For a subset K of X, $\bar{K}$, $\mathring{K}$, and ∂K denote its closure, interior, and boundary, respectively. By $b_r(x)$ we mean the open ball centered at x with radius $r \in \mathbb{R}^+$; b_1 denotes the open unit ball.

The semi-inner product $<\cdot,\cdot>_s : X \times X \to \mathbb{R}$ is defined by

$$<y,x>_s := \lim_{t \to 0+} \frac{|x+ty|^2 - |x|^2}{2t}, \quad (y,x) \in X \times X.$$

If F denotes the duality map on X - F is the subdifferential of $|\cdot|^2/2 : X \to \mathbb{R}$ - then

$$<y,x>_s = \max\{<y,w> : w \in F(x)\}.$$

It follows that $<\cdot,\cdot>_s$ is upper semi-continuous. If X is strictly convex then $F(x) \cap F(y) = \emptyset$ whenever $x \neq y$. If X is smooth then F is single-valued and $<y,x>_s$ is just the derivative of $|\cdot|^2/2$ at x in the direction y.

A set-valued map A from X into itself is called an operator on X. $D(A) := \{x \in X : (x,a) \in A\}$ and $R(A) := \{a \in X : (x,a) \in A\}$ denote the domain and range of A, respectively. $A^{-1} := \{(a,x) \in X \times X : (x,a) \in A\}$. If A^1 and A^2 are two operators on X, $A^1 + A^2 := \{(x,a^1+a^2) \in X \times X : (x,a^1) \in A^1$ and $(x,a^2) \in A^2\}$, and for $\lambda \in \mathbb{R}$ $\lambda A = \{(x,\lambda a) \in X \times X : (x,a) \in A\}$. The operator A is said to be accretive, if

$$\forall\ (x,a),(x',a') \in A \qquad 0 \leq \langle a - a', x - x' \rangle_s .$$

This is equivalent to the fact that

$\forall\ \lambda > 0 \qquad (I + \lambda A)^{-1}$ defines a contraction from X to X, i.e.,

$$\forall (x,a),(x',a') \in A \qquad |x - x'| \leq |(x + \lambda a) - (x' + \lambda a')| \qquad \forall\ \lambda \in \mathbb{R}^+ .$$

A is said to be m-accretive if A is accretive and if for all $\lambda \in \mathbb{R}^+$ $I + \lambda A$ is surjective. (The fact that $I + \lambda A$ is injective follows trivially from the accretiveness of A.) An m-accretive operator on X is a maximal element within the class of accretive operators on X.

Following E.H. Zarantonello [6] we call an operator A on X atone if A as well as -A are accretive, i.e.,

$$\forall\ (x,a),(x',a') \in A \qquad |x - x'| \leq |(x + \lambda a) - (x' + \lambda a')| \qquad \forall\ \lambda \in \mathbb{R}.$$

The following lemma about m-accretive operators is crucial for the proof of the theorem.

LEMMA 1. Let X be strictly convex and smooth. For an m-accretive operator A on X D(A) is convex, more precisely, it is the range of a contractive ray retraction.

For a proof see H. Brezis [1, Theorem 2.2] and R.C. Bruck [2].

If dim X = 2 and if the norm is strictly convex then by a result of L.A. Karlovitz the closed convex sets are exactly the contractive retracts.

Let us stay with two-dimensional spaces. By a subspace we mean a non trivial proper subspace, i.e., a straight line through the origin. Two subspaces (G_1, G_2) in X are called conjugate, if G_2 is parallel to one of the tangent lines of ∂b_1 at the points of $G_1 \cap \partial b_1$, and vice versa. There are always pairs of conjugate subspaces in the plane. Indeed,

LEMMA 2. There exist always two different pairs of conjugate subspaces in the plane.

For a proof see D. Laugwitz [4].

These conjugate spaces will be used to construct simple atone operators in the plane.

3. Proof of the Theorem

We endow ∂b_1 with the natural orientation (counterclockwise).

Let (G_1,G_2) and (G_1',G_2') be two pairs of conjugate subspaces whose existence is guaranteed by Lemma 2. We take (G_1,G_2) to be our coordinate system and select the unit vectors e_1 on G_1 and e_2 on G_2 in such a way that they are compatible with the orientation. With respect to the second pair, we may assume w.l.o.g. that G_1' intersects the first and third quadrant, and G_2' the second and fourth one, see Fig. 1.

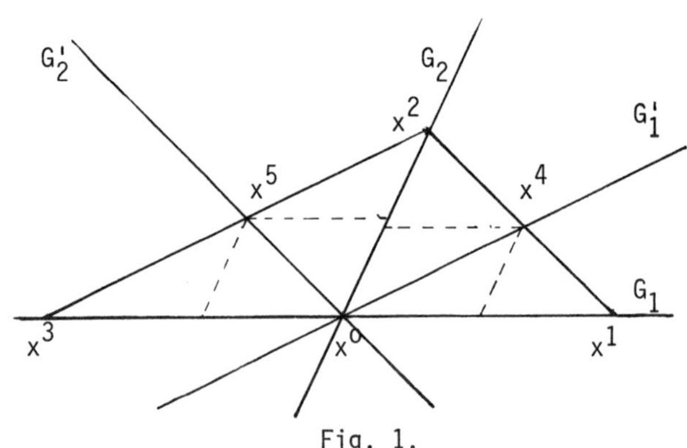

Fig. 1.

To construct an atone operator A on X we start out with the vectors $x^0 := (0,0)$, $x^1 := e_1 = (1,0)$, $x^2 := \{x^1 + G_2'\} \cap G_2 =: (0,\alpha)$, $\alpha > 0$, $x^3 := \{x^2 + G_1'\} \cap G_1 =: (-\beta,0)$, $\beta > 0$, and define A via $(x^0,x^0),(x^1,x^2)$, and (x^2,x^3). Obviously, A is atone.

Let B be an arbitrary m-accretive extension of A. Since X is strictly convex, by Lemma 1 the interior of the triangle C defined by x^0,x^1, and x^2 belongs to D(B). Taking further the smoothness of X into account, B is uniquely determined on $\mathring{C}$ via

$$(1)\quad \mathring{C} \ni z \to B(z) = T(z) \cap \{x^2 + T(z-x^1)\} \cap \{x^3 + T(z-x^2)\},$$

where $T(u)$, $u \neq 0$, denotes the subspace of X parallel to the tangent line of ∂b_1 at $u/|u|$. Indeed, if the three tangent lines would not intersect in a single point, either A or -A would not have an m-accretive extension. The mapping $B|_{\mathring{C}}$ is a continuous atone bijection of $\mathring{C}$ onto the interior of the triangle D defined by x^o, x^2, and x^3, see Fig. 2.

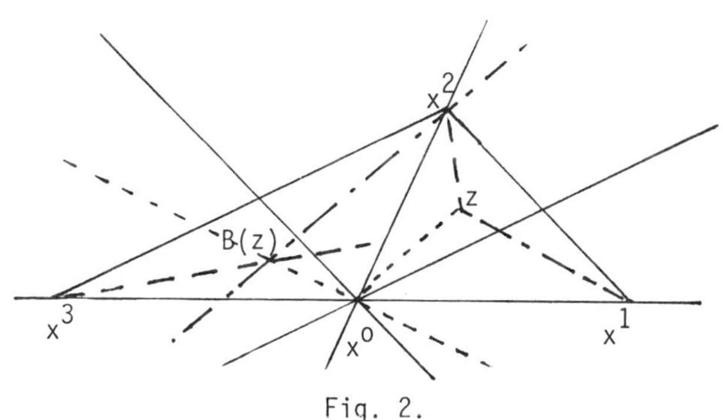

Fig. 2.

For a given $z \in \mathring{C}$ there exist uniquely determined $0 < \sigma, \tau < 1$ such that z is the intersection of the straight lines through x^2 and $(\sigma,0)$, and x^1 and $(0,\tau\cdot\alpha)$, respectively. The coordinates of z in terms of σ and τ are calculated as

$$(2) \qquad z = \left(\frac{\sigma(1-\tau)}{1-\sigma\tau}, \frac{\tau(1-\sigma)}{1-\sigma\tau} \cdot \alpha\right).$$

Let us define the mappings $\varphi,\psi : [0,1] \to [0,1]$ by $\varphi(0) := 0$, $\varphi(1) := 1$, and

$$(0,\varphi(\sigma)\cdot\alpha) := G_2 \cap \{x^3 + T((\sigma,0) - x^2)\},$$

and $\psi(0) := 0$, $\psi(1) := 1$, and

$$(-\psi(\tau)\cdot\beta,0) := G_1 \cap \{x^2 + T((0,\tau\cdot\alpha) - x^1)\},$$

respectively. Both functions are continuous and strictly increasing. From (1) we obtain

$$\mathring{C} \ni z \to B(z) = \{x^2 + T((0,\tau\cdot\alpha) - x^1)\} \cap \{x^3 + T((\sigma,0) - x^2)\},$$

i.e., B(z) is the point of intersection of the straight lines through x^3 and $(0,\varphi(\sigma)\cdot\alpha)$, and x^2 and $(-\psi(\tau)\cdot\beta,0)$, respectively, giving

$$(3) \qquad B(z) = \left(-\frac{\psi(\tau)(1-\varphi(\sigma))}{1-\varphi(\sigma)\psi(\tau)} \cdot \beta, \frac{\varphi(\sigma)(1-\varphi(\tau))}{1-\varphi(\sigma)\psi(\tau)} \cdot \alpha\right).$$

In order to simplify the notation, let us set

$$(4) \begin{cases} s := \sigma/(1-\sigma), & t := \tau/(1-\tau), \\ \Phi(s) := \varphi(\sigma)/(1-\varphi(\sigma)), & \Psi(t) := \psi(\tau)/(1-\psi(\tau)), \end{cases}$$

Φ and Ψ are continuous, strictly increasing functions on $[0,\infty]$ onto itself. Using (4), the equations (2) and (3) can be rewritten as

$$(5) \qquad z = \left(\frac{s}{1+s+t}, \frac{t}{1+s+t} \cdot \alpha\right),$$

$$(6) \quad B(z) = \left(-\frac{\Psi(t)}{1+\Phi(s)+\Psi(t)}\cdot\beta\ ,\ \frac{\Phi(s)}{1+\Phi(s)+\Psi(t)}\cdot\alpha\right).$$

We want to introduce one additional function $\Delta:[0,\infty]\to[0,\infty]$ by $\Delta(0):=0$, $\Delta(\infty):=\infty$, and $\Delta(u):=-B(z)_2/B(z)_1$, where $u=z_1/z_2$, $z\in\overset{\circ}{C}$. For $z\in X$, z_i, $i=1,2$, denotes its i-th coordinate. By (1) Δ is a well-defined, continuous and strictly increasing function.

It follows from the upper semi-continuity of B that for each $s,t,u\in[0,\infty]$ the ordered pairs

$$\left(\left(\frac{s}{1+s},0\right),\left(0,\frac{\Phi(s)}{1+\Phi(s)}\cdot\alpha\right)\right),\ \left(\left(0,\frac{t}{1+t}\cdot\alpha\right),\left(-\frac{\Psi(t)}{1+\Psi(t)}\cdot\beta,0\right)\right), \text{ and}$$

$$\left(\left(\frac{\alpha u}{1+\alpha u},\frac{1}{1+\alpha u}\cdot\alpha\right),\left(-\frac{\alpha}{\alpha+\beta\Delta(u)}\cdot\ ,\ \frac{\beta\Delta(u)}{\alpha+\beta\Delta(u)}\cdot\alpha\right)\right)$$

belong to B. Furthermore, since B is atone on C

$$(x^4,x^5):=\frac{1}{1+\beta}((\beta,\alpha),(-\beta,\alpha\beta))$$

as well as

$$((x_1^4,0),(0,x_2^5)),\text{ and }((0,x_2^4),(x_1^5,0))$$

belong to B, see Fig. 1. Since

$$x_1^4=\frac{\beta}{1+\beta}\quad\text{and}\quad x_2^4=\frac{1}{1+\beta}\cdot\alpha,$$

and consequently

$$x_1^5=-\frac{1}{1+\beta}\cdot\beta=-\frac{\Psi(1/\beta)}{1+\Psi(1/\beta)}\cdot\beta=-\frac{1}{1+(\beta/\alpha)\Delta(\beta/\alpha)}\cdot\beta$$

and

$$x_2^5=\frac{\beta}{1+\beta}\cdot\alpha=\frac{\Phi(\beta)}{1+\Phi(\beta)}\cdot\alpha=\frac{(\beta/\alpha)\Delta(\beta/\alpha)}{1+(\beta/\alpha)\Delta(\beta/\alpha)}\cdot\alpha,$$

we conclude that

$$(7) \quad \Phi(\beta)=\beta,\ \Psi(1/\beta)=1/\beta,\ \text{ and }\ \Delta(\beta/\alpha)=\alpha.$$

Let us go back to the definition of Δ. Using the representations (5) and (6), we obtain

$$(8) \quad \Delta(s/t\cdot\alpha)=\Phi(s)\cdot\alpha/(\Psi(t)\cdot\beta),\qquad 0<s,t<\infty.$$

Setting $s=t\cdot\beta$ and taking (7) into account, this leads to

(9) $\Psi(t)\cdot\beta = \Phi(t\cdot\beta), \qquad 0 \le t \le \infty,$

and

(10) $\Phi(1) = \Psi(1) = 1.$

Substituting (9) back into (8) and setting $t = 1$ gives us

$$\Delta(s/\alpha) = \Psi(s/\beta)\cdot\alpha, \qquad 0 \le s \le \infty,$$

and further

(11) $\Psi(s)\cdot\Psi(t) = \Psi(s\cdot t), \qquad 0 \le s,t \le \infty.$

Thus

(12) $\Psi(t) = t^{p-1}, \qquad 0 \le t \le \infty,$ for some $p > 1$.

The Δ-function then reads

(13) $\Delta(u) = (\alpha^p/\beta^{p-1})\, u^{p-1}, \qquad 0 \le u \le \infty.$

If x is a unit vector located in the first quadrant, then with $u = x_1/x_2$ $\Delta(u)$ determines the slope of ∂b_1 at x. Hence

$$\alpha^p x_1^{p-1}\, dx_1 + \beta^{p-1} x_2^{p-1}\, dx_2 = 0$$

for all $x \in \partial b_1$ for which $x_1, x_2 \ge 0$. By integration and by taking the initial value $(1,0)$ into account

$$\alpha^p x_1^p + \beta^{p-1} x_2^p = \alpha^p.$$

Since the vector $(0,1)$ also has to satisfy the equation we conclude that $\alpha^p = \beta^{p-1}$, giving finally

(14) $x_1^p + x_2^p = 1, \qquad x \in \partial b_1,\ x_1, x_2 \ge 0.$

In the Cartesian plane (G_1, G_2) and (G_1', G_2') are both orthogonal pairs of subspaces where G_1 and G_2 determine the x_1- and the x_2-axis, respectively.

Let $p \neq 2$. In this case it follows from (7) that $\alpha = \beta = 1$, and (G_1, G_2) and (G_1', G_2') are rotated about each other by 45°.

If we interchange the role of (G_1, G_2) and (G_1', G_2'), we get a second equation for ∂b_1 within the first quadrant with respect to (G_1', G_2'), say

(15) $x_1'^{p'} + x_2'^{p'} = 1, \qquad x' \in \partial b_1,\ x_1', x_2' \ge 0.$

Since both quadrants in (14) and (15) intersect in an open angle, this is only possible when $p = p' = 2$.

Thus in the Cartesian plane ∂b_1 is given by the Euclidean circumference.

3. Concluding Remarks

There are good reasons to believe that for the plane endowed with an arbitrary norm the following holds true:

Let X be a real, 2-dimensional, normed vector space. If every accretive operator in $X \times X$ has an m-accretive extension then the norm either generates an inner product or is square.

At this moment we are not able to prove our claim.

Concerning Crandall's and Liggett's result, they proved actually more than we stated in the Introduction. Indeed they proved

In $\ell_\infty(n)$, $n \in \mathbb{N}$, the class of maximally accretive operators coincides with the class of m-accretive ones.

We are able to give an alternative proof of this result which gives more insight into the toplogical structure of maximally accretive operators. The details will be given in a separate paper.

REFERENCES

[1] Brezis, H., *Opérateurs Maximaux Monotones.* Notas de Matemàtica (50). North-Holland Publishing Company, Amsterdam/London 1973.

[2] Bruck, R.C., *Nonexpansive projections on subsets of Banach spaces.* Pacific J. Math. 47 (1973), 341-355.

[3] Crandall, M.G. - Liggett, T.M., *A theorem and a counterexample in the theory of semigroups of nonlinear transformations.* Trans. Amer. Math. Soc. 160 (1971), 263-278.

[4] Laugwith, D., *Konvexe Mittelpunktsbereiche und normierte Räume.* Math. Z. 61 (1954), 235-244.

[5] Minty, G.J., *Monotone (nonlinear) operators in Hilbert space.* Duke Math. J. 29 (1962), 341-346.

[6] Zarantonello, E.H., *Piecewise atone interpolation of monotone operators.* Ann. Scuola Norm. Sup. Pisa. Serie IV, IX, 4 (1982), 497-528.

II Functional Analysis

International Series of
Numerical Mathematics, Vol. 65

INVARIANT FUNCTION SPACES CONNECTED WITH THE HOLOMORPHIC DISCRETE SERIES

Jaak Peetre[(1)]
Matematiska Institutionen
Lunds Universitet
S-220 07 Lund, Sweden

The theory of Möbius invariant spaces of holomorphic functions, as developed by Arazy, Fisher, and others, is briefly reviewed. Then an extension of it (connected with the discrete holomorphic series) is outlined. The main object of the lecture is however the advancement of the thesis that the systematic use of group invariant spaces is a useful point of view in Analysis, and we illustrate it on the hand of several concrete applications. In particular, we give a new simple proof of the trace ideal criterion for Hankel operators (Peller's theorem) in the case $1 < p < \infty$.

0. Introduction. This paper is a modest contribution to a vast general theme, to which I was directed through conversations with Jonathan Arazy (and later Stephen Fisher): the study of G-invariant function spaces where G is a (topological) group acting on functions defined on a given space Z. In the first place we are interested in Banach spaces and the group should act on them through isometries. In the second instance we may ask what happens if we only assume that G acts through (topological) isomorphims. One may also consider quasi-Banach spaces and more general (metric) spaces.

As a first example, let Z be any measurable space with measure m and let G be the group generated by the following two types of transformations T: either 1) T is multiplication by a unimodular measurable function a, say: $Tf(z) = a(z)f(z)$ a.e., $|a(z)| = 1$ a.e. on Z or 2) T is

(1) Joint work with Jonathan Arazy, Stephen Fisher, Svante Janson; I am solely responsible for any possible mistakes in this messy compilation, prepared on the occasion of the conference Functional Analysis and Approximation (Oberwohlfach, July 30 - Aug. 6, 1983). A somewhat more detailed version is available in the Lund report series.

induced by a measure preserving self-map of Z: $Tf(z) = f(b(z))$ a.e., $mE = mb(E)$ for any measurable subset $E \subset Z$. The spaces gotten then are the r.i. (rearrangement invariant) spaces. A remarkable feature in this case is the fact (the Calderón-Mitjagin theorem) that in this case all G-invariant spaces can be gotten by interpolation (in fact, via the K-method) from two "extremal" spaces, namely L^1 and L^∞.

The special case when Z is finite, say, $Z = \{1,\dots,n\}$, and m is the measure which to each point assigns the load 1. We are then dealing with r.i. spaces of finite sequences and the Calderón-Mitjagin theorem concerns interpolation between ℓ_n^1 and ℓ_n^∞; in this situation the result is more or less classical and touches upon a theme associated with a series of famous names: Schur, Karamata, Hardy-Littlewood-Polyá etc.; see e.g. [4] for some references. In this case the group is essentially the Coxeter group B_n (the semi-direct product of n copies of the cyclic group $\{1,-1\}$ with the symmetric group S_n). In a series of papers (see e.g. [16]) I.M.Zobin and/or B.G.Zobina have extended the Calderón-Mitjagin theorem to the setting of general Coxeter groups. There arises the question to what extent this work can be extended to the infinite dimensional case. Returning for a moment again to the finite dimensional case, a question which comes to ones mind is what happens if one lets the group act not on sequences (vectors) but on tensors, preferably tensors with prescribed symmetry properties. In the case of the symmetric group S_n proper (the Coxeter group A_n) this is in line with the classical theory of representations of that group (Frobenius, Schur, Young, H.Weyl).

Let us now turn to a fundamentally different situation, with G a special s.s. (semi-simple) Lie group, which will keep us busy for the rest of the time. More specifically, we let Z be the unit disk U, say, in the complex plane C and let G be the *Möbius group* (that is, the group of all holomorphic self-maps (automorphisms) of U); if $\omega \in G$ then there is a matrix $\begin{pmatrix} a & b \\ c & d \end{pmatrix}$ in $SU(1,1)$ (that is, the group of 2×2 complex unimodular matrices preserving the Hermitian form $||z||^2 = |z_1|^2 - |z_2|^2$ in C^2) such that $\omega(z) = (az+b)/(cz+d)$, we have thus $ad - bc = 1$, $c = \bar{b}$, $d = \bar{a}$, and the operator $T = T_\omega$ corresponding to ω is given by $T_\omega f(z) = f(\omega z)$.

In this case, among the G-invariant spaces - we are only concerned in the action of G on *holomorphic* functions - there appear a number of famous function spaces: H^∞, BMO, the Bloch space, the Dirichlet space etc. In fact, the last space is essentially the unique Hilbert space among the G-invariant Banach spaces. Furthermore, the last but one, the Bloch space, is the maximal G-invariant space and there is also a *minimal* such space, which turns out to be the predual of the Bloch space in the natural pairing given by the invariant inner product in the Dirichlet space. (Similarly, the pre-dual of BMO is essentially the Hardy class H^1, but only "covariantly"; H^1 has to be replaced by the Sobolev space W_1^1 ($f \in W_1^1 \iff$ $\iff \int_T |f'(z)||dz| < \infty$).)

Thus in a way we have isolated two extremal spaces (somehow analogous to L^∞ and L^1 in the previous example) but it is in no way clear if it is possible to reconstruct the general G-invariant space from these spaces; at least they are certainly not interpolation spaces. At any rate it is possible to associate rather general families (parametrized by a general r.i. space X) to any orbit of G under its action on clos U × clos U, which possesses an invariant measure. For instance, corresponding to the "open diagonal" $\{(z,z) | z \in U\}$, which set we may identify with U itself, we have Bergman (or Besov) type spaces A_X: $f \in A_X$ iff $(1-|z|^2)f'(z) \in X$ (with respect to $dxdy/(1-|z|^2)^2$, the invariant measure on the orbit in question). If $X = L^2$ we have the Dirichlet space and if $X = L^\infty$ the Bloch space; on the contrary, the minimal space does not arise in this way; cf. Sec. 3.4. For some more details about the theory outlined above see [1], [2], [3], [12].

REMARK. Recall that we always assume that the group acts via isometries. Actually, in [2] there is established that uniqueness prevails even if we permit isomorphisms only, imposing however a uniformity condition: $\sup_{\omega \in G} ||T||_\omega <$ $< \infty$. If we choose to drop the latter assumption there are of course plenty of non-equivalent invariant Hilbert spaces, and it is an interesting *open problem* to classify them. Arazy has also (see [1]) proved the interesting theorem that if a Möbius invariant Banach space (in the original sense, i.e. the group acting isometrically) has a 1-symmetric basis then it must be a Hilbert space (= Dirichlet space). This points to a radical difference between the two cases: G a group of the "Coxeter type" and G a s.s. Lie group.

However, we shall in this paper mostly be concerned (Sec. 1) with the corresponding theory in the more general case when G (the Möbius group) acts on holomorphic functions f on U (unit disk) via a "multiplier", defining the action T_ω of $\omega \in G$ by $(T_\omega f)(z) = f(\omega z)(cz+d)^{\mu-1}$ where μ is any given integer. Here our findings are in general, at least at the time being, far more fragmentary, in any case restricted essentially to the "botany of function spaces".

These actions are related to the so-called *holomorphic discrete series* for the group $SL(2,R)$ ($\approx SU(1,1)$); whence the fancy title of the paper. It seems therefore that the ultimate limits of the theory are the case when G is a s.s. Lie group possessing a maximal compact subgroup K which is itself not s.s. so that $Z = G/K$ can be realized as a symmetric domain in some C^n. E.g. Svante Janson and I (unpublished) have in some detail looked into the case when $G = SU(n,1)$ so that Z = unit ball in C^n. In this case one has an analogous decomposition into orbits as for $n = 1$ (the previous case), and to each orbit there corresponds a family of invariant spaces parametrized by r.i. spaces. It is likewise comparitively easy to get at least a rough idea of what is going on in the case of the polydisk U^n. Regarding the Bloch space in the context of general symmetric domains see [15].

In Sec. 2 we shall however outline the theory of maximal and minimal spaces in a completely abstract setting.

Sec. 3 is concerned with assorted applications (trace class Hankel operators (forms), rational approximation, the Fejér-Riesz inequality).

1. Holomorphic discrete series. In this Sec. G is thus the Möbius group (see Intr.) and we let it act on holomorphic functions via the formula

$$f(z) \to f(\omega z)(cz+d)^{\mu-1} \tag{1}$$

where $\omega \in G$, $\omega z = (az+b)/(cz+d)$, $c = \bar{b}$, $d = \bar{a}$, $ad - bc = 1$. μ is any integer; if $\mu \geq 1$ we must reckon modulo P_μ, the set of polynomials of degree $\leq \mu - 1$ (we also put $P_0 = 0$); in particular, if $\mu = 1$ we reckon modulo constants.

If $\mu < 0$ we define an invariant inner product by the formula

$$(f|f_1) = \int_U f(z)\overline{f_1(z)}(1-|z|^2)^{-\mu-1}dxdy; \tag{2}$$

the corresponding Hilbert spaces are the weighted Bergman spaces and the corresponding unitary representations make up the holomorphic discrete series.

Presently we shall however concentrate upon the case $\mu \geq 0$. In that case we put

$$(f|f_1) = -\frac{1}{2\pi i}\int_T \overline{f(t)t^{\mu-1}f_1^{(\mu)}(t)}\,\overline{dt}.$$

Clearly

$$(f|f_1) = \sum_{n\geq\mu}^{\infty} n(n-1)\ldots(n-(\mu-1))\hat{f}(n)\overline{\hat{f}_1(n)}$$

so that it is positive modulo P_μ.

CLAIM. This inner product is invariant under the group and is up to constants the only one with this property (group invariance).

PROOF. It suffices to compute $(f|f_1)$ when f and f_1 are in the orbit of z^μ, i.e.

$$f = \left(\frac{az+b}{cz+d}\right)(cz+d)^{\mu-1} \quad \text{and similarly for } f_1.$$

Write $(az+b)/(cz+d) = a/c - 1/c(cz+d)$. We then get

$$f = (-1)^\mu \frac{1}{c^\mu(cz+d)} \bmod P_\mu; \quad f^{(\mu)} = \mu!/(cz+d)^{\mu+1}, \tag{3}$$

similarly for f_1. It follows that

$$(f|f_1) = -\frac{1}{2\pi i}\int_T f(t)\,\overline{\frac{\mu!\,t^{\mu-1}}{(c_1t+d_1)^{\mu+1}}}\,\overline{dt} =$$

$$= \mu!/2\pi i \int_T f(t)\frac{1}{(a_1t+b_1)^{\mu+1}}\frac{dt}{t} = \frac{1}{a_1^{\mu+1}} f^{(\mu)}(-b_1/a_1),$$

whence (use again (3), now with f)

$$(f|f_1) = \frac{1}{a_1^{\mu+1}} \cdot \frac{\mu!}{(d-\frac{b_1}{a_1}c)^{\mu+1}} = \frac{\mu!}{(\bar{a}a_1-\bar{b}b_1)^{\mu+1}}.$$

Clearly this expression is invariant; in the denominator sits the Hermitian inner product in C^2 encountered in the Intr.

Conversely, given any invariant inner product $(\ |\ )$, since apparently then in particular $(f|z^{\mu}) = \text{const}\cdot\bar{a}^{-(\mu+1)}$, one concludes that $(f|f_1) = = \text{const}\cdot(\bar{a}a_1-\bar{b}b_1)^{-(\mu+1)}$. □

REMARK. The case $\mu = 0$ may be considered as a limiting case of the holomorphic discrete series, while the case $\mu > 0$ can be obtained via analytic continuation. To see this we have to replace G by its universal covering group $\tilde{G}$. Then the formula defining the group action is applicable for any complex number μ (not only for integers). For $\text{Re}\,\mu < 0$ formula (2) still can be used as definition of an invariant inner product. Using a subscript to indicate the μ dependence, it is then clear that the limit $\lim_{\mu\to 0} \mu\cdot(f|f_1)_{\mu}$ exists, provided, say, f and f_1 are analytic in a neighborhood of clos U. More generally, it is seen that $(f|f_1)_{\mu}$ has a holomorphic (meromorphic) continuation to all values of μ. The points $\mu = 0,1,2,\ldots$ are then simple poles and the residues can easily be computed. They turn out to define an inner product which is positive definite mod P_{μ}.

We can now begin to develop the analogue of the Arazy - Fisher theory, as outlined in the Intr.

There exists thus a unique Hilbert space H (analogue of the Dirichlet space) and $H \approx B_2^{\mu/2,2}$.

There exists further a m a x i m a l invariant Banach space B (analogue of Bloch) and a minimal i n v a r i a n t space B_* (analogue of pre-Bloch). They are in duality in the invariant pairing induced by the inner product in H. Also (as Besov spaces) $B \approx B_{\infty}^{(\mu-1)/2,\infty}$ and $B_* \approx B_1^{(\mu+1)/2,1}$.

B apparently is characterized by the property $|f^{(\mu)}(z)| \leq \text{const}\cdot \cdot (1-|z|^2)^{(\mu+1)/2}$, and the elements of B_* are those which admit the representation $f = \sum\lambda_k \cdot 1/c_k^{\mu}(c_k z+d_k) \bmod P_{\mu}$, with $\sum|\lambda_k| < \infty$; c_k and d_k are elements of corresponding matrices in the group.

More generally, one sees that the general Besov spaces $B_p^{s,p}$, $s = = (\mu-1)/2 + 1/p$, are invariant. They may also be gotten by interpolation from the "extremal" space : $B_p^{s,p} = (B_*,B)_{\theta p}$ with $\theta = 1/p$, s as before. We have further the invariant norm

$$||f||^2 = \iint_U ((1-|z|^2)^{(\mu+1)/2}|f^{(\mu)}(z)|)^p dxdy/(1-|z|^2)^2,$$

p r o v i d e d $1/p < (\mu+1)/2$ or $p > 2/(\mu+1)$. (In the present context we are willing to admit quasi-Banach spaces too!) Limiting case: W_p^μ, $1/p = (\mu+1)/2$ (cf. the case W_1^1 mentioned in the Intr.).

Instead of the μ th derivative $f^{(\mu)}$ one can also, in order to define invariant spaces, work with $f(z_1,z_2,\ldots,z_{\mu+1})$, the μth divided difference (in the sense of Newton).

Let g be the transform of f, i.e.

$$g(z) = f(\omega z)(cz+d)^{\mu-1}.$$

Then the μ th difference is transformed according to the formula

$$g(z_1.,,,.z_{\mu+1}) = \frac{f(\omega z_1,\ldots,\omega z_{\mu+1})}{(cz_1+d)\ldots(cz_{\mu+1}+d)} .$$

Notice that in the limit one obtains

$$g^{(\mu)}(z) = \frac{f^{(\mu)}(\omega z)}{(cz+d)^{\mu+1}} ,$$

which has been implicit in the preceding discussion.

For instance, one gets now the following equivalent expression for the inner product in H:

$$||f||^2 = \int_{T\times\ldots\times T} |f(z_1,\ldots,z_{\mu+1})|^2 \; |dz_1|^2 \ldots |dz_{\mu+1}|^2 : \prod_{i\neq k} |z_i-z_k|^{\frac{1}{\mu}-1} .$$

One can also let some of the points coalesce. In particular, this suggests a definition of "BMO":

$$f \in BMO_\mu \iff \forall z \in U \; \exists g_z \in P_\mu \quad \text{such that}$$

$$\int_T (|f(\zeta)-g_z(\zeta)|\cdot(\Pi_z(\zeta))^{(\mu-1)/2})^2 \Pi_z(\zeta)|d\zeta| \leq M < \infty,$$

where Π_z is the Poisson kernel at the point z; one can always take g_z to be the $(\mu-1)$th Taylor polynomial at the point z. Similar definition of "H^∞":

$$f \in H_\mu^\infty \iff \forall z\ \exists g_z \in P_\mu \quad \text{such that}$$

$$\sup_{\zeta \in T} (|f(\zeta)-g_z(\zeta)| \cdot (\Pi_z(\zeta))^{(\mu-1)/2}) \leq M < \infty.$$

2. Some abstract nonsense. We shall now outline a theory of maximal and minimal spaces in a general setting.

2.1. The highest level of abstraction. Let H be any Hilbert space, with scalar product $(\cdot|\cdot)$, and let $\Phi = \{\varphi_i\}_{i\in I}$ be a family *total* in H (i.e. finite linear combinations are dense in H). We can then consider the Banach space, denoted Min, generated by $\Phi : f \in \text{Min} \iff f = \sum \lambda_i \varphi_i$ where $\sum |\lambda_i| < \infty$ (i.e. $\{\lambda_i\} \in l^1(I)$). $||f||_{\text{Min}} = \inf \sum |\lambda_i|$ $(= ||\{\lambda_i\}||_1)$. Clearly Min is *dense* in H. Therefore, by general principles, H can be imbedded in Max, the dual of Min. If g is an element of Max and f is in Min, I still use the notation $(f|g)$ for the pairing between Min and Max. Also $||g||_{\text{Max}} = \sup |(g|\varphi_i)|$. We have thus the following situation:

$$\text{Min} \xrightarrow{\text{dense}} H \longrightarrow \text{Max and Min}^* = \text{Max}.$$

Let Max_0 be the closure of H in Max. Then, as clearly H is dense in Max_0, H can be imbedded in Max_0^*. One has always a canonical map $\text{Min} \longrightarrow \longrightarrow \text{Max}_0^*$, which to each φ_i assigns the linear function $g \longrightarrow (g|\varphi_i)$. *Under suitable assumptions* on (e.g. if the map $i \longrightarrow (f|\varphi_i)$, for $f \in H$ given, always is in $c_0(I)$), one can show that it is an isometry. One then has thus

$$\text{Max}_0^* = \text{Min}.$$

There is a general theorem (see [7]) which says that if E is a *reflexive* Banach space *densely* imbedded in a Hilbert space so that one has the situation

$$E \xrightarrow{\text{dense}} H \longrightarrow E^*,$$

then $(E,E^*)_{1/2,2} = H$ (real interpolation) isomorphically and $[E,E^*]_{1/2} = H$ (complex ditto) isometrically. Here E $(= \text{Max}_0)$ cannot be expected to be reflexive in general but it is easy to see that the conclusion of the theorem still holds true. Thus

$$[\mathrm{Max},\mathrm{Min}]_{1/2} = H \text{ isometrically}$$

and similarly with the real method.

2.2. The case of a general group representation. Now assume that there is given a (topological) group G and a unitary representation T of G in a Hilbert space H. If $\varphi_0 \in H$ is then a cyclic element for G, it is natural to take for Φ the orbit under G of φ_0, $\Phi = \{T_\omega\varphi_0\}_{\omega\in G}$ $(I = G)$. In this case thus:

$$||g||_{\mathrm{Max}} = \sup_G |(g|T_\omega\varphi_0)|,$$

$$||f||_{\mathrm{Min}} = \inf\{\sum|\lambda_i| : f = \sum\lambda_i T_{\omega_i}\varphi_0\},$$

and the previous theory is applicable.

2.3. The special case of a muliplier representaion. Finally, assume that H is a space of integrable functions (possible vector valued!) on a given measure space (Z,ν), $||f||^2 = \int_Z|f(z)|^2\,d\nu(z)$, and that G acts on H via a multiplier $A(\omega,z)$:

$$(T_\omega f)(z) = A(\omega,z)f(\omega z).$$

From $T_{\omega'} \circ T_\omega = T_{\omega\omega'}$ we readily obtain the "cocycle relation":

$$A(\omega\omega',z) = A(\omega',z)A(\omega,\omega' z).$$

In this case we have a natural choice for φ_0: namely $\varphi_0 = K_{z_0}$, the reproducing kernel of H at a given point $z_0 \in Z$: $f(z_0) = (f|K_{z_0})$ $(f \in H)$. Clearly: $T_{\omega^{-1}}K_{z_0}(\zeta) = A(\omega,z_0)K_z(\zeta)$ if $z = \omega z_0$. Thus now

$$||f||_{\mathrm{Max}} = \sup_{\omega\in G} |(g_\omega|T_\omega K_{z_0})| \;(= (T_{\omega^{-1}}f|K_{z_0})),$$

$$||f||_{\mathrm{Min}} = \inf\{\sum|\lambda_i| : f = \sum\lambda_i A(\omega_i,z_0)K_{z_i},\; z_i = \omega_i z_0\}.$$

EXAMPLE. $Z = U$ = unit circle, G = Möbius group, $(T_\omega f)(z) = f(\omega z)(cz+d)^{\mu-1}$, μ a given integer ≥ 0, if $\omega = \begin{pmatrix} a & b \\ c & d \end{pmatrix}$. This is the case of Sec. 1. To make the previous machinery work we have to identify f with $f^{(\mu)}$ (the μ th derivative), because we are counting modulo P_μ.

3. Assorted applications and auxiliary comments.

3.1. The trace ideal criterion for Hankel operators. The question when a given Hankel operator is of trace class p $(0 < p < \infty)$ is now completely settled: after Peller's pioneering work [9], the cases $p = 1$ and $1 < p < \infty$, the problem in the remaining case $0 < p < 1$ was quite recently solved independently by Semmes [13] and by Peller himself [10]. Below we offer a somewhat novel approach in the primitive case $1 < p < \infty$, which has the great virtue that it immediately extends to quite general situations (including the case of several (complex) variables).

To fix the ideas let us consider the situation in Sec. 1, i.e. G is the Möbius group acting on holomorphic functions defined in U via formula (1) of that Sec. but we concentrate now instead on the case when μ, still an integer, is < 0 (the limiting case $\mu = 0$ corresponding to the primitive situation of the classical Hankel operators (forms) is formally also permitted but requires a few modifications: BMO in the "left" endpoint etc.). On the Hilbert space (Bergman space) $H = B_2^{\mu/2,2}$ we consider bilinear forms on the type

$$\Gamma : (f_1,f_2) \to \int_U \bar{\varphi} f_1 f_2 (1-|z|^2)^{-2\mu}\, dxdy$$

(similar considerations can be made with multilinear forms) where φ is a holomorphic function which transforms according to the rule $\varphi(z) \to \varphi(\omega z) \cdot \cdot (cz+d)^{2\mu-2}$ (that is, as the product f_1, f_2; the previous parameter μ thus gets replaced by μ' given by $\mu' - 1 = 2\mu - 2$ or $\mu' = 2\mu - 1$); we say that φ is the (true) s y m b o l of Γ and write $\Gamma = \Gamma_\varphi$. We can also rewrite the preceding integral in the form (we omit an uninteresting numerical factor)

$$\int_T \bar{\Phi} f_1 f_2 |dz|;$$

we then say that Φ is the a p p a r e n t symbol of Γ and write $\Gamma = \Gamma_\Phi$; formally $\varphi = (R+1)\dots(R+1-2\mu)\Phi$ (the number of factors is $1-2\mu = -\mu'$) where $R = zd/dz$.

If $f_1, f_2 \in H$ then $f_1 f_2 \in B_1^{\mu,1}$ $(= B_1^{(\mu'+1)/2,1} = B_*')$. Therefore Γ_φ or Γ_Φ will be continuous if $\varphi \in B_\infty^{\mu-1,\infty}$ $(= B_\infty^{(\mu'-1)/2,\infty} = B')$ or $\Phi \in B_\infty^{-\mu,\infty}$. (Here we use a dash ' to remind of the change to the parameter μ'!) This gives us one endpoint result $(p = \infty)$; cf. [8].

To get the other one $(p = 1)$ we may argue as follows. Consider those symbols φ for which Γ_φ is of trace class 1. Then $\varphi \to ||\Gamma_\varphi||_{S_1}$ surely is invariant (for the μ'-action). Thus by minimality $||\Gamma_\varphi||_{S_1} \le c||\varphi||_{B_*'}$.

We can now interpolate between these two endpoint results. This gives an estimate of $||\Gamma_\varphi||_{S_p}$ for $1 < p < \infty$ in terms of an appropriate Besov norm (cf. Sec. 1). By duality we get the converse assertion. Namely, since we have the i s o m e t r i c imbedding $\Gamma : H' \to S_2$ (an easy computation!; alternatively, use the uniqueness of H'), we must have the formal relation $\Gamma^*\Gamma = \mathrm{id}$. Therefore we can argue as in [9] (cf. [8]). Thus we have

THEOREM. $\Gamma_\varphi \in S_p$ (<u>or</u> $\Gamma_\Phi \in S_p$) $\iff$ $\varphi \in B_p^{\mu+1/p-1,p}$ (<u>or</u> $\Phi \in B_p^{-\mu+1/p,p}$).□

REMARK. The case $\mu = 0$ is formally Peller's theorem 2.2 [9]; the present case $\mu < 0$ is the one treated in [8] in Peller's original spirit. Our proof thus trivializes the trace ideal criterion in the case $1 < p < \infty$. It would be interesting to have an analogous "trivialization" for $0 < p < 1$ too.

REMARK (press stop). The same ideas (group invariance) can also be used in connection with C a l d e r ō n - Z y g m u n d c o m m u t a t o r s (see [8] and the references given there, notably [6]). In fact, Janson and I have recently obtained the following generalization of the main result in [6]: Let $K_1, \dots, K_N$ be Calderōn-Zygmund operators in R^n and let b denote multiplication by a given function, likewise denoted b. Then $[K_1 \dots [K_N b] \dots] \in S_p$, $N > n/p$, iff $b \in B_p^{n/p,p}$. Still the case $1 < p < \infty$.

3.2. Rational approximation. As in the original Peller case [9] our result lends itself to an application to rational approximation (rate of convergence). However, since it is not clear whether there is a good analogue of the powerful AAK theorem here, we get only a result in one direction (an "inverse theorem"), and moreover the restriction $1 < p < \infty$ has the effect that we do not get the full range $(0,\infty)$ for the rate of approximation, having to restrict ourselves to the interval $(0,1)$. For details see [8].

3.3. Shields's Fejér-Riesz inequality. The classical Fejér-Riesz inequality [5] can be generalized and has been generalized in many directions. Shields in a recent paper [14] considers function f in the (classical) Dirichlet space, presently denoted by D $(f \in D \iff \int |f'(z)|^2 dxdy < \infty)$ normalized by the condition $f(0) = 0$, and proves for such functions the inequality

$$(1) \qquad \int_0^1 |f(x)|^2 x^2/(1-x)(\ln(1-x))^2 dx \le c^2 ||f||^2$$

and, more generally,

$$(2) \qquad \int_0^1 |f(x)|^2 d\tilde{\nu}(x) \le c^2 ||f||^2$$

where $\tilde{\nu}$ is any positive measure on the interval $(0,1)$ subject to the condition

$$(3) \qquad \tilde{\nu}([x,1)) = O(|\ln(1/(1-x))|^{-1}).$$

But he does not make any apparent use of the Möbius invariance of D. Passing to this point of view, a minute's reflextion reveals that the "correct" formulation of the inequality (1) is

$$(1') \qquad \int_0^1 \left(\frac{|f(x)|^2}{\frac{1}{2}\ln\frac{1+x}{1-x}}\right)^2 \frac{dx}{1-x^2} \le c^2 ||f||^2 \quad (f \in D,\ f(0) = 0).$$

For $1/2\cdot\ln(1+x)/(1-x)$ is nothing but the hyperbolic separation between 0 and x and $dx/(1-x^2)$ the corresponding line element. Similarly, we see that (2) is but a special case of the general inequality

(2')
$$\int_U \left(\frac{|f(z)-f(z_0)|}{d(z,z_0)}\right)^2 d\nu(z) \le c^2||f||^2 \quad (f \in D)$$

where ν now is any positive measure on U and z_0 a general point (we also drop the normalization, $f(z_0) = 0$), while to (3) there corresponds the condition

(3')
$$\nu(K_\alpha) \le c_1\alpha$$

where K_α is the disk with hyperbolic radius α about z_0.

At this point it is natural to ask whether there is a similar result for general Möbius invariant spaces. Indeed, this is the case and in this way we get also a generalization of Shields's inequality.

For the sake of simplicity, let us only consider the invariant Besov spaces $B_p \overset{def}{=} B_p^{sp}$, $s = 1/p$. If $f \in B_\infty$ $(= B)$ we have

$$|f(z)-f(z_0)| \le d(z,z_0)||f||_{B_\infty}$$

Similarly, since the elements of B_1 $(= B_*)$ are bounded we see that the inequality $|f(z)-f(z_0)| \ge \beta\, d(z,z_0)$ implies $d(z,z_0) \le c_2||f||/\beta$, i.e. $z \in K_\alpha$, $\alpha = c_2||f||/\beta$ ($||\cdot||$ is the B_1-norm). Using (3') we conclude that

$$\nu(|f(z)-f(z_0)| \ge \beta\, d(z,z_0)) \le c_3||f||_{B_1}/\beta \ (c_3 = c_2c_1).$$

Introducing the map $f(z) \to (f(z)-f(z_0))/d(z,z_0)$, call it T, we have thus proven that

$$T : B_1 \to L^1_{weak}(\nu),\ T : B_\infty \to L_\infty(\nu).$$

Since $B_p = (B_1,B_\infty)_{\theta p}$, $\theta = 1/p$ we get by interpolation

$$T : B_p \to L^p(\nu) \quad (1 < p \le \infty).$$

THEOREM. <u>Let ν be any positive measure on U which satisfies (3'). Then for $1 < p \le \infty$ and $f \in B_p$ we have the inequality</u>

$$\left[\int_U \left(\frac{|f(z)-f(z_0)|}{d(z,z_0)}\right)^p d\nu(z)\right]^{1/p} \le c||f||_{B_p},$$

with c *depending on* p *and* ν *only.* □

3.4. *On Carleson measure*. The considerations of the previous Sec. are intimately related to the idea of *Carleson measure*. In dealing with spaces of holomorphic functions, taken *modulo constants*, invariant or not, the following definition naturally comes to one's mind:

DEFINITION. *A measure* ν *on* U *is termed an X-Carleson measure iff for every* $f \in X$ *holds*

$$\int_U |f'(z)|d\nu(z) \leq c||f||_X.$$

We put also $||\nu||_{C(X)} = \inf c$ (the X-Carleson norm of ν).

One now also sees that, from a "higher" point of view, whereas these definitions are connected with the orbit $\Delta \cap (U \times U)$ of the action of the Möbius group on clos U × clos U, the one in Sec. 3.3 instead is connected with the action of the group on $U \times U$: in that case one deals with measures ρ on $U \times U$ subject to inequalities of the form

$$\iint_{U\times U} |f(z_1)-f(z_2)|d\rho(z_1,z_2) \leq c||f||_X;$$

the special case considered in Sec. 3.3 is thus concerned with measures of the special form $\rho = \nu \otimes \delta_{z_0}$.

It is plain that (*) if $X \subset Y$ then ν Y-Carleson $\Longrightarrow$ ν X-Carleson.

If X is a given (Möbius) invariant space we associate with it another one $\hat{X}$ as follows: $f \in \hat{X} \iff \sup \int |f'(z)|d\nu < \infty$ where the sup ranges over all ν with $||\nu||_{C(X)} \leq 1$. Clearly $X \subset \hat{X}$.

Let us consider B_* $(= B_1^{11})$, our minimal space, and the Sobolev space $W \overset{\text{def}}{=} W_1^{11}$.

PROPOSITION. *The following conditions are equivalent.*

1) ν *is Carleson* (in the usual sense).

2) ν *is W-Carleson.*

3) ν *is* B_**-Carleson.*

PROOF. 1) $\Longrightarrow$ 2). Obvious (definition of Carleson measure).

2) $\Longrightarrow$ 3). Obvious, in view of $B_* \subset W$ and the above observation (*).

3) $\Longrightarrow$ 1). Instead of the usual realization in terms of the disk U let us employ the upper halfplane P. Assume that ν is B_*-Carleson. We then have to prove that $\nu(Q) \leq c|I|$ for all intervals $I \subset \mathbf{R}$, Q the square in P with one side I. But thus is obvious by homogeneity. □

COROLLARY. $\hat{B}_* = W$. □

This shows that the space B_* cannot be defined using the first derivative only (cf. Intr.); one has to invoke the second derivative too. More precisely, the corollary shows that in a way the space W ($\approx H^1$) is the *minimal* space with this property (susceptible to a definition in terms of first derivatives).

REMARK. A result similar to the above proposition but in a different formulation is due to Vinogradov (see Peller's in many respects highly interesting paper [11]; it contains several items of interest from the Möbius point of view).

REFERENCES

[1] Arazy, J., Some aspects of the minimal Möbius-invariant space of analytic functions in the unit disc. Informal Conf. on the Theory of Interpolation Spaces and Allied Questions of Analysis, Lund, Aug.29 - Sept. 1, 1983.

[2] Arazy, J. - Fisher, S.D., The uniqueness of the Dirichlet space among Möbius-invariant Hilbert spaces. Technion Report Series No. MT-593, Haifa, 1983.

[3] Arazy, J. - Fisher, S.D. - Peetre, J., The minimal Möbius invariant spaces and related topics. In preparation.

[4] Beckenbach, E.T. - Bellman, R., Inequalities. (Ergebnisse 30.) Springer-Verlag, Berlin - Göttingen - Heidelberg, 1961.

[5] Fejér, L. - Riesz, F., Über einige funktionen-theoretische Ungleichungen. Math. Z. 11 (1921), 305-314.

[6] Janson, S. - Wolff, T., Schatten classes and commutators of singular integral operators. Ark. Mat. 20 (1982), 50-73.

[7] Lions, J.-L. - Peetre, J., Sur une classe d'espaces d'interpolation. Inst. Hautes Etudes Sci. Publ. Math. 19 (1964), 5-68.

[8] Peetre, J., Hankel operators, rational approximation, and allied questions of Analysis. Edmonton conference, June 1982. (To appear.)

[9] Peller, V.V., Hankel operators of class S_p and their applications (rational approximation, Gaussian processes, the majorant problem for operators). Mat. Sb. 113 (1980), 538-581. [Russian.]

[10] Peller, V.V., Hankel operators of the Schatten - von Neumann class γ_p, $0 < p < 1$. LOMI preprints E-6-82, Leningrad, 1982.

[11] Peller, V.V., Estimates of functions of power bounded operators on Hilbert spaces. J. Operator Theory 7 (1982), 341-372.

[12] Rubel, L.A. - Timoney, R.M., An extremal property of the Bloch space. Proc. Amer. Math. Soc. 75 (1979), 45-49.

[13] Semmes, S., Trace ideal criterion for Hankel operators, $0 < p < 1$. (To appear in : Integral Equations Operator Theory.)

[14] Shields, A., An analogue of the Fejér - Riesz theorem for the Dirichlet space. In: W.Beckner - A.P.Calderón - R.Fefferman - P.W.Jones (ed.), Conference in Harmonic Analysis in honor of Antoni Zygmund. Vol. II, pp. 810-820. Wadsworth, Belmont, 1983.

[15] Timoney, R.M., Bloch functions in several complex variables I. Bull. London Math. Soc. 12 (1980), 421-267.

[16] Zobin, I.M. - Zobina, V.G., Interpolation in spaces with given symmetries. Finite dimension. Izv. Vysš. Učebn. Zaved. Matematika Nr. 4, 1981, 19-28. [Russian.]

International Series of
Numerical Mathematics, Vol. 65

DERIVATIONS ON CERTAIN CSL ALGEBRAS II

Frank L. Gilfeather*
Department of Mathematics
University of Nebraska-Lincoln

Let A be a CSL algebra on a Hilbert space. Sufficient conditions are given so that all derivations of A into A are inner. The conditions are determined by an order relationship between the core projection of A. The proofs are based on a result of E.C. Lance which shows that the nth cohomology spaces for a nest algebra are trivial. Reflexive algebras whose invariant subspace lattice is an infinite ordinal direct sum of commutative subspace lattices satisfy the sufficient conditions. A nest subalgebra A of a von Neumann algebra which is a CSL algebra is also shown to have only inner derivations.

All derivations from a CSL algebra A of operators on Hilbert space into itself were shown by E. Christensen to be norm continuous [1]. For nest algebras they are inner and in fact the n th Hochschild cohomology group $H^n(A,\mathcal{B})$ is trivial for all n and all ultraweakly closed subalgebras $\mathcal{B}$ of $\mathcal{B}(H)$ containing A [5]. In [3] these results were extended to a CSL algebra which is finite intersection of nest algebras with mutually commuting nests. These latter algebras include reflexive algebras whose invariant subspace lattices are finite tensor products of nests. Also in [3] it was shown that $H^1(A,A)$ need not be trivial even when A is the intersection of two nest algebras. The obstruction to $H^1(A,A)$ being trivial in the case A is a CSL algebra containing an atomoic mass is studied in [2].

In this note we pursue sufficient conditions on a CSL algebra A for $H^n(A,\mathcal{B})$ to be trivial whenever $\mathcal{B}$ is an ultraweakly closed algebra of $\mathcal{B}(H)$ containing A. The main result will extend the above cited results in which $H^n(A,\mathcal{B})$ is trivial. The main ingredient here as in [5] and [3] is the existence of sufficient rank one operators in A which have common domain and or range. In addition we show that a CSL algebra A which is a nest subalgebra

* This research was partly supported by the NSF (USA) and the SERC (Great Britain).

of a von Neumann algebra has trivial cohomology spaces.

For general results on the cohomology of Banach algebras we refer to [7]. For details and definitions needed here we refer to Lance's paper [5]. Actually our proofs are confined to the case of the first cohomology space. For algebras A, and A modules $\mathcal{B}$, the space $H^1(A,\mathcal{B})$ consists of the derivations of A into $\mathcal{B}$ modulo the inner derivations. $H^1(A,\mathcal{B})$ is trivial if each derivation δ of A to $\mathcal{B}$ is inner. A derivation δ of A into $\mathcal{B}$ is a linear map with the property that $\delta(RS) = R\delta(S) + \delta(R)S$ and δ is inner if there exists a $T \in \mathcal{B}$ such that $\delta(R) \equiv \delta_T(R) = RT - TR$ for all $R \in A$.

In this paper all Hilbert spaces will be separable. Subspace lattices will all be commutative and closed in the strong operator toplogy. Where there is no confusion we shall identify the subspace and the orthogonal projection onto it. For a subspace lattice L we let C_L denote the core, the von Neumann algrebra generated by L or equivalently the double commutant L'' of L. An algebra A is a CSL algebra if $\text{Lat}A = L$ is a commutative subspace lattice. For core projections E and F with respect to a CSL algebra A we say that E and F are strictly ordered, denoted by $E << F$, whenever $E\mathcal{B}(H)F \subseteq A$. Whenever A contains a strictly ordered pair then A contains many rank one operators.

We define the orbit $O(E)$ of a nonzero core projection E to be the join of all core projections F for which $E << F$ or $F << E$. If $E = 0$ then one would have $O(E) = I$ so we shall only define $O(E)$ whenever $E \neq 0$.

In a nest algebra the orbit of a core projection can easily be described. Let $E \neq 0$ be a core projection for the nest N and let $N_1 = \sup\{N \in N \mid NE = 0\}$ and $N_2 = \inf\{N \in N \mid NE = E\}$. Then $N_1 << E$ and $E << N_2^{\perp}$ and thus $O(E) \supseteq N_1 + N_2^{\perp}$. A short argument shows that in fact $O(E)H = N_{1+} \vee N_{2-}^{\perp}$. For a lattice L a projection P is called a comparable element of L if for all $Q \in L$ either $Q \subseteq P$ or $P \subseteq Q$. Since $P << P^{\perp}$ it follows that $P^{\perp} \leqslant O(P)$ and $P \leqslant O(P^{\perp})$.

The main theorem gives sufficient conditions on a CSL algebra A so that every derivation δ of A into $\mathcal{B}(H)$ is inner, that is, $\delta = \delta_T$ for some $T \in \mathcal{B}(H)$. As in [3] and [5] this implies that $H^1(A,\mathcal{B}) = 0$ for any ultraweakly closed algebra in $\mathcal{B}(H)$ containing A. Moreover following the proofs in [3] or [5] one can extend this result to $H^n(A,\mathcal{B})$. The sufficient condition involving the orbits of core projections is easily verfied for the cases studied in [3] and [5] and the arguments below are based on those in [5]. Moreover the results below are satisfied whenever L is the infinite ordinal direct sum of commutative subspace lattices.

LEMMA 1. $0(E) = F_1 + F_2 - F_1F_2$ where $F_1 >> E >> F_2$. Moreover $F_1 \in L^{\perp}$ and $F_2 \in L$.

PROOF. Let $F_1 = \vee\{F \in Cl\, F >> E\}$ and $F_2 = \vee\{F \in Cl\, F << E\}$. Clearly $F_1 + F_2 - (F_1F_2) \leq 0(E)$. Conversely $0(E) = \vee\{F | F >> E$ or $F << E\} \subseteq F_1 \vee F_2$. Let $[AF_2H] = R$ and $T = RTE$. We claim $T \in A$ hence $R << E$ and thus $R \leq F_2$ and $F_2 \in L$. If $T \notin A$ then there exists a $P \in L$ so that $P^{\perp}TP \neq 0$. Then $y = P^{\perp}Tx \neq 0$ for some $x \in PH$. However $y \in R$ and given $\varepsilon \leq \frac{||y||}{2}$ there exists $A_1 \in A$, $z \in F_2H$ so that $||y - A_1z|| < \varepsilon$. Let $A_2 \equiv z \otimes x$ where $||x|| = 1$ and $x \in E$. Now $A_2 \in A$ since $z \in F_2H$ and $A_1A_2x = A_1z$. However $A_1A_2x = A_1A_2Px = PA_1A_2x \subseteq PH$ and $y \in P^{\perp}H$. Thus $||y|| \leq ||y - A_1z|| \leq \frac{||y||}{2}$ a contradiction.

Since $F_2 \in L^{\perp}$ one has $F_1^{\perp} \in L$ so that $E_2 \equiv F_2 - F_1F_2 \in L$ also. Similarly one has $E_1 \equiv F_2 - F_1F_2 \in L$.

LEMMA 2. Let $E >> F$ and δ be a derivation on A with $\delta|_{\mathcal{D}} = 0$. Then there exists $T \in \mathcal{B}(H)$ so that $T = FTF$ and $\delta(A) = \delta_T(A)$ for $A \in FAF$.

PROOF. Let $x \in EH$ with $||x|| = 1$ and $y \in F$. Define $Ty = \delta(y \otimes x)(x)$. Since δ is linear and bounded $T \in \mathcal{B}(FH)$ and is extended to $\mathcal{B}(H)$ so that $T = FTF$. If $A \in FAF$ then $\delta(A(y \otimes x))(x) = \delta(Ay \otimes x)(x) - TAy$ by definition. However using the derivation property of δ we have $\delta(A(y \otimes x)) = A\,(y \otimes x) + \delta(A)(y \otimes x)$. Applying x and equating with TAy we have $\delta_T(A)(y) = \delta(A)y$ for all $y \in FH$.

We have determined an operator T which depends on a vector $x \in EH$ via the formula $\delta(y \otimes x)(x) = Ty$. From this it follows that $\delta(y \otimes x)(x \otimes w) = Ty \otimes w$ and that $||T|| \leq ||\delta||$.

PROPOSITION 3. Let $E \neq 0$ be in the core of A and δ a derivation on A for which $\delta|_{\mathcal{D}} = 0$. There exists $T \in \mathcal{B}(H)$ such that $T = 0(E)T0(E)$ and $\delta(A) = \delta_T(A)$ for all $A \in 0(E)A0(E)$. Moreover $||T|| \leq ||\delta||$.

PROOF. By Lemma 1, $0(E) = F_1 + F_2 - F_1F_2$. First assume $F_1F_2 = 0$. Now by Lemma 2 there exists a $T \in \mathcal{B}(H)$ with $T = F_2TF_2$ and $\delta = \delta_T$ on F_2AF_2. Next we note that $\hat{\delta}(B) \equiv -\delta(B^*)^*$ for $B \in A^*$ defines a derivation on A^* and applying Lemma 2 to this case we have an $S \in \mathcal{B}(H)$ so that $\hat{\delta}(B) = \hat{\delta}_S(B)$ for B in $F_1A^*F_1$ and $S = F_1SF_1$. Thus $\delta(A) = \delta_{S^*}(A)$ for $A \in F_1AF_1$. Hence if we set $T = S^*$ on F_1H we have $\delta(A) = \delta_T(A)$ for A in F_1AF_1 or F_2AF_2. In defining T on F_1H and F_2H we may use the same $x \in EH$ with $\|x\| = 1$.

If $y \in F_2H$ and $z \in H$ then $(z \otimes x)\delta(x \otimes y) = z \otimes \delta(x \otimes y)^*(x) = -z \otimes \hat{\delta}(y \otimes x)x = -z \otimes Sy = -(z \otimes y)S^* = -(z \otimes y)T$.

Now assume $y \in F_2H$ and $z \in F_1H$ since $F_2 \gg E \gg F_1$ we have $z \otimes y \in A$. Next factor $z \otimes y$ as $z \otimes y = (z \otimes x)(x \otimes y)$. Then $\delta(z \otimes y) = (z \otimes x)\delta(x \otimes y) + \delta(z \otimes x)(x \otimes y) = T(z \otimes y) - (z \otimes y)T = \delta_T(z \otimes y)$. Recall that $F_1AF_2 = F_1\mathcal{B}(H)F_2$ and that restricted to the unit ball of A, δ is weakly continuous [2]. Now δ agrees with δ_T on rank one operators in F_1AF_2 and hence δ agrees with δ_T on F_1AF_2. Also $F_2AF_1 = 0$ so δ agrees with δ_T on $O(E)AO(E)$ under the assumption that $F_1F_2 = 0$.

When $F_1F_2 \neq 0$ we have $F_1F_2 \ll E \ll F_1F_2$ and thus $F_1F_2AF_1F_2$ is in the diagonal $\mathcal{D}$ of A. Recall that $\delta|_{\mathcal{D}} = 0$. From Lemma 2 and the remark after it we have $T = 0$ on F_1F_2H. Since the diagonal of A is the same as the diagonal of A* it follows that T is well defined on F_1F_2H and thus δ agrees with δ_T on $O(E)AO(E)$.

The last paragraph of the above proof in fact shows that if E and F are core projections with $E \neq 0$ and $F \ll E \ll F$ then $FAF \subseteq \mathcal{D}$.

LEMMA 4. <u>Let I be in the strong closure of the orbits O(E) for nonzero core projections E. Then the span of the rank one operators in the unit ball of A are weakly dense in the unit ball of A.</u>

PROOF. Let A be in the unit ball of A. Assume E_n are core projections such that $O(E_n)A\,O(E_n) \to A$ strongly. Fix E and let $O(E) = F_1 + F_2 - F_1F_2$ as in Lemma 1. Since $F_1F_2AF_1F_2 = F_1F_2\mathcal{B}(H)F_1F_2$, $F_1AF_2 = F_1\mathcal{B}(H)F_2$ and $F_2AF_1 = F_2AF_1F_2 = F_2\mathcal{B}(H)F_1F_2$ we have that the span of the rank one operators in $O(E)AO(E)$ are weak operator dense in $O(E)AO(E)$. Thus $O(E_n)AO(E_n)$ is in the closure of the span of the rank one operators in the unit ball of A. Hence A is also.

Using the results of Laurie and Longstaff, Lemma 4 implies that LatA is completely distributive whenever the hypotheses of Lemma 4 are satisfied [6]. The converse is not true in that the algebras A_n in [3] can consist entirely of finite rank operators yet not satisfy the conclusion of the following theorem.

Theorem 5. Let A be a CSL algebra for which the identity belongs to the strong operator closure of the orbits of the nonzero core projections. Every derivation of A into $\mathcal{B}(H)$ is inner.

PROOF. Since $\delta|\mathcal{D}$ is a derivation of a von Neumann algebra there exists $S \in \mathcal{C} \ni \delta|\mathcal{D} = \delta_S|\mathcal{D}$. Thus $\hat{\delta} = (\delta-\delta_S)|\mathcal{D} = 0$ and we may assume simply that $\delta|\mathcal{D} = 0$. If there exists a nonzero E in the core of A for which $I = O(E)$, then the result reduces to the preceding proposition. Assume there exists a sequence E_n of core projections with $O(E_n) \to I$ in the strong operator topology. For each n let $T_n \in \mathcal{B}(H)$ so that $\delta(A) = \delta_{T_n}(A)$ for $A \in O(E_n)AO(E_n)$. Now $\|T_n\| \leq \|\delta\|$ so there exists a subsequence $\{T_{n_k}\}$ which converges weakly to T on H. Let $x,y \in H$ so that $A \equiv x \otimes y \in A$, then there exists an N so that $x,y \in O(E_{n_k})$ for all $k > N$. Thus if $k > N$ $\delta(A) = \delta_{T_{n_k}}(A) = T_{n_k}A - AT_{n_k} \to TA - AT$. Thus δ agrees with δ_T on the set of rank one operators in A and hence by Lemma 4 on A itself.

The next result concerns derivations on nest subalgebras of von Neumann algebras (nsva). Specifically we show that if A is a nsva and a CSL algebra then $H^1(A,\mathcal{B}(H)) = 0$. As above these results can be extended to $H^n(A,\mathcal{B})$ whenever $\mathcal{B}$ is an ultraweakly closed algebra with $A \subseteq \mathcal{B} \subseteq \mathcal{B}(H)$. There is a nice necessary and sufficient condition for a CSL algebra to be a nsva. Let A be a CSL algebra and $L = \mathrm{Lat}A$. A is a nsva iff for all $P,Q \in L$ there exists $R,R^\perp \in L$ such that $PQ^\perp \leq R$ and $P^\perp Q \leq R^\perp$. In particular the CSL nsva algebras are precisely direct sums and integrals of nest algebras (c.f. §3 in [4]). We refer to [4] for terminology and especially for references to the measurable selection techniques employed here.

THEOREM 6. Let A be a nsva which is a CSL algebra. Then every derivation of A into $\mathcal{B}(H)$ is inner.

PROOF. Let $A = \int \oplus A_\lambda\, \mu(d\lambda)$ where $A_\lambda = \mathrm{alg}N(\lambda)$ given by (4.1) in [4]. Let $\{A_n\}$ be a dense set of selectors for $\lambda \to A_\lambda$ so that $\{A_n\}$ and $\{A_n(\lambda)\}$ respectively are weakly dense in the unit balls of A and A_λ. Thus δ determines derivations δ_λ on A_λ by $\delta(A_n)(\lambda) \equiv \delta_\lambda(A_n(\lambda))$. Hence there exists T_λ in $\mathcal{B}(H_0)$ so that $\delta_\lambda(A) = \delta_{T_\lambda}(A)$ whenever $A \in A_\lambda$. (We may assume that $H_0 = H_\lambda$ for all λ in the measure space Λ.) We must now use measurable selection to show $\lambda \to T_\lambda$ can be chosen measurably so that $T = \int \oplus T_\lambda$ induces T.

Let $G = \{(\lambda, B) | B \in \mathcal{B}(H_\lambda)$ and $\delta_\lambda(A) = \delta_B(A)$ for all $A \in A_\lambda\}$. It is easy to see that $G = \cap_n G_n$ where $G_n = \{(\lambda, B) | B \in \mathcal{B}(H_\lambda)\}$ and $A_n(\lambda)B - BA_n(\lambda) = \delta(A_n)(\lambda)\}$. The sets G_n are measurable in $\Lambda \times \mathcal{B}(H_0)$ since they are the null sets of the measurable maps $(\lambda, B) \to A_n(\lambda)B - BA_n(\lambda) - \delta(A_n)(\lambda)$. Thus by measurable selection $\lambda \to T_\lambda$ is measurable so that $(\lambda, T_\lambda) \in G$ μ-a.e. Hence $T = \int \oplus T_\lambda \; \mu(d\lambda)$ implements δ on A.

Complete and careful details of the above proof can be assertained by comparing it with certain proofs in [4], e.g. (3.7) and (5.4). The theorem can be extended easily to the case where A is not a CSL algebra but still a nsva of a tpye I von Neumann algebra. The corresponding questions are open to our knowledge for type II and III algebras.

REFERENCES

[1] Christensen, E., Derivations of Nest Algebras, Math. Ann. 229 (1977), 155-161.

[2] Gilfeather, F. Derivations on Certain CSL Algebras, J. London Math.. Soc., to appear.

[3] Gilfeather, F. Hopenwasser, A., Larson, D.R., Reflexive algebras with finite width lattices: J. Functional Analysis, to appear.

[4] Gilfeather, F., Larson, D.R., Nest-subalgebras of von Neumann algebras, Advances in Math. 46 (1982), 171-199.

[5] Lance, E.C. Cohomology and perturbations of nest algebras, Proc. London Math. Soc. 43 (1981), 334-356.

[6] Laurie, C., and Longstaff, W., A note on rank one operators in reflexive algebras, Proc. Amer. Math. Soc., to appear.

[7] Johnson, B.,Cohomology in Banach algebras, Memoirs of the Amer. Math. Soc. 127, Providence, R.I. (1972).

International Series of
Numerical Mathematics, Vol. 65

IDEALS IN C(X)

C.B. Huijsmans
Mathematical Institute
Leiden State University
Leiden

The main topic of this paper is the connection between the various types of order ideals and algebraic ideals in the Riesz space and ring $C(X)$ of all real continuous functions on some Tychonoff space X .

Since this conference is more or less in honour of "the men of seventy", I dedicate this article to one of them, my former teacher professor Zaanen. Almost as a matter of course, the subject of this paper is closely related to the theory of Riesz Spaces, his main field of interest during the last 15 years. One of the first things I remember from the time I was a student of professor Zaanen and was trying to master the basic theory of Riesz spaces, is that he always impressed upon me that the theory of order ideals in Riesz spaces is in a way much nicer than the theory of algebraic ideals in commutative rings. His favourite example in illustration was the fact that an order ideal in an order ideal of a Riesz space is an order ideal again, whereas the corresponding result for algebraic ideals in commutative rings in general does not hold. At that time (the early seventies) professor Zaanen was very interested in the connection between the several kinds of order ideals and algebraic ideals in structures which are both a Riesz space and a commutative ring. He restricted his considerations to $C(X)$, the collection of all realvalued continuous functions on some compact Hausdorff space X . In his book "Riesz Spaces I" ([6], written together with W.A.J. Luxemburg) section 34 is completely devoted to these problems.

We refer the reader to this monograph for the notations, terminology and elementary Riesz space theory. The principal order ideal generated by f is henceforth denoted by I_f , i.e., $I_f = \{g : |g| \le n_g |f| \text{ for some } n_g \in \mathbb{N}\}$, whereas the principal algebraic ideal generated by f (consisting of all ring multiples of f) is denoted by (f) . The disjoint complement $\{f\}^d$ of f is defined by $\{f\}^d = \{g : |g| \wedge |f| = 0\}$ and the annihilator $\{f\}^\perp$ of f is the set $\{g : gf = 0\}$. The function e in C(X) stands for the function defined by e(x) = 1 for all $x \in X$. In the case $X = \mathbb{R}$ (or a subinterval of $\mathbb{R}$) we define i by $i(x) = x$ for all x .

In appendix 9 of [7] and section 34 of [6] several results on the connection between the various sets of ideals are produced. We adhere to the notations of [7] in this respect. Denote the sets of all order ideals and all algebraic ideals in C(X) by $\mathcal{O}$ and $\mathcal{A}$ respectively, and the set of all (proper) order prime ideals and all (proper) algebraic prime ideals by $\mathcal{OP}$ and $\mathcal{AP}$. It is shown in [7] that in C(X) (with X compact and Hausdorff)

$$\mathcal{AP} \subset \mathcal{OP} \subset \mathcal{O} \subset \mathcal{A} .$$

The proof of the inclusion $\mathcal{AP} \subset \mathcal{OP}$ is erroneously attributed to Fremlin. It is, in fact, due to Kohls ([5], theorem 2.1) and holds even for arbitrary topological spaces. Because of the simple and elegant proof, we repeat it here for the reader's convenience.

THEOREM 1. $\mathcal{AP} \subset \mathcal{OP}$.

PROOF. The main difficulty lies in showing that $\mathcal{AP} \subset \mathcal{O}$. To this end, let $P \in \mathcal{AP}$. Take $f \in P$, $g \in C(X)$ such that $|g| \le |f|$. Put

$$h(x) = \begin{cases} \dfrac{g^2(x)}{f(x)} & \text{if } x \notin Z(f) \\ 0 & \text{if } x \in Z(f) , \end{cases}$$

where Z(f) denotes the zero-set of f . Evidently, $h \in C(X)$ and $g^2 = hf \in P$. Hence $g \in P$. It remains to show that P is order prime. According to [6], theorem 33.2, it suffices to prove that $f \wedge g = 0$ implies $f \in P$ or $g \in P$. Since $f \wedge g = 0$ is equivalent to $fg = 0$, we have $fg \in P$. Therefore, $f \in P$ or $g \in P$ and the proof is complete.

Observe that in the above proof we only used that $P = \sqrt{P}$ and that P is pseudoprime (i.e., $fg = 0$ implies $f \in P$ or $g \in P$). However, these two properties characterize the elements of AP ([1], 4.2a).

Contrary to $AP \subset OP$, the inclusion $O \subset A$ no longer holds if X is not compact. By way of example, consider I_i in $C(\mathbb{R})$. Since $i^2 \notin I_i$, the order ideal I_i is not an algebraic ideal.

The main purpose of the present paper is a systematic investigation of the connection between the various kinds of ideals in $C(X)$ with, if necessary, X a Tychonoff (= completely regular Hausdorff) space. Most of the results can be generalized, with some effort, to certain classes of f-algebras. Some of the results are jointly due to B. de Pagter and myself.

THEOREM 2. $O \subset A \Longleftrightarrow X$ <u>is pseudocompact.</u>

PROOF. "$\Rightarrow$": The order ideal I_e of all bounded functions in $C(X)$ is an algebraic ideal. Hence, $f = f.e \in I_e$ for all $f \in C(X)$, so X is pseudocompact.

"$\Leftarrow$": If $I \in O$, then it follows from $f \in I$, $r \in C(X)$ and $|rf| \leq M|f|$ for some $M > 0$ that $rf \in I$. This implies that $I \in A$.

In general, an algebraic ideal in $C(X)$ also need not be an order ideal. Take for instance in $C([0,1])$ the algebraic ideal (i) and define $g \in C([0,1])$ by

$$g(x) = \begin{cases} x \sin \frac{1}{x} & 0 < x \leq 1 \\ 0 & x = 0 . \end{cases}$$

Since $|g| \leq i$, $g \notin (i)$, it is evident that (i) is not an order ideal.

In this connection we mention the following theorem (see [1], theorem 6.2, [2] ; theorem 14.25 and [3], theorem 10.5).

THEOREM 3. <u>The following statements are equivalent</u> .

(i) $A \subset O$.

(ii) X <u>is an F-space (i.e., every bounded continuous function on a cozero-</u>

set of X has a continuous extension to the whole of X).

(iii) $C(X) = \{f^+\}^d + \{f^-\}^d = \{f^+\}^\perp + \{f^-\}^\perp$ for all $f \in C(X)$.

(iv) $(f) = (|f|)$ for all $f \in C(X)$.

(v) $(f,g) = (|f|+|g|)$ for all $f,g \in C(X)$, where (f,g) denotes the algebraic ideal generated by f and g .

(vi) Every algebraic ideal in $C(X)$ is an intersection of (algebraic) pseudoprime ideals.

(vii) Every Finitely generated algebraic ideal is principal.

(viii) $C(X)$ has the σ-interpolation property (i.e., if $f_n \uparrow \leq g_n \downarrow$, then there exists $h \in C(X)$ such that $f_n \leq h \leq g_n$ for all n).

The letter F in F-space comes from Finitely.

Denote by OM (AM) the set of all order maximal ideals (algebraic maximal ideals) in $C(X)$ respectively. As is well-known, compactness of X implies $OM = AM$. Every such ideal is fixed in this case, that is, of the form $M_x = \{f \in C(X) : f(x)=0\}$ (in fact, X is compact if and only if every member of AM is fixed).

In general, only the inclusion $OM \subset AM$ holds. This can be shown in several ways (amongst others by using quotients), but we prefer the following simple direct proof.

THEOREM 4. $OM \subset AM$.

PROOF. The proof is divided in several steps. Take $M \in OM$.
Step 1. Suppose that $0 \leq u$, $v \in C(X)$ are algebraic units with $u \notin M$, $v \in M$. Obviously, $u \vee v$ is an algebraic unit as well, which is not a member of M . Consequently, the order ideal $I(M, u\vee v)$ generated by M and $u \vee v$ is the whole of $C(X)$. Hence, there exists $0 \leq m \in M$, $k \in \mathbb{N}$ such that

$$(u\vee v)^2 v^{-1} \leq m + k(u\vee v) .$$

It follows that $u \vee v \leq m(u\vee v)^{-1}v + kv \leq m + kv$, so $u \vee v \in M$, a contradiction. The conclusion is that M contains all algebraic units as soon as M contains one.
Step 2. Suppose that M contains one algebraic unit and take $f \in C(X)$.

Since $|f| \vee e$ is an algebraic unit, Step 1 implies that $|f| \vee e \in M$ and so $f \in M$. Hence, $M = C(X)$, which is absurd. We have thus proved that M does not contain any algebraic unit.

Step 3. $M \in A$. Indeed, suppose on the contrary that there exist $0 \leq m \in M$, $0 \leq u \in C(X)$ with $mu \notin M$. The equality $I(M,mu) = C(X)$ implies

$$e \leq n + kmu \leq (n+m)(e \vee ku)$$

for some $0 \leq n \in M$ and $k \in \mathbb{N}$. Therefore, M contains the algebraic unit $n + m$, which is impossible by Step 2.

Step 4. $M \in AM$, for M is, as a proper algebraic ideal, contained in some algebraic maximal ideal N. Since $N \in AP \subset OP$, it follows that N is a proper order ideal containing M and so $M = N \in AM$. The proof is complete.

Combining theorems 1 and 4 we get

$$OM \subset AM \subset AP \subset OP .$$

In general, all inclusions are proper. We illustrate this with some examples.

EXAMPLES. (i) $AP \subsetneq OP$. The order ideal I_i in $C(\mathbb{R})$ does not contain i^2. There exists therefore ([6], theorem 33.4) $P \in OP$ such that $i \in P$, $i^2 \notin P$. Hence P is not even an algebraic ideal.

(ii) $AM \subsetneq AP$. Consider in $C([0,1])$ the ℓ-ideal

$$I = \left\{f \in C([0,1]) : 0 \in \operatorname{int} Z(f)\right\}$$

(an ℓ-ideal is a set which is both an order ideal and an algebraic ideal). Then I is contained in the unique algebraic (and order) maximal ideal $M_0 = \{f \in C([0,1]) : f(0)=0\}$. Since $i \notin \sqrt{I}$ and $\sqrt{I} = \cap(P \in AP : P \supset I)$, there exists $P_0 \in AP$, $P_0 \supset I$ such that $i \notin P_0$. Therefore, $P_0 \neq M_0$ and thus $P_0 \notin AM$ (cf. [2], section 2.8).

(iii) $OM \subsetneq AM$. Consider $X = \mathbb{N}$ with the discrete topology. The proper ℓ-ideal

$$I = \left\{f \in C(\mathbb{N}) : f(n)=0 \text{ for all but finitely many } n\right\}$$

is contained in some $M \in \mathcal{AM}$. Obviously, $e = (1,1,1,\ldots) \notin M$. It is straightforward to prove that $u = (1,2,3,\ldots) \notin I(M,e)$, showing that $M \notin \mathcal{OM}$.

In the next theorems necessary and sufficient conditions are given for equality in one of the above inclusions.

THEOREM 5. The following statements are equivalent.

(i) $\mathcal{AM} = \mathcal{AP}$.

(ii) X is a P-space (i.e., Z(f) is open for all $f \in C(X)$).

(iii) Every continuous function on a cozero-set of X has a continuous extension to the whole of X.

(iv) C(X) is von Neumann regular (for every $f \in C(X)$ there exists $g \in C(X)$ such that $f = f^2 g$).

(v) $(f) = (f^2)$ for all $f \in C(X)$.

(vi) $(f,g) = (f^2 + g^2)$ for all $f,g \in C(X)$.

(vii) Every algebraic ideal in C(X) is an intersection of algebraic prime ideals (i.e., $I = \sqrt{I}$ for all $I \in A$).

(viii) Every finitely generated algebraic ideal is generated by an idempotent.

(ix) C(X) is σ-laterally complete.

For the proof we refer to [2], 4 J and theorem 14.29. The letter P in P-space comes from Prime (every algebraic Prime ideal is maximal). Notice the analogy between theorems 3 and 5. In view of the remarks at the beginning of this paper it is perhaps nice to observe that the condition "every algebraic ideal in an algebraic ideal of C(X) is again an algebraic ideal" is equivalent to each of the statements of theorem 5 ([4], theorem 6.1). In this context we mention another example to show that order ideal theory in Riesz spaces is nicer than algebraic ideal theory in commutative rings. It is an elementary property in Riesz space theory that an order ideal maximal with respect to the property of not containing a given element is an order prime ideal ([6], theorem 33.4). The corresponding property for rings certainly does not hold in general. In fact, a moment's reflection shows that the commutative ring R is von Neumann regular if and only if every algebraic ideal in R, maximal with respect to the property of not containing an element of R, is an algebraic prime ideal.

THEOREM 6 ([3], corollary 7.4; [2], theorem 5.8).

$$\mathcal{OM} = \mathcal{AM} \Longleftrightarrow X \text{ is pseudocompact.}$$

The equality $\mathcal{AP} = \mathcal{OP}$ seldom occurs. We shall show that this happens only in the trivial case that X is finite. We need some preparations.

As is well-known, the commutative ring R is von Neumann regular if and only if $R = (f) \oplus \{f\}^{\perp}$ for all $f \in R$. Accordingly, the Riesz space L is said to be hyper-Archimedean whenever $L = I_f \oplus \{f\}^d$ for all $f \in L$. The terminology is justified by the fact that this property is equivalent to the statement that every quotient of L is Archimedean ([6], theorem 37.6). For $C(X)$ we have the following theorem.

THEOREM 7. The following are equivalent.

(i) $C(X)$ is hyper-Archimedean.

(ii) $\mathcal{OM} = \mathcal{OP}$.

(iii) $C(X)$ is von Neumann regular and X is pseudocompact.

(iv) X is finite (i.e., $C(X)$ is a coordinatewise ordered $\mathbb{R}^n$ for some n).

PROOF. See [2], 4 K.2; [3] theorem 10.2 and [6], theorem 37.6.

THEOREM 8. $\mathcal{AP} = \mathcal{OP} \Longleftrightarrow X$ is finite.

PROOF. "$\Leftarrow$": $\mathcal{OM} = \mathcal{OP}$ certainly implies $\mathcal{AP} = \mathcal{OP}$.

"$\Rightarrow$": The proof is for ease of survey divided in several steps.

Step 1. If $I \in \mathcal{O}$, then the equality $I = \cap(P \in \mathcal{OP} : P \supset I)$ and the hypothesis imply that I is an intersection of algebraic ideals, hence $I \in \mathcal{A}$. By theorem 2, X is pseudocompact.

Step 2. If I is an ℓ-ideal, i.e., $I \in \mathcal{A} \cap \mathcal{O}$, then

$$\begin{aligned}\sqrt{I} &= \cap(P \in \mathcal{AP} : P \supset I)\\ &= \cap(P \in \mathcal{OP} : P \supset I) = I .\end{aligned}$$

Step 3. Take an arbitrary $0 \le u \in C(X)$. The element $v = \sqrt{u}$ satisfies $v^2 \in \langle u\rangle$, the ℓ-ideal generated by u, i.e.,

$$\langle u\rangle = \left\{f\in C(X) : |f| \le ru, 0\le r\in C(X)\right\} .$$

It follows from step 2 that $v \in \sqrt{\langle u\rangle} = \langle u\rangle$, in other words $0 \le v \le ur$ for some $0 \le r \in C(X)$. Hence, $0 \le u \le u^2r^2$. Now

$$u - u^2r^2 = (u-u^2r^2)^+ - (u-u^2r^2)^- = -(u-u^2r^2)^-$$
$$= -u(ur^2-e)^+ = -uw^2 ,$$

with $w = \sqrt{(ur^2-e)^+}$. This implies that $u(e+w^2) = u^2r^2$, so $u = u^2t$ with $t = r^2(e+w^2)^{-1}$.

Step 4. Take $f \in C(X)$ arbitrary. By step 3, $|f| = f^2g$ for some $g \in C(X)$. Write for the moment $k = fg$. We show that $|f| = kf$ implies that $f = k|f|$. Indeed,

$$f^+ + f^- = kf^+ - kf^-$$

yields $f^+(e-k) = -f^-(e+k)$. Since the left hand term and the right hand term are disjoint, we infer $f^+ = kf^+$, $f^- = -kf^-$. Hence

$$f = f^+ - f^- = kf^+ + kf^- = k|f| .$$

It follows that $f = f^2gk$. By definition, $C(X)$ is von Neumann regular; since X is also pseudocompact, X must be finite by theorem 7.

To complete the survey, we mention one more theorem which is proved in [3], lemma 10.1. Let us denote the sets of all minimal order prime ideals and all minimal algebraic prime ideals in $C(X)$ by *MOP* and *MAP* respectively.

THEOREM 9. *MOP = MAP* .

The study of $C(X)$ considered as a Riesz space is in many respects more natural than regarding $C(X)$ as a ring. Let us give some examples in illustration. In the first place it seems somewhat mysterious why the collection of all algebraic prime ideals containing a fixed algebraic prime ideal is a chain with respect to inclusion ([2], 14.8), since this is certainly not always the case in arbitrary rings. However, on account of theorem 1 and the fact that any two order prime ideals of a Riesz space, containing a given order prime ideal, are comparable ([6], theorem 33.4), the above result suddenly becomes completely clear. Another example is supplied by the real compactification (Hewitt-compactification) υX of a Tychonoff space X ([2], 8.4-8.8), which is, in my opinion somewhat artificially, introduced by means of the set of all so-called real ([2], 5.6) algebraic (maximal) ideals in $C(\beta X)$, where βX is the Stone-Čech-compactification of X . However, υX can be realized more naturally. It is not difficult to show that the set OM of all order maximal ideals in $C(X)$, equipped with its hull-kernel topology ([6], section 35), is actually also a model for υX, completely analogous to the fact that the set AM (with its hull-kernel topology) is a model for βX . As observed before, X is compact (equivalently, $X = \beta X$) if and only if every member of AM is fixed. Similarly, X is realcompact if and only if each element of OM is fixed (equivalently, $X = \upsilon X$) . It follows immediately from

$$X \hookrightarrow \upsilon X = OM \subset \beta X = AM$$

and theorem 6 that X is compact if and only if X is realcompact and pseudocompact, a well-known topological theorem (see e.g. [2], 5 H.2).

Hopefully, these two examples make it clear that the order-theoretical aspects in the study of the $C(X)$ cannot be neglected. I like to end therefore by quoting professor Zaanen, who wrote some years ago in a book review ([8]): "it is evident that textbook authors cannot ignore order structures in vector spaces any longer".

REFERENCES

[1] Gillman, L. - Kohls, C.W., Convex and Pseudoprime Ideals in Rings of Continuous Functions, Math. Z. 72 (1959/60), 399-409.

[2] Gillman, L. - Jerison, M., Rings of Continuous Functions, Springer Verlag, New York/Heidelberg/Berlin 1976.

[3] Huijsmans, C.B. - de Pagter, B., On z-ideals and d-ideals in Riesz Spaces II, Indag. Math. 42 (= Proc. Neth. Ac. Sci. 83) (1980), 391-408.

[4] Kohls, C.W., Ideals in Rings of Continuous Functions, Fund. Math. 45 (1958), 28-50.

[5] Kohls, C.W., Prime Ideals in Rings of Continuous Functions, Ill. J. Math. 2 (1958), 505-536.

[6] Luxemburg, W.A.J. - Zaanen, A.C., Riesz Spaces I, North Holland Publishing Company, Amsterdam/London 1971.

[7] Zaanen, A.C., Ideals in Riesz Spaces, Troisième Colloque sur l'Analyse Fonctionelle (Liège, 1970), Vander, Louvain (1971), 137-146.

[8] Zaanen, A.C., M.R. 54 (1977), #11023.

International Series of
Numerical Mathematics, Vol. 65

GENERALISED FUNCTIONS AS LINEAR FUNCTIONALS ON GENERALIZED FUNCTIONS

Tom H. Koornwinder
Centrum voor Wiskunde
en Informatica (CWI)
Amsterdam

and

Jan J. Lodder
FOM-Instituut voor
Plasmafysica
Nieuwegein, Netherlands

We give a sketch of a rigorous foundation for the model for a symmetrical theory of generalised functions introduced earlier by the second author. On starting with a suitable subspace *PC* of the space *S'* of tempered distributions, we introduce a space *SGF* of "new" generalised functions as a space of linear functionals on *PC*. Both on *PC* and *SGF* we have all the usual operations including a product. On *PC* this product operation is somewhat arbitrary but on *SGF* it is canonical and much nicer. Finally, *PC* and *SGF* are put together into a space *GF* of linear functionals on *SGF*.

1. Introduction

Distribution theory arose out of the need to give a rigorous foundation to objects such as the delta function, which were used before in a heuristic way. In order to apply Fourier techniques, the space S' of tempered distributions was introduced. When S' is compared with other spaces invariant under the Fourier transform like S or $L^2(\mathbb{R})$ then some simple formal properties are missing in the theory of S' like a scalar product $S' \times S' \to \mathbb{C}$ or an ordinary product $S' \times S' \to S'$. These shortcomings are sometimes bothersome in applications of distribution theory in mathematics or physics.

In [7] a symmetrical theory of generalised functions was designed by the second author in order to combine the desirable features of distribution theory and L^2-theory. Here by "symmetrical" we mean that there is no longer a distinction between test functions and distributions, but that a scalar product exists on the space of generalised functions constructed in [7]. Applications of this theory to quantum electrodynamics were given in [8]. While the presentation of the theory in [7] was heuristic, here we give a sketch of a rigorous approach. Proofs are omitted; these will appear in a later paper.

The construction proceeds in several steps. In order to show the similarities and differences with distribution theory the subspace *SGF* of "new" generalised functions is introduced as a space of linear functionals on a suitable subspace *PC* of *S'*, in such a way that it is closed under the usual operators. On *PC* we define a non-associative product following Keller [4], [5], [6]. (This was earlier done in [7], but there the point singularities remained unspecified because of indeterminacy.) On *SGF*, being a bidual of *S*, a canonical product is inherited from *S*. The formal properties of the product on *SGF* are much nicer than on *PC*. There is also a lot of arbitrariness in the choice of the product on *PC*. The paper concludes with a synthesis of *PC* and *SGF* into a space *GF* of linear functionals on *SGF*. The theory of the space *GF*, when viewed as its own dual, may be shown to coincide with the symmetrical theory of generalised functions in [7]. Throughout the paper, "distributions" will be understood in the sense of Schwartz.

2. The Preliminary Class *PC*

Let *S* be the space of rapidly decreasing C^∞-functions on $\mathbb{R}$, equipped with the usual topology. Below we list a number of continuous linear endomorphisms of *S* by their action on elements ϕ of *S*:

(2.1) $$(D\phi)(x) := \frac{d\phi(x)}{dx},$$

(2.2) $$(X\phi)(x) := x\phi(x),$$

(2.3) $$(e^{aD}\phi)(x) := \phi(x+a), \quad a \in \mathbb{R},$$

(2.4) $$(e^{ibX}\phi)(x) := e^{ibx}\phi(x), \quad b \in \mathbb{R},$$

(2.5) $$(S_c\phi)(x) := \phi(cx), \quad c > 0,$$

(2.6) $$(F\phi)(x) := \int_{-\infty}^{\infty} \phi(\xi)e^{-i\xi x}d\xi,$$

(2.7) $$(P\phi)(x) = \check{\phi}(x) := \phi(-x),$$

(2.8) $$(M_\psi\phi)(x) = (\psi\phi)(x) := \psi(x)\phi(x), \quad \psi \in S,$$

(2.9) $$(C_\psi\phi)(x) = (\psi*\phi)(x) := \int_{-\infty}^{\infty} \psi(y)\phi(x-y)dy, \quad \psi \in S.$$

In (2.3) and (2.4) the power series $\Sigma_k\, a^k\, D^k\phi/k!$ and $\Sigma_k(ib)^k\, X^k\, \phi/k!$ do not converge in S for all ϕ, only on a dense subspace of analytic functions.

There are many well-known identities involving the operators defined above. Here we only mention:

(2.10) $DX - XD = I,$

(2.11) $FD = iXF,$

(2.12) $F^2 = 2\pi P,$

(2.13) $F^{-1} = (2\pi)^{-1}PF,$

(2.14) $F(\phi*\psi) = (F\phi)(F\psi),$

(2.15) $D(\phi\psi) = (D\phi)\psi + \phi(D\psi).$

Consider also the integration functional I and the evaluation functional E, both continious on S:

(2.16) $I(\phi) := \int_{-\infty}^{\infty} \phi(\xi)d\xi,$

(2.17) $E(\phi) := \phi(0).$

They satisfy

(2.18) $I(\phi) = E(F\phi),$

(2.19) $I(\phi\psi) = I(F\phi)(F^{-1}\psi).$

Let S' be the space of tempered distributions, i.e. of all continuous linear functionals on S. Generally, if V is a linear space and V' its dual space then we will write $<f,\phi>$ for the linear functional $f \in V'$ evaluated at $\phi \in V$. There is an embedding $S \to S'$ such that

(2.20) $<\psi,\phi> = \int_{-\infty}^{\infty} \psi(x)\phi(x)dx, \quad \psi, \quad \phi \in S.$

If A is any of the operators defined by (2.1)-(2.9) then there is a unique continuous linear operator $A'\colon S \to S$ such that

(2.21) $\langle A\psi,\phi\rangle = \langle\psi,A'\phi\rangle, \quad \psi, \quad \phi \in S,$

and there is an extension of A to S' (also denoted by A) such that

(2.22) $\langle Af,\phi\rangle = \langle f,A'\phi\rangle, \quad f \in S', \quad \phi \in S.$

For $\alpha \in \mathbb{C}$, $q \in \mathbb{Z}_+$ we define the element $x_+^\alpha(\log x_+)^q$ of S' as a Hadamard finite part:

(2.23) $\langle x_+^\alpha(\log x_+)^q,\phi\rangle := \mathrm{Res}_{\lambda=0}\lambda^{-1}\mathrm{AC} \int_0^\infty \phi(x)x^{\alpha+\lambda}(\log x)^q dx,$

where AC means analytic continuation and $\mathrm{Res}_{\lambda=0}$ the residue at $\lambda = 0$. Also:

(2.24) $x_-^\alpha(\log x_-)^q := P(x_+^\alpha(\log x_+)^q),$

(2.25) $\langle\delta^{(k)},\phi\rangle := (-1)^k(D^k\phi)(0) = (-1)^k E(D^k\phi), \quad k \in \mathbb{Z}_+.$

The linear span of the elements $x_\pm^\alpha(\log x_\pm)^q$ and $\delta^{(k)}$ is invariant under D, X, S_c, F, P (see [3] for explicit formulas).

Let the *preliminary class* PC be the smallest linear subspace of S' which contains all elements $x_\pm^\alpha(\log x_\pm)^q$ and $\delta^{(k)}$ and which is invariant under the operators defined by (2.1)-(2.9). We will rather use the following equivalent characterization as a definition:

DEFINITION 2.1. *The space* PC *consists of all finite linear combinations of the elements*

(2.26) $e^{aD}\delta^{(k)} \quad (k \in \mathbb{Z}_+, \ a \in \mathbb{R}),$

(2.27) $\phi e^{aD}x_\pm^\alpha(\log x_\pm)^k \quad (\alpha \in \mathbb{C}, \ k \in \mathbb{Z}_+, \ a \in \mathbb{R}, \ \phi \in S),$

(2.28) $\phi * e^{ibX}x_\pm^\alpha(\log x_\pm)^k \quad (\alpha \in \mathbb{C}, \ k \in \mathbb{Z}_+, \ b \in \mathbb{R}, \ \phi \in S).$

The class PC defined above is somewhat smaller than the preliminary class in [7]. This is done for convenience, but the results of this paper will remain valid with respect to the larger class.

More structure can be given to PC by using the spaces O_M of multipliers for S and O_C' of convolutors for P, as introduced by Schwartz [9]:

$$(2.29)\qquad \mathcal{O}_M := \{f \in \mathcal{S}' \mid \phi f \in \mathcal{S} \text{ for all } \phi \in \mathcal{S}\},$$

$$(2.30)\qquad \mathcal{O}'_C := \{f \in \mathcal{S}' \mid \phi * f \in \mathcal{S} \text{ for all } \phi \in \mathcal{S}\}.$$

Note that all elements of $\mathcal{O}_M$ are C^∞-functions and that $\mathcal{O}_M = \mathcal{F}(\mathcal{O}'_C)$, $\mathcal{S}.\mathcal{S}' \subset \mathcal{O}'_C$, $\mathcal{S}*\mathcal{S}' \subset \mathcal{O}_M$. If $f \in \mathcal{O}_M$, $g \in \mathcal{S}'$ then we can define $M_f g = fg \in \mathcal{S}'$ by

$$(2.31)\qquad \langle fg,\phi\rangle = \langle g,\phi f\rangle, \quad \phi \in \mathcal{S},$$

and if $f \in \mathcal{O}'_C$, $g \in \mathcal{S}'$ then we define $C_f g = f*g \in \mathcal{S}'$ by

$$(2.32)\qquad \langle f*g,\phi\rangle = \langle g,\phi*\check{f}\rangle, \quad \phi \in \mathcal{S}.$$

Now let:

$$(2.33)\qquad \mathcal{PC}_M := \mathcal{PC} \cap \mathcal{O}_M, \quad \mathcal{PC}_C := \mathcal{PC} \cap \mathcal{O}'_C.$$

PROPOSITION 2.2. $\mathcal{PC} = \mathcal{PC}_M + \mathcal{PC}_C$; $\mathcal{PC}_M \cap \mathcal{PC}_C = \mathcal{S}$; $\mathcal{PC}_M$ _is the linear span of the elements given by_ (2.28); $\mathcal{PC}_C$ _is the linear span of the elements given by_ (2.27).

Thus $\mathcal{PC}_C$ consists of piecewise C^∞-functions on $\mathbb{R}$ whith only finitely many singularities around which they have a quite specific asymptotic behaviour apparent from (2.26), (2.27). Furthermore, as $x \to \pm\infty$ they behave as rapidly decreasing C^∞-functions. The space $\mathcal{PC}_M$ can be characterized in a different way as follows:

PROPOSITION 2.3. $f \in \mathcal{PC}_M$ _if and only if_ $f \in C^\infty(\mathbb{R})$ _and, near_ $\pm\infty$, f _is a linear combination of functions_

$$x \mapsto e^{ibx}\,|x|^\alpha\,(\log|x|)^k\,h_\pm(x),$$

where $b \in \mathbb{R}$, $\alpha \in \mathbb{C}$, $k \in \mathbb{Z}_+$ _and_ $h_\pm \in C^\infty(\mathbb{R})$ _with asymptotic expansion of the form_

$$(2.34)\qquad h_\pm(x) \sim \sum_{j=0}^{\infty} c_{j,\pm}\,|x|^{-j}, \quad x \to \pm\infty.$$

Here (2.34) _means that for all_ $n,m \in \mathbb{Z}_+$ _we have:_

$$(\frac{d}{dx})^m(h_\pm(x) - \sum_{j=0}^{n} c_{j,\pm}|x|^{-j}) = O(|x|^{-n-m-1}) \text{ as } x \to \pm\infty.$$

3. A Product on PC

If $f \in PC_M$ then M_f sends both PC_M and PC_C into itself. If $f \in PC_C$, $g \in PC_M$ then we may define f.g as $M_g f$. However, it remains a problem to give a meaning to f.g if both f and g are in PC_C with common singular points. Similarly, we can ask for the meaning of $f*g$ if $f,g \in PC_M$. There have been many attempts in literature to find a reasonable definition for the product of two distributions on suitable subclasses (see for instance the references in [5]). In our opinion, the best definition for our purposes has been given by Keller [4], [5], [6]. We will adapt his approach in order to define the product on PC.

The point of departure is an extension to PC of the evaluation functional E, defined on S by (2.17).

DEFINITION 3.1. *An evaluation functional* E *is a linear functional on* PC *such that* $E(f)=f(0)$ *if* $f \in PC$ *and* f *is continuous at* 0.

For each choice of E we can define an integration functional I on PC by

$$(3.1) \qquad I(f) := E(Ff), \quad f \in PC.$$

Then $I(f) = \int_{-\infty}^{\infty} f(\xi)d\xi$ if $f \in PC \cap L^1(\mathbb{R})$. Note that we can fix any evaluation functional E by an arbitrary choice for $E(\delta^{(k)})$ $(k \in \mathbb{Z}_+)$, $E(x_\pm^\alpha(\log x_\pm)^q)$ $(\mathrm{Re}\,\alpha \le 0,\ \alpha \ne 0,\ q \in \mathbb{Z}_+$ or $\alpha = 0,\ 0 < q \in \mathbb{Z}_+)$, $E(x \mapsto \mathrm{sign}(x))$.

The following theorem is closely related to Keller's results, cf. Theorem 4.3 in part II of [6].

THEOREM 3.2. *For each choice of* E *there is a unique bilinear mapping* $(f,g) \to f.g$: $PC \times PC \to PC$ *such that*:

(i) $f.g = M_f g$ *if* $f \in PC_M$, $g \in PC$;

(ii) $f.(\phi g) = \phi(f.g)$ *if* $\phi \in S$, $f, g \in PC$ (*(S)-semi-associativity*);

(iii) $I(P(f.g)) = I(Ff.F^{-1}g)$ *if* $f,g \in PC$ (*Parseval formula*).

This mapping has the additional properties:

(iv) *If* $f,g \in PC$ *are continuous at* x *then* f.g *is continuous at* x *and* $(f.g)(x) = f(x)g(x)$;

(v) $D(f.g) = (Df).g + f.(Dg)$;

(vi) If $f \in PC_M$, $g \in PC$ then $f.g = g.f = M_{fg}$.

Now we can define a convolution product on PC (again depending on the choice of E) by

$$(3.2) \qquad f*g := F^{-1}(Ff.Fg), \quad f,g \in PC.$$

A large numbers of further remarks can be made:

a) If $f,g \in PC_C$ then f.g as a linear functional on S is given by

$$\langle f.g,\phi\rangle = I((Ff)(F^{-1}\phi*F^{-1}g)) = E(\check{f}*\phi g), \quad \phi \in S.$$

b) If $f,g,h \in PC$ and $h(x) = f(x)g(x)$ at the common regular points x of f and g then $f.g - h$ is a finite linear combination of elements $e^{aD}\delta^{(k)}$, where $k \in \mathbb{Z}_+$, a is a singular point of f or g. Thus, in order to evaluate f.g it is sufficient to compute the coefficients occurring in these finite linear combinations.

c) If $f,g \in PC$ are boundary values in the sense of S' of analytic functions F,G, respectively, on a strip $\{z \in \mathbb{C} \mid 0 < \operatorname{Im} z < b\}$ then f.g is the boundary value of FG. If $f,g \in PC$ have support bounded away from $-\infty$ then $f*g$ as defined by (3.2) coincides with the usual convolution product for such distributions.

d) Whatever the choice of E may be, the multiplication on PC can never be associative or commutative. For the nonassociativity this follows by the example in Schwartz [9]:

$$(\delta.x).x^{-1} = 0.x^{-1} = 0 \neq \delta = \delta.1 = \delta.(x.x^{-1}).$$

For the noncommutativity observe that

$$x^{-1}.\delta = -E(x^{-1})\delta \neq E(x^{-1})\delta - \delta' = \delta.x^{-1}.$$

We may always pass to a commutative algebra with new product

$$f\odot g := \tfrac{1}{4}(f.g+(\check{f}.\check{g})^{\vee} + g.f + (\check{g}.\check{f})^{\vee}),$$

which no longer satisfies property (ii) of Theorem 2.2. Note that $f \circledast g = \frac{1}{2}(f.g+g.f)$ if $E(\check{f}) = E(f)$ for all $f \in PC$.

e) The bilinear form $(f,g) \mapsto I(f.g)$ on $PC \times PC$ is nondegenerate for each choice of E. The Hermitian form

$$(f,g) \to \tfrac{1}{2}(I(f.g^*)+I(g^*.f))$$

on $PC \times PC$ can never be positive definite. Indeed, for real-valued $\phi \in S$

$$I((\delta+\phi).(\delta+\phi)) = E(\delta) + 2\phi(0) + \int_{-\infty}^{\infty} \phi(x)^2 dx$$

and, for given E, ϕ can always be chosen such that the right hand side is negative.

f) There is no preferred choice of E. Indeed, starting with a given E, the evaluation functionals $S_c^{\prime} E$ and $e^{ibX}E$ $(c>0,\ b\in\mathbb{R})$ defined by

$$(S_c^{\prime}E)(f) := E(S_c f), \quad f \in PC,$$

$$(e^{ibX}E)(f) := E(e^{ibX}f), \quad f \in PC,$$

also satisfy Definition 3.1 and we have

$$(S_c^{\prime}E)(\log|x|) = E(\log|x|) + \log c,$$

$$(e^{ibX}E)(x^{-1}) = E(x^{-1}) + ib.$$

More generally, we may transform E by multiplication with a smooth function which equals 1 at 0 or by a smooth transformation of the independent variable which leaves 0 fixed. Still we can impose an important restriction on the freedom of choice for E such that this restriction is invariant under all the above-mentioned transformations of E, namely:

(3.3) $$E(\delta^{(k)}) = 0 (k \in \mathbb{Z}_+) \text{ and } E(x_{\pm}^{\alpha,q}) = 0 (-\alpha \notin \mathbb{Z}_+,\ q \in \mathbb{Z}_+).$$

In particular this will imply that $\delta^{(k)}.\delta^{(\ell)} = 0$ for all $k,\ell \in \mathbb{Z}_+$. From now on we will assume that (3.3) holds.

g) As pointed out by Keller [6], a particular nice choice for E is

$$E(f) := \operatorname{Res}_{\lambda=0} \lambda^{-1} \, AC \; E\left(f * \frac{|x|^{\lambda-1}}{2\Gamma(\lambda)\cos\frac{1}{2}\pi\lambda}\right), \quad f \in \mathcal{PC}_C, \tag{3.4}$$

which is equivalent to the choice for I made in [7]:

$$I(g) := \operatorname{Res}_{\lambda=0} \lambda^{-1} \, AC \; I(|x|^{-\lambda} g), \quad g \in \mathcal{PC}_M. \tag{3.5}$$

Note that (3.5) is in the spirit of the Hadamard finite part (cf. (2.23)).

4. A Canonical Product on the Dual of $\mathcal{PC}$

In the previous section we introduced a far from canonical product on $\mathcal{PC}$. However, by using a simple extension principle first observed by Arens [1], [2]*) we can define a canonical associative product on a suitable space of linear functionals on $\mathcal{PC}$.

Let V be an algebra over $\mathbb{C}$, V' its algebraic linear dual space and V" its bidual. Then we can define bilinear mappings

$(\phi,f) \mapsto \phi f$: $V \times V' \to V'$,

$(F,f) \mapsto Ff$: $V'' \times V' \to V'$,

$(F,G) \mapsto FG$: $V'' \times V'' \to V''$ as follows:

$$\langle \phi f,\psi\rangle = \langle f,\phi\psi\rangle, \quad f \in V', \quad \phi,\psi \in V, \tag{4.1}$$

$$\langle Ff,\psi\rangle = \langle F,f\psi\rangle, \quad F \in V'', \quad f \in V', \quad \psi \in V, \tag{4.2}$$

$$\langle FG,f\rangle = \langle F,Gf\rangle, \quad F,G \in V'', \quad f \in V'. \tag{4.3}$$

V is naturally embedded in V" and the product on V" restricted to V gives back the original product on V. If the product on V is associative then the same holds on V", but if the product on V is commutative then this is not necessarily true for the product on V" (cf. R. Arens [2]). Of course, the above construction remains true if V' is replaced by a subspace X of V' and V" by a subspace Y of X', provided $V \times X \subset X$, $Y \times X \subset X$, $Y \times Y \subset Y$.

Let us apply this construction to the case that $V = \mathcal{S}$, $X = \mathcal{PC}$. Then

*) We thank C.B. Huijsmans for providing us these references.

$(\phi,f) \mapsto \phi f\colon \mathcal{S} \times \mathcal{PC} \to \mathcal{PC}$ coincides with the usual action of $\mathcal{S}$ on $\mathcal{PC}$. If $f \in \mathcal{PC}$ then we can define an element F_f of $\mathcal{PC}'$ by

$$(4.4) \qquad \langle F_f, g\rangle = I(f.g), \quad g \in \mathcal{PC}.$$

Of course, the mapping $f \to F_f$ depends on the choice of E. Now it follows from (4.2) that

$$(4.5) \qquad F_f g = f.g, \quad f,g \in \mathcal{PC},$$

and from (4.3) that $F_f F_g$ ($f.g \in \mathcal{PC}$) is the element of $\mathcal{PC}'$ defined by

$$(4.6) \qquad \langle F_f F_g, h\rangle = I(f.(g.h)), \quad h \in \mathcal{PC}.$$

Thus, if $f,g \in \mathcal{PC}$ then

$$(4.7) \qquad \langle F_f F_g - F_{f.g}, h\rangle = I(f.(g.h)-(f.g).h), \quad h \in \mathcal{PC}.$$

The left hand side of (4.7) vanishes whenever f and g are regular on the support of h. In order to describe $F_f F_g - F_{f.g}$ when acting on h with support on some of the singularities of f and g we have to introduce some further elements of $\mathcal{PC}'$: $\eta_{a,+}^{(\alpha,q)}$, $\eta_{a,-}^{(\alpha,q)}$, $\eta_{\infty,b}^{(\alpha,q)}$, $\eta_{-\infty,b}^{(\alpha,q)}$ ($\alpha \in \mathbb{C}$, $q \in \mathbb{Z}_+$, $a,b \in \mathbb{R}$), $\theta_a^{(k)}$ ($k \in \mathbb{Z}_+$, $q \in \mathbb{R}$):

$$(4.8) \qquad \langle \eta_{a,\pm}^{(\alpha,q)}, f\rangle := \text{coefficient of } e^{aD} x_\pm^\alpha (\log x_\pm)^q \text{ in asymptotic series of } f \text{ as } \pm(x-a) \downarrow 0,$$

$$(4.9) \qquad \langle \eta_{\pm\infty,b}^{(\alpha,q)}, f\rangle := \text{coefficientof } e^{-ibX} x_\pm^\alpha (\log x_\pm)^q \text{ in asymptotic series of } f \text{ as } x \to \pm\infty,$$

$$(4.10) \qquad \langle \theta_a^{(k)}, f\rangle := \text{coefficient of } \frac{(-1)^k}{k!} e^{aD} \delta^{(k)} \text{ in } f.$$

(The normalisation in (4.8), (4.9) is slightly different from the one in [7].) Now it is clear that $F_f F_g - F_{f.g}$ is a (possibly infinite) linear combination of elements of $\mathcal{PC}'$ of the type (4.8), (4.9), (4.10).

If $F \in \mathcal{PC}'$ and A is one of the operators given by (2.1)-(2.9) then define $AF \in \mathcal{PC}'$ by

(4.11) $\langle AF,f\rangle := \langle F,A'f\rangle, \quad f \in PC,$

where $\langle A'f,\phi\rangle := \langle f,A\phi\rangle$ $(f \in PC,\ \phi \in S)$.

DEFINITION 4.1. Let the space *SGF* of *special generalised functions* consist of all finite linear combinations of the elements

(4.12) $F_f (f \in PC)$

(4.13) $\sum_{p,q=0}^{\infty} c_{p,q}\, \eta_{a,\pm}^{(\alpha-p,q)} \quad (c_{p,q} \in \mathbb{C},\ a \in \mathbb{R},\ \alpha \in \mathbb{C}),$

(4.14) $\sum_{p,q=0}^{\infty} c_{p,q}\, \eta_{\pm\infty,b}^{(\alpha+p,q)} \quad (c_{p,q} \in \mathbb{C},\ b \in \mathbb{R},\ \alpha \in \mathbb{C}),$

(4.15) $\sum_{k=0}^{\infty} c_k\, \theta_q^{(k)} \quad (c_k \in C,\ a \in \mathbb{R}).$

Note that an infinite sum like (4.13), when tested against an element of *PC*, yields only finitely many nonzero terms.

THEOREM 4.2.

a) *SGF* is invariant under the operators inherited from (2.1)-(2.9).
b) $SGF \times PC \subset PC$ with product defined by (4.2).
c) $SGF \times SGF \subset SGF$ with product defined by (4.3).
d) The product on *SGF* is associative.

It might seem from Definition 4.1 that the definition of *SGF* depends on the choice of E. However, we can define another embedding $f \to G_f$ of *PC* in *PC'*, not depending on E, as follows. If f has no singularities on [a,b] except possibly at one interior point c then put

(4.16) $\langle G_f,g) := \operatorname{Res}_{\lambda=0} \lambda^{-1}\, AC \langle g,|x-c|^{\lambda} f\rangle,$

whenever $g \in PC$ with support inside [a,b]. (Note that $\langle g,|x-c|^{\lambda} f\rangle$ is well-defined for $\mathrm{Re}\,\lambda$ sufficiently large because g is a distribution of finite order.) Similarly, if f has no singularities at finite points in $[a,\infty)$ then put

(4.17) $\langle G_f,g\rangle := \operatorname{Res}_{\lambda=0} \lambda^{-1}\, AC \langle g,|x|^{-\lambda} f\rangle$

whenever $g \in PC$ with support inside $[a,\infty)$. The definition of G_f as $x \to -\infty$ is analogous to (4.17). Now, for each $f \in PC$, $F_f - G_f$ is a finite linear combination of elements of the form (4.13), (4.14), (4.15) and $\langle F_f - G_f, h\rangle = 0$ if $h \in PC$ with support outside the singularities of f.

Let the mapping $F \to f_F$ of SGF onto PC be defined by

$$(4.18) \qquad \langle f_F, \phi\rangle = \langle F, \phi\rangle, \quad \phi \in S,$$

where at the right hand side ϕ is considered as an element of PC. This mapping sends both F_f and G_f back to f and if satisfies

$$(4.19) \qquad f_{F_f F_g} = f.g, \quad f.g \in PC.$$

Summarizing, we see that SGF is a much nicer algebra than PC. The reason is that SGF has much more elements with point support ((4.13),(4.14),(4.15)) than PC (only (2.26)). These new elements admit enough freedom to carry information in order to have a product which is associative, behaves nicely under dilatation, and so on.

There is one final step to be made in order to get the full picture of [7]. In [7] the elements of PC and SGF live together in one bigger algebra of generalised functions which we denote here by GF. We might achieve this in our present approach by applying the construction of the beginning of this section once more, such that the algebra now equals PC with product obtained by a choice of E. Then we can realize both PC and SGF as subalgebras of the dual of SGF: PC by putting $\langle f,F\rangle := \langle F,f\rangle$ if $f \in PC$, $F \in SGF$, and SGF by putting $\langle F,G\rangle := \langle FG,1\rangle$ if $F,G \in SGF$. The details, in particular a minimal choice of GF as a subspace of SGF', have yet to be worked out.

Acknowledgement. The contribution of one of the authors JJL was performed under the Euratom-FOM association agreement with financial support of ZWO and Euratom.

REFERENCES

[1] Arens, R., Operations induced in function classes, Monatsh. Math. 55 (1951), 1-19.

[2] Arens, R., The adjoint of a bilinear operation, Proc. Amer. Math. Soc. 2 (1951), 839-848.

[3] Gel'fand, I.M-Shilov, G.E., Generalized functions, Vol. I. Academic Press, New York 1964.

[4] Keller, K., Konstruktion von Produkten in einer für Feldtheorien wichtigen Klasse von Distributionen. Thesis, Aachen, July 1974.

[5] Keller, K., Analytic regularizations, finite part prescriptions and products of distributions, Math. Ann. 236 (1978), 49-84.

[6] Keller, K., Irregular operations in quantum field theory I, II, Rep. Math. Phys. 14 (1978), 285-309; 16 (1979), 203-231.

[7] Lodder, J.J., A simple model for a symmetrical theory of generalized functions I-V, Phys. A 116 (1982), 45-58, 59-73, 380-391, 392-403, 404-410.

[8] Lodder, J.J., Quantum electrodynamics without renormalization I-IV, Phys. A 120 (1983), 1-29, 30-42, 566-578, 579-586.

[9] Schwartz, L., Théorie des Distributions. Hermann, Paris 1966.

International Series of
Numerical Mathematics, Vol. 65

SOME EMBEDDING THEOREMS FOR MODULAR CLASSES

Julian Musielak
Institute of Mathematics
A. Mickiewicz University
Poznan

In 1961 there were obtained results [1] concerning connections between the sets $\bigcap_{i=1}^{\infty} L^{\varphi_i}\langle a,b\rangle$, $\bigcup_{i=1}^{\infty} L^{\varphi_i}\langle a,b\rangle$ and $L^{\psi}\langle a,b\rangle$, where $L^{\varphi_i}\langle a,b\rangle$ and $L^{\psi}\langle a,b\rangle$ are Orlicz classes. These results were successively generalized in [2], [3] and [4], based on the notion of σ-absorbed resp. σ-absorbing families of sets. Here, we observe that the latter result may be extended by replacing families of sets by families of functionals.

1. Introduction

We consider three measure spaces $(E,\mathcal{E},\mu)$, $(\Xi,\mathcal{X},m)$ and $(H,\mathcal{Y},n)$ with $\mu(E)<\infty$, such that for any sequence (ε_i) of positive numbers for which $\varepsilon_1+\varepsilon_2+\ldots<\mu(E)$ there exists a sequence of pairwise disjoint sets $E_i \in \mathcal{E}$ satisfying $\mu(E_i)=\varepsilon_i$ for $i=1,2,\ldots$ We denote by S_1 (resp. S_2) the families of nonnegative $\mathcal{X}$-measurable functions Ξ on finite m-a.e. (resp. $\mathcal{Y}$-measurable functions on H finite n-a.e.).

Let A_i be a family of functionals defined over S_i, $i=1,2$, satisfying the following conditions for all $F_i \in A_i$, $i=1,2$:

1) $0 \leqslant F_i(s_i) \leqslant \infty$ for $s_i \in S_i$,

2) $F_i(0)=0$ and if $s_i>0$ a.e., then $F_i(s_i)>0$,

3) $s_i \leqslant s_i'$ a.e. implies $F_i(s_i) \leqslant F_i(s_i')$ for $s_i, s_i' \in S_i$,

4) if $s_i^n \to \infty$ as $n\to\infty$, then $F_i(s_i^n)\to\infty$ as $n\to\infty$.

The family A_1 will be called σ-*absorbed*, if there exists an increasing sequence of functionals (F_1^n), $F_1^n \in A_1$, such that for every $F_1 \in A_1$ there is an index k for which $F_1(s_1) \leqslant F_1^k(s_1)$ for all $0 \leqslant s_1 \in S_1$. The family

A_2 will be called σ-absorbing, if there exists a sequence of functionals (F_2^n), $F_2^n \in A_2$, such that for every $F_2 \in A_2$ there is an index k_0 for which $F_2^k(s_2) \leqslant F_2(s_2)$ for all $k \geqslant k_0$ and $0 \leqslant s_2 \in S_2$.

Let us consider two functions $\varphi : \Xi \times R_+ \to R_+$ and $\varphi : H \times R_+ \to R_+$, $R_+ = \langle 0,\infty)$, such that

1^0 $\varphi(\cdot,u)$ is X-measurable with respect to the variable $\xi \in \Xi$, and $\psi(:,u)$ is Y-measurable with respect to the variable $\eta \in H$ for all $u \geqslant 0$,

2^0 $\varphi(\xi,\cdot)$ is a φ-function (see e.g.[2]) for m-a.e. $\xi \in \Xi$ and $\psi(\eta,\cdot)$ is a φ-function for n-a.e. $\eta \in H$,

3^0 $F_1(a\ \varphi(\cdot,u)) < \infty$ and $F_2(a\ \psi(:,u)) < \infty$ for all $a,u > 0$, $F_1 \in A_1$, $F_2 \in A_2$.

Let $\alpha = (F_1,F_2,c,u_0)$ for $F_i \in A_i$, $i = 1,2$; $c,u_0 > 0$. The function ψ will be called α-weaker than φ if $F_2(2\mu(E)\ \psi(:,u)) \leqslant c\ F_1(\mu(E)\ \varphi(\cdot,u))$ for all $u \geqslant u_0$.

Let X be the set of all E-measurable, real-valued functions x on E such that $\varphi(\xi,|x(t)|)$ and $\psi(\eta,|x(t)|)$ are $\Xi \times E$-measurable on $E \times E$ or $Y \times E$-measurable on $H \times E$, respectively. Let us denote for $x \in X$:

$$g_\varphi(\xi,x) = \int_E \varphi(\xi,|x(t)|)\, d\mu \quad , \quad g_\psi(\eta,x) = \int_E \psi(\eta,|x(t)|)\, d\mu,$$

$$\rho^a_{F_1}(x) = F_1(a g_\varphi(\cdot,x)) \quad \text{and} \quad \rho^b_{F_2}(x) = F_2(b g_\psi(:,x)) \quad \text{for} \quad a,b > 0 .$$

In place of $\rho^1_{F_i}$ we shall simply write ρ_{F_i}, $i = 1,2$.

2. Embedding Theorems

The following lemma will be of importance.

LEMMA. *If $F_1 \in A_1$ is concave, $F_2 \in A_2$ is convex and ψ is α-weaker than φ with $\alpha = (F_1,F_2,c,u_0)$, then*

$$\rho^1_{F_2}(x) \leqslant \frac{1}{2}\, c\ \rho^1_{F_1}(x) + \frac{1}{2} F_2(2\ \mu(E)\ \psi(:,u_0)) .$$

PROOF. Let $T = \{t \in E : |x(t)| \geqslant u_0\}$; then

$$\rho_{F_2}^{1}(x) = F_2(g_\psi(:,x)) \leqslant \frac{1}{2} F_2(\int_T 2\psi(:,|x(t)|)d\mu) + \frac{1}{2} F_2(\int_{E\setminus T} 2\psi(:,|x(t)|\, du).$$

But

$$F_2(\int_T 2\,\psi(:,|x(t)|)d\mu) \leqslant \frac{1}{\mu(E)} \int_T cF_1(\mu(E)\,\varphi(\cdot,|x(t)|)\,d\mu$$

$$\leqslant cF_1(\int_T \varphi(\cdot,\, |x(t)|)d\mu) = c\,\rho_{F_1}(x)$$

and

$$F_2(\int_{E\setminus T} 2\psi(:,|x(t)|)d\mu) \leqslant F_2(2\,\mu(E)\psi(:,u_o)).$$

Hence there follows the desired inequality.

Now, we may establish an embedding theorem between modular classes

$$X^o_{\rho^a_{F_1}} = \{x \in X : \rho^a_{F_1}(x) < \infty\} \quad \text{and} \quad X^o_{\rho^b_{F_2}} = \{x \in X : \rho^b_{F_2}(x) < \infty\}.$$

Let us denote by B_i (resp. C_i) the class of all concave (resp. convex) functionals from A_i, $i = 1,2$. Then

THEOREM 1. 1) _If_

(a) _there exist_ $F_{1,0} \in B_1$, $F_{2,0} \in C_2$, $c,u_o > 0$ _such that_ ψ is α-_weaker_ than φ with $\alpha = (F_{1,0},F_{2,0},c,u_o)$, _then_

(a')
$$\bigcap_{F_1 \in B_1} X^o_{\rho_{F_1}} \subset \bigcup_{F_2 \in C_2} X^o_{\rho_{F_2}}$$

2) _If_

(b) _for every_ $F_2 \in C_2$ _there exist_ $F_{1,0} \in B_1, c,u_o > 0$ _such that_ ψ _is_ α-_weaker than_ φ _with_ $\alpha = (F_{1,0},F_2,c,u_o)$, _then_

(b')
$$\bigcap_{F_1 \in B_1} X^0_{\rho_{F_1}} \subset \bigcap_{F_2 \in C_2} X^0_{\rho_{F_2}}.$$

3) *If*

(c) *for every* $F_1 \in B_1$ *and* $F_2 \in C_2$ *there exist* $c, u_0 > 0$ *such that* ψ *is* α-*weaker than* φ *with* $\alpha = (F_1, F_2, c, u_0)$, *then*

(c')
$$\bigcup_{F_1 \in B_1} X^0_{\rho_{F_1}} \subset \bigcap_{F_2 \in C_2} X^0_{\rho_{F_2}},$$

4) *If*

(d) *for every* $F_1 \in B_1$ *there exist* $F_2 \in C_2, c, u_0 > 0$ *such that* ψ *is* α-*weaker than* φ *with* $\alpha = (F_1, F_2, c, u_0)$, *then*

(d')
$$\bigcup_{F_1 \in B_1} X^0_{\rho_{F_1}} \subset \bigcup_{F_2 \in C_2} X^0_{\rho_{F_2}}.$$

PROOF. This follows easily from the Lemma. Indeed, considering for example the case 1) and assuming x to belong to the left-hand side of the inclusion (a'), we have $\rho_{F_1}(x) < \infty$ for all $F_1 \in B_1$; in particular, $\rho_{F_{1,0}}(x) < \infty$. Hence by the Lemma and by 3^0,

$$\rho_{F_{2,0}}(x) \leqslant \frac{1}{2} c \, \rho_{F_{1,0}}(x) + \frac{1}{2} F_2(2 \mu(E) \psi(:, u_0)) < \infty.$$

Thus, x belongs to the right-hand side of the inclusion (a').

A theorem partially converse to Theorem 1 requires the assumption of σ-absorbicity of the families of functionals. It reads as follows.

THEOREM 2. 1) *If* A_1 *is* σ-*absorbed*, A_2 *is* σ-*absorbing*, *and*

(a")
$$\bigcap_{F_1 \in C_1} X^0_{\rho_{F_1}} \subset \bigcup_{F_2 \in B_2} X^{0}_{\rho^2_{F_2}},$$

then there holds the condition (a).

2) <u>If</u> A_1 <u>is</u> σ-<u>absorbed</u>, <u>and</u>

(b")
$$\bigcap_{F_1 \in C_1} X^o_{\rho_{F_1}} \subset \bigcap_{F_2 \in B_2} X^o_{\rho^2_{F_2}} ,$$

<u>then there holds the condition</u> (b).

3) <u>If</u>

(c")
$$\bigcup_{F_1 \in C_1} X^o_{\rho_{F_1}} \subset \bigcap_{F_2 \in B_2} X^o_{\rho^2_{F_2}} ,$$

<u>then there holds the condition</u> (c).

4) <u>If</u> A_2 <u>is</u> σ-<u>absorbing</u>, <u>and</u>

(d")
$$\bigcup_{F_1 \in C_1} X^o_{\rho_{F_1}} \subset \bigcup_{F_2 \in B_2} X^o_{\rho^2_{F_2}} ,$$

<u>then there holds the condition</u> (d).

PROOF. We define a function of the form $x(t) = u_i$ if $t \in E_i$, $x(t) = 0$ if $t \notin E_i$ for $i = 1,2,\ldots$, where the numbers $u_i > 0$ and pairwise disjoint sets $E_i \in \mathcal{E}$ are chosen so that $\mu(E_i) = \varepsilon_i$, where u_i and ε_i are selected in each of the cases (a")-(d") in such a manner that x belongs to the left-hand side of the inclusion and does not belong to the right-hand one. It is easily seen that $x \in X$.

In order to prove 1), let (F^n_1) and (F^n_2) be defined by the assumption that A_1 is σ-absorbed and A_2 is σ-absorbing, and let us suppose that (a) does not hold. Then there exists a sequence $u_n \uparrow \infty$ such that

$$F^n_2(2\ \mu(E)\ \psi(:,u_n)) > 2^n F^n_1(\mu(E)\ \varphi(\cdot,u_n)) > 2^n$$

For $n = 1,2,\ldots$ Taking

(*)
$$\varepsilon_n = \frac{\mu(E)}{2^n F^n_1(\mu(E)\ \varphi(\cdot,u_n))} , \qquad n = 1,2,\ldots,$$

we have $\varepsilon_1 + \varepsilon_2 + \ldots < \mu(E)$, and so there exists a sequence (E_n) of pairwise disjoint sets $E_n \in \mathcal{E}$ such that $\mu(E_n)=\varepsilon_n$. Defining the function x as above, and taking an arbitrary $F_1 \in C_1$, we may find an index k such that $F_1 \leq F_1^n$. Since $F_1^i \geq F_1^n$ for $i \geq n$, we have

$$\rho_{F_1}(x) \leq \sum_{i=1}^{\infty} \frac{\mu(E_i)}{\mu(E)} F_1(\mu(E)\ \varphi(\cdot, u_i))$$

$$\leq \sum_{i=1}^{n-1} \frac{\mu(E_i)}{\mu(E)} F_1^n(\mu(E)\ \varphi(\cdot, u_i)) + \sum_{i=n}^{\infty} \frac{\mu(E_i)}{\mu(E)} F_1^i(\mu(E)\ \varphi(\cdot, u_i))$$

$$= \sum_{i=1}^{n-1} \frac{\mu(E_i)}{\mu(E)} F_1^n(\mu(E)\ \varphi(\cdot, u_i)) + \sum_{i=n}^{\infty} \frac{1}{2^i} < \infty,$$

i.e., $x \in X^o_{\rho_{F_1}}$. Now, let $F_2 \in B_2$ be arbitrary, and let n_o be such that $F_2^n \leq F_2$ for $n \geq n_o$. Then

$$\rho^2_{F_2}(x) \geq \sum_{i=n_o}^{\infty} \frac{\mu(E_i)}{\mu(E)} F_2^i(2\ \mu(E)\ \psi(:, u_i))$$

$$\geq \sum_{i=n_o}^{\infty} \frac{\mu(E_i)}{\mu(E)} 2^i F_1^i(\mu(E)\ \varphi(\cdot, u_i)) = \infty\ ;$$

whence $x \notin X^o_{\rho^2_{F_2}}$ for every $F_2 \in B_2$. Thus, (a") does not hold.

We are going now to prove 3). Supposing that (c) does not hold, one may find an $F_1 \in C_1$, an $F_2 \in B_2$ and a sequence $u_n \uparrow \infty$ such that

$$F_2(2\ \mu(E)\ \psi(:, u_n)) \geq 2^n\ F_1(\mu(E)\ \varphi(\cdot, u_n)) > 2^n$$

for $n = 1,2,\ldots$ Taking

$$(**) \qquad \varepsilon_n = \frac{\mu(E)}{2^n\ F_1(\mu(E)\ \varphi(\cdot, u_n))}, \qquad n = 1,2,\ldots$$

we have again $\varepsilon_1 + \varepsilon_2 + \ldots < \mu(E)$, and we may define the respective pairwise disjoint sets $E_n \in \mathcal{E}$ with $\mu(E_n) = \varepsilon_n$. Taking x as above, we have

$$\rho_{F_1}(x) \leq \sum_{i=1}^{\infty} \frac{\mu(E_i)}{\mu(E)} F_1(\mu(E)\ \varphi(\cdot, u_i)) \leq \sum_{i=1}^{\infty} \frac{1}{2^i} < \infty\ ,$$

and

$$\rho^2_{F_2}(x) \geq \sum_{n=1}^{\infty} \frac{\mu(E_n)}{\mu(E)} F_2(2\,\mu(E)\,\psi(:,\,u_n))$$

$$\geq \sum_{n=1}^{\infty} \frac{\mu(E_n)}{\mu(E)} 2^n F_1(\mu(E)\,\varphi(\bullet,u_n)) = \infty \quad ,$$

i.e. $x \in X^0_{\rho_{F_1}}$, but $x \notin X^0_{\rho^2_{F_2}}$. This contradicts (c").

Parts 2) and 4) are proved analogously, choosing ε_n according to (*) in case 2), and according to (**) in case 4); the proofs are left to the reader.

3. Applications

In the special case when the functionals from the families A_1, A_2 are all linear, we have $B_i = C_i = A_i$ for $i = 1,2$. Also we have then obviously $\rho^a_{F_i}(x) = a\rho_{F_i}(x)$ for every $a > 0$. So Theorems 1 and 2 yield the following

COROLLARY. _If the functionals from_ A_i, $i = 1,2$, _are linear, then under the assumptions of_ σ-_absorbicity from Theorem 2 there hold the equivalencies_ (a)$\Leftrightarrow$(a'), (b)$\Leftrightarrow$(b'), (c)$\Leftrightarrow$(c'), _and_ (d)$\Leftrightarrow$(d').

In particular, taking $F_1(s_1) = \int_Z s_1(\xi)\,dm$ and $F_2(s_2) = \int_{Z*} s_2(\eta)\,dn$, where $Z \in \mathcal{Z}$, $Z* \in \mathcal{Z}*$, and $\mathcal{Z}$ (resp. $\mathcal{Z}*$) is a family of subsets of Ξ with $0 < m(Z) < \infty$ (resp. of H with $0 < n(Z*) < \infty$), the Corollary gives the results of [4].

We shall give now a more general example. Taking two families of sets $\mathcal{Z}$ and $\mathcal{Z}*$ as above, we call $\mathcal{Z}$ σ-absorbed after [4], if there exists an increasing sequence (Z_n) of sets, $Z_n \in \mathcal{Z}$, such that for every $Z \in \mathcal{Z}$ there is a k with $Z \subset Z_k$, and we call $\mathcal{Z}*$ σ-absorbing, if there exists a sequence (Z^*_n) of sets, $Z^*_n \in \mathcal{Z}*$, such that for every $Z* \in \mathcal{Z}*$ there is a k_0 such that $Z^*_k \subset Z*$ for every $k \geq k_0$. We take a concave φ-function Φ_1 and a convex φ-function Ψ_2 and we put

$$F_1(s) = \int_Z \Phi_1(s(\xi))\,dm \quad , \quad F_2(s) = \int_{Z*} \Psi_2(s(\eta))\,dn.$$

By A_1 (resp. A_2) we understand the family of all F_1 with $Z \in \mathcal{Z}$ (all F_2 with $Z^* \in \mathcal{Z}^*$). Then $B_1 = A_1$, $C_2 = A_2$, and σ-absorbicity properties of $\mathcal{Z}$ (resp. $\mathcal{Z}^*$) correspond to σ-absorbicity properties of A_1 (resp. A_2). Moreover, the condition $x \in X^o_{\rho_{F_1}}$ is equivalent to the condition $g_\varphi(\cdot,x) \in L^{\Phi_1}(Z,\mathcal{X},m)$, and the condition $x \in X^o_{\rho_{F_2}}$ to $g_\psi(\cdot,x) \in L^{\Psi_2}(Z^*,\mathcal{Y},n)$, where L^{Φ_1} and L^{Ψ_2} mean the respective Orlicz classes. Thus, Theorem 1 becomes an embedding theorem for Orlicz classes with mixed functions.

Another application is obtained if we take

$$F_1(s) = \int_Z s(\xi)\, dm \quad \text{and} \quad F_2(s) = \operatorname*{supess}_{\eta \in Z^*} s(\eta).$$

For example, Theorem 1, Part 1) states then that if there exist sets $Z \in \mathcal{Z}$, $Z^* \in \mathcal{Z}^*$ and numbers $c, u_0 > 0$ such that

$$\operatorname*{supess}_{\eta \in Z^*} \psi(\eta,u) \leqslant c \int_Z \varphi(\xi,u)\, dm \quad \text{for} \quad u \geqslant u_0 ,$$

then $\int_Z \int_E \varphi(\xi,|x(t)|)\, d\mu\, dm < \infty$ for every $Z \in \mathcal{Z}$ implies that $\operatorname{supess}_{\eta \in Z^*} \int_E \psi(\eta,|x(t)|)\, d < \infty$ for some $Z^* \in \mathcal{Z}^*$.

REFERENCES

[1] Matuszewska, W., Przestrzenie funkcji φ-calkowalnvch, I. Wlasności ogólne φ-funkcji i klas funkcji φ-calkowalnych. Prace Mat. 6 (1961), 121-139.

[2] Matuszewska, W. - Orlicz, W., A note on modular spaces, VI, Bull. Acad. Polon. Sci., Sér. Sci. Math., Astr. et Phys. 11 (1963), 449-454.

[3] Musielak, J. - Waszak, A., On some families of functions integrable with a parameter, Publ. Elektr. Fak. Univ. u. Beogradu 480 (1974), 127-137.

[4] Musielak, J. - Waszak, A., Some remarks on families of Orlicz classes. Comment. Math. Prace Mat. 19 (1977), 297-303.

International Series of
Numerical Mathematics, Vol. 65

NONTANGENTIAL MAXIMAL FUNCTIONS AND BOUNDED LOWER OSCILLATION

Colin Bennett
Department of Mathematics and Statistics
University of South Carolina
Columbia

Characterizations of the space BLO of functions of bounded lower oscillation are obtained in terms of the nontangential maximal function. As a corollary, a description is obtained of the harmonic functions in the upper half-space whose traces lie in BLO.

We work throughout with real-valued locally integrable functions f on R^n. The mean value of f over a bounded set E of positive measure will be denoted by $E(f)$ and the essential infimum by f_E :

(1) $$E(f) = \frac{1}{|E|} \int_E f(x)dx, \quad f_E = \operatorname*{ess\,inf}_{x \varepsilon E} f(x).$$

We shall use B to denote balls in R^n and Q to denote cubes with sides parallel to the coordinate axes.

The space BMO of functions of bounded mean oscillation consists of all locally integrable f on R^n for which

(2) $$||f||_{BMO} = \sup_Q Q(|f-Q(f)|)$$

is finite (cf. [5],[6]). When constant functions are factored out, BMO is a Banach space under the norm in (2). If the mean value $Q(f)$ in (2) is replaced by the essential infimum f_Q, the resulting functional defines the

Research supported in part by NSF Grant MCS-8301360

the space BLO of functions of bounded lower oscillation. Thus f belongs to BLO if and only if

(3) $$||f||_{BLO} = \sup_Q (Q(f) - f_Q)$$

is finite (cf. [2]). Although this functional is subadditive it is not invariant under changes of sign and so is far from being a norm. We use the norm notation simply to reflect the analogy with (2).

It follows immediately from (3) that each BLO-function is bounded below on each cube Q. However, BLO-functions may tend to $-\infty$ as $|x| \to \infty$, as the example $f(x) = -\log |x|$ shows. Elementary computations show that every bounded function f belongs to BLO, with

(4) $$||f||_{BLO} \leq 2\, ||f||_{L^\infty} \qquad (f \in L^\infty) ,$$

and that every BLO-function f belongs to BMO, with

(5) $$||f||_{BMO} \leq 2\, ||f||_{BLO} .$$

One of the deeper results in the area is that every BMO-function may be represented as the difference of two BLO-functions, that is, BMO = BLO-BLO (cf. [2]). There are several ways of characterizing BLO-functions, in terms of A_1-weights [2,6, p. 281], in terms of Hardy-Littlewood maximal functions [2], and in terms of "one-sided" Hardy-Littlewood maximal functions [1], that is $f \to \sup_{Q \ni x} Q(f)$ (with no absolute values).

In this paper we shall obtain characterizations of BLO in terms of the "one-sided" nontangential maximal operator. These results are similar in spirit to those in [1] although the formulation and the arguments are somewhat complicated by the fact that the maximal operator in this case involves a kernel (the Poisson kernel) whose support is not compact. We shall also obtain a characterization of the harmonic functions on R^+_{n+1} whose traces lie in $BLO(R^n)$. This result complements the corresponding theorem of Fabes, Johnson and Neri [3,4] for BMO.

For each $y > 0$, the Poisson kernel $P_y(x)$ is defined by

$$P_y(x) = \frac{\gamma_n y}{(|x|^2+y^2)^{\frac{n+1}{2}}} \qquad (x \in R^n) ,$$

where the normalizing constant γ_n depends only on the dimension n (cf.[8,p.61]). The Poisson integral

$$f(x,y) = (f * P_y)(x) = \int_{R^n} f(t)P_y(x-t)dt$$

is then well-defined on $R_+^{n+1} = \{(x,y): x \in R^n, y > 0\}$ for each $f \in L^1(dt/(1+|t|^{n+1}))$ and hence in particular for each $f \in BMO(R^n)$. We shall use the fact that if $f \in BMO$, then $f(x,y) \to f(x)$ nontangentially a.e., which follows from, for example, the finiteness a.e. of the truncated square function [7, Lemma 2.13] and [8,p.206], or from [8,p.213]. We have used the same letter f to denote both $f(x)$ and its Poisson integral $f(x,y)$ so it will be convenient to extend the notation established in (1) to subsets E of $R_+^{n+1} \cup R^n$ in the obvious way (with respect to the $(n+1)$-dimensional Lebesgue measure on these sets). Let us also observe here that the BMO-norm in (2) is equivalent to the following expression (for any p satisfying $1 \leq p < \infty$) involving the Poisson integral:

$$(6) \qquad ||f||_{BMO} \sim \sup_{(x,y) \in R_+^{n+1}} \left(\int_{R^n} |f(t) - f(x,y)|^P P_y(x-t)dt\right)^{1/P} .$$

A proof may be found in [6,pp.224,234] for $p=1,2$ and $n=1$. The proof for general p and n is similar.

For each $x \in R^n$ and each $\alpha > 0$, denote by $\Gamma_\alpha(x)$ the cone

$$\Gamma_\alpha(x) = \{(t,y) \in R_+^{n+1} : |x-t| \leq \alpha y\} .$$

The one-sided nontangential maximal function $N_\alpha f$ of $f \in BMO$ is defined by

$$N_\alpha f(x) = \sup_{(t,y)\,\varepsilon\,\Gamma_\alpha(x)} f(x,y) \ .$$

The absence of absolute values in this expression will be crucial. It will reflect the asymmetry in BLO which results from the fact that BLO fails to be invariant under changes of sign.

Observe that since the Poisson integrals $f(x,y)$ of BMO-functions $f(x)$ converge nontangentially a.e. to $f(x)$ as $y \to 0$, we have $N_\alpha f(x) \geq f(x)$ a.e. Furthermore, since $|N_\alpha f| \leq N_\alpha(|f|)$, which is the familiar nontangential maximal function of $|f|$, and since the latter operator is bounded on $L^2(R^n)$ (cf. [6,p.28], [8,p.197]), the same is therefore true of the maximal operator N_α.

In order to state our first result we shall need the following notation. For each $\alpha > 0$ and $(x,y) \,\varepsilon\, R_+^{n+1}$, denote by $B_\alpha(x,y)$ the ball

$$B_\alpha(x,y) = \{t \,\varepsilon\, R^n : |t-x| \leq \alpha y\}$$

and by $C_\alpha(x,y)$ the inverted cone

$$C_\alpha(x,y) = \{(t,u) \,\varepsilon\, R_+^{n+1} : |t-x| \leq \alpha(y-u)\} \ .$$

Note that $B_\alpha(x,y)$ is the projection of $C_\alpha(x,y)$ on R^n.

THEOREM 1. *Suppose* $f \in BMO(R^n)$. *Then the following assertions are equivalent*:

(i) $f \,\varepsilon\, BLO$;

(ii) $N_\alpha f(x) - f(x) \leq c$ *for almost all* x in R^n ;

(iii) $f(x,y) - f_{B_\alpha(x,y)} \leq c$ *for all* (x,y) *in* R_+^{n+1} ;

(iv) $f(x,y) - f_{C_\alpha(x,y)} \leq c$ *for all* (x,y) *in* R_+^{n+1} ,

where the constants c *depend only on* α,n *and* f. *The least constant* $c = ||N_\alpha f-f||_\infty$ *for which* (ii) *holds is also the least constant* c *for which either of* (iii) or (iv) *holds. Furthermore, there are constants* c_1 *and*

c_2, _depending only on_ α _and_ n, _such that_

(7) $$c_1 \,||f||_{BLO} \leq ||f||_{BMO} + ||N_\alpha f - f||_\infty \leq c_2 \,||f||_{BLO} .$$

Proof. Suppose first that (ii) holds. Fix $(x,y) \in R^{n+1}_+$ and let $(t,u) \in C_\alpha(x,y)$. Then, for any $s \in R^n$, the point $(x-s, y-u)$ lies in the cone $\Gamma_\alpha(t-s)$ so we have $f(x-s,y-u) - f(t-s) \leq (N_\alpha f)(t-s) - f(t-s) \leq c$ for almost all s. Hence,

$$f(x,y) - f(t,u) \leq \int_{R^n} [f(x-s,y-u) - f(t-s)]\, P_u(s)ds \leq c,$$

so taking the infimum over all $(t,u) \in C_\alpha(x,y)$ we see that (iv) holds for the same value of c.

That (iv) implies (iii) is clear because $f(x,y) \to f(x)$ nontangentially a.e. as $y \to 0$ and $B_\alpha(x,y)$ is merely the projection of the cone $C_\alpha(x,y)$ onto R^n. Hence $f_{B_\alpha(x,y)} \geq f_{C_\alpha(x,y)}$ so that (iii) holds when-whenever (iv) does and with the same value of c.

Next we show (iii) implies (ii). To this end, let x be any Lebesgue point of f in R^n and choose $(t,y) \in \Gamma_\alpha(x)$. Then $x \in B_\alpha(t,y)$ and so

$$f(t,y) - f(x) \leq f(t,y) - f_{B_\alpha(t,y)} \leq c .$$

Passing to the supremum over all $(t,y) \in \Gamma_\alpha(x)$, we obtain (ii) with the same constant as in (iii).

This establishes the equivalence of (ii), (iii), and (iv) and shows that the least constant c for which the inequality holds is the same in all three cases. Before passing to the remainder of the proof, let us observe that in defining BLO it is immaterial whether we use an oscillation condition over cubes, as in (3), or over balls B in R^n. Thus

(8) $$||f||_{BLO} \sim \sup_B \,(B(f) - f_B) .$$

To see this, suppose first that $f \varepsilon$ BLO and let B be any ball in R^n. Then the smallest cube Q containing B has volume proportional to the volume of B with the constant of proportionality depending only on the dimension n.
Hence

$$B(f) - f_B \leq B(f) - f_Q = \frac{1}{|B|}\int_B (f-f_Q) \leq \frac{1}{|B|}\int_Q (f-f_Q) \leq c\ ||f||_{BLO}$$

for all balls B. The proof of the converse is similar.

We now show that (i) implies (iii). Suppose $f \varepsilon$ BLO and let B^0 be the ball $B_\alpha(x,y) = \{t: |s-t| \leq \alpha y\}$. For each k=1,2,..., let B^k denote the ball $B_\alpha(x,2^k y) = \{t: |x-t| \leq 2^k\alpha y\}$. Then we may write

$$(9) \qquad f(x,y) - f_{B_\alpha(x,y)} = \int_{R^n} [f(s) - f_{B^0}]\ P_y(x-s)ds$$

$$= \int_{B^0} [f(s) - f_{B^0}]\ P_y(x-s)ds + \sum_{k=1}^{\infty} \int_{B^k\setminus B^{k-1}} [f(s) - f_{B^0}]P_y(x-s)ds\ .$$

But

$$(10) \qquad P_y(x-s) \leq \begin{cases} \dfrac{\gamma_n}{y^n}, & s \ \varepsilon\ B^0 \\[2ex] \dfrac{\gamma_n}{\alpha^{n+1} y^n 2^{(k-1)(n+1)}}, & s\ \varepsilon\ B^k\setminus B^{k-1}\ . \end{cases}$$

Using this and the fact that $f_{B^0} \geq f_{B^k}$ (k=1,2,...), we obtain from (9),

$$f(x,y) - f_{B_\alpha(x,y)} \leq \frac{\gamma_n}{y^n} \left\{\int_{B^0} [f(s) - f_{B^0}]ds + \sum_{k=1}^{\infty} (\tfrac{2}{\alpha})^{n+1} 2^{-k(n+1)} \int_{B^k}[f(s)-f_{B^k}]ds\right\}$$

$$\leq c \left\{[B^0(f) - f_{B^0}] + \sum_{k=1}^{\infty} 2^{-k}[B^k(f) - f_{B^k}]\right\} ,$$

with c depending only on α and n. Hence, from (8),

$$f(x,y) - f_{B_\alpha(x,y)} \leq c\ ||f||_{BLO}(1 + \sum_{k=1}^{\infty} 2^{-k}) \leq c\ ||f||_{BLO}$$

for all balls $B_\alpha(x,y)$ in R^n. This shows that (iii) holds and hence also that $||N_\alpha f-f||_\infty \leq c\ ||f||_{BLO}$. Combining this with (5), we therefore obtain the second of the inequalities in (7).

Finally, we show that (iii) implies (i). Any ball B in R^n may be expressed in the form $B = B_\alpha(x,y)$ for suitable $(x,y) \varepsilon R_+^{n+1}$ and we may write

$$B(f) - f_B = B(f) - f(x,y) + f(x,y) - f_B$$

$$\leq \frac{1}{|B|} \int_B |f(s) - f(x,y)|\, ds + f(x,y) - f_B .$$

But for all $s \varepsilon B$ we have $P_y(x-s) \geq cy^{-n}$, with c depending only on α and n, so the previous estimate gives

$$B(f) - f_B \leq c \int_{R^n} |f(s) - f(x,y)| P_y(x-s)ds + f(x,y) - f_B .$$

The integral term is, by (6), dominated by the BMO-norm of f. Since $B = B_\alpha(x,y)$, the remaining term can be estimated by (iii). The resulting constant c, however, as we have observed above, may be taken to be $||N_\alpha f-f||_\infty$. Hence, altogether we have

$$B(f) - f_B \leq c(||f||_{BMO} + ||N_\alpha f-f||_{L^\infty})$$

for all balls B in R^n. By virtue of (8), this shows that $f \varepsilon$ BLO and that the first of the inequalities in (7) holds. This completes the proof.

As a corollary, we derive the following characterization of the harmonic functions in R_+^{n+1} whose traces lie in $BLO(R^n)$.

COROLLARY 2. _Let_ $u(x,y)$ _be a harmonic function in_ R_+^{n+1}. _Then_ u _is the Poisson integral_ $u(x,y) = (f*P_y)(x)$ _of a function_ $f \varepsilon BLO(R^n)$ _if and only if_ $y|\nabla u|^2 dxdy$ _is a Carleson measure and_

(11) $$\sup_{(x,y)\varepsilon R_+^{n+1}} (u(x,y) - u_{C_\alpha(x,y)}) < \infty .$$

Proof. If $u(x,y) = (f*P_y)(x)$ for some $f \varepsilon$ BLO, then since $f \varepsilon$ BMO, a theorem of Fefferman and Stein [5] shows that $y|\nabla u|^2 dxdy$ is a Carleson measure. Furthermore, since $u(x,y) = f(x,y)$ and $f \varepsilon$ BLO, Theorem 1 shows that (11) holds.

Conversely, suppose $y|\nabla u|^2 dxdy$ is a Carleson measure. Then a result of Fabes, Johnson, and Neri [3,4] shows that $u(x,y) = (f*P_y)(x)$ for some function $f \varepsilon BMO(R^n)$. Then $u(x,y) = f(x,y)$ so if (11) holds, then Theorem 1 shows that $f \varepsilon BLO(R^n)$, as desired.

Next, we explore the action of the maximal operator N_α on BMO.

LEMMA 3. _Suppose_ $f \varepsilon BMO(R^n)$ _and_ $\alpha > 0$. _Then there is a constant_ c _depending only on_ n _and_ α _such that_

(12) $$B(N_\alpha f) \leq c\ ||f||_{BMO} + \inf_B N_\alpha f\ .$$

Proof. Fix a ball B in R^n and write $B = B_\alpha(x_0,y_0) = \{x \varepsilon R^n : |x-x_0| \leq \alpha y_0\}$. Let A be the concentric ball with three times the radius: $A = B_\alpha(x_0,3y_0)$. Suppose $f \varepsilon BMO(R^n)$ and write

$$f = \{[f-f(x_0,y_0)]\chi_A\} + \{f(x_0,y_0)\chi_A + f\chi_{A^c}\} = g+h,$$

say, where χ_A denotes the characteristic function of A.

Since N_α is bounded on $L^2(R^n)$, we have

$$B(N_\alpha g) = \frac{1}{|B|}\int_B N_\alpha g(x)dx \leq |B|^{-\frac{1}{2}}\left(\int_B |N_\alpha g(x)|^2 dx\right)^{\frac{1}{2}}$$

$$\leq c|B|^{-\frac{1}{2}}\left(\int_{R^n}|g(x)|^2 dx\right)^{\frac{1}{2}} = c|B|^{-\frac{1}{2}}\left(\int_A |f(x) - f(x_0,y_0)|^2 dx\right)^{\frac{1}{2}}\ .$$

However, $P_{y_0}(x-x_0) \geq c/y_0^n$ whenever $x \varepsilon A$ so, using (6) (with p=2), we obtain

$$B(N_\alpha g) \leq c\left(\int_{R^n} |f(x) - f(x_0,y_0)|^2\ P_y(x-x_0)dx\right)^{\frac{1}{2}} \leq c\ ||f||_{BMO}\ .$$

Because of the subadditivity of N_α, the desired estimate (12) will therefore be established if we can show that

$$B(N_\alpha h) \leq c\ ||f||_{BMO} + \inf_B N_\alpha f\ .$$

In fact, we shall prove the stronger result that, for every $x \varepsilon B$, the quantity $N_\alpha h(x)$ is dominated by the expression on the right of the above inequality. Equivalently,

(13) $$h(t,y) \leq c\ ||f||_{BMO} + \inf_B N_\alpha f,$$

whenever $(t,y) \varepsilon \Gamma_\alpha(x)$ and $x \varepsilon B$. We consider two cases.

CASE I. We assume $y \leq y_0$ so our hypotheses are

$$|x - x_0| \leq \alpha y_0,\ |t-x| \leq \alpha y \leq \alpha y_0\ .$$

From the definition of h, we have

$$h(t,y) = \int_{R^n} h(s)P_y(t-s)ds = f(x_0,y_0) \int_A P_y(t-s)ds + \int_{A^c} f(s)P_y(t-s)ds$$

$$= f(x_0,y_0) + \int_{A^c}[f(s) - f(x_0,y_0)]P_y(t-s)ds,$$

so

(15) $$h(t,y) \leq f(x_0,y_0) + \int_{A^c}|f(s) - f(x_0,y_0)|P_y(t-s)ds.$$

Now every $s \varepsilon A^c$ satisfies $|s-x_0| \geq 3\alpha y_0$ so it follows from (14) that $|s-t| \geq \alpha y_0$ and hence that $|s-t|^2 + y_0^2 \leq c|s-t|^2$. In particular, we have $P_y(t-s) \leq cP_{y_0}(t-s)$ for all $s \varepsilon A^c$. However, since $|s-t| \geq |s-x_0|/3$, we also have $P_{y_0}(t-s) \leq cP_{y_0}(x_0-s)$ for all $s \varepsilon A^c$. Hence, these estimates taken together with (15) give

(16) $$h(t,y) \leq f(x_0,y_0) + c \int_{A^c}|f(s) - f(x_0,y_0)|P_{y_0}(x_0-s)ds.$$

Appealing once again to (6) we see that the integral term is dominated by a multiple of the BMO-norm of f. Furthermore, by virtue of (14), the point (x_0,y_0) lies in the cone $\Gamma_\alpha(\xi)$ for every $\xi \in B$ and so $f(x_0,y_0) \leq N_\alpha f(\xi)$ for all $\xi \in B$. With these observations, it is clear that (16) produces the desired estimate (13).

CASE II. Now we assume $y > y_0$ so our hypotheses are

(17) $$|x-x_0| \leq \alpha y_0 < \alpha y, \quad |t-x| \leq \alpha y.$$

Define z by

$$\alpha z = \max(\alpha y, \; |x_0-t| + \alpha y_0).$$

Then elementary computations involving (17) give

(19) $$y \leq z \leq 3y.$$

We write

(20) $$h(t,y) = (h*P_y)(t) = [h-f(t,z)] * P_y(t) + f(t,z).$$

Now any $\xi \in B$ satisfies $|\xi-t| \leq \alpha z$ so $(t,z) \in \Gamma_\alpha(\xi)$ and hence $f(t,z) \leq (N_\alpha f)(\xi)$ for all $\xi \in B$. Thus, in order to establish the desired estimate (13), we see from (20) that it will suffice to show that

$$I = [h-f(t,z)] * P_y(t) \leq c \; ||f||_{BMO}.$$

It follows from (19) that $P_y \leq cP_z$ so using this and the definition of h we have

$$I \leq c \int_{R^n} |h(s) - f(t,z)| P_z(t-s)ds$$

$$= c \; \{\int_A |f(x_0,y_0) - f(t,z)| P_z(t-s)ds + \int_{A^c} |f(s) - f(t,z)| P_z(t-s)ds\}$$

$$\leq c \{\int_A |f(x_0,y_0) - f(s)|P_z(t-s)ds + \int_{R^n} |f(s) - f(t,z)|P_z(t-s)ds\}.$$

By (6), the latter integral is dominated by a multiple of the BMO-norm of f. Hence, it remains only to show that the same is true of the first term. But we see from (19) that $P_z \leq \gamma_n/z^n \leq \gamma_n/y^n$ while on the other hand we have $P_{y_0}(x-s) \geq c/y^n$ for all $s \in A$. Hence, by (6),

$$\int_A |f(x_0,y_0) - f(s)| P_z(t-s)ds \leq c \int_A |f(x_0,y_0) - f(s)|P_{y_0}(x -s)ds$$

$$\leq c \,||f||_{BMO} ,$$

and the proof is complete.

THEOREM 4. *Suppose* $\alpha > 0$. *If* $f \in BMO(R^n)$, *then either* $N_\alpha f$ *is identically infinite or* $N_\alpha f$ *belongs to* $BLO(R^n)$ *and*

$$||N_\alpha f||_{BLO} \leq c \,||f||_{BMO} , \tag{21}$$

where c *is a constant depending only on* α *and* n.

Proof. If $N_\alpha f(x) < \infty$ at one point $x \in R^n$, then Lemma 3 shows that the mean value of $B(N_\alpha f)$ is finite for any ball B in R^n that contains x. In that case, $N_\alpha f$ is finite a.e. on B, hence a.e. on R^n since B is arbitrary. Since $\inf_B N_\alpha f$ is then finite for any ball B in R^n, it may be subtracted from each side of (12) to give

$$B(N_\alpha f) - \inf_B N_\alpha f \leq c \,||f||_{BMO} ,$$

and this, together with (8), gives (21).

Finally, we obtain the following characterization of BLO-functions as maximal functions of BMO-functions (modulo bounded functions).

THEOREM 5. *Suppose* $\alpha > 0$ *and* f *is a locally integrable function on* R^n.

Then $f \in BLO$ *if and only if there is a* BMO-*function* F, *with* $N_\alpha F \not\equiv \infty$, *and a bounded function* h *such that*

$$f = N_\alpha F + h. \tag{22}$$

Furthermore, when $f \in BLO$, *there are constants* c_1 *and* c_2, *depending only on* n *and* α, *such that*

$$c_1 \,||f||_{BLO} \le \inf(\,||F||_{BMO} + ||h||_{L^\infty}) \le c_2 \,||f||_{BLO}, \tag{23}$$

where the infimum extends over all representations (22) *of* f.

Proof. If f is represented as in (22), it follows from Theorem 4 that $f \in BLO$. The first of the estimates in (23) then follows at once from (4) and (21). Conversely, if $f \in BLO$, we represent f as in (22) by choosing $F = f$ and $h = f - N_\alpha f$. It is clear that $F \in BMO$ and, from Theorem 1, that $h \in L^\infty$ (in particular, $N_\alpha F = N_\alpha f \not\equiv \infty$). The second inequality in (23) follows directly from (7).

REFERENCES

[1] Bennett, C. *'Another characterization of BLO'*, Proc. Amer. Math. Soc., 85 (1982), 552-556.

[2] Coifman, R. R. and Rochberg, R. *'Another characterization of BMO'*, Proc. Amer. Math. Soc., 79 (1980), 249-254.

[3] Fabes, E. B., Johnson, R. L. and Neri, U., *'Green's formula and a characterization of the harmonic functions with BMO traces'*, Ann. Univ. Ferrara, 21 (1975), 147-157.

[4] __________, *'Spaces of harmonic functions representable by Poisson integrals of functions in BMO and* $L_{p,\lambda}$*'*, Indiana Univ. Math. J., 25 (1976), 159-170.

[5] Fefferman, C. and Stein E. M., *'H^p spaces of several variables'*, Acta Math., 129 (1972), 137-193.

[6] Garnett, J. B., *'Bounded Analytic Functions'*, Academic Press, New York, 1981.

[7] Neri, U., 'Some properties of functions with bounded mean oscillation', Studia Math., 61 (1977), 63-75.

[8] Stein, E. M., 'Singular Integrals and Differentiability Properties of Functions', Princeton University Press, 1970.

III Abstract Approximation

International Series of
Numerical Mathematics, Vol. 65

NEGATIVE RESULTS IN CONNECTION WITH FAVARD'S PROBLEM ON THE COMPARISON OF APPROXIMATION PROCESSES

Rolf Joachim Nessel and Erich van Wickeren *)
Lehrstuhl A für Mathematik
Rheinisch-Westfälische Technische Hochschule
Aachen

Twenty years ago, at the first Oberwolfach conference "On Approximation Theory", Favard posed the problem of the comparison of approximation processes. In connection with the classical uniform boundedness principle he pointed out that one feature of a solution to this complex problem would certainly consist in deriving significant negative results. The aim of the present note is to discuss the matter on the basis of our previous q u a n t i t a t i v e uniform boundedness and condensation principles.

Let X be a (complex) Banach space (with norm $\|\circ\|_X$), and X^* be the class of functionals T on X which are sublinear, i.e.,

$$|T(f+g)| \leq |Tf| + |Tg| \quad , \quad |T(af)| = |a|\ |Tf|$$

for all $f,g \in X$ and scalars a, and which are bounded, i.e.,

$$\|T\|_{X^*} := \sup\{|Tf|\,;\ \|f\|_X = 1\} < \infty.$$

Given a sequence $\{T_n\} \subset X^*$ (e.g., remainder functionals of some approximation process), the classical uniform boundedness principle (UBP) states ($n \in \mathbb{N}$, the set of natural numbers):

THEOREM 1. _Strong boundedness of_ T_n _on_ X, _i.e._, $|T_n f| = \mathcal{O}_f(1)$ _for each_ $f \in X$, _implies uniform boundedness_, _i.e._, $\|T_n\|_{X^*} = \mathcal{O}(1)$.

For our purposes it is appropriate to stress the negative character of

*) Supported by Deutsche Forschungsgemeinschaft Grant No. Ne 171/5-1.

this principle. Indeed, negation of the latter assertion together with the very definition of the norm of a functional immediately leads to the following (equivalent) formulation as a resonance principle (cf. [11, p.20]).

THEOREM 2. *Suppose that for* $T_n \in X^*$ *there are elements* $h_n \in X$ *with*

(1) $$\|h_n\|_X \leq C_1 \qquad (n \in \mathbb{N}),$$

(2) $$|T_n h_n| \neq \mathcal{O}(1) \qquad (n \to \infty).$$

Then there exists $f_o \in X$ *such that*

(3) $$|T_n f_o| \neq \mathcal{O}(1) \qquad (n \to \infty).$$

Given two sequences $\{R_n\}$, $\{V_n\} \subset X^*$, the(strong) comparison problem asks for an estimate of type

$$|R_n f| = \mathcal{O}_f(|V_n f|) \qquad (n \to \infty)$$

to be valid for each f of a prescribed class. In this connection it was Favard [9,10] who applied Theorem 1,2 to the ratio $R_n/\|V_n\|_{X^*}$ in order to deduce

THEOREM 3. *Suppose that for* R_n, $V_n \in X^*$ *there are elements* $h_n \in X$ *satisfying* (1) *and* $|R_n h_n| \neq \mathcal{O}(\|V_n\|_{X^*})$. *Then there exists* $f_o \in X$ *such that*

(4) $$|R_n f_o| \neq \mathcal{O}(|V_n f_o|) \qquad (n \to \infty).$$

The conditions of this first contribution depend upon the quantity $\|V_n\|_{X^*}$, i.e., upon the supremum of the values $|V_n f|$ over the unit ball in X. From the point of view of applications this may be too large. In fact, it would be desirable (cf. (2-4)) to replace $|R_n h_n| \neq \mathcal{O}(\|V_n\|_{X^*})$ by $|R_n h_n| \neq \mathcal{O}(|V_n h_n|)$. But the latter condition seems to be too weak. Therefore we consider suitable subspaces $Y \subset X$ for which the corresponding functional norm $\|V_n\|_{Y^*}$ essentially depends only upon the values $|V_n h_j|$, $1 \leq j \leq n$. Combining this aspect with the classical UBP one obtains a result which extends both, Theorem 2 and 3, in a way suitable for the following.

THEOREM 4. Suppose that for $T_n, R_n, V_n \in X^*$ there are elements $h_n \in X$ satisfying (1,2) as well as

(5) $$|R_n h_n| \neq \mathcal{O}(1) \qquad (n \to \infty),$$

(6) $$|V_n h_n| \leqslant C_2, \qquad \limsup_{m\to\infty} |V_m h_n| \leqslant C_3 \qquad (n \in \mathbb{N}).$$

Then there exists $f_o \in X$ satisfying (3,4).

PROOF. Select a subsequence $\{n_k\} \subset \mathbb{N}$ such that for $k \geqslant 2$ (cf. (2,5,6))

(7) $$\max\{k, \max_{1\leqslant j\leqslant k-1} \|V_{n_j}\|_{X^*}\} \leqslant \begin{cases} |T_{n_k} h_{n_k}|^{1/2}, & k \text{ even} \\ |R_{n_k} h_{n_k}|^{1/2}, & k \text{ odd} \end{cases},$$

(8) $$\max_{1\leqslant j\leqslant k-1} |V_{n_k} h_{h_j}| \leqslant C_3 + 1 .$$

Then h_{n_j} belongs to the Banach space

$$Y := \{f \in X; \, |V_{n_k} f| = \mathcal{O}_f(1)\}, \quad \|f\|_Y := \|f\|_X + \sup_{k\in\mathbb{N}} |V_{n_k} f|$$

and satisfies (cf. (1,6-8))

$$\|h_{n_j}\|_Y \leqslant C_1 + \sup_{k\in\mathbb{N}} \begin{Bmatrix} C_3 + 1, & k > j \\ C_2 \;, & k = j \\ C_1 \|V_{n_k}\|_{X^*}, & k < j \end{Bmatrix} \leqslant C \begin{cases} |T_{n_j} h_{n_j}|^{1/2}, & j \text{ even} \\ |R_{n_j} h_{n_j}|^{1/2}, & j \text{ odd.} \end{cases}$$

Consequently, one has by (7) for n even

$$\|T_{n_k}\|_{Y^*} \geqslant |T_{n_k} h_{n_k}| / \|h_{n_k}\|_Y \geqslant k/C,$$

thus $\|T_{n_k}\|_{Y^*} \neq \mathcal{O}(1)$, and analogously $\|R_{n_k}\|_{Y^*} \neq \mathcal{O}(1)$. Hence by Theorem 1 there exist $f_1, f_2 \in Y$ such that $|T_{n_k} f_1| \neq \mathcal{O}(1)$, $|R_{n_k} f_2| \neq \mathcal{O}(1)$. Setting

$$f_0 := \begin{cases} f_1 , & \text{if also } |R_{n_k} f_1| \neq \mathcal{O}(1) \\ f_2 , & \text{if also } |T_{n_k} f_2| \neq \mathcal{O}(1) \\ f_1 + f_2 , & \text{otherwise,} \end{cases}$$

this yields the assertion (note that $f_0 \in Y$). □

Let us turn to quantitative results, thereby surveying our previous UBP's with rates (cf. [2;3] and the literature cited there) in the light of the present comparison problem. To this end, let ω be a function, continuous on $(0,\infty)$ such that

(9) $$0 < \omega(s) \leqslant \omega(t), \quad \omega(t)/t \leqslant \omega(s)/s \qquad (0 < s \leqslant t)$$

(abstract modulus of continuity (cf. [13,p. 96ff])),sometimes additionally satisfying

(10) $$\text{(i)} \lim_{t\to 0+} \omega(t) = 0, \qquad \text{(ii)} \lim_{t\to 0+} \omega(t)/t = \infty.$$

Moreover, let $U \subset X$ be a seminormed linear subspace with seminorm $|\circ|_U$. Consider the Peetre K-functional

$$K(t,f) = K(t,f;X,U) := \inf\{\|f-g\|_X + t|g|_U ;\ g \in U\},$$

a standard measure of smoothness for elements of abstract spaces. Then intermediate spaces (abstract Lipschitz classes) are defined by

$$X_\omega := \{f \in X;\ K(t,f;X,U) = \mathcal{O}_f(\omega(t)),\ t \to 0+\}.$$

Obviously, X_ω is a Banach space under the norm

$$\|f\|_{X_\omega} := \|f\|_X + \sup_{t>0} K(t,f;X,U)/\omega(t).$$

In the following $\{\varphi_n\}$ always denotes a sequence of (strictly) positive numbers, tending to zero. In these terms the next result states a quantitative version of Theorem 4, equipped with large $\mathcal{O}$-rates.

THEOREM 5. <u>Suppose that for</u> $T_n, R_n, V_n \in X^*$ <u>there are elements</u> $h_n \in U$ <u>satisfying</u> (1,2,5) <u>as well as</u> $(n \in \mathbb{N})$

(11) $$\|h_n\|_U \leqslant C_4/\varphi_n,$$

(12) $$|V_n h_n| \leqslant C_2 \quad , \quad \limsup_{m\to\infty} |V_m h_n| /\varphi_m \leqslant C_{3,n}.$$

Then for each ω with (9, 10 (ii)) there exists $f_\omega \in X_\omega$ such that

(13) $$|T_n f_\omega| \neq O(\omega(\varphi_n)),$$

(14) $$|R_n f_\omega| \neq O(|V_n f_\omega|).$$

Concerning a proof, apply Theorem 4 to the Banach space X_ω, the functionals $T_n/\omega(\varphi_n)$, $R_n/\omega(\varphi_n)$, $V_n/\omega(\varphi_n)$, and the elements $g_n = \omega(\varphi_n)h_n$. In view of (1,9,11) one has

$$K(t,h_n) \leqslant \min\{C_1, C_4 t/\varphi_n\} \leqslant C\,\omega(t)/\omega(\varphi_n),$$

thus (1) for g_n. Moreover, (2,5) are obviously satisfied, whereas (6) follows by (cf.(10(ii),12))

$$\limsup_{m\to\infty} \frac{|V_m g_n|}{\omega(\varphi_m)} = \omega(\varphi_n) \limsup_{m\to\infty} \frac{|V_m h_n|}{\varphi_m} \lim_{m\to\infty} \frac{\varphi_m}{\omega(\varphi_m)} = 0.$$

Hence (3,4) imply (13,14) .

When dealing with rates, however, one may even replace large O-rates by small o-ones.

THEOREM 6. Suppose that for T_n, R_n, $V_n \in X^*$ there are elements $h_n \in U$ satisfying (1,11,12) as well as (instead of (2,5))

(15) $$\limsup_{n\to\infty} |T_n h_n| \geqslant C_5 > 0,$$

(16) $$\limsup_{n\to\infty} |R_n h_n| \geqslant C_6 > 0.$$

Then for each ω with (9,10) there exists $f_\omega \in X_\omega$ such that $(n \to \infty)$

(17) $$|T_n f_\omega| \neq o(\omega(\varphi_n)) ,$$

(18) $$|R_n f_\omega| \neq o(|V_n f_\omega|).$$

PROOF. We proceed essentially as in [4] via the gliding hump method. Thus, starting with $n_1 = 1$, $\delta_1 = 1$ one may construct sequences $\{n_k\} \subset \mathbb{N}$, $\{\delta_k\} \subset \{0,1\}$ such that for $k \geqslant 2$

(19) $$n_k > n_{k-1}, \qquad \varphi_{n_k} < \min\{1/k,\ \varphi_{n_{k-1}}\},$$

(20) $$\omega(\varphi_{n_k}) \max\{1, \|T_{n_{k-1}}\|_{X^*}, \|R_{n_{k-1}}\|_{X^*}, \|V_{n_{k-1}}\|_{X^*}\} \leq \omega(\varphi_{n_{k-1}})/k,$$

(21) $$\sum_{j=1}^{k-1} \omega(\varphi_{n_j})/\varphi_{n_j} \leq \omega(\varphi_{n_k})/\varphi_{n_k},$$

(22) $$|V_{n_k} h_{n_j}| \leq \omega(\varphi_{n_k}) \qquad (1 \leq j \leq k-1),$$

(23) $$\omega(\varphi_{n_k}) \leq 10 \begin{cases} |T_{n_k} g_k|/C_5 , & k \text{ even} \\ |R_{n_k} g_k|/C_6 , & k \text{ odd}, \end{cases}$$

$$g_k := \sum_{j=1}^{k} \delta_j \, \omega(\varphi_{n_j}) \, h_{n_j}.$$

Indeed, if the first k-1 elements of the sequences, and thus $g_{k-1} \in U$, are given, consider, for k even,

$$M_{k-1} := \limsup_{n\to\infty} |T_n g_{k-1}| / \omega(\varphi_n).$$

In case $M_{k-1} \leq C_5/5$ there exists $N_k \in \mathbb{N}$ such that for $n \geq N_k$ (cf. (12))

(24) $$|T_n g_{k-1}| \leq (M_{k-1} + C_5/5)\omega(\varphi_n) \leq (2/5)\, C_5\, \omega(\varphi_n),$$

(25) $$\max_{1 \leq j \leq k-1} |V_n h_{n_j}| / (C_{3,n_j} + 1) \leq \varphi_n .$$

Then choose $n_k \geq N_k$ large enough to satisfy (19-21) (cf. (9,10)) as well as (cf. (10,15))

$$\max_{1 \leq j \leq k-1} (C_{3,n_j} + 1)\, \varphi_{n_k} \leq \omega(\varphi_{n_k}) \quad , \qquad |T_{n_k} h_{n_k}| \geq C_5/2,$$

thus (22) in view of (25). Setting $\delta_k = 1$, the element g_k is well-defined, and

$$|T_{n_k} g_k| \geqslant \omega(\varphi_{n_k})\, |T_{n_k} h_{n_k}| - |T_{n_k} g_{k-1}| \geqslant (C_5/10)\, \omega(\varphi_{n_k}),$$

i.e., (23) holds true. On the other hand, if $M_{k-1} > C_5/5$, take $\delta_k = 0$, thus $g_k = g_{k-1}$. Then of course one may find n_k large enough to satisfy (19-23). If k is odd, one proceeds analogously, replacing T_n by R_n.

Since X is complete and (cf. (1,20))

$$\sum_{j=k}^{\infty} \delta_j \omega(\varphi_{n_j}) \| h_{n_j} \|_X \leqslant C_1 \sum_{j=k}^{\infty} \omega(\varphi_{n_j}) \leqslant 2\, C_1\, \omega(\varphi_{n_k}), \tag{26}$$

the following element is well-defined in X:

$$f_\omega := \sum_{j=1}^{\infty} \delta_j \omega(\varphi_{n_j})\, h_{n_j}. \tag{27}$$

Moreover, $f_\omega \in X_\omega$. Indeed, since for $0 < t \leqslant \varphi_1$ there exists $k \in \mathbb{N}$ such that $\varphi_{n_{k+1}} < t \leqslant \varphi_{n_k}$ (cf. (19)), one obtains by (9,11,21,26)

$$\begin{aligned} K(t,f_\omega) &\leqslant \| f_\omega - g_k \|_X + t\, |g_k|_U \\ &\leqslant \sum_{j=k+1}^{\infty} \omega(\varphi_{n_j}) \| h_{n_j} \|_X + C_4\, t \sum_{j=1}^{k} \omega(\varphi_{n_j})/\varphi_{n_j} \\ &\leqslant 2\, C_1\, \omega(\varphi_{n_{k+1}}) + 2\, C_4\, t\, \omega(\varphi_{n_k})/\varphi_{n_k} = \mathcal{O}(\omega(t)). \end{aligned} \tag{28}$$

Now (20,23,26) deliver for k even

$$\begin{aligned} |T_{n_k} f_\omega| &\geqslant |T_{n_k} g_k| - \| T_{n_k} \|_{X^*} \| f_\omega - g_k \|_X \\ &\geqslant (C_5/10)\, \omega(\varphi_{n_k}) - 2\, C_1 \| T_{n_k} \|_{X^*} \omega(\varphi_{n_{k+1}}) \\ &\geqslant \omega(\varphi_{n_k})(C_5/10 - 2\, C_1/(k+1)), \end{aligned} \tag{29}$$

thus (17), and analogously for k odd

$$|R_{n_k} f_\omega| \geqslant \omega(\varphi_{n_k})(C_6/10 + o(1)). \tag{30}$$

Moreover, in view of (12,20,22,26) one has

$$(31)\qquad |V_{n_k} f_\omega| \leq \omega(\varphi_{n_k})|V_{n_k} h_{n_k}| + \Big(\sum_{j=1}^{k-1} + \sum_{j=k+1}^{\infty}\Big)\omega(\varphi_{n_j})|V_{n_k} h_{n_j}|$$

$$\leq \omega(\varphi_{n_k})\,[C_2 + \sum_{j=1}^{k-1}\omega(\varphi_{n_j})] + 2C_1\|V_{n_k}\|_{X^*}\omega(\varphi_{n_{k+1}})$$

$$\leq \omega(\varphi_{n_k})\,[C_2 + 2\,\omega(\varphi_1) + 2C_1],$$

thus (18) by (30). □

Let us mention that Theorem 6 subsumes Theorem 5 in the sense that the rate of divergence in (13,14) may be specified. In fact, the proof of Theorem 6 also delivers

COROLLARY 7. *Under the assumptions of Theorem* 5 (*plus* (10(i))) *there exists* $f_\omega \in X_\omega$ *satisfying*

$$|T_n f_\omega| \neq o(|T_n h_n|\,\omega(\varphi_n)) \quad , \quad |R_n f_\omega| \neq o(|R_n h_n|\,|V_n f_\omega|).$$

Note that an application of Theorem 6 (and 5) is also possible in those situations where only one of the assertions, e.g. (17), is needed. Indeed, setting $R_n = T_n$, $V_n = 0$, one obtains

COROLLARY 8. *If* $T_n \in X^*$, $h_n \in U$ *satisfy* (1,11,15), *then for each* ω *with* (9,10) *there exists* $f_\omega \in X_\omega$ *with* (17).

Moreover, conditions (1,11,15) immediately imply lower bounds for the norms of T_n on X and U, respectively, namely ($\|g\|_U := \|g\|_X + |g|_U$)

$$(32)\qquad \limsup_{n\to\infty}\|T_n\|_{X^*} > 0, \qquad \limsup_{n\to\infty}\|T_n\|_{U^*}/\varphi_n > 0.$$

But this has to be realized by the same sequence $\{h_n\}$, so that (32) is not equivalent to (1,11,15). However, this can be prevented by considering the spaces $U_n = U$, now normed by

$$\|g\|_{U_n} := \|g\|_X + \varphi_n\,|g|_U.$$

Then one may reformulate Corollary 8 in a way similar to Theorem 1.

COROLLARY 9. _With_ ω, _subject to_ (9,10), _strong convergence of_ $T_n/\omega(\varphi_n) \in X^*$ _on_ X_ω, _i.e.,_ $|T_n f| = o_f(\omega(\varphi_n))$ _for each_ $f \in X_\omega$, _implies the "uniform convergence on_ U_n", _i.e.,_ $\|T_n\|_{U_n^*} = o(1)$.

Instead of the K-functional it is often more convenient in the applications to employ other appropriate functionals $S_t \in X^*$, $t \to 0+$, as a measure of smoothness, e.g., moduli of continuity or the error of best approximation. Indeed,

THEOREM 10. _Suppose that for_ $S_t, T_n, R_n, V_n \in X^*$ _there are a positive function_ $\sigma(t)$ _on_ $(0,\infty)$ _and elements_ $h_n \in X$ _satisfying_ (1,12,15,16) _as well as (instead of_ (11))

$$(33) \qquad |S_t h_n| \leqslant C_7 \min\{1, \sigma(t)/\varphi_n\} \qquad (n \in \mathbb{N},\ t > 0).$$

Then for each ω _with_ (9,10) _there exists_ $f_\omega \in X$ _satisfying_ (17,18) _as well as (instead of_ $f_\omega \in X_\omega$)

$$(34) \qquad |S_t f_\omega| = O(\omega(\sigma(t))) \qquad (t \to 0+).$$

PROOF. Apply Theorem 6 to the Banach space

$$Y := \{f \in X,\ |S_t f| = O_f(1)\}, \quad \|f\|_Y := \|f\|_X + \sup_{t>0} |S_t f|$$

and to the subspace

$$U := \{g \in Y;\ |S_t g| = O_g(\sigma(t))\}, \quad |g|_U := \sup_{t>0} |S_t g|/\sigma(t).$$

Then (11) holds true in view of (33), and consequently there exists $f_\omega \in Y_\omega$ satisfying (17,18). Since

$$|S_t f_\omega| \leqslant \inf_{g \in U} \{|S_t(f_\omega - g)| + |S_t g|\}$$

$$\leqslant \inf_{g \in U} \{\|f_\omega - g\|_Y + \sigma(t)\,|g|_U\} = O(\omega(\sigma(t))),$$

this completes the proof. □

On the other hand, starting from Theorem 10, one may interpret Theorem 6 as the particular case $S_t f := K(t,f)$, $\sigma(t) = t$, since then (1,11) automatically imply (33).

Let us mention that Theorem 10 is also valid for more than two sequences of functionals $\{T_n\}$, $\{R_n\}$ satisfying (15,16). Indeed, a countable number of sequences of functionals, thus a double sequence $\{T_{n,p}\}_{n,p\in\mathbb{N}}$, is admissible. To be more precise, one obtains the following result which subsumes the condensation principles with rates given in [4;5] (see also [6]).

THEOREM 11. _Suppose that for_ $S_t, T_{n,p}, V_n \in X^*$ _there are a positive function_ $\sigma(t)$ _on_ $(0,\infty)$ _and elements_ $h_n \in X$ _satisfying_ (1,12,33) _as well as_ (cf. (15, 16))

$$\limsup_{n\to\infty} |T_{n,p} h_n| \geq C_{8,p} > 0 \qquad (p\in\mathbb{N}). \tag{35}$$

Then for each ω _with_ (9,10) _there exists_ $f_\omega \in X$ _satisfying_ (34) _as well as_ (cf. (17,18))

$$|T_{n,p} f_\omega| \neq o_p(\omega(\varphi_n)), \tag{36}$$

$$|T_{n,p} f_\omega| \neq o_p(|V_n f_\omega|). \tag{37}$$

PROOF. As in [4] let the sequence $\{p_k\}\subset\mathbb{N}$ be given via $p_k = 2^l - m$, if $k = 2^l + m$, $0 \leq m < 2^l$ $(m, l \in \mathbb{N}\cup\{0\})$. It follows that for each $p\in\mathbb{N}$ there are infinitely many $k\in\mathbb{N}$ with $p_k = p$. Proceeding as in the proof of Theorem 6, construct sequences $\{n_k\}\subset\mathbb{N}$, $\{\delta_k\}\subset\{0,1\}$ satisfying (19,21,22) and

$$\omega(\varphi_{n_k}) \max\{1, \|T_{n_{k-1},p_{k-1}}\|_{X^*}, \|V_{n_{k-1}}\|_{X^*}\} \leq \omega(\varphi_{n_{k-1}})/k, \tag{20*}$$

$$\omega(\varphi_{n_k}) \leq (10/C_{8,p_k})\, |T_{n_k,p_k} g_k|. \tag{23*}$$

Then the element $f_\omega \in X$ (cf. (27)) satisfies (34) since for $\varphi_{n_{k+1}} < \sigma(t) \leq \varphi_{n_k}$ one has in view of (9,21,26,33) (cf. 28))

$$|S_t f_\omega| \leq C_7 \sum_{j=k+1}^{\infty} \omega(\varphi_{n_j}) + C_7\, \sigma(t) \sum_{j=1}^{k} \omega(\varphi_{n_j})/\varphi_{n_j}$$

$$\leq 2\,C_7\, \omega(\varphi_{n_{k+1}}) + 2\,C_7 \sigma(t)\omega(\varphi_{n_k})/\varphi_{n_k} \leq 4\,C_7\, \omega(\sigma(t)),$$

whereas for $\sigma(t) > \varphi_1$ one has by (9,26,33)

$$|S_t f_\omega| \leq 2\,C_7\, \omega(\varphi_1) \leq 2\,C_7 \omega(\sigma(t)).$$

Analogously to (29) one derives

$$|T_{n_k,p_k} f_\omega| \geq \omega(\varphi_{n_k})\ (C_{8,p_k}/10 - 2\,C_1/(k+1)),$$

which yields (36), and thus (37) in connection with (31). □

So far, nothing has been said about applications. In fact, it was the main purpose of our previous investigations [6-8;12] to stress the wide applicability of the present abstract approach, in particular of Theorem 6,10. This also includes (cf. [6;12]) a detailed discussion of a question raised by S.B. Steckin on the occasion of the conference on "Constructive Function Theory" at Blagoevgrad, Bulgaria, in 1977. In contrast to the more theoretical merits of Favard's suggestions in 1963, Steckin posed his problem in very concrete terms, concerned with the comparison of the modulus of continuity with the functional of best trigonometric approximation on the space $C_{2\pi}$. Steckin's problem was solved by Boman [1] whose concrete reasoning was the starting point of our general approach which in particular enabled us in the present note to work out the interconnection with the original contributions [9;10] of Favard.

REFERENCES

[1] Boman, J., A problem of Steckin on trigonometric approximation. In: Constructive Function Theory'77 (Proc. Conf. Blagoevgrad, 1977, Eds. Sendov, Bl. - Vacov, D.). Publ. House Bulg. Acad. Sci., Sofia 1980, 269-273.

[2] Dickmeis, W. - Nessel, R.J., On uniform boundedness principles and Banach-Steinhaus theorems with rates. Num. Funct. Anal. Optim. 3 (1981), 19-52.

[3] Dickmeis, W.-Nessel, R.J., Quantitative Prinzipien gleichmäßiger Beschränktheit und Schärfe von Fehlerabschätzungen. Forschungsberichte des Landes NRW, 3117, Westdeutscher Verl., Opladen 1982.

[4] Dickmeis, W. - Nessel, R.J., Condensation principles with rates. Studia Math. 75 (1982), 55-68.

[5] Dickmeis, W. - Nessel, R.J., On condensation principles with rates II. In: Constructive Function Theory'81 (Proc. Conf. Varna, 1981, Eds. Sendov, Bl. - et.al.). Publ. House Bulg. Acad. Sci., Sofia 1983, 275-278.

[6] Dickmeis, W. - Nessel, R.J. - van Wickeren, E., On the sharpness of estimates in terms of averages. Math. Nachr. (in print).

[7] Dickmeis, W. - Nessel, R.J. - van Wickeren, E., A general approach to quantitative negative results in approximation theory.(in print).

[8] Dickmeis, W. - Nessel, R.J. - van Wickeren, E., A general approach to counterexamples in numerical analysis. Numer. Math. (in print).

[9] Favard, J., Sur la comparaison des procédés de summation. In: On Approximation Theory (Proc. Conf. Oberwolfach, 1963, Eds. Butzer, P.L. - Korevaar, J., ISNM 5), Birkhäuser, Basel 1964, 4-11.

[10] Favard, J., On the comparison of the processes of summation. SIAM J. Numer. Anal. 1 (1964), 38-52.

[11] Kaczmarz, S. - Steinhaus, H., Theorie der Orthogonalreihen. Chelsea Publ., New York 1951.

[12] Nessel, R.J. - van Wickeren, E., Some negative results in connection with Marchaud-type inequalities. In: General Inequalities 4 (Proc. Conf. Oberwolfach, 1983). Birkhäuser, Basel (in print).

[13] Timan, A.F., Theory of Approximation of Functions of a Real Variable. Pergamon Press, New York 1963.

International Series of
Numerical Mathematics, Vol. 65

A REMARK ON QUANTITATIVE GLIDING HUMP METHODS

Werner Dickmeis
Lehrstuhl A für Mathematik
Rheinisch-Westfälische Technische Hochschule Aachen
Aachen

The usual gliding hump method (e.g., for the proof of the uniform boundedness principle) consists in a succesive selection of a subsequence $\{n_k\}$ of indices, to obtain certain resonance properties. In this note it is shown that under appropriate additional assumptions one may even select a geometrical subsequence, i.e., $n_k = M^k$ for some well-determined natural M. Thus, in constrast to a sucessive selection, this method has the advantages of a constructive approach.

Let X be a Banach space (with norm $\|\cdot\|_X$) and X* the class of sublinear (i.e, subadditive and positive homogeneous) functionals on X with finite (usual) operator norm $\|\cdot\|_{X^*}$. Let A and B be arbitrary index sets, $\sigma_y(t) = o(1)$, $t \to 0+$, be a positive function on [0,1] for each $y \in A$, and $\tau_n = o(1)$, $n \to \infty$, be a positvc scqucnce (with $n \in \mathbb{N}$, the set of natural numbers). Furthermore, ω denotes an abstract modulus of continuity, thus a continuous function with (cf. [9, p. 96ff])

$$0 = \omega(0) < \omega(t) \leq \omega(s), \qquad \omega(s)/s \leq \omega(t)/t \qquad (0 < t \leq s).$$

Additionally we suppose that $\omega(t) \neq O(t)$, thus $t/\omega(t) = o(1)$, $t \to 0+$.

In continuation of previous investigations on uniform boundedness and condensation principles with rates (see e.g. [3], [4],[6]), the following result was proved in [2].

THEOREM 1. <u>For</u> $S_{t,y}$, $T_{n,x} \in X^*$ <u>there exist constants</u> C <u>and elements</u> $g_n \in X$ <u>satisfying for each</u> $0 \leq t \leq 1$, $y \in A$, $n \in \mathbb{N}$, $x \in B$, $j < n$, <u>respectively</u>,

(1) $$\|g_n\|_X \leq C_1,$$

(2) $$|S_{t,y}g_n| \leqslant C_2 \min\{1,\sigma_y(t)/\tau_n\},$$

(3) $$\|T_{n,x}\|_{X^*} \leqslant C_{3,n},$$

(4) $$|T_{n,x}g_j| \leqslant C_{4,x}\, C_{5,j}\, \tau_n,$$

(5) $$|T_{n,x}g_n| \geqslant C_{6,x} > 0.$$

Then for each modulus ω there exists a subsequence $\{n_k\}_{k\in\mathbb{N}} \subset \mathbb{N}$ such that the element $f_\omega := \sum_{k=1}^{\infty} \omega(\tau_{n_k})g_{n_k}$ is well-defined in X and satisfies for each $y \in A$, and $x \in \mathcal{B}$

(6) $$|S_{t,y}f_\omega| = \mathcal{O}(\omega(\sigma_y(t))) \qquad (t \to 0+),$$

(7) $$\limsup_{n\to\infty} |T_{n,x}f| / \omega(\tau_n) \geqslant C_{6,x}.$$

This quantitative condensation principle is a simplified version of a resonance theorem, given in [2], which in particular includes the case of varying index sets $\mathcal{B}_n$ (instead of one fixed set $\mathcal{B}$). On the other hand, it may also be considered as an extended version of the condensation principle in [5], in as much as a rather arbitrary family of functionals $\{S_{t,y}\}$ (instead of a Peetre-K-functional) is admitted as a measure of smoothness. In fact, Theorem 1 shows by means of the counterexample f_ω, that for the process $\{T_{n,x}\}$ a (hypothetic) rate $\mathcal{O}_f(\omega(\tau_n))$ for all f satisfying the smoothness condition (6) cannot be improved to the rate $o_f(\omega(\tau_n))$.

The proof of Theorem 1 consists in a quantitative gliding hump method. One successively chooses indices n_k to construct the counterexample $f_\omega \in X$. Of course, it is the main purpose of such resonance theorems to establish the (pure) existence of counterexamples under rather general assumptions, thus - in so far as gliding hump methods are used - to establish the (pure) existence of such a subsequence $\{n_k\}$. Many concrete counterexamples in the literature, however, are based upon a geometrical subsequence $\{M^k\}_{k\in\mathbb{N}}$, which sometimes even allows more detailed information about the element f_ω. See e.g. [7, p. 55] concerning best approximation, [1] concerning the trapezoidal quadrature rule, or [8] concerning difference schemes.

It is the purpose of this note to show that under some restrictions concerning the admissible rates of convergence it is possible to obtain such (explicit) geometrical subsequences also in the framework of (abstract) resonance principles.

THEOREM 2. _Under the hypotheses of Theorem_ 1, _with_

$$\sigma_y(t) = t^{\alpha},\ \alpha > 0, \qquad \tau_n = n^{-\beta},\ \beta > 0,$$

(_independent of_ $y \in A$) and with constants satisfying

$$C_{3,n} \leqslant C_3, \qquad C_{4,x}\, C_{5,j} \leqslant C_4\, j^{\beta}, \qquad C_{6,x} \geqslant 3\, C_6 > 0$$

(_independent of_ $n \in \mathbb{N}$, $x \in B$, _respectively_), _for each modulus_ $\omega_\delta(t) = t^{\delta}$, $0 < \delta < 1$, _there exists a natural number_ M _such that for the geometrical subsequence_ $n_k = M^k$ $(k \in \mathbb{N})$ _the assertions of Theorem_ 1 _hold true_.

PROOF. Choose $M \in \mathbb{N}$ such that

$$M \geqslant \max\{(1 + C_1\, C_3/C_6)^{1/\beta\delta},\ (1 + C_5/C_6)^{1/\beta(1-\delta)}\}.$$

Then for each $k \in \mathbb{N}$

$$\sum_{j=k+1}^{\infty} M^{-\beta\delta j} = M^{-\beta\delta(k+1)}/(1 - M^{-\beta\delta}) \leqslant (C_6/C_1\, C_3)\, M^{-\beta\delta k},$$

$$\sum_{j=1}^{k-1} M^{\beta(1-\delta)j} = (M^{\beta(1-\delta)k} - 1)/(M^{\beta(1-\delta)} - 1) \leqslant (C_6/C_5)\, M^{\beta(1-\delta)k}.$$

Thus the element $f_\delta := f_{\omega_\delta} := \sum_{k=1}^{\infty} M^{-\beta\delta k}\, g_{M^k}$ is well-defined in X since X is complete.

For each $t \in (0, M^{-\alpha\beta}]$ one may find $k \in \mathbb{N}$ such that $M^{-\beta(k+1)} < t^{\alpha} \leqslant M^{-\beta k}$. Herewith one obtains (6) since

$$|S_{t,y} f_\omega| \leqslant C_2 \sum_{j=1}^{k-1} M^{\beta(1-\delta)j}\, t^{\alpha} + C_2\, M^{\beta(1-\delta)k}\, t^{\alpha} + C_2 \sum_{j=k+1}^{\infty} M^{-\beta\delta j}$$

$$\leqslant C_2\,[(C_6/C_5+1)\,M^{\beta(1-\delta)k}\,t^{\alpha}+(C_6/C_1C_3)\,M^{-\beta\delta k}]$$

$$\leqslant C_2[C_6/C_5+1+M^{\beta\delta}\,C_6/C_1C_3]t^{\alpha\delta}=\mathcal{O}(t^{\alpha\delta}).$$

On the other hand, for each $k\in\mathbb{N}$ one has

$$|T_{M^k,x}f_\delta|\geqslant M^{-\beta\delta k}\,|T_{M^k,x}g_{M^k}|-C_5\sum_{j=1}^{k-1}M^{\beta(1-\delta)j-\beta k}-C_1C_3\sum_{j=k+1}^{\infty}M^{-\beta\delta j}$$

$$\geqslant M^{-\beta\delta k}\{3C_6-C_6-C_6\},$$

which yields (7),completing the proof. □

As a first example consider the Fejér means on $X=C_{2\pi}$, the space of 2π-periodic, continuous functions on the real line $\mathbb{R}$ (with the usual sup-norm $\|\cdot\|_C$). Let $f^\wedge(k)=(1/2\pi)\int_{-\pi}^{\pi}f(u)\,e^{-iku}\,du$ denote the k-th Fourier coefficient of f. Then the Fejér means of f are defined by

$$F_{n-1}(f;x)=\sum_{|k|\leqslant n}(1-|k|/n)\,f^\wedge(k)\,e^{ikx}\qquad(x\in\mathbb{R}).$$

If f satisfies a Lipschitz condition of order δ, $0<\delta<1$, i.e.,

$$f\in\mathrm{Lip}\,\delta:=\{f\in C_{2\pi};\ \|f(u+t)-f(u)\|_C=\mathcal{O}_f(t^\delta)\},$$

then it is well-known that

$$|F_{n-1}(f;x)-f(x)|=\mathcal{O}_f(n^{-\delta})\qquad(n\to\infty)$$

uniformly for all $x\in\mathbb{R}$.

By the condensation principles given above, one may indeed show that this result is sharp, even in a pointwise sense:

CORLOLLARY 3. _For each_ $0<\delta<1$ _there exists a function_ $f_\delta\in\mathrm{Lip}\delta$, _e.g._,

$$f_\delta(x)=\sum_{k=1}^{\infty}M^{-k\delta}\,e^{iM^kx}\quad,\quad M\geqslant\max\{7^{1/\delta},\ 4^{1/(1-\delta)}\}$$

such that uniformly for each $x \in \mathbb{R}$

$$\limsup_{n\to\infty} n^{\delta} \, |F_{n-1}(f_{\delta};x) - f_{\delta}(x)| \geq 1/3.$$

PROOF. Consider the functionals ($x \in B = \mathbb{R}$, $A = \{1\}$)

$$S_{t,1}f := \|f(u+t) - f(u)\|_C \quad , \quad T_{n,x}f = |F_{n-1}(f;x) - f(x)| \,,$$

and the functions $g_n(x) = e^{inx}$. Then one has

$$S_{t,1}g_n = \|e^{inx}(e^{int} - 1)\|_C \leq 2 \min\{1,nt\},$$

$$T_{n,x}g_j - |e^{ijx}| \, (j/n) \qquad \text{for} \quad j \leq n.$$

Thus conditions (1)-(5) are satisfied with $C_1 = 1$, $C_2 = 2$, $\sigma_1(t) = t$, $\tau_n = 1/n$ (i.e., $\alpha = \beta = 1$), $C_3 = 2$, $C_4 = 1$, and $C_6 = 1/3$. Hence an application of Theorem 2 completes the proof. □

Of course, an approach via Theorem 1 would also yield the existence of counterexamples f_{ω} for each modulus $\omega(t) \neq O(t)$, but without any concrete representation since then the underlying subsequence $\{n_k\}$ is only implicitely known.

The author would like to express his sincere gratitude to Prof. R.J. Nessel and E. van Wickeren for a critical reading of the manuscript and many valuable suggestions.

REFERENCES

[1] Brass, H., Der Wertebereich des Trapezverfahrens. In: Hämmerlin, G., ed., Numerische Integration. (ISNM, vol. 45) Birkhäuser Verlag, Basel 1979, pp. 89-108.

[2] Dickmeis, W., Ein quantitatives Resonanzprinzip und Schärfe von punktweisen Fehlerabschätzungen fast überall (in print).

[3] Dickmeis, W. - Nessel, R.J., Quantitative Prinzipien gleichmäßiger Beschränktheit und Schärfe von Fehlerabschätzungen. (Forschungsberichte des Landes Nordrhein-Westfalen 3117) Westdeutscher Verlag, Opladen 1982.

[4] Dickmeis, W. - Nessel, R.J., Condensation principles with rates. Studia Math. 75 (1982), 55-68.

[5] Dickmeis, W. - Nessel, R.J., On condensation principles with rates II. In: Sendov, Bl. - et al., Constructive Function Theory '81. Publishing House of the Bulgarian Academy of Sciences, Sofia 1983, pp. 275-278.

[6] Dickmeis, W. - Nessel, R.J. - van Wickeren, E., On the sharpness of estimates in terms of averages. Math. Nachr. (in print).

[7] Feinermann, R.P. - Newman, D.J., Polynomial Approximation. William & Wilkins, Baltimore 1974.

[8] Hedstrom, G.W., The rate of convergence of some difference schemes. SIAM J. Numer. Anal. 5 (1968), 363-406.

[9] Timan, A.F., Theory of Approximation of Functions. Pergamon Press, New York 1963.

International Series of
Numerical Mathematics, Vol. 65
© 1984 Birkhäuser Verlag Basel

INTERPOLATION OF H^1 AND H^∞

Robert Sharpley
Department of Mathematics and Statistics
University of South Carolina
Columbia

The interpolation spaces for H^1 and H^∞ are characterized as the Hardy spaces of the interpolation spaces for L^1 and L^∞. This description is provided by recent work of Peter Jones on constructive solutions of $\bar{\partial}$ problems and of Brudnyǐ - Krugljak in the general theory of interpolation.

A Banach space X is called an interpolation space of a Banach couple (X_0,X_1) if each linear operator T, whose restriction to X_i is a bounded operator from X_i into itself $(i=0,1)$, is also a bounded operator on X. In [4] Calderón characterized the interpolation spaces of L^1 and L^∞ as the spaces X of measurable functions whose norms satisfy a rearrangement condition; specifically, there must exist a constant $c > 0$ so that

$$f \in X \text{ and } g \prec f \to \|g\|_X \leq c\,\|f\|_X . \tag{1}$$

Here $g \prec f$ (the Hardy-Littlewood-Polya preorder) means

$$\int_0^t g^*(s)ds \leq \int_0^t f^*(s)ds \quad , \text{ all } t > 0$$

and g^* denotes the decreasing rearrangement of $|g|$ [4]. We call such spaces X rearrangement-invariant function spaces. In this brief note we will provide a description of the interpolation spaces for the Banach couple of Hardy spaces (H^1,H^∞). For simplicity we work on R and the upper half plane R^2_+, but similar results hold for T and the disc. The Hardy space

Research supported in part by NSF Grant MCS-8301360.

$H(X)$ of a rearrangement-invariant function space X over R is defined to be the collection of functions f in X which have analytic extensions into the upper half plane and whose norm is given by

(2) $$||f||_{H(X)} = ||\ |f|\ ||_X \ .$$

The standard notation of H^p $(=H^p(R))$ will be used for $H(L^p)$. The main ingredients of the proof are identification of the K-functional for (H^1,H^∞) by Peter Jones [5], using rather deep constructive results for $\bar\partial$ problems, together with recent work in general interpolation theory by Brudnyǐ and Kruglјak [3].

THEOREM. _A necessary and sufficient condition for a space_ Y _to be an interpolation space for the pair_ (H^1,H^∞) _is that_ Y _be equal (with equivalent norms) to a Hardy space_ $H(X)$ _for some interpolation space_ X _of the pair_ (L^1,L^∞) _(i.e., for some rearrangement-invariant function space_ X_)._

Proof. From the proof of Theorem 3 in Jones [5] it follows immediately that the Peetre K-functional (see [3], [7] page 261) for the pair (H^1,H^∞) can be estimated by

(3) $$c_1\ K(f,t) \le \int_0^t (Nf)^*(s)ds \le c_2\ K(f,t) \quad ,\ \text{all } t > 0$$

for some fixed positive constants c_i $(i=1,2)$. Here Nf is the nontangential maximal function of f in R^2_+ ; i.e. if F is the harmonic extension of f into R^2_+ , then Nf is defined by

$$Nf\ (x) = \sup\{|F(t,y)| : (t,y) \in R^2_+,\ |x-t| \le y\} \ .$$

For our purposes, a slight improvement of (3) is required, namely

(4) $$c_1\ K(f,t) \le \int_0^t f^*(s)ds \le c_2\ K(f,t) \quad ,\ \text{all } t > 0 \ .$$

The right hand inequality follows immediately from (3) since $|f| \le Nf$ a.e. The inequality is evident directly as well since $\int_0^t f^*$ is a subadditive functional of f, $\int_0^t g^* \le ||g||_{L^1} = ||g||_{H^1}$ and $\int_0^t h^* \le t||h||_{L^\infty} = t||h||_{H^\infty}$.

Hence

$$\int_0^t f^*(s)ds \leq \inf_{f=g+h} \{||g||_{H^1} + t||h||_{H^\infty}\} = K(f,t) .$$

For the left hand inequality in (4), let F denote the analytic extension of f into the upper half plane. Factor F as BG^2 where B is a Blaschke product and G is a zero-free analytic function in R^2_+. Let g be the function on R of boundary values of G, then $Nf \leq (Ng)^2$. Hence

$$\int_0^t (Nf)^*(s)ds \leq \int_0^t (Ng)^*(s)^2ds$$

(5)

$$\leq c\int_0^t (Mg)^*(s)^2ds$$

since Ng is no larger than a constant multiple of the Hardy-Littlewood maximal function Mg (see page 197 of [8]). In addition, Herz's inequality (see, for example,[1]) states that $(Mg)^*(s) \leq \frac{5}{s} \int_0^s g^*(r)dr$, so using the specialized Hardy inequality

$$\int_0^t [\int_0^s g^*(r)dr/s]^2ds \leq 4 \int_0^t g^*(s)^2ds$$

(obtained from integration by parts), we obtain

$$\int_0^t (Mg)^*(s)^2ds \leq c \int_0^t g^*(s)^2ds.$$

Together with (5) and the fact that $|g|^2 = |f|$ a.e. this shows that for all $t > 0$

$$\int_0^t (Nf)^*(s)ds \leq c \int_0^t f^*(s)ds$$

for each f belonging to $H^1 + H^\infty$. This inequality together with Jones' inequality (3) establishes (4).

Suppose now that X is a rearrangement-invariant function space and $H(X)$ is its corresponding Hardy space. If T is a bounded linear operator on both H^1 and H^∞, then obviously

$$K(Tf,t) \leq A\ K(f,t) \qquad \text{all} \quad t > 0$$

where A is the maximum of the two operator norms. If the estimate (4) is applied to each side of the last inequality, then Calderón's result (1) shows that $H(X)$ is an interpolation space. Conversely, let Y be an interpolation space for the pair (H^1,H^∞). Jones has shown [6] that if both f and g belong to $H^1 + H^\infty$ and $g \prec f$, then there exists a linear operator T such that $Tf = g$ and T is bounded on both H^1 and H^∞. Applying Corollary 3 of [3], Y is equal to the space $K_\Phi(H^1,H^\infty)$ and

$$||f||_Y = \Phi(K(f,\cdot))$$

where Φ is the function norm of some interpolation space for the pair $(L^\infty,L^\infty_{1/t})$. It follows from (1) that $\Phi(\int_0^{(\cdot)} g^*(s)ds) =: ||g||_X$ is a rearrangement-invariant function norm. Hence by the estimates in (4),

$$c_1||f||_Y \leq ||f||_X \leq c_2\, ||f||_Y \quad .$$

This result may be rephased in terms of real Hardy spaces in an obvious way. Recall that ReH^p is the space of functions in L^p whose Hilbert transforms also belong to L^p. For $1 < p < \infty$, Riesz's theorem shows that ReH^p is L^p with an equivalent norm. In general, for a rearrangement-invariant function space X the real Hardy space of X is defined by

$$Re\ H(X) = \{f \in X : Hf \in X\}$$

with norm

$$||f||_{ReH(X)} = ||f||_X + ||Hf||_X \ .$$

Here Hf denotes the Hilbert transform of f.

COROLLARY. <u>The K-functional for the pair (ReH^1,ReH^∞) is equivalent within constants to the expression</u>

(6) $$\int_0^t f^*(s)ds + \int_0^t (Hf)^*(s)ds \quad .$$

<u>The collection of interpolation spaces for this pair are precisely the real</u>

Hardy spaces of the interpolation spaces of (L^1, L^∞). A necessary and sufficient condition for a rearrangement-invariant Banach function space X to be an interpolation space for $(\mathrm{Re}H^1, \mathrm{Re}H^\infty)$ is that the Boyd indices of X satisfy $0 < \beta_X \leq \alpha_X < 1$.

Proof. Only the last statement requires verification, but in view of (6), this is precisely the content of [2].

REFERENCES

[1] Bennett, C. and Sharpley, R., Weak type inequalities for H^p and BMO. Proc. Symp. Pure Math. 35.1 (1979), 201-229.

[2] Boyd, D. W., The Hilbert transform on rearrangement invariant spaces. Canad. J. Math. 19 (1967), 599-616.

[3] Brudnyĭ, Ju. A. and Krugljak, N. Ja., Real interpolation functors. Soviet Math. Dokl. 23 (1981), 5-8.

[4] Calderón, A. P., Spaces between L^1 and L^∞ and the theorem of Marcinkiewicz. Studia Math. 26 (1966), 273-299.

[5] Jones, P., L^∞ estimates for the $\bar{\partial}$ problem in a half-plane. to appear Acta Math.

[6] ________, personal communication.

[7] Kreĭn, S. G., Petunin, Ju.I. and Semenov, E. M., Interpolation of Linear Operators. (Translations Math. Monographs, Vol. 54) Amer. Math. Soc., Providence, R.I., 1982.

[8] Stein, E. M., Singular Integrals and Differentiability Properties of of Functions. Princeton University Press, Princeton, N.J., 1970.

International Series of
Numerical Mathematics, Vol. 65

n-WIDTHS FOR $\mathcal{C}_p^\alpha$ SPACES

R.A. DeVore and R.C. Sharpley
Department of Mathematics
University of South Carolina
Columbia

and

S.D. Riemenschneider
Department of Mathematics
University of Alberta
Edmonton

The smoothness spaces $\mathcal{C}_p^\alpha$, introduced by DeVore and Sharpley in [5], coincide with the Sobolev spaces W_p^α for integer α and $p > 1$. Considering them as a natural extension of the Sobolev spaces for fractional α and values of $p > 0$, we compute the n-widths $d_n(U(\mathcal{C}_p^\alpha), L_q)$ for $\alpha > N/p - N/q$, $0 < p \leq \infty$, $1 \leq q \leq +\infty$.

1. Introduction

In the last several years smoothness spaces, defined using in some way the L_p norm for $0 < p < 1$, have played an increasingly important role in approximation theory. For example, Brudnyi [2] described the functions that can be approximated in L_q by rational functions of degree n to the order $n^{-\alpha}$ in terms of Besov and Lipschitz spaces defined using $p = (\alpha + 1/q)^{-1}$. Likewise, spaces of generalized bounded variation V_p^λ, $0 < p \leq \infty$, arose naturally in the study of approximation by splines with free knots (see e.g. [1,2,3]). DeVore [4] used the Hardy-Littlewood maximal function as a mapping from L_1 to L_p, $1/2 < p < 1$, to give a short elementary proof of Popov's Theorem on rational approximation. This led DeVore to use other maximal functions, $f_{\alpha,p}^b$, and their associated spaces $\mathcal{C}_p^\alpha$ $(f_{\alpha,p}^b \in L_p)$ to study rational functions. All of these spaces are in some sense a natural replacement for Sobolev spaces when $0 < p < 1$. An advantage of the $\mathcal{C}_p^\alpha$ spaces is that the maximal function $f_{\alpha,p}^b$ behaves like a (fractional) derivative and is relatively easy to use.

There has also been a resurgence of research in the theory of n-widths for smoothness spaces due largely to the deep results of Kashin [10-14] and Gluskin [6-7]. If the unit ball $U(X)$ of some quasi-normed space X is compactly embedded in the Banach space Y, then the Kolmogorov n-width of

$U(X)$ is the quantity

$$(1.1)\qquad d_n(U(X),Y) = \inf_{Y_n} \sup_{x\varepsilon U(X)} \inf_{y\varepsilon Y_n} \|x-y\|_Y ,$$

where Y_n ranges over all possible n-dimensional subspaces of Y. The problem is to determine $d_n(U(X),Y)$ as a function of n, at least up to constants.

When $X = \ell_p^m$, $1 \le p \le \infty$, and $Y = \ell_q^m$, $1 \le q \le \infty$, the results of Kashin and Gluskin complete the asymptotic determination of $d_n(U(\ell_p^m),\ell_q^m)$ as a function of m,n. By discretization techniques dating back to Maiorov [16], the finite dimensional results permit the determination of the n-widths $d_n(U(W_p^\alpha),L_q)$ and $d_n(U(B_p^{\alpha,r}),L_q)$ for Sobolev and Besov spaces on $\Omega = [0,1]^N$ (see [8-15]). The orders of $d_n(U(W_p^\alpha),L_q)$ and $d_n(U(B_p^{\alpha,r}),L_q)$ are the same as those for $d_n(U(\mathscr{C}_p^\alpha),L_q)$, $1 \le p \le \infty$, given in Theorem 1. (Some of the upper estimates on $d_n(U(W_p^\alpha),L_q)$ appearing in the literature have an additional power of $\log n$ that, in light of Gluskin's subsequent results [7], can be removed for all but two values of α.) It is the purpose of the present note to extend these results to $0 < p < 1$ using $\mathscr{C}_p^\alpha$ spaces.

2. Spaces $\mathscr{C}_p^\alpha$ and their Widths

The spaces $\mathscr{C}_p^\alpha = \mathscr{C}_p^\alpha[0,1]^N$, $0 < p \le \infty$, $0 < \alpha$, are defined by means of certain maximal functions. For any cube $Q \subset \Omega = [0,1]^N$, and a function $f \in L_p(Q)$, let $P_Q f$ be any best approximant to f in the $L_p(Q)$ norm from the subspace $\mathbf{P}_{(\alpha)}$ of polynomials with total degree $\le (\alpha)$, where (α) is the greatest integer $< \alpha$. Define

$$(2.1)\qquad \begin{aligned} f^b_{\alpha,p}(x) &:= \sup_{\substack{Q\\ x\varepsilon Q\subset\Omega}} |Q|^{-\alpha/N}\Big(\frac{1}{|Q|}\int_Q |f-P_Q f|^p\Big)^{1/p} \\ &= \sup_{\substack{Q\\ x\varepsilon Q\subset\Omega}} \inf_{\pi\varepsilon\mathbf{P}_{(\alpha)}} |Q|^{-\alpha/N}\Big(\frac{1}{|Q|}\int_Q |f-\pi|^p\Big)^{1/p} .\end{aligned}$$

The space $\mathscr{C}_p^\alpha$ is the set of functions for which the quantity

(2.2) $$\|f\|_{\mathcal{C}_p^\alpha} := \|f\|_{L_p(\Omega)} + |f|_{\mathcal{C}_p^\alpha}, \quad |f|_{\mathcal{C}_p^\alpha} := \|f^b_{\alpha,p}\|_{L_p(\Omega)}$$

is finite. When $1 \le p \le \infty$, $\|f\|_{\mathcal{C}_p^\alpha}$ is a norm, but for $0 < p < 1$ it is only a quasi-norm.

The spaces $\mathcal{C}_p^\alpha$ are smoothness spaces in the sense that the statement "$f \in \mathcal{C}_p^\alpha$" implies smoothness or differentiability properties on f. Indeed, if $f \in L_p(\Omega)$ and $f^b_{\alpha,p}(x) < +\infty$, then the Peano derivative $D_\nu f$ exists at x for any multi-index ν with $|\nu| < \alpha$. Moreover, when $p > 1$, the weak derivatives of order ν, $|\nu| < \alpha$ exist, and we even have the relation

(2.3) $$W_p^\alpha = \mathcal{C}_p^\alpha \quad \text{if} \quad \alpha \in Z^+ \text{ and } 1 < p \le \infty.$$

When $p = 1$, the proper inclusion $\mathcal{C}_1^\alpha \subset W_1^\alpha$ holds for $\alpha \in Z^+$.

For non-integral α, the spaces $\mathcal{C}_p^\alpha$ are related to other smoothness spaces of fractional order. We need the embedding with Besov spaces

(2.4) $$B_p^{\alpha,p} \subset \mathcal{C}_p^\alpha \subset B_p^{\alpha,\infty}, \quad \alpha \text{ not an integer, } 0 < p \le \infty.$$

These embeddings are proper and unimprovable within the scale of Besov spaces.

Various spaces $\mathcal{C}_p^\alpha$ are related by the embeddings

(2.5) $$\mathcal{C}_p^\alpha \subset \mathcal{C}_q^\beta, \quad 0 < p \le q \le \infty \text{ and } 0 \le \beta \le \alpha - \frac{N}{p} + \frac{N}{q}.$$

(When $\alpha = 0$, $\mathcal{C}_p^\alpha := L_p$.) Of importance for n-widths is the compact embedding

(2.6) $$\mathcal{C}_p^\alpha \subset L_q, \quad \alpha > \frac{N}{p} - \frac{N}{q}, \quad 0 < p \le q \le \infty.$$

A detailed discussion of the spaces $\mathcal{C}_p^\alpha$, including the proofs of the above remarks, can be found in [5].

The above discussion indicates that the $\mathcal{C}_p^\alpha$ spaces form a natural framework for the extension of the n-width results for Sobolev spaces to the case $0 < p < 1$. To describe the results we divide the parameters (p,q)

into four regions (see figure 1):

I: $1 \leq q \leq p \leq \infty$; II: $0 < p \leq q \leq 2$, $1 \leq q$

III: $2 \leq p < q \leq \infty$; IV: $0 < p \leq 2 \leq q \leq \infty$.

THEOREM 1. *The asymptotic order of the n-widths* $d_n(U(\mathscr{C}_p^\alpha), L_q)$ *are as follows: For* $(p,q) \in I$,

$$n^{-\alpha/N}, \text{ if } \alpha > \frac{N}{p} - \frac{N}{q}; \tag{2.7}$$

for $(p,q) \in II$,

$$n^{-(\frac{\alpha}{N} - \frac{1}{p} + \frac{1}{q})}, \text{ if } \alpha > \frac{N}{p} - \frac{N}{q} \tag{2.8}$$

for $(p,q) \in III$,

$$\begin{aligned} &n^{-\frac{\alpha}{N}}, \text{ if } \alpha > (\tfrac{N}{p} - \tfrac{N}{q})/(1 - \tfrac{2}{q}) \text{ or} \\ &n^{-\frac{q}{2}(\frac{\alpha}{N} - \frac{1}{p} + \frac{1}{q})}, \text{ if } \frac{N}{p} - \frac{N}{q} < \alpha < (\tfrac{N}{p} - \tfrac{N}{q})/(1 - \tfrac{2}{q}); \end{aligned} \tag{2.9}$$

and for $(p,q) \in IV$,

$$\begin{aligned} &n^{-(\frac{\alpha}{N} - \frac{1}{p} + \frac{1}{2})}, \text{ if } \alpha > \frac{N}{p} \text{ or} \\ &n^{-\frac{q}{2}(\frac{\alpha}{N} - \frac{1}{p} + \frac{1}{q})}, \text{ if } \frac{N}{p} - \frac{N}{q} < \alpha < \frac{N}{p}. \end{aligned} \tag{2.10}$$

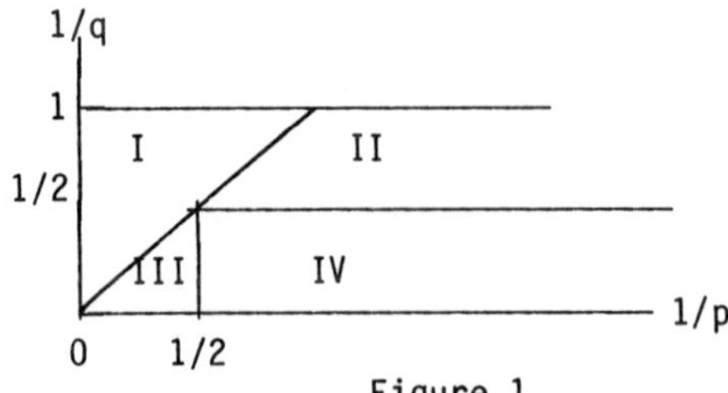

Figure 1

As mentioned in the introduction, the orders given in Theorem 1 hold for $d_n(U(B_p^{\alpha,r}),L_q)$ and $d_n(U(W_p^{\alpha}),L_q)$, when $1 \leq p,q \leq \infty$, $0 < r \leq \infty$. When $\alpha = (\frac{N}{p} - \frac{N}{q})/(1 - \frac{2}{q})$ for $(p,q) \in$ III and $\alpha = \frac{N}{p}$ for $(p,q) \in$ IV, the lower estimate for the n-widths has the order given in the theorem but the upper estimate contains an additional power of $\log n$. The closing of this gap is the remaining major problem.

When $1 \leq p,q \leq \infty$, Theorem 1 is an immediate consequence of the basic fact

$$(2.11) \qquad X \subset Z,\ W \subset Y \Rightarrow d_n(U(X),Y) \leq O(1)d_n(U(Z),W)$$

together with the embeddings (2.3), (2.4) and the known orders for $d_n(U(W_p^{\alpha}),L_q)$, $d_n(U(B_p^{\alpha,r}),L_q)$. Throughout the remainder of the paper $O(1)$ will denote a generic constant, independent of n, which may be different at each occurance.

3. Upper Estiamtes for $0 < p < 1$

When $0 < p < 1$, we can successfully use the embeddings (2.5) with the fact (2.11). For $(p,q) \in$ IV, $0 < p < 1$, we can transfer to the case $p \geq 1$ by the embedding $\mathscr{C}_p^{\alpha} \subset \mathscr{C}_2^{\beta}$ where $\beta = \alpha - N/p + N/2$. Furthermore, $\beta > N/2$ iff $\alpha > N/p$, and $N/2 - N/q < \beta < N/2$ iff $N/p - N/q < \alpha < N/p$. Substituting $\beta = \alpha - N/p + N/2$ into the upper estimates for $d_n(U(\mathscr{C}_2^{\beta}),L_q)$ and using (2.11), we obtain the order (2.10) for the upper estimate of $d_n(U(\mathscr{C}_p^{\alpha}),L_q)$.

Similarly, the estimate in region II can be reduced to the case $p = q$. Indeed, for $(p,q) \in$ II, we have $\mathscr{C}_p^{\alpha} \subset \mathscr{C}_q^{\beta}$, $\beta = \alpha - N/p + N/q$, and the upper estimate follows from the estimate for $d_n(U(\mathscr{C}_q^{\beta}),L_q)$.

If the order of $d_n(U(\mathscr{C}_p^{\alpha}),L_q)$ were known, then from the embeddings (2.3), (2.4) together with the fact that the Besov spaces, $B_p^{\alpha,r}$ are interpolation spaces between pairs $\mathscr{C}_p^{\alpha_0}$, $\mathscr{C}_p^{\alpha_1}$ (see [5]), it would be possible to determine the orders of $d_n(U(W_p^{\alpha}),L_q)$ and $d_n(U(B_p^{\alpha,r}),L_q)$. A direct proof of the upper estimates in Theorem 1 uses the standard discretization techniques and Lemma 1 below.

Let $\{Q_j\}$ be a partition of $\Omega = [0,1]^N$ into n equal cubes of volume 1/n. Let $\mathbf{PP}_n$ denote the space of all functions f such that $f|_{Q_j} \in \mathbf{P}_{(\alpha)}$. Clearly, dim $\mathbf{PP}_n = O(1)n$.

LEMMA 1. For $f \in \mathscr{C}_p^\alpha$, there is an $S \in \mathbf{PP}_n$ such that for any q with $0 < p \le q \le \infty$ and $\alpha > N/p - N/q$, we have

$$\text{(3.1)} \qquad \|f - S\|_{L_q(\Omega)} \le O(1)n^{-(\frac{\alpha}{N} - \frac{1}{p} + \frac{1}{q})}\|f\|_{\mathscr{C}_p^\alpha}.$$

We only sketch the proof. Select $S \in \mathbf{PP}_n$ so that $S|_{Q_j} = P_{Q_j}f$, and use the local weak type inequality relating the decreasing rearrangement $[(f-S)\chi_{Q_j}]^*$ and $f^b_{\alpha,p}$ in Lemma 4.2 of [5]. From this inequality and Hardy's inequality, we obtain $\|f - S\|_{L_q(Q_j)} \le O(1)|Q_j|^{\frac{\alpha}{N} - \frac{1}{p} + \frac{1}{q}}\|f^b_{\alpha,p}\|_{L_p(Q_j)}$. To pass to the norms on Ω, we use the fact that $\ell_p \subset \ell_q$ when $p \le q$.

4. Lower estimates

For the lower estimates, we follow the approach of Höllig [9]. The method requires the embedding of ℓ_p^m into $\mathscr{C}_p^\alpha$ through the coefficients of some spline expansion.

Let $M(x)$ be the tensor product of the one dimensional cardinal B-spline, normalized so that supp $M(x) = [0,1]^N = \Omega$. The degree of the cardinal B-spline is chosen so large that $M \in \mathscr{C}_p^\alpha(R^N)$.

Let $\{Q_j\}$ be the decomposition of $[0,1]^N$ into cubes of volume m^{-1}. Choose $x_j \in Q_j$ so that $Q_j - x_j = m^{-1/N}\Omega$, and define $M_j(x) = M(m^{1/N}(x-x_j))$. Then supp $M_j = Q_j$ and $\|M_j\|_{L_p(\Omega)} = m^{-1/p}\|M\|_{L_p(R^N)}$.

Let $S_m := \{S(x) = \sum c_j M_j(x) \colon \{c_j\} \in \ell_p^m\}$. Then dim $S_m = m$ and $S_m \subset \mathscr{C}_p^\alpha$. We need the estimates

$$\text{(4.1)} \qquad \|S\|_{L_r(\Omega)} \le O(1)m^{-1/r}\|\{c_j\}\|_{\ell_r^m}, \quad 0 < r \le \infty, \; S \in S_m$$

(4.2) $$\|\{c_j\}\|_{\ell_r^m} \le O(1)m^{1/r}\|S\|_{L_r(\Omega)}\ , \quad 0 < r \le \infty,\ S \in \mathcal{S}_m$$

and

(4.3) $$\|S\|_{\mathcal{C}_p^\alpha} \le O(1)m^{\frac{\alpha}{N} - \frac{1}{p}}\|\{c_j\}\|_{\ell_p^m}\ , \quad 0 < p \le \infty,\ \alpha > 0.$$

Inequalities (4.1) and (4.2) are straightforward (the second uses the fact that $M_j^r(x) \ge C^r > 0$ on a cube $Q_j^* \subset Q_j$ with $2|Q_j^*| = |Q_j|$). Before proving (4.3), we use it to derive the lower bounds.

Assume $0 < p < 1$ and $1 \le q \le \infty$. Let P be a bounded projection from $L_q(\Omega)$ onto $\mathcal{S}_m \cap L_q(\Omega)$, for example, $Pf(x) = \sum_j a_j(\int_{Q_j} f)M_j(x)$, where $a_j = (\int_{Q_j} M_j(x)dx)^{-1}$. Then

(4.4) $$d_n(U(\mathcal{C}_p^\alpha) \cap \mathcal{S}_m, L_q \cap \mathcal{S}_m) \le \|P\| d_n(U(\mathcal{C}_p^\alpha) \cap \mathcal{S}_m, L_q) \le O(1)d_n(U(\mathcal{C}_p^\alpha), L_q).$$

Factoring the identity map $I: \ell_p^m \to \ell_q^m$ as

$$I: \ell_p^m \xrightarrow{T_1} \mathcal{C}_p^\alpha \cap \mathcal{S}_m \xrightarrow{J} L_q \cap \mathcal{S}_m \xrightarrow{T_2} \ell_q^m$$

where J is the embedding operator, we obtain by (4.2)-(4.4)

(4.5) $$\begin{aligned} d_n(U(\ell_p^m), \ell_q^m) &\le \|T_1\|\|T_2\| d_n(U(\mathcal{C}_p^\alpha) \cap \mathcal{S}_m, L_q \cap \mathcal{S}_m) \\ &\le O(1)m^{\frac{\alpha}{N} - \frac{1}{p} + \frac{1}{q}} d_n(U(\mathcal{C}_p^\alpha), L_q)\ . \end{aligned}$$

In [7], Gluskin gave the lower bounds,

(4.6) $$d_n(U(\ell_p^m), \ell_q^m) \ge O(1) \begin{cases} m^{1/q - 1/p} & , (p,q) \in II \\ m^{1/q - 1/2}(\frac{m}{n} - 1)^{1/2} & , (p,q) \in IV \end{cases}$$

valid for $n^{q/2} \ge m \ge 2n$. We take $m = 2n$ in (4.5) and (4.6) to obtain

(2.8) and the first case in (2.10), and $m \sim n^{q/2}$ to obtain the second case in (2.10). It should be noted that Gluskin does not state (4.6) for $0 < p < 1$, but his proof does give this case.

It remains to prove (4.3). Let $0 < p < 1$. We have to estimate $|S|_{\mathscr{C}_p^\alpha}$, $S = \sum c_j M_j$, in terms of $\|\{c_j\}\|_{\ell_p^m}$. Let x and Q be given with $x \in Q \subset \Omega$. Set $Q_j^* := \{m^{1/N}(x-x_j) : x \in Q\}$. Then $|Q_j^*| = m|Q_j|$. Choose $P_j^* \in \mathbf{P}_{(\alpha)}$ to be a best approximation in $L_p(Q_j^*)$ to $M(x)$. Then $P_j(x) = P_j^*(m^{1/N}(x-x_j))$, $x \in Q$, is a best approximation in $L_p(Q)$ to $M_j(x)$. Let $P = \sum_j c_j P_j$. Then

$$\begin{aligned}\int_Q |S-P|^p &\le \sum_j |c_j|^p \int_Q |M_j - P_j|^p = m^{-1}\sum_j |c_j|^p \int_{Q_j^*} |M - P_j^*|^p \\ &\le m^{\alpha p/N}|Q|^{\alpha p/N+1}\sum_j |c_j|^p |Q_j^*|^{-\alpha p/N-1}\int_{Q_j^*} |M - P_j^*|^p \\ &\le m^{\alpha p/N}|Q|^{\alpha p/N+1}\sum_j |c_j|^p (\inf_{Q_j^*} M_{\alpha,p}^b)^p \\ &\le m^{\alpha p/N}|Q|^{\alpha p/N+1}\sum_j |c_j|^p (M_{\alpha,p}^b(m^{1/N}(x-x_j)))^p.\end{aligned}$$

Therefore,

$$S_{\alpha,p}^b(x) \le m^{\alpha/N}\{\sum |c_j|^p (M_{\alpha,p}^b(m^{1/N}(x-x_j)))^p\}^{1/p}, \tag{4.7}$$

and

$$\begin{aligned}|S|_{\mathscr{C}_p^\alpha} = \|S_{\alpha,p}^b\|_{L_p(\Omega)} &\le m^{\alpha/N}\{\sum_j |c_j|^p \|M_{\alpha,p}^b(m^{1/N}(\cdot - x_j))\|_{L_p(\mathbf{R}^N)}^p\}^{1/p} \\ &\le O(1) m^{\alpha/N-1/p}\{\sum_j |c_j|^p\}^{1/p}.\end{aligned}$$

Since the proof of (4.3) for $1 \le p \le \infty$ is more technical and this case is not needed here, we give only a brief sketch of the argument. As above we arrive at (4.7) with $p = 1$; which is all that is required since $\|S_{\alpha,1}^b\|_{L_p(\Omega)} \sim \|S_{\alpha,p}^b\|_{L_p(\Omega)}$ (see [5]). However, since the supports of $M_{\alpha,1}^b(m^{1/N}(x-x_j))$ overlap significantly we cannot take the power and integration through the sum and pull out a common $\|M^b_{\alpha,1}\|_{L_p(R^N)}$ factor. The trick is to look at the integral over a fixed cube Q_{j_0}, and to separate the sum into the cubes touching Q_{j_0} (good terms) and the rest (bad terms). For the small number of cubes touching Q_{j_0} we can take the power and integration through the sum at the expense of a constant. For the remaining Q_j, we use the fact that the maximal function $M^b_{\alpha,1}(u)$ can be estimated in terms of the dist(u, supp M) to obtain $M^b_{\alpha,1}(m^{1/N}(x-x_j)) \leq O(1)[m\ \mathrm{dist}(Q_j,Q_{j_0})]^{-\alpha-N}$ for $x \in Q_{j_0}$. After summing over j_0, the resulting estimate for the bad terms can be viewed as the convolution of $\{c_j\}$ with a ℓ_1-kernel.

REFERENCES

[1] Bergh, J. - Peetre, J., On the spaces V_p $(0<p<\infty)$. Bolletino U.M.I. (4) 10(1974), 632-648.

[2] Brudnyi, Yu. A., Rational approximation and exotic Lipschitz spaces. In Quantitative Approximation, R.A. DeVore and K. Scherer editors, Academic Press, New York 1980, 25-30.

[3] Burchard, H.G. - Hale, D.F., Piecewise polynomial approximation on optimal meshes. J. Approximation Theory 14(1975), 128-147.

[4] DeVore, R.A., Maximal functions and their application to rational approximation. In Second Edmonton Conference on Approximation Theory, Z. Ditzian et al editors, CMS Conference Proceedings vol. 3, American Mathematical Society, Providence 1983, 143-155.

[5] DeVore, R.A - Sharpley, R.C., Maximal functions measuring smoothness. Mem. Amer. Math. Soc., to appear.

[6] Gluskin, E.D., On some finite-dimensional problems in the theory of widths. (Russian) Vestnik Leningrad. Univ. Mat. Mekh. Astronom. (1981), 5-10.

[7] Gluskin, E.D., Norms of stochastic matrices and widths of finite dimensional sets. (Russian) Mat. Sbornik 120 (162) (1983), 180-189.

[8] Höllig, K., Approximationzahlen von Sobolev-Einbettungen. Math. Ann. 242(1979), 273-281.

[9] Höllig, K., Diameters of classes of smooth functions. In Quantitative Approximation, R.A. DeVore and K. Scherer editors, Academic Press, New York 1980, 163-175.

[10] Kashin, B.S., The diameters of certain finite dimensional sets and classes of smooth functions. (Russian) Izv. Akad. Nauk SSSR, Ser. Mat., 41(1977), 334-351.

[11] Kashin, B.S., On estimates of diameters. In Quantitative Approximation, R.A. DeVore and K. Scherer editors, Academic Press, New York 1980, 177-184.

[12] Kashin, B.S., Diameters of Sobolev classes with low order of smoothness. (Russian) Vestnik Moskov. Univ. Ser. I Mat. Meh. 36(1981), 50-54.

[13] Kashin, B.S., On some properties of matrices as bounded operators from ℓ_2^n into ℓ_2^m. (Russian) Akad. Nauk Armjan. SSR Dokl. 15(1980), 379–394.

[14] Kulanin, E.D., Bounds on widths of Sobolev classes with low smoothness. (Russian) Vestnik Moskov. Univ. Ser. I Mat. Meh. 38(1983), 24-30.

[15] Lubitz, C., Weylzahlen von Diagonaloperatoren und Sobolev-Einbettungen. Bonner Mathematische Schriften Nr. 144, Bonn 1982.

[16] Maiorov, V.E., Discretization of the problem of diameters. (Russian) Uspekhi Mat. Nauk 30(1975), 179-180.

International Series of
Numerical Mathematics, Vol. 65

THE N-WIDTH OF BV ∩ Lip α

Ronald A. DeVore
Department of Mathematics and Statistics
University of South Carolina
Columbia

We determine asymptotically the n-width of the space of functions BV ∩ Lip α as a subset of C. Since the n-width behaves like $n^{-(\alpha+1)/2}$, it is seen that this class of functions has significantly smaller n-width than Lip α.

There are many theorems of analysis which show that the space $S_\alpha := BV \cap \text{Lip}\,\alpha$ is small when compared to Lip α. For example, we know that each $f \in S_\alpha$ has an absolutely convergent Fourier series if $\alpha > 0$ but this is not the case for Lip α unless $\alpha > \frac{1}{2}$. In this note, we shall measure the size of S_α by computing the n-width of the unit ball of S_α in the space $C_{2\pi}$ of 2π periodic continuous functions.

One way to measure the size of a compact set K of a Banach space X is through the notion of n-widths. If X_n is an n-dimensional subspace of X then

$$E(f,X_n) := \inf_{x_n \in X_n} \|f-x_n\|_X$$

is the error in approximating f by elements of X_n and

$$E(K,X_n) := \sup_{f \in K} E(f,X_n)$$

is the error in approximating the compact set K. The n-width of K looks for best n-dimensional subspaces:

Research supported in part by NSF Grant MCS-8101661

(1) $$d_n(K)_X := \inf_{X_n} E(K,X_n) .$$

The asymptotic behavior of $d_n(K)_X$ gives some idea of the size of K in X. For example if $C_{2\pi}$ denotes the space of continuous 2π periodic functions, then the unit ball $U(\text{Lip }\alpha)$ of $\text{Lip }\alpha$ has n-width

(2) $$d_n(U(\text{Lip }\alpha))_{C_{2\pi}} \approx n^{-\alpha}, \quad n \to \infty .$$

Actually, there is a much larger set than $U(\text{Lip }\alpha)$ which also has n-width $n^{-\alpha}$. It was shown by B. S. Kashin [1] that the Sobolev space W_2^α satisfies:

$$d_n(U(W_2^\alpha))_{C_{2\pi}} \approx n^{-\alpha}, \quad \alpha > \tfrac{1}{2} .$$

Since $W_2^\alpha = B_2^{\alpha,2}$ (the Besov space), it follows from an interpolation of linear operators argument that

(3) $$d_n(U(B_2^{\alpha,r})) \approx n^{-\alpha}, \quad \alpha > \tfrac{1}{2}; \quad r > 0 .$$

Since $\text{Lip }\alpha \underset{\neq}{\subset} B_2^{\alpha,r}$, (3) is a (substantial) improvement over (2).

The main result of this note is to determine the n-width of the set $K_\alpha := \{f \in S_\alpha : \max(\|f\|_{BV}, \|f\|_{\text{Lip }\alpha}) \le 1\}$ in $C_{2\pi}$. The following result should be compared with (2).

Theorem 1. *For each* $0 < \alpha \le 1$, *there are constants* $C_1, C_2 > 0$ *such that*

(4) $$C_1 n^{-(\alpha+1)/2} \le d_n(K_\alpha)_{C_{2\pi}} \le C_2 n^{-(\alpha+1)/2} .$$

Proof. The upper estimate follows from an embedding of K_α into a Besov space. If $f \in K_\alpha$ and $h > 0$, we have

(5) $$\int_0^{2\pi} |f(x+h) - f(x)|^2 dx \le h^\alpha \int_0^{2\pi} |f(x+h) - f(x)| \, dx \le h^{\alpha+1}$$

The last inequality is apparent for absolutely continuous functions and follows for any f in K_α since such an f is the uniform limit of absolutely continuous functions from K_α.

It follows from (5) that $K_\alpha \subset U(B_2^{\beta,\infty})$, $\beta := (\alpha+1)/2$ and so from (3),

$$d_n(K_\alpha) \leq d_n(U(B_2^{\beta,\infty})) \leq C\, n^{-\beta} = n^{-(\alpha+1)/2}$$

which is the right hand side of (4).

To prove the lower estimate in (4), we shall use an idea of W. Rudin [2] and construct sufficiently many functions in S_α which are mutually orthogonal. In part, the orthogonality will come from making supports disjoint and in part, by using Latin squares to guarantee orthogonality where supports are not disjoint.

It is enough to prove (4) when n is a 2 power. Let $N := 2n$ and choose m as the largest 2 power such that

$$2m \leq (2N/\pi)^\alpha$$

It follows that

$$m \geq \tfrac{1}{4}(2N/\pi)^\alpha \tag{6}$$

Also $m \leq N$.

Now divide $[0,2\pi]$ into N/m intervals $I_\nu := [x_\nu, x_{\nu+1}]$ $\nu = 0,1,\ldots,(N/m)-1$: $x_\nu := 2\nu\pi\, m/N$. For each ν we choose points $x_{\mu\nu} := x_\nu + (2\mu-1)\pi/N$, $\mu = 1,\ldots, m$ from I_ν.

We will also need a Latin square

$$A := (a_{ij}) \qquad a_{ij} = \pm 1$$

of dimension $m \times m$. Let $v_1,\ldots,v_m$ denote the columns of A. Then, these vectors are orthogonal.

Here is how we construct our orthogonal functions. If $1 \leq i \leq N$, we write

$$i = km + \ell;\ 1 \le \ell \le m\ .$$

Define g_i as the piecewise linear function

$$g_i : = \sum_{\mu=1}^{m} a_{\mu\ell}\ \psi_{\mu k}$$

with $\psi_{\mu k}$ the "hat function " which is supported on the interval $[x_{\mu k} - \pi/N,\ x_{\mu k} + \pi/N]$ has value zero at the end points of this interval and 1 at the point $x_{\mu k}$.

It is easy to see that the g_i are orthogonal. Indeed, suppose that $i_1 = k_1 m + \ell_1$ and $i_2 = k_2 m + \ell_2$. If $k_1 \neq k_2$ then the g_i are orthogonal because they have disjoint supports. If $k_1 = k_2$, then

$$\int_0^{2\pi} g_{i_1} g_{i_2} = \frac{\pi}{6N}\, v_{\ell_1} \cdot v_{\ell_2} = 0$$

because the columns of A are orthogonal.

The functions g_i have L_2 norm $(\frac{2\pi m}{3N})^{\frac{1}{2}}$. Therefore, the normalized functions

$$f_i : = (\frac{3N}{2\pi m})^{\frac{1}{2}}\, g_i$$

form an orthonormal system.

Now suppose X_n is any n-dimensional subspace of $C_{2\pi}$ and let $\phi_1,\ldots,\phi_n$ be an orthonormal basis for X_n. We extend this orthonormal system by adjoining functions $\phi_{n+1}, \ldots$ to arrive at a complete orthonormal system. Now consider the matrix $B: = (b_{ij})$ with $b_{ij}: = (f_i,\phi_j)^2$ The sum of the entries in a given row of B is one and the sum of the entries in a given column is at most one. Hence

$$\sum_{i=1}^{N} E(f_i,X_n)^2 = \sum_{i=1}^{N} \sum_{j=n+1}^{\infty} (f_i,\phi_j)^2 = N - \sum_{i=1}^{N} \sum_{j=1}^{n} (f_i,\phi_j)^2$$

$$\geq N - n = n.$$

As a result

$$(7) \qquad \max_{1 \leq i \leq N} E(f_i, X_n)^2 \geq \tfrac{1}{2}$$

It is easy to check that $\|g_i\|_{BV} = 2m$ and $\|g_i\|_{\mathrm{Lip}\,\alpha} = (\frac{N}{\pi})^{\alpha}$.
Hence the function $\tilde{f}_i := (\frac{\pi}{N})^{\alpha} g_i = (\frac{\pi}{N})^{\alpha} (\frac{2\pi m}{3N})^{\frac{1}{2}} f_i$ satisfy

$$(8) \qquad \|\tilde{f}_i\|_{BV} \leq \|\tilde{f}_i\|_{\mathrm{Lip}\,\alpha} = 1$$

so that $\tilde{f}_i \in K_{\alpha}$

According to (7) and (8)

$$\sqrt{2\pi}\, E(K_{\alpha}, X_n)_{C_{2\pi}} \geq E(K_{\alpha}, X_n)_{L_2} \geq \max_{1 \leq i \leq n} E(\tilde{f}_i, X_n)_{L_2} \geq (\frac{1}{\sqrt{2}})(\frac{\pi}{N})^{\alpha}(\frac{2\pi m}{3N})^{\frac{1}{2}}$$

$$\geq C\, N^{-\alpha-\frac{1}{2}}\, m^{\frac{1}{2}} \geq C\, N^{-(\alpha+1)/2} \geq C\, n^{-(\alpha+1)/2}.$$

Since X_n is arbitrary this gives the lower estimate in (4).

The approach above extends readily to the more general compact set

$$(9) \qquad K_{\alpha,\beta} := \{f : |f|_{B_p^{\alpha,r}} \leq 1 \ ; \ |f|_{B_q^{\beta,s}} \leq 1\} \ ; \ 1/p + 1/q = 1 \ ; \ p \leq q$$

with α, β, r, $s > 0$ and $|\cdot|_{B_p^{\alpha,r}}$, the usual Besov space semi-norms.
The space $B_p^{\alpha,r} \hookrightarrow B_q^{\beta,s}$ if $\alpha > \beta + 1/p - 1/q$ and $B_q^{\beta,s} \hookrightarrow B_p^{\alpha,r}$ if $\beta \geq \alpha$.
Thus, the only interesting cases are when

$$\beta < \alpha \leq \beta + 1/p - 1/q .$$

Theorem 2. *If* $K_{\alpha,\beta}$ *is defined as in* (9) *with* $1 \le p \le 2$, *and* $\beta < \alpha \le \beta + 1/p - 1/q$, *then there are constants* $C_1, C_2 > 0$ *such that*

$$C_1 n^{-(\alpha+\beta)/2} \le d_n(K_{\alpha,\beta})_{C_{2\pi}} \le C_2 n^{-(\alpha+\beta)/2} \tag{10}$$

Proof. The proof is essentially the same as Theorem 1 and so we only indicate the necessary changes. Let Δ_h^k denote the k-th difference with step size h. If $f \in K_{\alpha,\beta}$ and $k := [\alpha+\beta] + 1$, then

$$\int_0^{2\pi} |\Delta_h^k(f,x)|^2 dx \le (\int_0^{2\pi} |\Delta_h^k(f,x)|^p dx)^{1/p} (\int_0^{2\pi} |\Delta_h^k(f,x)|^q dx)^{1/q}$$

$$\le C_0 h^\alpha ||f||_{B_p^{\alpha,r}} h^\beta ||f||_{B_q^{\beta,s}} \le C_0 h^{\alpha+\beta}$$

with C_0 depending only on k. Hence $K_{\alpha,\beta}$ is contained in a ball of radius $\sqrt{C_0}$ in $B_2^{\alpha+\beta,\infty}$ with C_0 depending only on k. It then follows from (3) that

$$d_n(K_{\alpha,\beta}))_{C_{2\pi}} \le \sqrt{C_0}\, d_n(U(B_2^{\alpha,\infty}))_{C_{2\pi}} \le C n^{-(\alpha+\beta)/2}$$

which is the upper estimate in (10)

For the lower estimate, we argue similarly to Theorem 1 except that we need smoother functions f_i. Let ϕ be a non-negative function in C^∞ which is supported on $[-\frac{1}{2},\frac{1}{2}]$ with

$$\int_{-\infty}^{\infty} \phi^2 = 1 \; ; \; ||\phi||_\infty \le 2$$

We suppose as before that n is a 2 power, $N := 2n$ and we define m as the largest 2 power such that $m \le N^{\gamma+1}$ with $\gamma := (\beta-\alpha)/(1/p-1/q)$. Note that $-1 \le \gamma < 0$. The points $x_{\mu\nu}$ are defined as in Theorem 1. Finally we define $h_0 := 2\pi/N$.

The function $\phi_{\mu\nu} := m^{-\frac{1}{2}} h_0^{-\frac{1}{2}} \phi(\frac{(x-x_{\mu\nu})}{h_0})$ is supported on $[x_{\mu\nu} - \frac{1}{2}h_0, x_{\mu\nu} + \frac{1}{2}h_0]$. If $1 \le i \le N$, we write $i = km + \ell$ with

$1 \leq \ell \leq m$ and define

$$f_i := \sum_{\mu=1}^{m} a_{\mu\ell}\, \phi_{\mu k}$$

where $a_{\mu\nu}$ are the entries of the Latin square A.

It follows that the f_i are an orthogonal system and therefore as in (7) for any n-dimensional space X_n

$$\max_{1 \leq i \leq N} E(f_i, X_n)^2_{C_{2\pi}} \geq \tfrac{1}{2} . \tag{11}$$

We now check the norm of f_i in $B_p^{\alpha,r}$. Let $k > \alpha$. Now, $\| \phi_{\nu\mu}^{(k)} \|_\infty \leq Cm^{-\frac{1}{2}} h_0^{-k-\frac{1}{2}}$ and since the $\phi_{\mu\nu}$ have disjoint supports and $|a_{\mu\nu}| = 1$,

$$|\Delta_h^k(f_i, x)| \leq C \max\, (m^{-\frac{1}{2}} h_0^{-\frac{1}{2}} (h/h_0)^k, 1) \qquad \text{on support of } f_i.$$

Hence,

$$\| \Delta_h^k(f_i) \|_p \leq C \max\, (m^{-\frac{1}{2}} h_0^{-\frac{1}{2}} (h/h_0)^k, 1)\, (m/N)^{1/p}$$

It follows that the modulus of smoothness $\omega_k(f_i)_p$ satisfies

$$\omega_k(f_i, h)_p \leq C(m/N)^{1/p} \begin{cases} m^{-\frac{1}{2}} h_0^{-\frac{1}{2}} (h/h_0)^k, & h \leq h_0 \\ 1 \qquad , & h \geq h_0 \end{cases}$$

Hence,

$$|f_i|_{B_p^{\alpha,r}} = \left(\int_0^\infty (h^{-\alpha} \omega_k(f_i, h)_p^r \frac{dh}{h}\right)^{1/r} \leq C\, (m/N)^{1/p}\, [m^{-\frac{1}{2}} h_0^{-\frac{1}{2}} h_0^{-\alpha} + h_0^{-\alpha}]$$

$$\leq C\, (m/N)^{1/p}\, m^{-\frac{1}{2}}\, h_0^{-\frac{1}{2}}\, h_0^{-\alpha}$$

$$\leq CN^{\gamma(1/p-\frac{1}{2})}\, N^{\alpha} \leq CN^{(\alpha+\beta)/2}$$

The same computation shows that $|f_i|_{B_p^{\beta,s}} \leq C\, n^{(\alpha+\beta)/2}$.

The functions $g_i := C^{-1}\, n^{-(\alpha+\beta)/2}\, f_i$ are in $K_{\alpha,\beta}$ and according to (11) give the lower estimate in (10). □

Acknowledgement: During the visit of Professor G. G. Lorentz to Columbia, S.C., there were preliminary discussions concerning approximation theorems for BV $\cap$ Lip α. We thank Professor Lorentz for these initial discussions.

REFERENCES

[1] B. S. Kashin, The widths of certain finite dimensional sets and classes of smooth functions, Isv. Akad. Nauk SSSR Ser. Mat. 41 (1977), 334-351 (Russian).

[2] W. Rudin, L_2 approximation by partial sums of orthogonal developments, Duke Math. J. 19 (1952), 1-4.

IV Approximation in Real Domains

International Series of
Numerical Mathematics, Vol. 65

THE NECESSITY OF A NEW KIND OF MODULUS OF SMOOTHNESS

Vilmos Totik
Bolyai Institute
Attila József University
Szeged

A new type of modulus of smoothness is defined. It very much resembles the ordinary moduli of smoothness, only the increment in it varies togehter with the variable x. This new measurement provides a natural frame-work for the solution of several approximation-theoretical problems. Applications are given concerning best polynomial approximation and many aspects of operator approximation.

1. Introduction

Let (a,b) be a finite or infinite interval,

$$\Delta_h^1 f(x) = f(x+h) - f(x-h) \quad , \quad \Delta_h^r f(x) = \Delta_h^{r-1} f(x+h) - \Delta_h^{r-1} f(x-h)$$

the r-th symmetric difference of f, $1 \leq p \leq \infty$, and for the sake of brevity let L^∞ denote the Banach space of the set of bounded and continuous functions equipped with the supremum norm. The higher order modulus of smoothness mentioned in the title has the form

$$\omega_\varphi^r(f,\delta)_p := \sup_{h\leq\delta} \|\Delta_{h\varphi}^r f\|_{L^p} + \text{something depending on } \varphi$$

where φ is a weight function, and the additional term depends on the behaviour of φ around the endpoints of the interval (a,b) (if $x \pm rh\,\varphi(x) \notin (a,b)$, then let $\Delta_{h\varphi(x)}^r f(x) = 0$). This was found - with a slight modifiction - by Z. Ditzian [4] and the author (see e.g. [15]) independently. The main term is

$$\sup_{h\leq\delta} \|\Delta_{h\varphi}^r f\|_{L^p},$$

and the point in it is that the increment $h\varphi(x)$ varies together with x. We do not describe how the additional term is defined if φ is given, in subsesequent sections we shall see several concrete examples with fully defined ω_φ^r. If $\varphi\equiv 1$, then we obtain the ordinary higher order modulus of smoothness.

In trigonometric approximation every point on the real line (or on the one-dimensional torus) behaves similarly, and that is the reason why ordinary moduli of smoothness are so efficient in trigonometric approximation theory. However, if we work on an interval and we are interested in global results then this homogeneity disappears and, clearly, the position of the points inside the interval must somehow be reflected in the measurement of the "goodness" of the functions, e.g., as is done in the case of ω_φ^r. It has turned out that several problems in approximation theory that resisted being solved in terms of ordinary (weighted) moduli of smoothness can be answered by the aid of our new modulus. The aim of the article is to report on some recent results that show the wide applicability of ω_φ^r and some of its properties.

2. Best Approximation by Polynomials on Finite Intervals

In trigonometric approximation perhaps the two most important estimates are Jackson's theorem

$$E_n[f]_{L^p_{2\pi}} \leqslant K\, \omega^r(f,\tfrac{1}{n})_{L^p_{2\pi}} \tag{2.1}$$

and the opposite inequality of de la Vallée Poussin and S.B. Stechkin

$$\omega^r(f,\tfrac{1}{n})_{L^p_{2\pi}} \leqslant K n^{-r} \sum_{k=o}^{n} (k+1)^{r-1}\, E_k[f]_{L^p_{2\pi}}, \tag{2.2}$$

where $E_k[f]_{L^p_{2\pi}}$ denotes the best approximation of $f\in L^p_{2\pi}$ by trigonometric polynomials of order at most k, and $\omega^r(f,\delta)=\sup_{h\leqslant\delta}\|\Delta_h^r f\|_{L^p_{2\pi}}$ is the r-th modulus of smoothness of f. It is a fascinating and important problem how the analogues of (2.1) and (2.2) look like in polynomial approximation, especially if we take into account that, according to the celebrated results of Timan, Dzjadyk, Gopengauz and Teljakowskii, in C [-1,1]-space the approximation becomes better as we approach the endpoints - 1 and 1. There is an extensive literature about this subject, nevertheless a satisfactory solution has been obtained only very recently in two different settings.

First of all, P.L. Butzer, R.L. Stens and M. Wehrens have succeeded in refining their remarkable Legrendre transform method (cf. [2]) to a degree that in [3] they were able to solve this problem in the frame-work of the Legendre transform. The main idea of this method is to apply the translation

$$\tau_h f(x) = \frac{1}{\pi} \int_{-1}^{1} f(xh+u\sqrt{1-x^2}\sqrt{1-h^2})(1-u^2)^{-1/2}\,du$$

instead of the ordinary one and to use

$$(2.3)\qquad \chi_r(f,\delta)_p = \sup\{\|\Delta_{h_1}\dots\Delta_{h_r} f\|_p \quad , \quad 0 \leqslant 1-h_j \leqslant \delta \quad , \quad j=1,\dots,r\}$$

where, of course, $\Delta_h f := f-\tau_h f$. If $E_n[f]_p$ denotes the best approximation of $f \in L^p[-1,1]$ in L^p-norm by algebraic polynomials of degree at most n, then one of their results states that for $\alpha < r$, $E_n[f]_p = O(n^{-\alpha})$ and $\chi_r(f,\delta)_p = O(\delta^{\alpha/2})$ are equivalent.

Another complete solution for the above problem was found by K.G. Ivanov [5]. We shall return to his results in Section 9.

Now our new modulus provides a very close analogue of (2.1) and (2.2). Indeed, let

$$\omega_\varphi^r(f,\delta)_p = \sup_{h\leqslant\delta} \|\Delta_{h\varphi}^r f\|_p + \sup_{h\leqslant\delta^2} \|\Delta_h^r f\|_p$$

with $\varphi(x) = \sqrt{1-x^2}$ (note that on the right hand side in the additional second term we wrote δ^2!), and for this we have

THEOREM 1. _The inequalities_

$$E_n[f]_p \leqslant K\omega_\varphi^r(f,\tfrac{1}{n})_p \qquad (n>r)$$

and

$$\omega_\varphi^r(f,\tfrac{1}{n})_p \leqslant Kn^{-r} \sum_{k=o}^{n} (k+1)^{r-1} E_k[f]_p$$

hold true.

If one is interested in Lipschitz order approximation then the additional term can be dropped, that is, for $\alpha < r$

$$(2.4)\qquad E_n[f]_p = O(n^{-\alpha})$$

and

(2.5) $$\|\Delta^r_{h\varphi} f\|_p = O(h^\alpha) \qquad (\varphi(x) = \sqrt{1-x^2}),$$

are equivalent.

3. Uniform Approximation by Positive Linear Operators

Let us consider the so called Szász-Mirakjan operators

$$S_n f(x) := e^{-nx} \sum_{k=0}^{\infty} f(\tfrac{k}{n}) \frac{(nx)^k}{k!} \qquad (x \geqslant 0).$$

These are the $[0,\infty)$-analogues of the Bernstein polynomials but for them a new problem arises: determine those functions f for which $S_n f$ tends to f *uniformly* on the whole interval $[0,\infty)$. The same question can be raised for every other linear operator process and the answer is given by

THEOREM 2. *Let $\{L_n\}$ be a sequence of positive linear operators, and let*

$$L_n((t-x)^2,x) \sim \alpha_n^2\, \varphi^2(x) \qquad (\alpha_n \to 0).$$

With certain mild assumptions upon $\{L_n\}$ and φ (see [13]) we have that for an $f \in L^\infty$ the relations "$L_n f - f = o(1)$ uniformly" and

$$\omega^1_\varphi(f,\delta)_\infty = o(1) \quad \text{as} \quad \delta \to 0; \quad \omega^1_\varphi(f,\delta)_\infty = \sup_{h \leqslant \delta} \|\Delta^1_{h\varphi} f\|_\infty$$

are equivalent.

For example, $S_n f - f = o(1)$ holds uniformly on $[0,\infty)$ if and only if $\lim_{h\to 0} \sup_x |f(x+h\sqrt{x}) - f(x-h\sqrt{x})| = 0$, which is in turn equivalent to the uniform continuity of the function $f(x^2)$ on $[0,\infty)$.

4. Best Uniform Estimates on Positive Linear Operators

Let $\{L_n\}$ be again a sequence of positive linear operators. The following estimate of Shisha and Mond [8] is well-known:

$$|L_n f(x) - f(x)| \leq K v(f, \mu_n(x)) \quad , \quad \mu_n^2(x) := L_n((t-x)^2, x),$$

where v is the ordinary modulus of continuity. For example, if f belongs to Lip α, then we obtain for the Szász-Mirakjan operators the estimate $|S_n f(x) - f(x)| \leq K(x/n)^{\alpha}$, which is very rough if x is large. Now we want to give estimates on $L_n f - f$ which are global (that is hold uniformly on the whole interval) and best possible (can be reversed in the most important cases).

Let us suppose again that

$$L_n((t-x)^2, x) \sim \alpha_n^2 \varphi^2(x) \quad , \quad \omega_\varphi^2(f,\delta)_\infty = \sup_{h \leq \delta} \|\Delta_{h\varphi}^2 f\|_\infty \qquad (\alpha_n \to 0).$$

Then with some mild assumptions on $\{L_n\}$ and φ we have

THEOREM 3. _If_ $f \in L^\infty$, _then_

$$|L_n f - f| \leq K\, \omega_\varphi^2(f, \alpha_n)_\infty.$$

With some further assumptions (cf. [13]) we can state

THEOREM 4. _If_ $0 < \alpha \leq 1$ _and_ $f \in L^\infty$, _then_

$$L_n f - f = O(\alpha_n^{2\alpha}) \qquad \underline{\text{iff}} \qquad \omega_\varphi^2(f,\delta)_\infty = O(\delta^{2\alpha}).$$

For example, for the Bernstein polynomials

$$B_n f(x) = \sum_{k=0}^{n} f\left(\frac{k}{n}\right) \binom{n}{k} x^k (1-x)^{n-k}$$

we obtain that for any $f \in C[-1,1]$, $B_n f - f = O(n^{-\alpha})$ $(0 < \alpha \leq 1)$ and $\|\Delta_{h\varphi}^2 f\| = O(h^{2\alpha})$ $(\varphi(x) = \sqrt{x(1-x)})$ are equivalent (a result of Lorentz and Schumaker [7] and Ditzian, c.f. [4]). Comparing this with the equivalence of (2.4) and (2.5) (transformed to [0,1]) we can see that for $0 < \alpha < 1$ and $f \in C[0, 1]$

$$E_n[f]_\infty = O(n^{-2\alpha}) \tag{4.1}$$

as well as

(4.2) $$B_n f - f = O(n^{-\alpha})$$

are also equivalent (a result of Ivanov [6]). Let us mention that (4.1) and (4.2) are no longer equivalent for $\alpha = 1$. In fact, the function

$$f(x) = \sum_{k=1}^{\infty} (4k)^{-3} \cos[4k \arccos (2x-1)]$$

satisfies (4.1) with $\alpha = 1$, but since the derivative of f does not belong to Lip 1 on [1/4,3/4], (4.2) does not hold (see the saturation result of Lorentz and Schumaker [7]).

5. Approximation by Contraction Operators in L^p-spaces

This is another large topic where ω_φ^r is indispensable.

Since the Bernstein polynomials are not suitable for L^p-approximation, Kantorovich suggested the following modification

$$K_n f(x) = \sum_{k=o}^{n} ((n+1) \int_{k/(n+1)}^{(k+1)/(n+1)} f(u)\,du) \binom{n}{k} x^k (1-x)^{n-k}.$$

The saturation problem for them was solved by V. Maier and S.D. Riemenschneider. The problem of characterizing

(5.1) $$\| K_n f - f \|_{L^p} \leqslant K n^{-\alpha}$$

for an arbitrary $0 < \alpha < 1$ was open for some five years, and a final solution was found in terms of ω_φ^r (see [12,14]), namely

THEOREM 5. *Let* $1 \leqslant p < \infty$, $0 < \alpha < 1$, $f \in L^p(0,1)$ *and*

(5.2) $$\omega_\varphi^2(f,\delta)_p := \sup_{h \leqslant \delta} \| \Delta_{h\varphi}^2 f \|_p \quad , \qquad \varphi(x) = \sqrt{x(1-x)}.$$

Then (5.1) *and* $\omega^2(f,\delta)_p = O(\delta^{2\alpha})$ *are equivalent*.

A "best estimate" can be given with a modification of (5.2), namely with

(5.3) $$\bar{\omega}_{\varphi}^{2}(f,\delta) := \sup_{h\leqslant\delta}\|\Delta_{h\varphi}^{2}f\|_{p} + \sup_{h\leqslant\delta^{2}}\|\Delta_{h}^{2}f\|_{p} + \delta^{2}\|f\|_{p}.$$

THEOREM 6. _If_ $1\leqslant p<\infty$ _and_ $f\in L^{p}(0,1)$, _then_

$$\|K_{n}f-f\|_{p}\leqslant K\,\bar{\omega}_{\varphi}^{2}(f,\frac{1}{\sqrt{n}}),$$

and this is the best estimate in the sense that except for the case $p=\alpha=1$, (5.1) _is equivalent to_

(5.4) $$\bar{\omega}^{2}(f,\frac{1}{\sqrt{n}})\leqslant Kn^{-\alpha}$$

Let us mention that the case $p=\alpha=1$ is really an exception: (5.1) and (5.4) are not equivalent for $p=\alpha=1$.

Comparing Theorem 5 with the equivalence of (2.4) and (2.5) we can see that the L^{p} analogue of Ivanov's result also holds: for $0<\alpha<1$ and $1\leqslant p<\infty$ the relations $E_{n}[f]_{p}=\mathcal{O}(n^{-2\alpha})$ and $\|K_{n}f-f\|_{p}=\mathcal{O}(n^{-\alpha})$ are equivalent.

Several authors obtained quantitative estimates for positive contractions but neither of these results is exact, e.g. for the Kantorovich polynomials. A general global result which is best possible in the above sense has not been found as yet, nevertheless for the most important operators (such as the gamma operators, the Kantorovich-type modification of the Szász-Mirakjan, Baskakov and Meyer-König and Zeller operators) the perfect analogue of the results above have been proved (mostly with individual proofs), see [10,11,12].

6. Jackson and de la Vallée Poussin-Stechkin Type Estimates for Operators

In trigonometric approximation the estimates (2.1) and (2.2) play a fundamental role. Now the question arises if the same can be done in operator approximation. We know estimates of this type for the gamma operators

$$G_{n}f(x)=\frac{x^{n+1}}{n!}\int_{0}^{\infty}e^{-xu}u^{n}f(\frac{n}{u})\,du$$

as well as for the so-called Post-Widder operators. Indeed, let $1\leqslant p\leqslant\infty$, $f\in L^{p}(0,\infty)$, and

$$\omega_\varphi^2(f,\delta)_p = \sup_{h\leq\delta}\|\Delta_{h\varphi}^2 f\|_p \quad , \quad \varphi(x) = x.$$

Then with the notation $G_n[f]_p = \|G_nf - f\|_{L^p(0,\infty)}$ we have the following ([11]).

THEOREM 7. _The inequalities_

$$G_n[f]_p \leq K\,\omega_\varphi^2(f,\tfrac{1}{\sqrt{n}})_p \tag{6.1}$$

and

$$\omega_\varphi^2(f,\tfrac{1}{\sqrt{n}})_p \leq K\,n^{-1}\sum_{k=0}^{n} G_k[f]_p \tag{6.2}$$

hold true.

Of course (6.1) and (6.2) imply at once that $G_n[f]_p \leq Kn^{-\alpha}$ $(0<\alpha<1)$ and $\omega_\varphi^2(f,\delta)_p \leq K\,\delta^{2\alpha}$ are equivalent.

It is an open problem whether similar estimates hold for other operators, e.g. for the Kantorovich polynomials. All what we know is a much easier result: with (5.3) and $K_n[f]_p = \|f - K_nf\|_p$ we have $K_n[f]_p \leq K\,\overline{\omega}_\varphi^2(f,1/\sqrt{n})_p$ $(\varphi(x) = \sqrt{x(1-x)})$, and for every $\gamma < 1$

$$\overline{\omega}_\varphi^2(f,\tfrac{1}{\sqrt{n}})_p \leq K_\gamma n^{-\gamma}\sum_{k=0}^{n}(k+1)^{\gamma-1}K_k[f]_p\,. \tag{6.3}$$

A proof for (6.3) in the case $\gamma = 1$ would imply the interesting relations

$$K_n[f]_p \leq Kn^{-1/2}\sum_{k=0}^{\sqrt{n}} E_k[f]_p$$

and

$$E_n[f]_p \leq K\,n^{-2}\sum_{k=0}^{n^2} K_k[f]_p$$

between best approximation and the quantity $\|K_nf - f\|_p$.

7. The Reason for the Applicability of ω_φ^r

Now the question arises why our ω_φ^r occurs so "often" and what is the reason that it provides solutions for seemingly different types of problems. The key is the so-called Peetre K-functional

$$K(f,t^r)_p = \inf_g(\|f-g\|_p + t^r\|\varphi^r g^{(r)}\|_p),$$

for which it has turned out that the inequalities

$$\frac{1}{A}\,\omega_\varphi^r(f,t)_p \leq K(f,t^r)_p \leq A\,\omega_\varphi^r(f,t)_p$$

hold (see [12] and also [4]), and it is well-known that interpolation norms are very important in approximation theory.

8. Properties of ω_φ^r

The well-known and really essential properties of ordinary moduli of smoothness hold just as well for our new ones (see [14]). Indeed,

THEOREM 8.

(i) $\omega_\varphi^r(f,\lambda\delta)_p \leq K\,\lambda^r\,\omega_\varphi^r(f,\delta)_p \qquad (\lambda \geq 1);$

(ii) $\lim_{\delta\to 0}\ \omega_\varphi^r(f,\delta)_p/\delta^r = 0$ __iff__ f __is a polynomial of degree at most__ r-1;

(iii) __If__ $\psi \leq \varphi$, __then__ $\omega_\psi^r(f,\delta)_p \leq K\,\omega_\varphi^r(f,\delta)_p$;

(iv) $\omega_\varphi^r(f,\delta)_p = O(\delta^r)$ iff

$$\begin{cases} f^{(r-1)} \text{ is absolutely continuous and } \varphi^r f^{(r)} \in L^p \text{ if } 1 < p \leq \infty, \\ f^{(r-2)} \text{ is absolutely continuous and } \varphi^r f^{(r-1)} \in BV \text{ if } p = 1. \end{cases}$$

9. Possible Substitutes

It is easy to see that (with no additional term - which is almost always the case if $p=\infty$) $\omega_\varphi^r(f,\delta)_\infty = O(\delta^\alpha)$ is equivalent to

$$\sup_x (\varphi(x))^\alpha |\Delta_h^r f(x)| = O(h^\alpha) \qquad (h\to 0),$$

i.e., to a weighted condition. For example, we obtain from the equivalence of (2.4) and (2.5) that

$$E_n[f]_\infty \le K n^{-\alpha} \quad \text{and} \quad \sup_x (1-x^2)^{\alpha/2} |\Delta_h^r f(x)| \le K\, h^\alpha$$

($\alpha>0$, $r>\alpha$) are equivalent (for $r=2$ this is a result of Stens and Wehrens). Thus, if one is interested in Lipschitz order approximation then at least for $p=\infty$ the condition given by $\omega_\varphi^r(f,\delta)_p$ can be expressed by a weighted ordinary modulus of smoothness. This is definitely not true for $p<\infty$ (see e.g. [9]), and in this case there seems not to exist an equivalent measurement in terms of $\Delta_h^r f$ with constant (not varying) h.

Another quantity which resembles ω_φ^r was studied by K.G. Ivanov [5] and several other Bulgarian mathematicians. Their construction is similar to that of ours, only they take an integration first and omit the "sup" : let $1\le p\le\infty$, $\delta=\delta(x)$ be a function

$$v_r(f,x,\delta(x)) := (2\delta(x))^{-1} \int_{-\delta(x)}^{\delta(x)} |\Delta_u^r f(x)|\, du ,$$

and

$$\tau_r(f,\delta)_p := \| v_r(f,\cdot,\delta(\cdot))\|_{L^p}$$

(they even incorporated a weight function but for the sake of simplicity we omit it). The similarity between $\omega_\varphi^r(f,\delta)_p$ and $\tau_r(f,\delta)_p$ is clear, and there are certain signs that actually they are equivalent measurements.

For $\tau_r(f,\delta)_p$ Ivanov derived the following result concerning best approximation ([5]) : let $Q_n(x) = \sqrt{1-x^2}\,/\,n + 1/n^2$, $1\le p\le\infty$ and $f\in L^p[-1,1]$. Then

$$E_{n+r}[f]_p \le K\, \tau_r(f,Q_n)_p$$

and

$$\tau_r(f,Q_n)_p \leqslant Kn^{-r} \sum_{k=o}^{n} (k+1)^{r-1} E_k[f]_p$$

hold true. These inequalities also completely solve the problem of Section 2.

Of course a third possible substitute may be (2.3).

10. The Role of the Additional Terms

Generally, the additional term in our modulus of continuity is essential, it may strongly influence the behaviour of ω_φ^r (c.f. our second example below). On the other hand, there are important cases when it can simply be dropped. Let us consider e.g. the ω_φ^r defined in the second section and its main term

$$v_\varphi^r(f,\delta)_p = \sup_{h\leqslant\delta} \|\Delta_{h\varphi}^r f\|_p \quad , \quad \varphi(x) = \sqrt{1-x^2}.$$

We have seen that $\omega_\varphi^r(f,\delta)_p = O(\delta^\alpha)$ characterizes $E_n[f]_p = O(n^{-\alpha})$, and we have also mentioned that the simpler quantity $v_\varphi^r f$ can do the same, because in general we have

$$v_\varphi^r(f,\delta)_p \leqslant \omega_\varphi^r(f,\delta)_p \leqslant K \int_0^\delta \frac{v_\varphi^r(f,u)}{u}\, du.$$

To show the importance of the additional term let us consider the operators

$$R_n f(x) = (1+x)^{-n} \sum_{k=o}^{n} f\left(\frac{k}{n-k+1}\right) \binom{n}{k} x^k$$

introduced by G. Bleimann, P.L. Butzer and L. Hahn [1]. The transforms are rational functions and can be used to approximate functions that are continuous on $[0,\infty)$ and have finite limit ($f(\infty)$) at infinity (shortly $f \in C[0,\infty]$). Now $\{R_n\}$ is saturated with order $\{n^{-1}\}$ and has saturation class

$$\{f \mid f' \text{ abs. cont.}, \ x(1+x)^2 |f''(x)| \leqslant K_f \ \text{ a.e.}\}.$$

For non-optimal approximation we have

THEOREM 9. _Let_ $f \in C[0,\infty]$ _and_ $0 < \alpha < 1$. _Then_ $R_n f - f = O(n^{-\alpha})$ _uniformly on_ $[0,\infty)$ _if and only if_

(10.1) $$x^{\alpha}(1+x)^{2\alpha}|\Delta_h^2 f(x)| \leq Kh^{2\alpha} \qquad (0 < h \leq \tfrac{x}{2},\ x > 0)$$

and

(10.2) $$|f(x) - f(\infty)| \leq K x^{-\alpha} \qquad (x > 0)$$

are satisfied.

While (10.1) is a "normal" condition, (10.2) is rather a curious one since it describes the growth of $f - f(\infty)$ at the infinity and not its smoothness. Now the reason behind this phenomenon is that in our case an appropriate modulus of smoothness is, with $\varphi(x) = \sqrt{x(1+x)}$,

$$\omega_{\varphi}^2(f,\delta) = \sup_{h\leq\delta} \|\Delta_{h\varphi}^2 f\|_{L^{\infty}[0,1/(2h^2)]} + \sup_{x,x' \geq \delta^{-2}} |f(x) - f(x')| .$$

11. A Proof

In this section we give the outline of a typical proof that uses our new modulus of smoothness. Let us consider e.g. Theorem 6.

Let $\varphi(x) = \sqrt{x(1-x)}$,

$$\omega_{\varphi}^2(f,\delta)_p = \sup_{h\leq\delta} \|\Delta_{h\varphi}^2 f\|_p + \sup_{h\leq\delta^2} \|\Delta_{h\varphi}^2 f\|_p ,$$

$$\bar{\omega}_{\varphi}^2(f,\delta)_p = \omega_{\varphi}^2(f,\delta)_p + \delta^2\|f\|_p,$$

$D = \{g \mid g \in L^p(0,1),\ g' \text{ abs. cont.},\ x(1-x)\ g''(x) \in L^p(0,1)\}$,

$$Sg = (\varphi^2 g, g) \in L^p \times L^p \qquad ; \quad \|Sg\|_p = \|\varphi^2 g''\|_p + \|g\|_p ,$$

and

$$K(t^2,f)_p = \inf_{g\in D}(\|f-g\|_p + t^2\|\varphi^2 g''\|_p).$$

By section 7 $\omega_\varphi^2(f,t)_p$ and $K(t^2,f)_p$ have the same order.

The proof of Theorem 6 is divided into several steps, of which we do not prove the first two.

1) $$\|f'\|_p \leqslant K\|Sf\|_p \qquad (f \in D).$$

2) If $B_n f$ are the Bernstein polynomials associated with f, then

$$\|K_n f - B_n f\|_p \leqslant \frac{K}{n}\|f'\|_p \qquad (f \in D).$$

3) $$\|B_n f - f\|_p \leqslant \frac{K}{n}\|Sf\|_p \qquad (f \in D).$$

PROOF. By Taylor's formula we have

$$f(t) = f(x) + f'(x)(t-x) + \int_0^{t-x} \frac{t-x-\tau}{(x+\tau)(1-x-\tau)} (x+\tau)(1-x-\tau)f''(x+\tau)d\tau.$$

Now for $\tau \in (0,t-x)$, $t,x \in (0,1)$ it follows that

$$\left|\frac{t-x-\tau}{(x+\tau)(1-x-\tau)}\right| \leqslant \frac{|t-x|}{x(1-x)},$$

and so, noting that $B_n(t-x;x) \equiv 0$, it follows that

$$|B_n f(x) - f(x)| \leqslant B_n\left(\frac{|t-x|}{x(1-x)} \left|\int_0^{t-x} (x+\tau)(1-x-\tau)|f''(x+\tau)|d\tau\right|;x\right).$$

For $p > 1$ we use the maximal function $M(\cdot,\cdot)$ and the maximal inequality to obtain

$$\|B_n f - f\|_p \leqslant \left\|B_n\left(\frac{(t-\cdot)^2}{\varphi^2} M(\varphi^2 f'',\cdot);\cdot\right)\right\|_p =$$

$$= \frac{K}{n}\|M(\varphi^2 f'',\cdot)\|_p \leqslant \frac{K}{n}\|\varphi^2 f''\|_p \leqslant \frac{K}{n}\|Sf\|_p$$

where we also used the fact that

$$B_n((t-x)^2;x) = \frac{x(1-x)}{n} \qquad (x\in[0,1]).$$

For $p = 1$ Fubini's theorem yields (put $p_{n,-1} \equiv 0$, $p_{n,k}(x) = \binom{n}{k}x^k(1-x)^{n-k}$)

$$\|B_nf - f\|_{L^1(0,1/2)} \leq \int_0^{1/2} \{ \sum_{k=0}^{n} \frac{|k/n-x|}{x(1-x)} | \int_x^{k/n} u(1-u) |f''(u)| du| p_{n,k}(x)\} dx$$

$$\leq \int_0^1 \{ \sum_{k=0}^{n} \frac{|k/n-x|}{x} | \int_x^{k/n} u(1-u) | f''(u)| du| p_{n,k}(x)\} dx =$$

$$= \int_0^1 \{ \sum_{k=0}^{n} \frac{(k/n-x)}{x} \int_x^{k/n} u(1-u)|f''(u)| du) p_{n,k}(x)\} dx$$

$$= \int_0^1 \{ \sum_{k=0}^{n} \frac{k/n-x}{x} (\int_0^{k/n} u(1-u)| f''(u)| du) p_{n,k}(x)\} dx$$

$$= \sum_{k=0}^{n} (\int_0^{k/n} u(1-u)| f''(u)| du) \int_0^1 (p_{n-1,k-1}(x) - p_{n,k}(x)) dx$$

$$\leq \sum_{k=0}^{n} (\int_0^1 u(1-u)| f''(u)| du) (\frac{1}{n} - \frac{1}{n+1}) \leq \frac{K}{n} \|Sf\|_{L^1} .$$

A similar estimate holds on the interval (1/2,1), and so the proof is complete.

4) From steps 1-3 we obtain for any $f \in L^p$ and $g \in D$ the estimate

$$\|K_nf - f\|_p \leq \|K_n(f - g)\|_p + \|f - g\|_p + \|K_ng - g\|_p$$

$$\leq K(\|f - g\|_p + \frac{1}{n} \|\varphi^2g''\|_p) + \frac{K}{n} \|g\|_p .$$

Taking here the infinimum for all possible g and taking into account that we may restrict ourselves to g's with $\|g\|_p \leq 2 \|f\|_p$ (otherwise $g \equiv 0$ would be a better choice), we obtain

$$\|K_nf - f\|_p \leq K(K(\frac{1}{n}, f)_p + \frac{1}{n} \|f\|_p)$$

$$\leq K(\omega_\varphi^2(f, \frac{1}{\sqrt{n}})_p + (\frac{1}{\sqrt{n}})^2 \|f\|_p) \leq K \,\bar{\omega}_\varphi^2(f, \frac{1}{\sqrt{n}}),$$

and this is the first estimate in Theorem 6.

5) We shall prove the other half of Theorem 6 only for $0<\alpha<1$; to this end we shall use the inequalities

$$\|SK_nf\|_p \leq K_o n \|f\|_p \qquad (f\in L^p)$$

and

$$\|SK_nf\|_p \leq K_o \|Sf\|_p \qquad (f\in D)$$

(the proof of these are straightforward; we omit them).

We have to prove only that $\|K_nf-f\|_p = \mathcal{O}(n^{-\alpha})$ implies $K(t^2,f)=\mathcal{O}(t^{2\alpha})$ (see 4.). Let $0<c$ be so small that $c^\alpha K_o < 1/2$ holds. We have

$$K(\tfrac{1}{n},f) \leq \|K_nf-f\|_p + \tfrac{1}{n}\|SK_nf\|_p,$$

and here

$$\|SK_nf\|_p \leq \|SK_n(f-K_{[cn]}f)\|_p + \|SK_n K_{[cn]}f\|_p$$

$$\leq K_o n\|f-K_{[cn]}f\|_p + K_o \|SK_{[cn]}f\|_p$$

$$\leq K K_o c^\alpha n^{1-\alpha} + K_o(K_o(cn)\|f-K_{[c^2n]}f\|_p$$

$$+ K_o \|SK_{[c^2n]}f\|_p) \leq \ldots \leq$$

$$\leq K(K_oc^\alpha n^{1-\alpha} + (K_o c^\alpha)^2 n^{1-\alpha} + \ldots) \leq K n^{1-\alpha},$$

which completes the proof.

REFERENCES

[1] Bleimann, G. - Butzer, P.L. - Hahn, L., A Bernstein-type operator approximating continuous functions on the semi-axis. Indag. Math. 42 (1980), 255-262.

[2] Butzer, P.L. - Stens, R.L. - Wehrens, M., Approximation by algebraic convolution integrals. Approximation Theory and Functional Analysis

(ed. J.B. Prolla), North-Holland Publishing Company, Amsterdam/ New York 1979, 71-120.

[3] Butzer, P.L. - Stens, R.L. - Wehrens, M., Higher moduli of continuity based on the Jacobi translation operator and best approximation. C.R. Math. Rep. Acad. Sci. Canada 2 (1980), 83-88.

[4] Ditzian, Z., On interpolation of $L^p[a,b]$ and weighted Sobolev spaces. Pacific J. Math. 90 (1980), 307-323.

[5] Ivanov, K.G., Some characterization of best algebraic approximation in $L^p[-1,1]$ $(1 \leq p \leq \infty)$, C.R. Acad. Bulgare Sci. 34 (1981), 1229-1232.

[6] Ivanov, K.G., On Bernstein polynomials. C.R. Acad. Bulgare Sci. 35 (1982), 893-896.

[7] Lorentz, G.G. - Schumaker, L.L., Saturation of positive operators. J. Approx. Theory 5 (1972), 413-424.

[8] Shisha, O. - Mond, B., The degree of convergence of sequences of linear operators. Proc. Nat. Acad. Sci. USA 60 (1968), 1196-1200.

[9] Totik, V., Problems and solutions concerning Kantorovitch operators. J. Approx. Theory. 47 (1982), 51-68.

[10] Totik, V., Approximation by Meyer-König and Zeller type operators. Math. Z. 182 (1983), 425-446.

[11] Totik, V., The gamma operators in L^p spaces. Publ. Math. Debrecen. (submitted).

[12] Totik, V., An interpolation theorem and its applications to positive operators. Pacific J. Math. 110 (1983), No.1.

[13] Totik, V., Uniform approximation by positivie linear operators on infinite intervals. Analysis Math. (to appear)

[14] Totik, V., Some properties of a new kind of modulus of smoothness. Zeitschrift für Analysis und Ihre Anwendungen.(to appear)

[15] Totik, V., Approximation by Szász-Mirakjan-Kantorovich operators in L^p $(p > 1)$. Analysis Math. 9 (1983, 147-167.

International Series of
Numerical Mathematics, Vol. 65

APPROXIMATION OF FUNCTIONS OF TWO VARIABLES BY ALGEBRAIC POLYNOMIALS, I.

Kamen G. Ivanov
Center of Mathematics and Mechanics
Bulgarian Academy of Sciences
Sofia, Bulgaria

The aim of the paper is to introduce new moduli of continuity for functions of two variables, particularly for the case of best algebraic approximation. A Stečkin-type theorem as well as a converse theorem are established.

1. Introduction

Let D be the closure of a bounded domain in the plane $\mathbb{R}^2$, and let Γ stand for the boundary of D. We say that D belongs to $\check{C}^2$ iff Γ is the union of a finite number of curves and arcs with continuous curvature and the angles in the join-points of the arcs are positive and less than π. For example, polygons, disks and rings belong to $\check{C}^2$. Let us mention that the above definition permits the elements of $\check{C}^2$ to have some holes.

We denote the points in the plane by x,y,z and its coordinates (with respect to some fixed coordinate system) by subscripts - $x=(x_1,x_2)$.

The space $L_p(D)$ $(1\leqslant p<\infty)$ consists of all measurable functions f on D such that the norm $\|f\|_p=[\int_D |f(x)|^p dx]^{1/p}$ is finite. In the limit case we set $L_\infty(D)=C(D)$ - the space of all continuous functions f on D with the norm $\|f\|_\infty=\sup\{|f(x)| : x\in D\}$.

The best approximation of the function $f\in L_p(D)$ by means of algebraic polynomials of degree at most n is defined as

$$E_n(f)_p := \inf\{\|f-Q\|_p : Q\in H_n\},$$

$$H_n := \{Q : Q(x) = \sum_{i+j\leqslant n} a_{ij}\, x_1^i\, x_2^j \;,\quad a_{i,j}\in\mathbb{R}\}\ .$$

The aim of this paper, the structural description of the constructive classes $\{f \in L_p(D) : E_n(f)_p = O(n^{-\alpha})\}$ $(\alpha > 0)$, is considered in a few papers only for the uniform case $p = \infty$. Fuksman [3] describes these classes when the boundary of D is a level curve of a given polynomial. Konovalov [2] generalizes Fuksman's results for $D \in \check{C}^2$. In both works the structural classes are defined in terms of the derivatives of the functions when $\alpha > 1$, and the definitions are rather complicate.

In this article we consider new moduli for functions of two variables. A Stečkin's type theorem for $D \in \check{C}^2$ and $1 \leq p \leq \infty$ is given. The converse theorem is proved when $p = \infty$, $D \in \check{C}^2$ or $1 \leq p \leq \infty$ with special types of domains; we believe that the converse theorem is even true under the conditions of the direct theorem.

The moduli defined in Section 2 are analogous to the moduli of functions of one variable introduced in [4,5]. The local modulus concept is used when the global characteristic of functions is defined. We think that these moduli are a natural tool for estimating the best algebraic approximation because they take into account the improvement of the approximation near the boundary of the domain. But also the moduli introduced when modified properly are equivalent to the usual integral moduli of functions of two variables.

The problems are treated only for the case of two variables but we think that no principal difficulties should arise if functions of several variables are considered.

2. Moduli of Functions

Let $\|\cdot\|$ denote the Euclidean norm in $\mathbb{R}^2$. We define the distance between the point x and the set $V \subset \mathbb{R}^2$ by $\rho(x,V) = \inf\{\|x-y\| : y \in V\}$. Let $[x,y]$ denote the segment with the end-points x and y, i.e., $[x,y] = \{z \in \mathbb{R}^2 : z_i = (1-t)x_i + ty_i,\ 0 \leq t \leq 1,\ i = 1,2\}$.

We shall consider two types of metrical relations between the points in the plane. The first relation is the usual Euclidean metric. All integrals and measures are taken with respect to this metric. The second relation $g(x,y)$ will be used only to define the δ-neighborhood of a point. Nevertheless this metric plays a basic role because it takes into account the reflection of the approximating set properties in the geometry of the domain. So we can solve different problems by varying g. For example, if we consider approximations by means of trigonometric polynomials, then it is suitable to set

(1) $$g^*(x,y) = \|x-y\| .$$

In our case we set

(2) $$g(x,y) = \|x-y\| + |\sqrt{\rho(x,\Gamma)} - \sqrt{\rho(y,\Gamma)}| + \|x-y\| (\|x-y\| + \rho(x,\Gamma_0) + \rho(y,\Gamma_0))^{-1/2},$$

where Γ_0 is the set of corner points of D (if $\Gamma_0 = \phi$, then define $\rho(\cdot,\Gamma_0) \equiv 1$). One can show that g is a metric on the plane. We do not consider the problem of finding the best constants and so we are free to replace g by an equivalent metric relation g_1. Here "equivalent" means that there are constants $k_1,k_2 > 0$ such that

(3) $$k_1 g(x,y) \leqslant g_1(x,y) \leqslant k_2 g(x,y)$$

for each $x,y \in \mathbb{R}^2$. For example, we can set

(4) $$g_1(x,y) = \|x-y\| + \overset{y}{\underset{x}{V}} \sqrt{\rho(\cdot,\Gamma)} + \overset{y}{\underset{x}{V}} \sqrt{\rho(\cdot,\Gamma_0)} ,$$

where $V_x^y \varphi$ denotes the variation of the restriction of φ on [x,y]. By analogy with (4) we can easily write down a necessary form of the metrical relation for the case of several variables. Let us mention that g_1 is not a metric (the triangular rule is not valid) although g_1 satisfies (3).

Now we define the δ-neighborhood of $x \in \mathbb{R}^2$ $(\delta > 0)$ as

(5) $$U(\delta,x) = \{y \in \mathbb{R}^2 : g(x,y) \leqslant \delta\}.$$

The next lemma clarifies the form of the neighborhood. Let $x \in \mathbb{R}^2$ and $y \in \Gamma$ be such that $\rho(x,\Gamma) = \|x-y\|$. Then R(a,b;x) denotes the rectangle with center at x, with the first side parallel to the segment [x,y] and equal to a, and with the second side equal to b.

LEMMA 1. *There are constants* c_1 *and* c_2, *depending only on* D, *such that*

$$U(c_1\delta,x) \subset R(\delta\sqrt{\rho(x,\Gamma)} + \delta^2, \delta\sqrt{\rho(x,\Gamma_0)} + \delta^2;x) \subset U(c_2\delta,x)$$

for each $x \in D$, $\delta \leqslant 1$.

So to facilitate our imagination of a δ-neighbourhood we may consider them as some rectangles equivalent to these in Lemma 1. For example, if

a) $D = [-1,1] \times [-1,1]$, then for $x \in D$ $U(\delta,x)$ is like the rectangle

$$[x_1 - \delta\sqrt{1-x_1^2} - \delta^2,\ x_1 + \delta\sqrt{1-x_1^2} + \delta^2] \times [x_2 - \delta\sqrt{1-x_2^2} - \delta^2,\ x_2 + \delta\sqrt{1-x_2^2} + \delta^2].$$

b) $D = \{x \in \mathbb{R}^2 : \|x\| \leq 1\}$, then for $x \in D$ $U(\delta,x)$ is like the rectangle $R(\delta\sqrt{1-\|x\|} + \delta^2,\ \delta;\ x)$.

Now we are in a position to define moduli of functions of two variables.

The k-th finite differences with increment $h \in \mathbb{R}^2$ of the function f given on D is defined as

$$\Delta_h^k f(x) = \begin{cases} \sum_{i=0}^{k} (-1)^{k-i} \binom{k}{i} f(x+ih) & , \text{ if } [x,x+kh] \subset D, \\ 0 & , \text{ if } [x,x+kh] \not\subset D. \end{cases}$$

As usual $x + ih$ denotes the point $(x_1 + ih_1,\ x_2 + ih_2)$.

Using (5) we define the local q-modulus of order k of the function $f \in L_q(D)$ as

$$\text{(6)} \quad \omega_k(f,x;U(\delta,x))_q = \begin{cases} \left[\frac{1}{\mu U(\delta,x)} \int_{U(\delta,x)} |\Delta_{y-x}^k f(x)|^q \, dy\right]^{\frac{1}{q}} & , \text{ if } 1 \leq q < \infty, \\ \sup\{|\Delta_{y-x}^k f(x)| : y \in U(\delta,x)\} & , \text{ if } q = \infty . \end{cases}$$

Here μV denotes the Lebesgue measure of the set V. The definition (6) is correct if $U(x,\delta)$ has been given in an arbitrary way, but then the local modulus may not be a measurable function of x. But if we use $U(\delta,x)$ from (5) with g given in (1) or (2), and if $f \in L_p(D)$ $(p \geq q)$, then the local modulus belongs to $L_p(D)$ as a function of x.

Finally we define the q,p - modulus of order k of the function $f \in L_r(D)$ $(r = \max\ \{p,q\}\ )$ as

$$\text{(7)} \qquad \tau_k(f;\delta)_{q,p} = \|\omega_k(f,\ \cdot\ ;\ U(\delta,\cdot))_q\|_p .$$

We list only a few of the properties of τ_k.

LEMMA 2. <u>If</u> $f,\ f_1 \in L_r(D)$ $(r = \max\{p,q\},\ 1 \leq p,\ q \leq \infty)$, <u>then</u>

1) $\tau_k(f + f_1;\delta)_{q,p} \leq \tau_k(f;\delta)_{q,p} + \tau_k(f_1;\delta)_{q,p}$;

2) $\tau_k(\alpha f;\delta)_{q,p} = |\alpha| \; \tau_k(f;\delta)_{q,p}$ <u>for</u> $\alpha \in \mathbb{R}$;

3) $\tau_k(f;\delta)_{q,p} \leqslant c(k,D) \, \|f\|_p$ <u>for</u> $q \leqslant p$;

4) $\tau_k(f;\delta)_{q_1,p} \leqslant \tau_k(f;\delta)_{q_2,p}$ <u>for</u> $1 \leqslant q_1 \leqslant q_2 \leqslant \infty$;

5) $\tau_k(f;\delta)_{q;p_1} \leqslant [\mu D]^{1/p_1 - 1/p_2} \; \tau_k(f;\delta)_{q,p_2}$ <u>for</u> $1 \leqslant p_1 \leqslant p_2 \leqslant \infty$;

6) <u>Let</u> $\tau_k^*(f;\delta)_{q,p}$ <u>be defined by using</u> g^* <u>from</u> (1) <u>in</u> (5) <u>instead of</u> g <u>from</u> (2), <u>and let</u> $\omega_k(f;\delta)_p$ <u>be the usual integral modulus. Then</u>

$$c_3 \; \omega_k(f;\delta)_p \leqslant \tau_k^*(f;\delta)_{q,p} \leqslant \omega_k(f;\delta)_p$$

<u>for each</u> $1 \leqslant q \leqslant p$, <u>where</u> c_3 <u>depends only on</u> D <u>and</u> k.

3. Direct and Converse Theorems

Combining the ideas from [1,2] with these from [6,7] and using the moduli from (7) we can prove the following theorem of Stečkin's type for the best algebraic approximation.

THEOREM 1. <u>If</u> k <u>is a natural</u>, $D \in \check{C}^2$, $f \in L_p(D)$, $1 \leqslant q \leqslant p \leqslant \infty$, <u>then</u>

$$E_n(f)_p \leqslant c_4 \; \tau_k(f;\tfrac{1}{n})_{q,p} ,$$

<u>where</u> c_4 <u>depends on</u> k <u>and</u> D.

At present we can convert Theorem 1 for some cases.

THEOREM 2. <u>If</u> k <u>is a natural</u>, $f \in L_p(D)$, <u>and either</u> $D \in \check{C}^2$, $p = \infty$ <u>or</u> D <u>is a parallelogram or a disk and</u> $1 \leqslant p \leqslant \infty$, <u>then</u>

$$\tau_k(f;\tfrac{1}{n})_{p,q} \leqslant c_5 \; n^{-k} \sum_{s=0}^{n} \; (s+1)^{k-1} \; E_s(f)_p ,$$

<u>where</u> c_5 <u>depends only on</u> k <u>and</u> D.

As a consequence of Theorem 1 and Theorem 2 we obtain

COROLLARY 1. <u>Let</u> $0 < \alpha < k$. <u>Then under the conditions of Theorem</u> 2 <u>we have</u>

$$E_n(f)_p = O(n^{-\alpha}) \quad (n \to \infty) \iff \tau_k(f;\delta)_{p,p} = O(\delta^{\alpha}) \quad (\delta \to 0+).$$

Finally we shall give the trigonometrical analog of the above statements. Let $D^* = [0,2\pi] \times [0,2\pi]$, and let $E_n^*(f)_p$ be the best approximation of the periodic function $f \in L_p(D^*)$ by trigonometric polynomials of degree at most n. Then the following two theorems are true

THEOREM 1'. *If* k *is a natural*, $f \in L_p(D^*)$, $1 \leq p \leq \infty$, *then*

$$E_n^*(f)_p \leq c_6 \, \tau_k^*(f;\tfrac{1}{n})_{p,p} ,$$

where c_6 *depends on* k.

THEOREM 2'. *Under the conditions of Theorem 1' we have*

$$\tau_k^*(f;\tfrac{1}{n})_{p,p} \leq c_7 \, n^{-k} \sum_{s=0}^{n} (s+1)^{k-1} E_s^*(f)_p ,$$

where c_7 *depends on* k.

As a consequence of Theorem 1' and Theorem 2' we obtain

COROLLARY 1'. *Let* $0 < \alpha < k$. *Then under the conditions of Theorem 1' we have*

$$E_n^*(f)_p = O(n^{-\alpha}) \quad (n \to \infty) \iff \tau_k^*(f;\delta)_{p,p} = O(\delta^{\alpha}) \quad (\delta \to 0+).$$

REFERENCES

[1] Dzjadik, V.K. - Konovalov, V.N., *A method of partition of unity in domains with piecewise smooth boundary into a sum of algebraic polynomials of two variables that have certain kernel properties* (Russian). Ukrain. Mat. Ž. 25 (1973), 179-192.

[2] Konovalov, V.N., *On some constructive characteristics of some classes of functions of several variables* (Russian). Dissertation, Kiev, 1972.

[3] Fuksman, A.L., *The approximation of functions of real variables by algebraic polynomials in a closed region* (Russian). Dokl. Akad. Nauk SSSR 178 (1968), 1263-1266.

[4] Ivanov, K.G., Direct and converse theorems for the best algebraic approximation in C[-1,1] and $L_p[-1,1]$. C.R. Acad. Bulgare Sci. 33 (1980), 1309-1312.

[5] Ivanov, K.G., On a new characteristic of functions, I. Serdica 8 (1982), 262-279.

[6] Ivanov, K.G., On a new characteristic of functions, II. Pliska 5 (1983), 151-163.

[7] Ivanov, K.G., A constructive characteristic of the best algebraic approximation in L_p [-1,1] $(1 \leqslant p \leqslant \infty)$. Constructive Function Theory 81, Sofia (1983), 357-367.

International Series of
Numerical Mathematics, Vol. 65

POLYNOMIAL APPROXIMATION ON DISJOINT INTERVALS

József Szabados
Mathematical Institute
Hungarian Academy of Sciences
Budapest

The problem of polynomial approximation of certain classes of functions on sets with disjoint components is considered in the special case when the set is the union of two disjoint intervals of equal length. A recent result of C.K. Chui and M. Hasson [1] on the order of approximation of this kind is slightly generalized, and the behaviour of the best approximating polynomial in between the two intervals is investigated. A condition for the boundedness of these polynomials is given, and it is shown that this condition is not far from being best possible.

A classical theorem of J.L. Walsh and H.G. Russell from 1934 (see e.g. [3, p. 75]) states that if K is a closed, limited point set whose complement is connected and regular, and the function $f(z)$ is single valued and analytic on and within the equipotential locus $K_R = \{z=x+iy;\ G(x,y) = \log R > 0\}$, then there exist polynomials $p_n(z)$ of degree at most n such that

$$\max_{z\in K} |f(z) - p_n(z)| \leq MR^{-n} \qquad (n = 0,1,\dots),$$

where $M = M(R)$ is independent of n and z. Here $G(x,y)$ is Green's function of the complement of K.

This very general result also settles the problem of approximating functions on sets with several components. In a recent paper, W.H.J. Fuchs [2] gave upper and lower estimates for the order of approximation in the case when K consists of finitely many simply connected disjoint components, and the restriction of the function $f(z)$ to each of these components is an entire function. His estimate is of the form (under certain further restriction on K)

$$O(n^{-q-1/2}e^{-n\beta}) \quad (q = q(f) \geqslant 0 \text{ integer}, \ \beta > 0) ,$$

where β depends on the geometry of K.

In [1], C.K. Chui and M. Hasson gave a nice result for the special case when $K = [-b,-a] \cup [a,b]$ $(0 < a < b)$, and the functions

$$f_1(z) = f(z)|_{[a,b]} \text{ and } f_2(z) = f(z)|_{[-b,-a]}$$

are analytic in the domains

$$D_1 = D_1(a,b) = \left\{ z = \sqrt{\frac{b^2-a^2}{4}(w+\frac{1}{w}) + \frac{a^2+b^2}{2}} \ ; |w| = \frac{b-a}{b+a} \right\}$$

and

$$D_2 = D_2(a,b) = \{z; \ -z \in D_1(a,b)\} ,$$

respectively. Denoting $c = \frac{b-a}{b+a}$ and

$$E_n(f) = \inf_{p_n \in P_n} \max_{x \in K} |f(x) - p_n(x)| \qquad (n = 0,1,\ldots) ,$$

where P_n is the set of all polynomials of degree at most n, they show that

$$\overline{\lim_{n\to\infty}} (E_n(f))^{1/n} \leqslant \sqrt{c} .$$

In this paper, by imposing certain continuity conditions on the function on the boundaries, we sharpen the result of [1]. However, our main result will be the investigation of the behaviour of the best approximating polynomials between the intervals where they approximate. It will turn out that if the function is smooth enough on the boundaries of analyticity, then the best approximating polynomial will be bounded in the interval mentioned above.

Let $\omega(h)$ be any given modulus of continuity, and for a function g(z), continuous on D_k, let

$$\omega(g,h) = \sup_{\substack{z_1,z_2 \in D_k \\ |z_1 - z_2| \leq h}} |g(z_1) - g(z_2)| \qquad (k = 1 \text{ or } 2)$$

be its modulus of continuity. Let $D = D_1 \cup D_2$, and

$$AC(\omega)D = \{f; f_k = f\,|_{D_k} \text{ is analytic inside } D_k (k=1,2) \text{ and}$$

$$\omega(f,h) = \max_{k=1,2} \omega(f_k,h) = O(\omega(h)) \text{ as } h \to 0+\} .$$

THEOREM 1. _If_ $f(z) \in AC(\omega)D$, _then_

(1) $$E_n(f) = O(\omega(\tfrac{1}{\sqrt{n}}) + \tfrac{1}{\sqrt{n}})\, c^{n/2}$$

and

(2) $$\max_{|x| \leq a} |p_n^*(x)| = O\left(1 + \int_{1/\sqrt{n}}^{1} \frac{\omega(h)}{h}\, dh \right) ,$$

where $p_n^*(x)$ _is the best approximating polynomial of_ $f(x)$ _on_ K.

REMARKS. 1. It is shown in [1] that the factor $c^{n/2}$ in (1) cannot be replaced by a $d^{n/2}$ with $d < c$. If we take $f(x) = \operatorname{sgn} x$, then $\omega(h) \equiv 0$, and the estimate becomes $O(n^{-1/2} c^{n/2})$. By the result of Fuchs quoted above, the factor $n^{-1/2}$ cannot be improved, either.

2. The estimate (2) is not sharp: if we take $f(x) = 1/\log|x|$, then (2) gives only $O(\log \log n)$, while, in fact, a more careful calculation shows that it should be $O(1)$. The highest possible increase of the best approximating polynomials is expressed in the following obvious

COROLLARY. _If_ $f(z)$ _is analytic inside_ D _and continuous on_ D, _then for the best approximating polynomials_ $p_n^*(x)$ _on_ K _we have_

$$\max_{|x| \leq a} |p_n^*(x)| = o(\log n) .$$

PROOF OF THEOREM 1. The proof can be considered as a quantitative version of the corresponding proof given in [1]. Since $f_1(\sqrt{x})$ is analytic on $[a^2,b^2]$,

$$f_1(\sqrt{x}) = \sum_{k=0}^{\infty} a_k T_k\left(\frac{2x-a^2-b^2}{b^2-a^2}\right) \qquad (a^2 \leq x \leq b^2),$$

where $T_k(\cdot)$ is the Chebyshev polynomial of degree k normalized by $T_k(1)=1$, and

$$a_k = \frac{2b_k}{b^2-a^2} \int_{a^2}^{b^2} \frac{f_1(\sqrt{x})T_k\left(\frac{2x-a^2-b^2}{b^2-a^2}\right)}{\sqrt{1-\left(\frac{2x-a^2-b^2}{b^2-a^2}\right)^2}}\,dx = b_k \int_{-1}^{1} f_1\left(\sqrt{\frac{b^2-a^2}{2}x+\frac{a^2+b^2}{2}}\right) \times$$

$$\times \frac{T_k(x)}{\sqrt{1-x^2}}\,dx = \frac{b_k}{2}\int_{-\pi}^{\pi} f_1\left(\sqrt{\frac{b^2-a^2}{2}\cos t+\frac{a^2+b^2}{2}}\right)\cos kt\,dt =$$

$$= \frac{b_k}{4i}\int_{|z|=1} f_1(\sqrt{u(z)})(z^{k-1}+z^{-k-1})\,dz \qquad (k=0,1,2,\dots)$$

with

$$b_k = \begin{cases} 1/\pi & \text{if } k=0 \\ 2/\pi & \text{if } k=1,2,\dots \end{cases}$$

and

$$u(z) = \frac{b^2-a^2}{4}\left(z+\frac{1}{z}\right)+\frac{a^2+b^2}{2} = \frac{a^2+b^2}{2}(1-\cos t)+iab\sin t = iabt+O(t^2) \qquad (t\to 0).$$

Hence, with $z=-ce^{it}$,

$$a_k = \frac{b_k}{2i}\int_{|z|=c} f_1(\sqrt{u(z)})z^{k-1}dz = \frac{b_k}{2}(-c)^k\int_{-\pi}^{\pi} f_1(\sqrt{u(-ce^{it})})e^{ikt}dt =$$

$$= (-1)^k\frac{b_k}{2}c^k\int_{-\pi}^{\pi}\{f_1(\sqrt{u(-ce^{i(t+\frac{\pi}{k})})}) - f_1(\sqrt{u(-ce^{it})})\}e^{ikt}dt.$$

Here

$$|\sqrt{u(-ce^{i(t+\frac{\pi}{k})})} - \sqrt{u(-ce^{it})}| = O(\frac{1}{\sqrt{k}}) \min(1, \frac{t}{\sqrt{kt}}) .$$

Hence $|a_k| = O(c^k \omega(1/\sqrt{k}))$, $k = 1,2,\ldots$, and with the notation

$$S_m(f_1(\sqrt{x}),x) = \sum_{k=o}^{m} a_k T_k(\frac{2x-a^2-b^2}{b^2-a^2})$$

we obtain

$$(3) \quad |f_1(x) - S_m(f_1(\sqrt{x}),x^2)| \leqslant \sum_{k=m+1}^{\infty} |a_k| = O(\omega(\frac{1}{\sqrt{m}})c^m) \qquad (a \leqslant x \leqslant b , m = 1,2,\ldots) .$$

Similarly, we have

$$\frac{f_1(\sqrt{x})}{\sqrt{x}} = \sum_{k=o}^{\infty} a'_k T_k(\frac{2x-a^2-b^2}{b^2-a^2}) \qquad (a^2 \leqslant x \leqslant b^2) ,$$

where, as above,

$$|a'_k| \leqslant \frac{b^2-a^2}{4} c^k \int_{-\pi}^{\pi} | \frac{f_1(\sqrt{u(-ce^{i(t+\frac{\pi}{k})})})}{\sqrt{u(-ce^{i(t+\frac{\pi}{k})})}} - \frac{f_1(\sqrt{u(-ce^{it})})}{\sqrt{u(-ce^{it})}} | dt =$$

$$= O(c^k) \int_0^{\pi} \frac{\omega(\frac{1}{\sqrt{k}})}{\sqrt{t}} dt + \int_0^{\pi} \frac{\min(1,\frac{1}{\sqrt{kt}})}{\sqrt{k t(t+\frac{\pi}{k})}} dt = O(\omega(\frac{1}{\sqrt{k}}) + \frac{1}{\sqrt{k}})c^k \quad (k = 1,2,\ldots)$$

so that

$$|\frac{f_1(x)}{x} - S_m(\frac{f_1(\sqrt{x})}{\sqrt{x}}, x^2)| = O(\omega(\frac{1}{\sqrt{m}}) + \frac{1}{\sqrt{m}}) c^m \qquad (a \leqslant x \leqslant b) .$$

Let now

$$p_{n,1}(x) = \frac{1}{2}[S_{[n/2]}(f_1(\sqrt{x}),x^2) + xS_{[(n-1)/2]}(\frac{f_1(\sqrt{x})}{\sqrt{x}},x^2)] \in P_n .$$

Then it follows that

$$|f_1(x) - p_{n,1}(x)| = O(\omega(\frac{1}{\sqrt{n}}) + \frac{1}{\sqrt{n}})c^{n/2} \qquad (a \leqslant x \leqslant b) ,$$

and

$$\max_{-b\leq x\leq -a}|p_{n,1}(x)| = \frac{1}{2}\max_{a\leq x\leq b}|S_{[n/2]}(f_1(\sqrt{x}),x^2)-xS_{[(n-1)/2]}(\frac{f_1(\sqrt{x})}{\sqrt{x}},x^2)| =$$

$$= O(\omega(\frac{1}{\sqrt{n}}) + \frac{1}{\sqrt{n}})\, c^{n/2} \quad .$$

Similarly, we can find a $p_{n,2}(x) \in P_n$ such that

$$|f_2(x) - p_{n,2}(x)| = O(\omega(\frac{1}{\sqrt{n}})+\frac{1}{\sqrt{n}})\, c^{n/2} \qquad (-b\leq x\leq -a)\ ,$$

$$\max_{a\leq x\leq b}|p_{n,2}(x)| = O(\omega(\frac{1}{\sqrt{n}}) + \frac{1}{\sqrt{n}})\, c^{n/2} \quad .$$

Thus with $p_n(x) = p_{n,1}(x) + p_{n,2}(x)$ we have

$$(4) \quad |f(x) - p_n(x)| \leq |f(x) - p_{n,1}(x)| + |p_{n,2}(x)| = O(\omega(\frac{1}{\sqrt{n}}+\frac{1}{\sqrt{n}})\, c^{n/2}$$

if $x \in [a,b]$; the same holds if $x \in [-b,-a]$. This establishes (1).

To prove (2), we write

$$T_k(\frac{2x^2-a^2-b^2}{b^2-a^2}) = \frac{(-1)^k}{2}\left\{(\frac{\sqrt{a^2-x^2}+\sqrt{b^2-x^2}}{\sqrt{b^2-a^2}})^{2k} + (\frac{\sqrt{b^2-x^2}-\sqrt{a^2-x^2}}{\sqrt{b^2-a^2}})^{2k}\right\} =$$

$$= \frac{(-1)^k}{2}\{\varphi(x)^k + \varphi(x)^{-k}\} \qquad (|x|\leq a,\ k=0,1,\ldots)\ ,$$

where

$$1 \leq \varphi(x) = \frac{(\sqrt{a^2-x^2}+\sqrt{b^2-x^2})^2}{b^2-a^2} \leq \frac{1}{c} \qquad (|x|\leq a)\ .$$

Thus we have

$$|S_m(f_1(\sqrt{x}),x^2)| = \frac{1}{\pi}\left|\int_{|z|=c} f_1(\sqrt{u(z)})\frac{1-(-z\varphi(x))^{m+1}}{1+z\varphi(x)}\frac{dz}{z}\right| + O(1) =$$

$$= \frac{1}{\pi}\left|\int_{|z|=c,\,|\arg z|\geq 1/m}[f_1(\sqrt{u(z)}) - f_1(0)]\frac{1-(-z\varphi(x))^{m+1}}{1+z\varphi(x)}\frac{dz}{z}\right| + O(1) =$$

$$= O(1) \int_{1/m\leq|t|\leq\pi} \frac{\omega(\sqrt{|u(-ce^{it})|})}{|\frac{1}{\varphi(x)} - ce^{it}|} dt = O(1 + \int_{1/m}^{1} \frac{\omega(\sqrt{t})}{t} dt) \qquad (|x| \leq a) .$$

In order to estimate $xS_m(\frac{f_1(\sqrt{x})}{\sqrt{x}}, x^2)$, we may assume that $0 \leq x \leq a$. Then

$$|xS_m(\frac{f_1(\sqrt{x})}{\sqrt{x}}, x^2)| = x\frac{b^2-a^2}{8} \left| \int_{|z|=c} \frac{f_1(\sqrt{u(z)})}{\sqrt{u(z)}} \frac{1-(-z\varphi(x))^{m+1}}{1+z\varphi(x)} \frac{dz}{z} \right| + O(1) =$$

$$= O(1)x \left\{ \int_0^{1-c\varphi(x)} \frac{dt}{\sqrt{\sin\frac{t}{2}}\,(1-c\varphi(x))} + \int_{1-c\varphi(x)}^{\pi} \frac{dt}{t\sqrt{\sin\frac{t}{2}}} \right\} = O(1) \frac{x}{\sqrt{1-c\varphi(x)}} =$$

$$= O(1) \frac{x}{\sqrt{1-(1-\frac{x^2}{ab})^2}} = O(1) \frac{1}{\sqrt{\frac{2}{ab} - \frac{x^2}{a^2b^2}}} = O(1) .$$

Hence,

$$\max_{|x|\leq a} |p_{n,1}(x)| = O(1 + \int_{1/\sqrt{n}}^{1} \frac{\omega(t)}{t} dt),$$

and similarly,

$$\max_{|x|\leq a} |p_{n,2}(x)| = O(1 + \int_{1/\sqrt{n}}^{1} \frac{\omega(t)}{t} dt).$$

Thus we have the estimate

$$\max_{|x|\leq a} |p_n(x)| = O(1 + \int_{1/\sqrt{n}}^{1} \frac{\omega(t)}{t} dt). \tag{5}$$

In order to get the same estimate for the _best_ approximating polynomial $p_n^*(x)$, we need the following

LEMMA. _If for a_ $p(x) \in P_n$ _we have_ $\max_{a\leq|x|\leq b} |p(x)| \leq M$, _then_

$$\max_{|x|\leq a} |p(x)| \leq Mc^{-n/2} .$$

PROOF. This is just a special case of a lemma due to J. Walsh [3, p. 77] applied to the set K. Namely, the multi-valued function $w = w(z)$ mapping the complement of K onto the exterior of $|w| = 1$ is defined by

$$z^2 = \frac{b^2-a^2}{4}(w^2+w^{-2})+\frac{a^2+b^2}{2} .$$

So if $|w| = 1/\sqrt{c}$, then the corresponding level curve is D which contains the interval $[-a,a]$; so the lemma follows.

Turning to the proof of (2) of Theorem 1, we note that by (1)

$$|p_n^*(x) - p_n(x)| \leq |p_n^*(x) - f(x)| + |f(x) - p_n(x)| = O(\omega(\tfrac{1}{\sqrt{n}})+\tfrac{1}{\sqrt{n}})c^{n/2} \quad (a \leq |x| \leq b),$$

and thus by the Lemma and (5)

$$|p_n^*(x)| \leq |p_n^*(x) - p_n(x)| + |p_n(x)| = O(\omega(\tfrac{1}{\sqrt{n}})) + O(1 + \int_{1/\sqrt{n}}^{1} \frac{\omega(t)}{t}\, dt) =$$

$$= O(1 + \int_{1/\sqrt{n}}^{1} \frac{\omega(t)}{t}\, dt) \qquad (|x| \leq a) .$$

This completes the proof of Theorem 1.

The next theorem gives a lower estimate for the norm of best approximating polynomials in $[-a,a]$ in the class $AC(\omega)D$.

THEOREM 2. <u>For any modulus of continuity</u> $\omega(h)$, <u>there exists a function</u> $g(z) \in AC(\omega)D$ such that

$$\limsup_{n\to\infty} \frac{|p_n^*(0)|}{\omega(\frac{1}{n}) \log n} > 0 ,$$

$p_n^*(x)$ <u>being the best approximating polynomial of</u> $g(x)$ <u>on</u> K.

PROOF. The proof is an extension of the famous example of Fejér of a continuous function whose Fourier series diverges at a point. We may assume that

$$\limsup_{n\to\infty} \omega(\tfrac{1}{n}) \log n = \infty , \tag{6}$$

since otherwise the statement is trivial. Let

$$\varphi_n(z) = \sum_{\substack{k=1 \\ k\neq n+1}}^{2n+1} \frac{T_k(\frac{2z-a^2-b^2}{b^2-a^2})}{(n-k+1)T_k(-\frac{a^2+b^2}{b^2-a^2})} .$$

First we prove that $\max_{z\in D}|\varphi_n(z^2)| = O(1)$. Let $w = -ce^{it}$ and $z^2 = \frac{b^2-a^2}{4}(w+w^{-1}) + \frac{a^2+b^2}{2}$; then $\frac{2z^2-a^2-b^2}{b^2-a^2} = \frac{1}{2}(w+w^{-1})$, and

$$T_k(\frac{2z^2-a^2-b^2}{b^2-a^2}) = \frac{1}{2}(w^k+w^{-k}) = \frac{(-1)^k}{2}(c^k+c^{-k})\cos kt + i\frac{(-1)^k}{2}(c^k-c^{-k})\sin kt ,$$

$$T_k(-\frac{a^2+b^2}{b^2-a^2}) = \frac{(-1)^k}{2}(c^k+c^{-k}) .$$

Thus we have

$$\varphi_n(z^2) = \sum_{\substack{k=1 \\ k\neq n+1}}^{2n+1} (\frac{\cos kt}{n-k+1} - i\frac{\sin kt}{n-k+1} + i\frac{2c^k\sin kt}{(c^k+c^{-k})(n-k+1)}) .$$

Here the first sum is exactly the one for which Fejêr proved the boundedness. For the second sum we have

$$\left|\sum_{\substack{k=1 \\ k\neq n+1}}^{2n+1} \frac{\sin kt}{n-k+1}\right| = \left|\sum_{k=1}^{n} \frac{\sin(n-k+1)t-\sin(n+k+1)t}{k}\right| = \left|-2\cos(n+1)t \sum_{k=1}^{n} \frac{\sin kt}{k}\right| = O(1) ,$$

the latter estimate being well-known. Finally, the third sum can be estimated by

$$\sum_{\substack{k=1 \\ k\neq n+1}}^{2n+1} \frac{2c^k\sin kt}{(c^k+c^{-k})(n-k+1)} \leqslant 2\sum_{k=1}^{\infty} c^{2k} = O(1) .$$

Now it follows from the definition of $\varphi_n(z)$ that for any pair of integers m,n we have

$$(7) \qquad S_m(\varphi_n(z),0) \geqslant 0 \ , \ S_n(\varphi_n(z),0) = \sum_{k=1}^{n} \frac{1}{k} \geqslant \log n .$$

Let us define a sequence of indices $1 = n_1 < n_2 < \dots$ with the following properties

(8) $$\omega\left(\frac{1}{n_{k+1}}\right) \leq \frac{1}{2}\,\omega\left(\frac{1}{n_k}\right), \quad n_{k+1}\,\omega\left(\frac{1}{n_{k+1}}\right) \geq 2n_k\,\omega\left(\frac{1}{n_k}\right) \qquad (k = 1,2,\dots)$$

(the latter is possible because of (6)). Further let

$$g(z) = \sum_{k=1}^{\infty} \omega\left(\frac{1}{n_k}\right)\varphi_{n_k}(z^2) \qquad (z \in D)\,.$$

By (8), the series here converges uniformly on D; hence g(z) is analytic inside D. We show that $g(z) \in AC(\omega)D$. Let

$$1/n_{j+1} \leq h \leq 1/n_j\,, \quad z_1, z_2 \in D\,, \quad |z_1 - z_2| \leq h\,.$$

Then by the definition of D we have

$$z_k^2 = \frac{b^2-a^2}{4}\left(-ce^{it_k} - \frac{1}{c}\,e^{-it_k}\right) + \frac{a^2+b^2}{2} = (a^2+b^2)(1-\cos t_k) + iab\sin t_k \quad (k = 1,2)\,.$$

Hence

$$\sin\frac{t_1-t_2}{2} = \frac{z_1^2 - z_2^2}{2(a^2+b^2)\sin\frac{t_1+t_2}{2} + 2iab\cos\frac{t_1+t_2}{2}}\,,$$

i.e. $|t_1-t_2| = O(|z_1-z_2|) = O(h)$. Since $\varphi_n(z^2)$ is a trigonometric polynomial of degree $2n+1$ of the variable t, evidently

$$|\varphi_n(z_1^2) - \varphi_n(z_2^2)| \leq |t_1 - t_2|\left|\frac{d\varphi_n(z^2)}{dt}\right| \leq (2n+1)|t_1 - t_2|\max_t|\varphi_n(z^2)| = O(nh)\,.$$

Thus it follows that

$$|g(z_1) - g(z_2)| \leq \left(\sum_{k=1}^{j} + \sum_{k=j+1}^{\infty}\right)\omega\left(\frac{1}{n_k}\right)|\varphi_{n_k}(z_1^2) - \varphi_{n_k}(z_2^2)| =$$

$$= O(h)\sum_{k=1}^{j} n_k\,\omega\left(\frac{1}{n_k}\right) + O(1)\sum_{k=j+1}^{\infty}\omega\left(\frac{1}{n_k}\right) = O\left(hn_j\,\omega\left(\frac{1}{n_j}\right) + \omega\left(\frac{1}{n_{j+1}}\right)\right) = O(\omega(h))\,.$$

On the other hand, by (7)

$$S_{n_j}(g(\sqrt{x}),0) = \sum_{k=1}^{\infty} \omega(\frac{1}{n_k}) S_{n_j}(\varphi_{n_k}(x),0) \geq \omega(\frac{1}{n_j}) \log n_j .$$

Since $g(z) \in AC(\omega)D$, it is clear from the proof of (1) that

$$|g(x) - S_{n_j}(g(\sqrt{x}),x^2)| = O(\omega(\frac{1}{\sqrt{n_j}})c^{n_j}) \qquad (a \leq x \leq b)$$

(see (3)). Hence

$$|p^*_{2n_j}(x) - S_{n_j}(g(\sqrt{x}),x^2)| = O(\omega(\frac{1}{\sqrt{n_j}})c^{n_j}) \qquad (a \leq x \leq b) ,$$

i.e. applying the Lemma again,

$$|p^*_{2n_j}(0)| \geq |S_{n_j}(g(\sqrt{x}),0)| - |p^*_{2n_j}(0) - S_{n_j}(g(\sqrt{x}),0|$$

$$\geq \omega(\frac{1}{n_j}) \log n_j - O(\omega(\frac{1}{\sqrt{n_j}})c^{n_j}c^{-n_j}) \geq \frac{1}{2}\omega(\frac{1}{n_j}) \log n_j ,$$

which proves the theorem.

REFERENCES

[1] Chui, C.K. - Hasson, M., Degree of uniform approximation on disjoint intervals. Pacific J. Math. 105 (1983), 291 - 297.

[2] Fuchs, W.H.J., On the degree of Chebyshev approximation on sets with several components. Izv. Akad. Nauk Armjan. SSR Ser. Mat. 13 (1978), 396 - 404.

[3] Walsh, J.L., Interpolation and Approximation by Rational Functions in the Complex Domain. AMS Coll. Publ., Vol. XX, Providence 1960.

International Series of
Numerical Mathematics, Vol. 65

EXTREMALPOLYNOME IN DER L^1- UND L^2-NORM AUF ZWEI DISJUNKTEN INTERVALLEN

Franz Peherstorfer
Institut für Mathematik
Johannes Kepler Universität
Linz

Let $\alpha,\beta \in (-1,+1)$ with $\alpha \leq \beta$ and let P_n be a polynomial of degree n with leading coefficient one which deviates least from zero on $[-1,\alpha] \cup [\beta,1]$ with respect to the L^1-norm. We show that P_n can be represented as a product of two polynomials which are orthogonal with respect to weight functions vanishing on (α,β). For the recursion coefficients of those orthogonal polynomials a recurrence relation is given.

1. L^1-Extremalpolynome auf $[-1,\alpha] \cup [\beta,1]$

Sei $\alpha,\beta \in (-1,+1)$ mit $\alpha \leq \beta$. Wir betrachten folgendes L^1-Approximationsproblem: Sei $n \in \mathbb{N}$. Bestimme ein Polynom $P_n = x^n + \ldots$, so daß

$$\int_{-1}^{\alpha} |P_n(x)|\,dx + \int_{\beta}^{1} |P_n(x)|\,dx$$

$$= \min_{\substack{a_i \in \mathbb{R} \\ i \in \{1,\ldots,n\}}} \left\{ \int_{-1}^{\alpha} |x^n + a_1 x^{n-1} + \ldots + a_n|\,dx + \int_{\beta}^{1} |x^n + a_1 x^{n-1} + \ldots + a_n| \right\}$$

P_n nennen wir im folgenden L^1-Extremalpolynom.

Dieses Problem wurde zum erstenmal von N.I. Ahiezer, in [1] untersucht. Ahiezer stellte das Extremalpolynom mit Hilfe von elliptischen Funktionen dar. Wir wollen zeigen, daß sich P_n als Produkt zweier orthogonaler Polynome darstellen läßt.

LEMMA 1 (siehe [1]):

(a) P_n <u>ist Extremalpolynom</u> $\Longleftrightarrow$ $\int_{-1}^{\alpha} x^k \operatorname{sgn} P_n(x)dx + \int_{\beta}^{1} x^k \operatorname{sgn} P_n(x)dx = 0$ <u>für</u> $k = 0,\dots,n-1$.

(b) <u>Ist</u> P_n <u>Extremalpolynom, dann besitzt</u> P_n <u>lauter einfache Nullstellen, und mindestens</u> (n-1) <u>Nullstellen liegen in</u> $(-1,\alpha) \cup (\beta,1)$.

(c) <u>Es existiert immer ein Extremalpolynom</u> P_n, <u>das keine Nullstelle in</u> (α,β) <u>besitzt</u>.

(d) <u>Ist</u> P_n <u>Extremalpolynom und besitzt</u> P_n <u>alle Nullstellen in</u> $(-1,\alpha) \cup (\beta,1)$, <u>dann ist</u> P_n <u>eindeutig bestimmt</u>.

BEWEIS. (a), (b) und (c) wurden in [1] gezeigt.
Zu (d): Wegen (a) gilt

$$\int_{-1}^{\alpha} x^k \operatorname{sgn} P_n + \int_{\beta}^{1} x^k \operatorname{sgn} P_n = 0 \quad \text{für} \quad k = 0,\dots,n-1.$$

Annahme:es existiert ein weiteres Extremalpolynom $\overline{P}_n$. O.B.d.A. können wir annehmen, daß $x_n > y_n$, wobei x_n bzw. y_n die größte Nullstelle von P_n bzw. $\overline{P}_n$ bezeichnet. Wegen (a) gilt nun

$$(1) \qquad \int_{-1}^{\alpha} x^k (\operatorname{sgn} P_n - \operatorname{sgn} \overline{P}_n) + \int_{\beta}^{1} x^k (\operatorname{sgn} P_n - \operatorname{sgn} \overline{P}_n) = 0 \quad \text{für} \quad k = 0,\dots,n-1.$$

Da $\operatorname{sgn} P_n - \operatorname{sgn} \overline{P}_n$ höchstens (n-1) Vorzeichenwechsel auf [-1,+1] besitzt, folgt wegen (1), daß $\operatorname{sgn} P_n = \operatorname{sgn} \overline{P}_n$ auf [-1,+1].

BEZEICHNUNG. Es sei

$$w(x) = \begin{cases} \dfrac{1}{\pi\sqrt{1-x^2}} \sqrt{\dfrac{x-\alpha}{x-\beta}} & \text{für} \quad x \in (-1,\alpha) \cup (\beta,1) \\ 0 & \text{für} \quad x \in (\alpha,\beta) \end{cases}$$

und

$$\tilde{w}(x) = \begin{cases} \dfrac{1}{\pi\sqrt{1-x^2}} \sqrt{\dfrac{x-\beta}{x-\alpha}} & \text{für} \quad x \in (-1,\alpha) \cup (\beta,1) \\ 0 & \text{für} \quad x \in (\alpha,\beta). \end{cases}$$

$p_n = x^n + \dots$ ($\tilde{p}_n = x^n + \dots$) bezeichne das zur Gewichtsfunktion w ($\tilde{w}$) orthogonale Polynom vom Grad n mit Hauptkoeffizienten 1, d.h.

$$\int_{-1}^{+1} x^k p_n(x)w(x)dx = 0 \quad \text{für} \quad k = 0,\dots,n-1.$$

Man beachte, daß die Polynome $\{p_n\}$ $(\{\tilde{p}_n\})$ für $n \geq 2$ Rekurrenzrelationen der Form

$$p_n(x) = (x-\alpha_n)p_{n-1}(x) - \lambda_n p_{n-2}(x)$$

$$(\tilde{p}_n(x) = (x-\tilde{\alpha}_n)\tilde{p}_{n-1}(x) - \tilde{\lambda}_n \tilde{p}_{n-2}(x))$$

mit $p_1(x) = x-\alpha_1$ $(\tilde{p}_1(x) = x-\tilde{\alpha}_1)$ und $p_o(x) = \tilde{p}_o(x) = 1$ genügen.

Ist v ein auf $(-1,+1)$ nichtnegatives Polynom, dann bezeichnet p_n^v $(\tilde{p}_n^v)$ das zu vw $(v\tilde{w})$ orthogonale Polynom vom Grad n mit Hauptkoeffizient 1.

SATZ 1. Sei P_n L^1-Extremalpolynom, das keine Nullstelle in (α,β) besitzt. Dann gilt für $m \in \mathbb{N}$

$$P_{2m-1} = \begin{cases} p_m \cdot \tilde{p}_{m-1}^{(1-x^2)}, & \text{falls } \int_{-1}^{+1} p_m^2 w \leq \int_{-1}^{+1} \tilde{p}_m^2 \tilde{w}, \\ \tilde{p}_m \cdot p_{m-1}^{(1-x^2)}, & \text{falls } \int_{-1}^{+1} \tilde{p}_m^2 \tilde{w} \leq \int_{-1}^{+1} p_m^2 w, \end{cases}$$

und

$$\int_{-1}^{\alpha} |P_{2m-1}| + \int_{\beta}^{1} |P_{2m-1}| = 2 \min\{\int_{-1}^{+1} p_m^2 w, \int_{-1}^{+1} \tilde{p}_m^2 \tilde{w}\},$$

sowie

$$P_{2m} = \begin{cases} p_m^{(1+x)} \tilde{p}_m^{(1-x)}, & \text{falls } \int_{-1}^{+1} [p_m^{(1+x)}]^2 (1+x)w \leq \int_{-1}^{+1} [\tilde{p}_m^{(1+x)}]^2 (1+x)\tilde{w}, \\ \tilde{p}_m^{(1+x)} p_m^{(1-x)}, & \text{falls } \int_{-1}^{+1} [\tilde{p}_m^{(1+x)}]^2 (1+x)\tilde{w} \leq \int_{-1}^{+1} [p_m^{(1+x)}]^2 (1+x)w, \end{cases}$$

und

$$\int_{-1}^{\alpha} |P_{2m}| + \int_{\beta}^{1} |P_{2m}| = 2 \min\{\int_{-1}^{+1} [p_m^{(1+x)}]^2 (1+x)w, \int_{-1}^{+1} [\tilde{p}_m^{(1+x)}]^2 (1+x)\tilde{w}\}.$$

BEWEIS. Sei P_{2m-1} Extremalpolynom. Dann gilt wegen Lemma 1 (a), daß

$$\int_{-1}^{+1} U_k \operatorname{sgn} P_{2m-1} = \int_{\alpha}^{\beta} U_k \operatorname{sgn} P_{2m-1} \quad \text{für} \quad k = 0,\dots,2m-2,$$

wobei U_k das Tchebycheffpolynom 2. Art bezeichnet.

Aus [5, Lemma 6] folgt

$$P_{2m-1} = \begin{cases} p_m \tilde{p}_{m-1}^{(1-x^2)}, & \text{falls } \operatorname{sgn} P_{2m-1} = -1 \text{ auf } (\alpha,\beta), \\ \tilde{p}_m p_{m-1}^{(1-x^2)}, & \text{falls } \operatorname{sgn} P_{2m-1} = 1 \text{ auf } (\alpha,\beta). \end{cases}$$

Sei nun $R_{2m-1} = p_m \tilde{p}_{m-1}^{(1-x^2)}$ und $S_{2m-1} = \tilde{p}_m p_{m-1}^{(1-x^2)}$. Dann folgt mit [5, Lemma 6, Satz 1, Lemma 1 und Lemma 2]

$$\int_{-1}^{+1} U_k \operatorname{sgn} R_{2m-1} = - \int_\alpha^\beta U_k \quad \text{für } k = 0,\dots,2m-2$$

und

$$\int_{-1}^{+1} U_{2m-1} \operatorname{sgn} R_{2m-1} = - \int_\alpha^\beta U_{2m-1} + 2^{2m} \int_{-1}^{+1} p_m^2 w$$

sowie

$$\int_{-1}^{+1} U_k \operatorname{sgn} S_{2m-1} = \int_\alpha^\beta U_k \quad \text{für } k = 0,\dots,2m-2$$

und

$$\int_{-1}^{+1} U_{2m-1} \operatorname{sgn} S_{2m-1} = \int_\alpha^\beta U_{2m-1} + 2^{2m} \int_{-1}^{+1} \tilde{p}_m^2 \tilde{w}.$$

Ist nun $P_{2m-1} = R_{2m-1}$, also $\operatorname{sgn} P_{2m-1} = \operatorname{sgn} R_{2m-1} = -1$ auf (α,β), dann folgt

$$2 \int_{-1}^{+1} p_m^2 w = \int_{-1}^{\alpha} |R_{2m-1}| + \int_\beta^1 |R_{2m-1}| < \int_{-1}^{\alpha} |S_{2m-1}| + \int_\beta^1 |S_{2m-1}|$$

$$= 2 \int_{-1}^{+1} \tilde{p}_m^2 \tilde{w} + \int_\alpha^\beta S_{2m-1} - \int_\alpha^\beta |S_{2m-1}| \leq 2 \int_{-1}^{+1} \tilde{p}_m^2 \tilde{w}.$$

Analog folgt $\int_{-1}^{+1} \tilde{p}_m^2 \tilde{w} \leq \int_{-1}^{+1} p_m^2 w$, falls $P_{2m-1} = S_{2m-1}$. Ist $\int_{-1}^{+1} p_m^2 w = \int_{-1}^{+1} \tilde{p}_m^2 \tilde{w}$, dann folgt, daß sowohl R_{2m-1} als auch S_{2m-1} Extremalpolynom ist.

Für $n = 2m$ ergibt sich die Behauptung analog.

2. Bestimmung der Rekursionskoeffizienten

LEMMA 2.

(a) $\int_{-1}^{+1} \frac{w(t)}{(1-xt)}\,dt = \sqrt{\frac{1-\alpha x}{1-\beta x}} \cdot \frac{1}{\sqrt{1-x^2}}$ <u>für</u> $x \in (-1,+1)$.

(b) $\int_{-1}^{+1} \frac{\tilde{w}(t)}{(1-xt)}\,dt = \sqrt{\frac{1-\beta x}{1-\alpha x}} \cdot \frac{1}{\sqrt{1-x^2}}$ <u>für</u> $x \in (-1,+1)$.

BEWEIS. Zu (a). Sei

$$F(z) = \left(\frac{1-2\alpha z+z^2}{1-2\beta z+z^2}\right)^{1/2} \quad \text{für} \quad z \in \mathbb{C},\ |z| < 1. \tag{2}$$

Dann gilt: F ist analytisch im offenen Einheitskreis und $\lim_{r\to 1^-} \operatorname{Re} F(re^{i\varphi})$ $= \pi w(\cos\varphi)\sin\varphi$ f.ü. für $\varphi \in (0,\pi)$. Somit erhalten wir (siehe [4, p. 64]), daß

$$F(z) = \frac{1}{2}\int_{0}^{2\pi} \frac{e^{i\varphi}+z}{e^{i\varphi}+z}\, w(\cos\varphi)|\sin\varphi|\,d\varphi = (1-z^2)\int_{0}^{\pi} \frac{w(\cos\varphi)\sin\varphi}{1+z^2-2z\cos\varphi}\,d\varphi. \tag{3}$$

Für $z \in \mathbb{C}$ mit $|z| < 1$ sei nun $y = \frac{1}{2}(z + \frac{1}{z})$, d.h. $y \notin [-1,+1]$ u. $z = y-\sqrt{y^2-1}$. Dann folgt mit (2) und (3)

$$\left(\frac{y-\alpha}{y-\beta}\right)^{1/2} = \left(\frac{(1/2)(z+1/z)-\alpha}{(1/2)(z+1/z)-\beta}\right)^{1/2} = F(z) = \sqrt{y^2-1}\int_{-1}^{+1} \frac{w(t)}{y-t}\,dt.$$

Setzt man $x = \frac{1}{y}$, $y \in \mathbb{R}\setminus[-1,+1]$, so folgt die Behauptung

(b) ergibt sich analog.

BEZEICHNUNG. Sei $s_n(x) = \prod_{i=1}^{n}(x-x_i)$. Dann bez. $s_n^*(x) = x^n s_n(\frac{1}{x}) = \prod_{i=1}^{n}(1-x_i x)$ das zu s_n reziproke Polynom.

q_{n-1} ($\tilde{q}_{n-1}$) bezeichne das zu p_n ($\tilde{p}_n$) gehörige Polynom zweiter Art; d.h.

$$q_{n-1}(x) = \int_{-1}^{+1} \frac{p_n(x)-p_n(t)}{x-t}\, w(t)\,dt.$$

LEMMA 3.

(a) Für $n \in \mathbb{N}$ gilt:

$$(x-\alpha)p_n^2(x) - (x-\beta)(x^2-1)q_{n-1}^2(x) = A_n x + B_n \quad \text{mit}$$

$$A_n = 2\int_{-1}^{+1} p_n^2 w \quad \text{und} \quad B_n = -(\alpha+\beta)\int_{-1}^{+1} p_n^2 w + 2\int_{-1}^{+1} t p_n^2 w.$$

(b) Für $n \in \mathbb{N}\setminus\{1\}$ gilt:

$$(x-\alpha)p_n(x)p_{n-1}(x) - (x-\beta)(x^2-1)q_{n-1}(x)q_{n-2}(x) = C_n x^2 + D_n x + E_n$$

$$C_n = \int_{-1}^{+1} p_{n-1}^2 w, \quad D_n = -\frac{(\alpha+\beta)}{2}\int_{-1}^{+1} p_{n-1}^2 w \quad \text{für} \quad n \geq 2, \text{ und}$$

$$E_2 = 2\int_{-1}^{+1} p_2^2 w - [1-\lambda_2 + (\beta^2-\alpha^2)/4]\int_{-1}^{+1} p_1^2 w \quad \text{sowie}$$

$$E_n = 2\int_{-1}^{+1} p_n^2 w - [1/2 + (\beta-\alpha)^2/8]\int_{-1}^{+1} p_{n-1}^2 w \quad \text{für} \quad n \geq 3.$$

BEWEIS. Sei

$$\sqrt{\frac{1-\alpha x}{1-\beta x}} \cdot \frac{1}{\sqrt{1-x^2}} = \sum_{i=o}^{\infty} c_i x^i .$$

Dann ergibt eine einfache Rechnung, daß

$$c_1 = \frac{\beta-\alpha}{2} \quad \text{und} \quad c_2 = \frac{3}{8}\beta^2 - \frac{\alpha\beta}{4} - \frac{\alpha^2}{8} + \frac{1}{2}. \tag{4}$$

Im folgenden sei nun das lineare Funktional $c: \mathbb{P} \to \mathbb{R}$, $\mathbb{P}$ Vektorraum der reellen Polynome, definiert durch $c(x^i) = \int_{-1}^{+1} t^i w(t)dt = c_i$ für $i \in \mathbb{N}_o$. Dann folgt mit Lemma 2 (a) (siehe z.B. [3, Theorem 1.17])

$$\sqrt{\frac{1-\alpha x}{1-\beta x}} \cdot \frac{1}{\sqrt{1-x^2}} = \frac{q_{n-1}^*(x)}{p_n^*(x)} + \frac{x^{2n}}{p_n^{*2}(x)} \; c\left(\frac{p_n^2(t)}{1-xt} \right) \tag{5}$$

für $x \in (-1,+1)$ und $n \in \mathbb{N}$.

Zu (a). Aus (5) folgt unmittelbar

$$\frac{(1-\alpha x)}{(1-\beta x)(1-x^2)} = \frac{q_{n-1}^{*2}(x)}{p_n^{*2}(x)} + \frac{2x^{2n}}{p_n^{*2}(x)} \frac{q_{n-1}^*(x)}{p_n^*(x)} \; c\left(\frac{p_n^2(t)}{1-xt} \right) + O(x^{4n}),$$

woraus sich mit $c\left(\frac{p_n^2(t)}{1-xt}\right) = \int_{-1}^{+1} p_n^2 w + \left(\int_{-1}^{+1} t p_n^2 w\right)x + \left(\int_{-1}^{+1} t^2 p_n w\right)x^2 + \ldots$

$$(1-\alpha x)p_n^{*2}(x) - (1-\beta x)(1-x^2)q_{n-1}^{*2}(x) = \left(2\int_{-1}^{+1} p_n^2\right)x^{2n}$$

$$+ 2\left((c_1-\beta)\int_{-1}^{+1} p_n^2 + \int_{-1}^{+1} t p_n\right)x^{2n+1} + O(x^{2n+2}),$$

ergibt. Beachtet man nun, daß in der obigen Formel links ein Polynom vom Grad $(2n+1)$ steht, so folgt (a) mit Hilfe von (4) sofort.

Zu (b). Aus (5) erhalten wir für $n \geqq 2$

$$\frac{(1-\alpha x)}{(1-\beta x)(1-x^2)} = \frac{q_{n-2}^*(x)}{p_{n-1}^*(x)}\,\frac{q_{n-1}^*(x)}{p_n^*(x)} + \frac{x^{2n-2}}{p_{n-1}^{*2}(x)}\,\frac{q_{n-1}^*(x)}{p_n^*(x)}\; c\left(\frac{p_{n-1}^2(t)}{1-xt}\right)$$

$$+ \frac{x^{2n}}{p_n^{*2}(x)}\,\frac{q_{n-2}^*(x)}{p_{n-1}^*(x)}\; c\left(\frac{p_n^2(t)}{1-xt}\right) + O(x^{4n-2}),$$

woraus folgt

(6) $$(1-\alpha x)p_n^*(x)p_{n-1}^*(x) - (1-\beta x)(1-x^2)q_{n-1}^*(x)q_{n-2}^*(x)$$

$$= x^{2n-2}(1-\beta x)(1-x^2)\,\frac{q_{n-1}^*(x)}{p_{n-1}^*(x)}\; c\left(\frac{p_{n-1}^2(t)}{1-xt}\right)$$

$$+ x^{2n}(1-\beta x)(1-x^2)\,\frac{q_{n-2}^*(x)}{p_n^*(x)}\; c\left(\frac{p_n^2(x)}{1-xt}\right) + O(x^{4n-2}).$$

Sei nun

$$\frac{q_{n-1}^*(x)}{p_{n-1}^*(x)} = 1 + \tilde{c}_1 x + \tilde{c}_2 x + O(x^3) \quad \text{für} \quad x \in (-1,+1).$$

Für $n = 2$ gilt wegen

$$\frac{q_1^*(x)}{p_1^*(x)} = \frac{1-\alpha_2 x}{1-\alpha_1 x} = 1 + (\alpha_1-\alpha_2)x + (\alpha_1^2-\alpha_2\alpha_1)x^2 + O(x^3)$$

und $c_1 = \alpha_1$, daß

(7) $$\tilde{c}_1 = c_1-\alpha_2 \quad \text{und} \quad \tilde{c}_2 = c_1^2-\alpha_2 c_1.$$

Für $n \geq 3$ erhalten wir

$$\frac{q^*_{n-1}(x)}{p^*_{n-1}(x)} = (1-\alpha_n x)\,\frac{q^*_{n-2}(x)}{p^*_{n-1}(x)} - \lambda_n x^2\,\frac{q^*_{n-3}(x)}{p^*_{n-1}(x)}$$

$$= 1 + (c_1-\alpha_n)x + (c_2-\lambda_n-\alpha_n c_1)x^2 + O(x^3),$$

also

(7') $$\tilde{c}_1 = c_1-\alpha_n \quad \text{und} \quad \tilde{c}_2 = c_2-\lambda_n-\alpha_n c_1 \quad \text{für} \quad n \geq 3.$$

Aus (6) ergibt sich nun

(8) $$(1-\alpha x)p^*_n(x)p^*_{n-1}(x) - (1-\beta x)(1-x^2)q^*_{n-1}(x)q^*_{n-2}(x)$$

$$= \left(\int_{-1}^{+1} p^2_{n-1}\right)x^{2n-2} + \left[(\tilde{c}_1-\beta)\int_{-1}^{+1} p^2_{n-1} + \int_{-1}^{+1} tp^2_{n-1}\right]x^{2n-1}$$

$$+ \left[\int_{-1}^{+1} p^2_n + (\tilde{c}_2-1-\beta\tilde{c}_1)\int_{-1}^{+1} p^2_{n-1} + (\tilde{c}_1-\beta)\int_{-1}^{+1} tp^2_{n-1} + \int_{-1}^{+1} t^2p^2_{n-1}\right]x^{2n} + O(x^{2n+1})$$

$$= C_n x^{2n-2} + D_n x^{2n-1} + E_n x^{2n}.$$

Die Formel für C_n ist somit bewiesen. Beachten wir, daß

(9) $$\lambda_n = \int_{-1}^{+1} p^2_{n-1} \Big/ \int_{-1}^{+1} p^2_{n-2} \quad \text{und} \quad \alpha_n = \int_{-1}^{+1} tp^2_{n-1} \Big/ \int_{-1}^{+1} p^2_{n-1},$$

so folgt mit (8), (7), (7') und (9), daß für $n \geq 2$

$$D_n = -\frac{(\beta+\alpha)}{2}\int_{-1}^{+1} p^2_{n-1} - \alpha_n \int_{-1}^{+1} p^2_{n-1} + \int_{-1}^{+1} tp^2_{n-1} = -\frac{(\beta+\alpha)}{2}\int_{-1}^{+1} p^2_{n-1}.$$

Zur Berechnung von E_n benötigen wir noch die Beziehung

(10) $$\int_{-1}^{+1} p^2_n = \int_{-1}^{+1} t^2p^2_{n-1} - \alpha_n \int_{-1}^{+1} tp^2_{n-1} - \lambda_n \int_{-1}^{+1} p^2_{n-1}.$$

Für $n = 2$ folgt somit aus (8), (7) und (10)

$$E_2 = 2\int_{-1}^{+1} p^2_2 + (c_1^2-1-\beta c_1+\lambda_2)\int_{-1}^{+1} p^2_1,$$

woraus mit (4) die Behauptung folgt.

Für $n \geq 3$ folgt aus (8), (7') und 10)

$$E_n = 2\int_{-1}^{+1} p_n^2 + (c_2-1-\beta c_1-\alpha_n(c_1-\beta))\int_{-1}^{+1} p_{n-1}^2 + (c_1-\beta)\int_{-1}^{+1} tp_{n-1}^2$$

$$= 2\int_{-1}^{+1} p_n^2 + (c_2-1-\beta c_1)\int_{-1}^{+1} p_{n-1}^2,$$

wobei das zweite Gleichheitszeichen mit Hilfe von (9) folgt. Mit (4) erhalten wir die Formel für E_n.

SATZ 2. Die Orthogonalpolynome $\{p_n\}$ genügen für $n \geq 2$ einer Rekurrenzrelation der Form

$$p_n(x) = (x-\alpha_n)p_{n-1}(x) - \lambda_n p_{n-2}(x)$$

mit $p_1(x) = x-(\beta-\alpha)/2$ und $p_0(x) = 1$, wobei die Rekursionskoeffizienten α_n, λ_n für $n \geq 3$ der folgenden Rekurrenzrelation genügen:

$$\lambda_{n+1} = d\alpha_n - \alpha_n^2 - \lambda_n + (1+c^2)/2,$$

$$\alpha_{n+1} = \lambda_n(\alpha_n+\alpha_{n-1}-d)/\lambda_{n+1} - \alpha_n+d$$

mit $c = (\beta-\alpha)/2$, $d = (\beta+\alpha)/2$ und

$$\lambda_2 = (1-\alpha^2+d^2)/2, \qquad \alpha_2 = -(\beta+\alpha c^2)/2\lambda_2 + d,$$

$$\lambda_3 = d\alpha_2-\alpha_2^2 + (1-\lambda_2+c^2)/2, \quad \alpha_3 = \lambda_2(\alpha_2-\alpha)/2\lambda_3 + d-\alpha_2.$$

BEWEIS. Für $n \geq 2$ gilt

$$(x-\alpha)p_n(x)p_{n-1}(x) - (x-\beta)(x^2-1)q_{n-1}(x)q_{n-2}(x)$$

$$= (x-\alpha_n)[(x-\alpha)p_{n-1}^2(x) - (x-\beta)(x^2-1)q_{n-2}^2(x)]$$

$$- \lambda_n[(x-\alpha)p_{n-1}(x)p_{n-2}(x) - (x-\beta)(x^2-1)q_{n-2}(x)q_{n-3}(x)],$$

d.h. $$C_2x^2 + D_2x + E_2 = (x-\alpha_2)(A_1x+B_1) - \lambda_2(x-\alpha)(x-\alpha_1)$$

und für $n \geq 3$

$$C_n x^2 + D_n x + E_n = (x-\alpha_n)(A_{n-1}x+B_{n-1}) - \lambda_n(C_{n-1}x^2+D_{n-1}x+E_{n-1}).$$

Koeffizientenvergleich und $\alpha_1 = c$ ergibt daher

(11) $\quad E_2 = -\alpha_2 B_1 - \lambda_2 \alpha c$

sowie für $n \geq 3$

(11') $\quad E_n + \lambda_n E_{n-1} = -\alpha_n B_{n-1}.$

Mit Lemma 3 und $S = (1+c^2)/2$ ergibt sich aus (11')

$$2\int_{-1}^{+1} p_n^2 - S\int_{-1}^{+1} p_{n-1}^2 + \lambda_n\{2\int_{-1}^{+1} p_{n-1}^2 - S\int_{-1}^{+1} p_{n-2}^2\}$$

$$= -2\alpha_n\{\int_{-1}^{+1} tp_{n-1}^2 - d\int_{-1}^{+1} p_{n-1}^2\}.$$

Dividiert man nun diese Gleichung durch $\int_{-1}^{+1} p_{n-1}^2$ und beachtet man, daß $\lambda_n = \int_{-1}^{+1} p_{n-1}^2 / \int_{-1}^{+1} p_{n-2}^2$, so erhält man für $n \geq 3$ die gewünschte Rekursionsformel

$$\lambda_{n+1} + \lambda_n - S = -\alpha_n(\alpha_n - d)$$

λ_3 erhält man analog aus (11) und Lemma 3.

Aus Lemma 3 und der Rekurrenzrelation für $\{p_n\}$ folgt für $n \geq 2$

$$A_n x + B_n = (x-\alpha)p_n^2(x) - (x-\beta)(x^2-1)q_{n-1}^2(x)$$

$$= (x-\alpha_n)^2[(x-\alpha)p_{n-1}^2(x) - (x-\beta)(x^2-1)q_{n-2}^2(x)]$$

$$+ \lambda_n^2[(x-\alpha)p_{n-2}^2(x) - (x-\beta)(x^2-1)q_{n-3}^2(x)]$$

$$- 2\lambda_n(x-\alpha_n)[(x-\alpha)p_{n-1}(x)p_{n-2}(x) - (x-\beta)(x^2-1)q_{n-2}(x)q_{n-3}(x)]$$

$$= (x-\alpha_n)^2(A_{n-1}x+B_{n-1}) + \lambda_n^2(A_{n-2}x+B_{n-2})$$

$$- 2\lambda_n(x-\alpha_n)(C_{n-1}x^2+D_{n-1}x+E_{n-1}).$$

Setzen wir $x = \alpha_n$, so ergibt sich für $n = 2$

$$A_2\alpha_2+B_2 = \lambda_2^2(\alpha_2-\alpha)$$

und für $n \geqq 3$

$$A_n\alpha_n+B_n = \lambda_n^2(A_{n-2}\alpha_n+B_{n-2}). \tag{12}$$

Wegen Lemma 3(a) und (9) gilt für $n \geqq 1$

$$B_n/A_n = -d+\int_{-1}^{+1} tp_n^2/\int_{-1}^{+1} p_n^2 = -d+\alpha_{n+1}. \tag{13}$$

Einsetzen in (12) ergibt

$$2\int_{-1}^{+1} p_n^2(\alpha_n+\alpha_{n+1}-d) = 2\lambda_n^2\int_{-1}^{+1} p_{n-2}^2(\alpha_n+\alpha_{n-1}-d)$$

woraus mit (9) für $n \geqq 3$ folgt

$$\lambda_{n+1}(\alpha_n+\alpha_{n+1}-d) = \lambda_n(\alpha_n+\alpha_{n-1}-d),$$

Analog erhält man α_3.

Aus der Beziehung

$$(x-\alpha)(x-c)^2 - (x-\beta)(x^2-1) = A_1x+B_1$$

erhält man A_1 und B_1, woraus sich mit (13) und $A_1 = \lambda_2/2$ die Startwerte α_2, λ_2 ergeben.

BEMERKUNG. Mit Hilfe der Beziehung $\int_{-1}^{+1} p_n^2 w = \prod_{i=2}^{n+1} \lambda_i$ und Satz 2 läßt sich also die Minimalabweichung bestimmen.

KOROLLAR 1. *Die Orthogonalpolynome* $\{\tilde{p}_n\}$ *genügen für* $n \geqq 2$ *einer Rekurrenzrelation der Form*

$$\tilde{p}_n(x) = (x-\tilde{\alpha}_n)\tilde{p}_{n-1}(x) - \tilde{\lambda}_n\tilde{p}_{n-2}(x)$$

mit $\tilde{p}_1(x) = x-(\alpha-\beta)/2$ *und* $\tilde{p}_0(x) = 1$, *wobei die Rekursionskoeffizienten* $\tilde{\alpha}_n$, $\tilde{\lambda}_n$ *für* $n \geqq 3$ *der gleichen Rekurrenzrelation wie die* α_n, λ_n *genügen. Die Anfangswerte berechnen sich folgendermaßen:*

$$\tilde{\lambda}_2 = (1-\beta^2+d^2)/2, \qquad \tilde{\alpha}_2 = -(\alpha+\beta c^2)/2\tilde{\lambda}_2+d,$$

$$\tilde{\lambda}_3 = d\tilde{\alpha}_2-\tilde{\alpha}_2^2+(1-\tilde{\lambda}_2+c^2)/2, \quad \tilde{\alpha}_3 = \tilde{\lambda}_2(\tilde{\alpha}_2-\beta)/2\tilde{\lambda}_3+d-\tilde{\alpha}_2.$$

BEWEIS. Wegen Lemma 1 gilt

$$\sqrt{\frac{1-\beta x}{1-\alpha x}} \cdot \frac{1}{\sqrt{1-x^2}} = \frac{\tilde{q}^*_{n-1}(x)}{\tilde{p}^*_n(x)} + O(x^{2n}),$$

d.h. man braucht in Satz 2 nur α durch β und β durch α zu ersetzen.

BEISPIEL. Für den einfachen Fall $\alpha = -\beta$, $\beta \in [0,1)$, erhalten wir mittels Induktion $\alpha_n = (-1)^{n+1}\beta$ für $n \in \mathbb{N}$, $\lambda_2 = (1-\beta^2)/2$ und $\lambda_n = (1-\beta^2)/4$ für $n \in \mathbb{N}\setminus\{2\}$. Weiters gilt $\tilde{\alpha}_n = -\alpha_n$ und $\tilde{\lambda}_n = \lambda_n$ für $n \in \mathbb{N}$. Somit erhalten wir das bekannte Ergebnis (siehe [2]), daß $p_{2n}(x) = \tilde{p}_{2n}(x) = K.\ T_n\left(\frac{2x^2-\beta^2-1}{1-\beta^2}\right)$, wobei $K \in \mathbb{R}^+$.

REFERENCES

[1] Ahiezer, N. I., Verallgemeinerung einer Korkine-Zolotareffschen Minimum Aufgabe. Soobsc. Naucno-Issled. Inst. Mat. Meh. i Har'kov. Mat. Obsc. 13 (4) (1936), 3-14.

[2] Barkov, G. I., Systems of polynomials orthogonal on two symmetric intervals (in Russian). Izv. Vysh. Uceb. Zaved. Matematika 4 (17) (1960), 3-16.

[3] Brezinski, C., Padé-Type Approximation and general orthogonal polynomials. (ISNM, vol. 50) Birkhauser Verlag, Basel/Stuttgart (1980).

[4] Geronimus, Ya. L., Polynomials orthogonal on a circle and their applications. AMS Translations, Ser. 1, 3 (1962), 1-78.

[5] Peherstorfer, F., Orthogonal polynomials in L^1-approximation. Manuskript.

International Series of
Numerical Mathematics, Vol. 65

SHORTEST PATH ALGORITHMS FOR THE APPROXIMATION BY NOMOGRAPHIC FUNCTIONS

Manfred v. Golitschek
Institut für Angewandte Mathematik und Statistik
der Universität Würzburg
Würzburg

Continuous functions of two real variables are approximated on compact domains in the uniform norm by the classes NOM of nomographic functions which appear as approximation subspaces in bivariate approximation theory, in the theory of integral equations, functional equations, scalings of matrices, Goursat-type problems for the wave equation. The approximation problem is converted into the negative cycle problem in a properly chosen family of weighted directed graphs, and a version of the Ford-Bellman algorithm for finding the shortest paths leads to constructive proofs of new characterization and existence theorems.

1. Introduction

Throughout this paper D is a compact set in $\mathbb{R}^2$, S and T are the smallest real compact sets for which $D \subseteq S \times T$, $u \in C(S)$ and $v \in C(T)$ are given positive continuous functions, $g \in C(\mathbb{R})$ is a given strictly increasing continuous function. We shall approximate continuous functions f, $f \in C(D)$, by the class of nomographic functions

$$\mathrm{NOM} := \{ w \in \ell_\infty(D) \mid w(s,t) = g(\,x(s)v(t) + u(s)y(t)) \,, \quad (s,t) \in D \,, \quad x \in \ell_\infty(S) \text{ and } y \in \ell_\infty(T) \text{ are arbitrary bounded functions} \} \,,$$

and shall use the uniform norm $\|\cdot\| = \|\cdot\|_D$ on D.

This is our main problem:

PROBLEM A. Let $f \in C(D)$ be given. Find $w^* \in NOM$ for which the infimum

$$dist(f,NOM,D) := \inf_{w \in NOM} \|f - w\|_D$$

is attained.

Special cases of nomographic functions NOM have appeared in the literature in various guises and forms since more than thirty years, see Section 2. The conversion of Problem A into the negative cycle problem in a properly chosen family of weighted directed graphs (Sections 3 and 4) enables us to apply the efficient shortest path algorithms for solving Problem A : See Sections 5 and 7. One of these algorithms, our MAIN ALGORITHM in Section 5 , leads to constructive proofs of new characterization theorems (Theorem 2.2, Theorem 4.3, Remark 4.1) for the infimum dist(f,NOM,D) and of a new existence theorem (Theorem 6.1) for the general Problem A which provides new sufficient conditions for the domain D under which minimal bounded or minimal continuous solutions $w^* \in NOM$ exist.

2. Applications: Special Cases of Problem A.

2.1. Approximation by Univariate Functions. In 1951, Diliberto and Straus [8] suggested an alternating direction method for Problem A where $S = T = [0,1]$, $D = S \times T$, and

$$\text{(2.1)} \qquad NOM = \{ w \in \ell_\infty(D) \mid w(s,t) = x(s) + y(t) \ , \ x \in \ell_\infty(S),\ y \in \ell_\infty(T) \}.$$

Using the same method Aumann [1] proved the existence of continuous minimal solutions $x^*(s) + y^*(t) \in NOM$ for any $f \in C(D)$ in the above special case. See also [3; 4; 5; 17; 18; 25] .

Our Theorem 6.1 guarantees the existence of bounded or continuous minimal solutions under very general assumptions on D.

In 1972, Collatz [6] investigated our Problem A for the domain $D = S \times T$, $S = T = [0,1]$ and the approximation subspace

$$NOM = \{ w \in \ell_\infty(D) \mid w(s,t) = x(s)v(t) + u(s)y(t) \ , \ x \in \ell_\infty(S),\ y \in \ell_\infty(T) \}.$$

in order to compute approximants $w \in NOM$ for the kernel function $f(s,t)$ of a given Fredholm or Hammerstein integral equation. See also [13] .

If $u = 1$, $v = 1$, and $g = \exp$, our Problem A is equivalent to the problem where $f \in C(D)$, $D \subseteq S \times T$, and

$$NOM = \left\{ X(s)Y(t) \;\middle|\; X \in \ell_\infty(S)\,,\; Y \in \ell_\infty(T)\,,\; X > 0\,,\; Y > 0 \right\}.$$

For more details see [14] .

2.2. Asymmetric Scalings of Matrices. The preconditioning of matrices is of considerable importance in matrix calculations. See [10; 12; 13; 15; 16; 26; 27] .
An elementary preconditioning technique is to multiply the given matrix by diagonal matrices:

SCALING PROBLEM. *Let* $B = (b_{ij})$ *be an* (m,n)*-matrix. Find positive numbers* u_i $(i=1,\dots,m)$ *and* v_j $(j=1,\dots,n)$ *such that the infimum*

$$\gamma := \inf_{u_i, v_j} \frac{\max_{(i,j)\in D} |b_{ij} u_i v_j|}{\min_{(i,j)\in D} |b_{ij} u_i v_j|}$$

is attained, where $D := \left\{(i,j) \;\middle|\; b_{ij} \neq 0 \right\}$.

Putting $a_{ij} := \log |b_{ij}|$, $x_i := -\log |u_i|$, and $y_j := \frac{1}{2}\log\gamma - \log|v_j|$ the scaling problem is converted into a Problem A where $S = \{1;2;\dots;m\}$, $T = \{1;2;\dots;n\}$, and where NOM is of the form (2.1). See [12] and [13] .

The symmetric scaling problem where $m = n$ and $v_j = 1/u_j$, $j=1,\dots,n$, has been studied by Rothblum and Schneider [26] , Saunders and Schneider [27] , Schneider and v.Golitschek [16] .

2.3. Application to Functional Equations. In 1972, Buck [2] studied the following problem (in a somewhat more general setting) :

Suppose that $\beta_0, \beta_1, \ldots, \beta_n$ are continuous real-valued functions on $[0,1]$. Consider the system of functional equations

$$\varphi(\beta_{i-1}(s)) - \varphi(\beta_i(s)) = w_i(s) \ , \quad s \in [0,1] \ , \ i=1,2,\ldots,n \ , \tag{2.2}$$

where we ask for a continuous solution φ when the functions $w_i \in C[0,1]$ are given.

In general, (2.2) cannot have a solution φ unless

$$w_k(s)+w_{k+1}(s)+\ldots+w_m(s) = 0 \quad \text{for} \quad s \in \Gamma_{k-1,m} \tag{2.3}$$

for $k=1,2,\ldots,n$, $m=1,2,\ldots,n$, $k \leq m$, where Γ_{ij}, $i<j$, is the set of all $s \in [0,1]$ with $\beta_i(s) = \beta_j(s)$.

Let D be the compact set in $\mathbb{R}^2$ consisting of the graphs of the n+1 mappings β_i, i.e.

$$D = \bigcup \left\{ (s,\beta_i(s)) \;\middle|\; s \in [0,1] \ , \ i=0,1,\ldots,n \right\} .$$

Buck describes the solvability of (2.2) as follows.

THEOREM 2.1. <u>The system</u> (2.2) <u>admits arbitrarily good approximants</u> φ <u>for every choice of functions</u> $w_i \in C[0,1]$, $i=1,2,\ldots,n$, <u>obeying</u> (2.3) <u>if and only if</u>

$$\operatorname{dist}(f,\mathrm{NOM}^*,D) = 0 \quad \underline{\text{for each function}} \quad f \in C(D) \tag{2.4}$$

<u>where</u>

$$\mathrm{NOM}^* := \left\{ w \in C(D) \;\middle|\; w(s,t) = x(s)+y(t) \ , \ x \ \underline{\text{and}} \ y \ \underline{\text{continuous}} \right\} . \tag{2.5}$$

<u>2.4. Goursat-type Problems for the Wave-Equation.</u> The set $W := \mathrm{NOM}^* \cap C^2(\mathbb{R}^2)$ is identical with the set of functions $w \in C^2(\mathbb{R}^2)$ which solve the wave equation $w_{st}(s,t) = 0$ on $\mathbb{R}^2$.

PROBLEM. *Let* $D \subseteq S \times T$ *be a compact subset of* $\mathbb{R}^2$. *For which functions* $f \in C(D)$ *is*

$$(2.6) \qquad \inf_{w \in W} \|f - w\|_D = 0 \quad ?$$

The answer is obvious: (2.6) is valid if and only if $\text{dist}(f,\text{NOM}^*,D) = 0$.

The methods developed in this paper lead to the following results which are very helpful for the investigation of the problems in the sections 2.3 and 2.4 :

THEOREM 2.2. *Let* $D \subseteq S \times T$ *be a compact subset of* $\mathbb{R}^2$. *Let* NOM *and* NOM^* *be defined as in* (2.1) *and* (2.5). *Then, for any* $f \in C(D)$,

$$(2.7) \qquad \inf_{w \in \text{NOM}^*} \|f - w\|_D = \text{dist}(f,\text{NOM},D) \quad ,$$

$$(2.8) \qquad \text{dist}(f,\text{NOM},D) = \lim_{k \to \infty} \frac{1}{2k} \sup_{p \in W_k} L^*(p,f)$$

where W_k *is the set of all sequences* p ,

$$(2.9) \qquad p = \{(s_1,t_0),(s_1,t_1),(s_2,t_1),\ldots,(s_k,t_{k-1}),(s_k,t_k)\} \subseteq D \quad ,$$

of not necessarily distinct $2k$ *points in* D , *and*

$$L^*(p,f) := \sum_{j=1}^{k} (f(s_j,t_j) - f(s_j,t_{j-1})) \ .$$

The proof of Theorem 2.2 will be published elsewhere. We conjecture that the relation (2.8) can even be replaced by

$$(2.10) \qquad \text{dist}(f,\text{NOM},D) = \sup_{p \in P^*} \frac{L^*(p,f)}{|p|}$$

if the set P^*,

$$P^* := \left\{ p \in W_k \;\middle|\; \text{all } |p| = 2k \text{ points of } p \text{ are distinct and } t_0 = t_k \right\},$$

of all "simple closed paths in D" is non-empty. We also conjecture that

$$\text{(2.11)} \qquad \operatorname{dist}(f, NOM, D) = 0 \qquad \text{for all } f \in C(D)$$

if and only if the set P^* is empty.

3. The Negative Cycle Problem.

By $G = [V,E,L]$ we denote a weighted directed graph with the vertices $V = \{1; 2; \ldots; N\}$, the arcs $E \subseteq V \times V$, and the weight function $L: E \to \mathbb{R}$. The sequences $\beta = \{(i_r, i_{r+1}) \mid r = 1,2,\ldots,k\} \subseteq E$, $k \geq 1$, are called paths in G of the length $|\beta| = k$ and of the weight $L(\beta) := \sum_{r=1}^{k} L(i_r, i_{r+1})$, and Y denotes the set of all cycles (= closed paths) in the graph G., i.e. $\beta \in Y$ iff $i_{k+1} = i_1$. The number $\lambda^* := \min\{L(\beta)/|\beta| : \beta \in Y\}$ is called the minimum cycle mean of G.

The negative cycle problem (i.e. the problem if G contains cycles of negative weight) and the problem (3.1) below are equivalent. (See Lawler [21], v.Golitschek and Schneider [16], a.o.):

THEOREM 3.1. <u>There is no cycle β of negative weight $L(\beta)$ in $G = [V,E,L]$ if and only if the inequalities</u>

$$\text{(3.1)} \qquad z_r + L(r,q) - z_q \geq 0 \,, \qquad (r,q) \in E$$

<u>have a solution $z = (z_1, z_2, \ldots, z_N) \in \mathbb{R}^N$.</u>

<u>In particular, the minimum cycle mean λ^* of G is the largest of all numbers λ for which the inequalities</u>

$$\text{(3.2)} \qquad z_r + L(r,q) - z_q \geq \lambda \,, \qquad (r,q) \in E$$

<u>have a solution $z \in \mathbb{R}^N$.</u>

There are important algorithms which solve the negative cycle problem, in particular the shortest path algorithms of Ford and Bellman, and of Yen. See Lawler [23] , Yen [28] .

Recently, Karp [20] suggested an efficient algorithm for the computation of the minimum cycle mean λ^* . Its first version, independently suggested a little later in [11-13] , will play an essential role in this paper. This version computes the numbers

$$(3.3) \qquad z_{jk} := \min\left\{0 \; ; \; L(\beta) \;\middle|\; \begin{array}{l}\beta \text{ runs through the set of paths in } G \text{ with} \\ \text{the final vertex } j \text{ and with } |\beta| \leq k\end{array}\right\}$$

by the recurrence $z_{jo} := 0$, $j \in V$,

$$(3.4) \qquad z_{jk} := \min\left\{ z_{j,k-1} \; ; \; L(i,j) + z_{i,k-1} \;\middle|\; i : (i,j) \in E \right\}$$

for all $j \in V$ and $k=1,2,\ldots,N$.

The following theorem is then the main tool for the computation of λ^* , and can also be used for solving the negative cycle problem; see [12 ; 13] :

THEOREM 3.2. *If* $\lambda^* \geq 0$, *there exists an index* k , $k \leq N$, *so that the set* $J_k := \{ j \in V \mid z_{jk} < z_{j,k-1} \}$ *is empty, and the numbers* $z_j := z_{jk}$, $j \in V$, *solve the inequalities* (3.1) .

If $\lambda^* < 0$, *the sets* $J_1, J_2, \ldots, J_N$ *are non-empty and*

$$(3.5) \qquad \lambda^* = \min_{j \in V} \max_{0 \leq k < N} \frac{z_{jN} - z_{jk}}{N - k} .$$

4. Problem A and the Negative Cycle Problem.

We first observe that the infimum of Problem A can be characterized similarly as the minimum cycle mean λ^* in Theorem 3.1 :

THEOREM 4.1. 1. dist(f,NOM,D) *is the infimum of all numbers* d *for which*

(4.1) $$h_d(s,t) \leq X(s) + Y(t) \leq H_d(s,t) \quad , \quad (s,t) \in D$$

<u>has a solution</u> $X \in \ell_\infty(S)$, $Y \in \ell_\infty(T)$, <u>where we put</u>

(4.2) $$h_d(s,t) := \frac{g^{-1}(f(s,t)-d)}{u(s)v(t)} \quad , \quad H_d(s,t) := \frac{g^{-1}(f(s,t)+d)}{u(s)v(t)} \quad .$$

2. $X \in \ell_\infty(S)$, $Y \in \ell_\infty(T)$ <u>solve</u> (4.1) <u>if and only if</u>

(4.3) $$x(s) := u(s)X(s) \quad , \quad y(t) := v(t)Y(t)$$

<u>solve</u>

(4.4) $$\|f(s,t) - g(x(s)v(t)+u(s)y(t))\|_D \leq d \quad .$$

We now suppose that $D \subseteq S \times T$ is a finite point set, say

(4.5) $$S = \{\sigma_1; \sigma_2; \ldots; \sigma_m\} \quad , \quad T = \{\tau_1; \tau_2; \ldots; \tau_n\} \quad .$$

For any real number d we define the weighted directed graph $G_d = [V,E,L_d]$ where

(4.6)
$$V := \{1;2;\ldots;m+n\}$$
$$E := \{(i,m+j) \; ; \; (m+j,i) \mid (\sigma_i,\tau_j) \in D\}$$
$$L_d(i,m+j) := H_d(\sigma_i,\tau_j) \quad , \quad L_d(m+j,i) := -h_d(\sigma_i,\tau_j) \quad .$$

(This idea has also been used by Saunders and Schneider [27; Section 3] and v.Golitschek and Schneider [16; Section 7] .)

The relations between Problem A and the negative cycle problem for the weighted graphs G_d are now obvious :

THEOREM 4.2. <u>Let</u> $D \subseteq S \times T$ <u>satisfy</u> (4.5).

1. G_d <u>has no cycle</u> β <u>of negative weight</u> $L_d(\beta)$ <u>if and only if</u> $\mathrm{dist}(f,NOM,D) \leq d$.
2. <u>The inequalities</u>

$$(4.7) \qquad z_r + L_d(r,q) - z_q \geq 0 \quad , \quad (r,q) \in E$$

have a solution $z \in \mathbb{R}^{m+n}$ if and only if $\mathrm{dist}(f,NOM,D) \leq d$.
3. If $(z_1,z_2,\ldots,z_{m+n})$ solves (4.7) then the functions x and y ,

$$(4.8) \qquad \begin{aligned} x(\sigma_i) &:= -\,u(\sigma_i)z_i \quad , \quad i=1,2,\ldots,m \\ y(\tau_j) &:= \;\; v(\tau_j)z_{m+j} \quad , \quad j=1,2,\ldots,n \end{aligned}$$

are solutions of (4.4) , and conversely .

PROOF. One only has to verify that any solution z of (4.7) leads via (4.8) to a solution of (4.4), and conversely : This is Part 1-2. Part 3 follows from Theorem 3.1. ◻

In Theorem 4.2, the cycles in G_d play an essential role. Hence one has to consider the corresponding point sequences in D. It turns out that there is a one-to-one correspondence between the cycles in G_d and the closed paths in D which are defined as follows.

DEFINITION. Let D be any compact set in $\mathbb{R}^2$, $D \subseteq S \times T$. Any sequence p of $2k$ not necessarily distinct points in D of the form

$$(4.9) \qquad p = \{(s_1,t_0),(s_1,t_1),(s_2,t_1),\ldots,(s_k,t_{k-1}),(s_k,t_k)\} \quad , \quad k \geq 1$$

is called a path in D of length $|p| = 2k$. p is called a closed path in D if $t_k = t_0$. The set of all paths p in D is denoted by W , and of all closed paths in D by P .

If $D \subseteq S \times T$ satisfies (4.5), there is a one-to-one correspondence between the cycles $\beta = \{(m+j_0,i_1),(i_1,m+j_1),\ldots,(m+j_{k-1},i_k),(i_k,m+j_0)\}$ in E and the closed paths

$$p(\beta) = \{(\sigma_{i_1},\tau_{j_0}),(\sigma_{i_1},\tau_{j_1}),\ldots,(\sigma_{i_k},\tau_{j_{k-1}}),(\sigma_{i_k},\tau_{j_0})\}$$

in D . We notice that the weight of β in G_d is given by

$$L_d(\beta) = \sum_{(r,q) \in \beta} L_d(r,q) = \sum_{l=1}^{k} (H_d(\sigma_{i_l}, \tau_{j_l}) - h_d(\sigma_{i_l}, \tau_{j_{l-1}})) .$$

Hence we introduce the weights for a path p in D as follows.

DEFINITION. _Let_ D _be any compact set in_ $\mathbb{R}^2$.

1. _For_ $d \in \mathbb{R}$ _and any path_ p _in_ D _of the form_ (4.9) ,

$$(4.10) \qquad L^*_d(p,f) := \sum_{l=1}^{k} (H_d(s_l,t_l) - h_d(s_l,t_{l-1})) .$$

2. _If_ g = identity, _we put_

$$(4.11) \qquad L^*_0(p,f) := \sum_{l=1}^{k} \left(\frac{f(s_l,t_l)}{u(s_l)v(t_l)} - \frac{f(s_l,t_{l-1})}{u(s_l)v(t_{l-1})} \right) ,$$

$$(4.12) \qquad U^*(p) := \sum_{l=1}^{k} \left(\frac{1}{u(s_l)v(t_l)} + \frac{1}{u(s_l)v(t_{l-1})} \right) .$$

The next theorem is an immediate corollary of Theorem 4.2 and Theorem 3.1. It states that the "worst closed path" in D determines the infimum of Problem A.

THEOREM 4.3. _Let_ $D \subseteq S \times T$ _satisfy_ (4.5) _and let_ P _be non-empty._

1. _The infimum of_ Problem A _is characterized by_

$$(4.13) \qquad \mathrm{dist}(f,NOM,D) = \sup_{p \in P} \mathrm{dist}(f,NOM,p) .$$

2. _The inequality_ $\mathrm{dist}(f,NOM,D) \leq d$ _holds if and only if_

$$(4.14) \qquad L^*_d(p,f) \geq 0 \quad \text{for all } p \in P .$$

3. _For each_ $p \in P$, _the infimum_ $d = \mathrm{dist}(f,NOM,p)$ _is uniquely determined by the relation_ $L^*_d(p,f) = 0$.

If g = identity , *then*

(4.15) $\quad \mathrm{dist}(f,\mathrm{NOM},p) = |L_0^*(p,f)| / U^*(p) \quad .$

REMARK 4.1. In Theorem 4.3 , the domain D is a finite point set. We conjecture that Theorem 4.3 is valid for every compact domain $D\subset\mathbb{R}^2$, but we can only prove that the relations (4.13) and (4.14) are valid if the compact domain $D\subset\mathbb{R}^2$ has the property P1 :

PROPERTY P1. *There exists a positive integer* m *so that any path* p *in* D *of the form* (4.9) *can be supplemented to a closed path* p' *in* D *by affixing at the last point of* p *at most* 2m *appropriate points of* D. (Notice that the paths and closed paths consist of an even number of not necessarily distinct points of D .)

EXAMPLE 4.1. The domain $D \subseteq S \times T$ has the property P1 if there exist finitely many numbers $t_1, t_2,\ldots, t_n$ in T such that

$$S = \bigcup \{ S_{t_j} \mid j=1,2,\ldots,n \}$$

where we put $S_t := \{ s \in S \mid (s,t) \in D \}$, $T_s := \{ t \in T \mid (s,t) \in D \}$.

5. The Main Algorithm and its Properties.

In 1979, at a conference at Oberwolfach [11] , I suggested a new algorithm for the scaling problem of Section 2.2 without noticing its close relation to the shortest path algorithms in graph theory. Later [12 ; 13] I realized that my algorithm in [11] , if formulated in the setting of graph theory, is a version of Karp's algorithm [20] for the minimum cycle mean problem. In what follows below, we shall proceed in the opposite direction: Using the special structure of the directed graphs G_d and the relations (4.3) and (4.8), we convert the recurrence (3.4) into the setting of Problem A and thus obtain our Main Algorithm below which determines if a number d satisfies the inequality $d > \mathrm{dist}(f,\mathrm{NOM},D)$: see Theorem 5.2.

Our Main Algorithm can be used for any (general) Problem A; it is identical with the algorithm in [11] if it is applied to the scaling problem of Section 2.2 where g = identity, $u = 1$, $v = 1$, and D is a finite point set: see Theorem 5.3.

MAIN ALGORITHM. Let a real number d and a Problem A be given with a compact (not necessarily discrete) domain $D \subseteq S \times T$ and $f \in C(D)$.

Step 0. Put $X_0(s,d) := 0 \quad , \quad s \in S \quad ,$

$$(5.1) \qquad Y_0(t,d) := \inf \{ H_d(s,t) \mid s \in S_t \} , \quad t \in T ,$$

where H_d and h_d are defined by (4.2).

Step 1. For $k=1,2,3,\ldots$ compute

$$(5.2) \qquad X_k(s,d) := \sup \{ X_{k-1}(s,d) \; ; \; h_d(s,t) - Y_{k-1}(t,d) \mid t \in T_s \}$$

for $s \in S$ and put $I_k(d) := \{ s \in S \mid X_k(s,d) > X_{k-1}(s,d) \}$;

$$(5.3) \qquad Y_k(t,d) := \inf \{ Y_{k-1}(t,d) \; ; \; H_d(s,t) - X_k(s,d) \mid s \in S_t \}$$

for $t \in T$ and put $J_k(d) := \{ t \in T \mid Y_k(t,d) < Y_{k-1}(t,d) \}$;

(5.4) S T O P if $J_k(d)$ is empty.

It follows from (5.1)-(5.3), the continuity of h_d and H_d, and the compactness of D that the suprema and infima in (5.1)-(5.3) are attained. More detailed, we have

LEMMA 5.1. For any $s \in S$, $t \in T$, $d \in \mathbb{R}$, $k=0,1,2,\ldots,$

$$X_k(s,d) = \lim_{\varepsilon \to 0+} (\sup \{ X_k(s',d) \mid s' \in S , \; |s - s'| < \varepsilon) ,$$

$$Y_k(t,d) = \lim_{\varepsilon \to 0+} (\inf \{ Y_k(t',d) \mid t' \in T , \; |t - t'| < \varepsilon) .$$

The next theorem states that the Main Algorithm terminates for any Problem A after finitely many iterations if $d > \mathrm{dist}(f,\mathrm{NOM},D)$.

THEOREM 5.2. _We apply the_ Main Algorithm _to a given_ Problem A.
1. _If_ $J_k(d)$ _is empty for some_ k _then_ $d \geq \text{dist}(f,\text{NOM},D)$.
2. _If_ $d > \text{dist}(f,\text{NOM},D)$ _then there exists an index_ k _for which the set_ $J_k(d)$ _is empty_ .

PROOF. 1. If $J_k(d)$ is empty, i.e. if $Y_k(\ ,d) = Y_{k-1}(\ ,d)$, the functions $x := u\ X_k(\ ,d)$ and $y := v\ Y_k(\ ,d)$ solve (4.4).

2. For any $d' < d$ there exists a constant $M = M(f,d,d') > 0$ so that

$$(5.5)\qquad H_d(s,t)-H_{d'}(s,t) \geq M \quad , \quad h_{d'}(s,t)-h_d(s,t) \geq M \quad , \quad (s,t) \in D\ .$$

We put $d^* := \text{dist}(f,\text{NOM},D)$ and suppose that $d > d^*$ and $J_k(d) \neq \phi$ for al indices k . Because of Lemma 5.1, we find for each $k \geq 1$ and $t_k \in J_k(d)$ a path $p_k = p_k(t_k,d) \in W$ of the form (4.9) satisfying

$$(5.6)\qquad Y_k(t_k,d) - Y_r(t_r,d) = \sum_{j=r+1}^{k} (H_d(s_j,t_j) - h_d(s_j,t_{j-1}))$$

for $r=0,1,\ldots,k-1$. We put $d' := (d+d^*)/2$. There exist two functions $X \in \ell_\infty(S)$ and $Y \in \ell_\infty(T)$ such that

$$(5.7)\qquad h_{d'}(s,t) \leq X(s) + Y(t) \leq H_{d'}(s,t) \quad , \quad (s,t) \in D\ ,$$

see Theorem 4.1. Hence, by (5.5)-(5.7),

$$\begin{aligned} Y_k(t_k,d)-Y_0(t_0,d) &\geq 2kM + \sum_{j=1}^{k} (H_{d'}(s_j,t_j)-h_{d'}(s_j,t_{j-1})) \\ &\geq 2kM + Y(t_k) - Y(t_0) \longrightarrow +\infty \quad \text{for } k \longrightarrow \infty \end{aligned}$$

But this is a contradiction since

$$Y_k(t_k,d) - Y_0(t_0,d) \leq 2\ \|Y_0(\ ,d)\|_D \quad . \ \square$$

We obtain an analogue of Theorem 3.2 if the set T consists of finitely many real numbers. In particular, the relation (5.9) below is the translation of (3.5) into the setting of Problem A.

THEOREM 5.3. _Suppose that our_ Main Algorithm _is applied to a_ Problem A _where the domain_ $D \subseteq S \times T$ _satisfies_

$$(5.8) \qquad T = \{\tau_1; \tau_2; \ldots; \tau_n\} .$$

1. _All sets_ $J_1(d), \ldots, J_n(d)$ _are non-empty if and only if_ $\mathrm{dist}(f,NOM,D) > d$.
2. _If_ g = identity, $u = 1$, $v = 1$, _and_ $0 \leq d \leq \mathrm{dist}(f,NOM,D)$ _then_

$$(5.9) \qquad \mathrm{dist}(f,NOM,D) = d + \max_{t \in T} \min_{0 \leq k < n} \frac{-Y_n(t,d) + Y_k(t,d)}{2(n-k)} .$$

PROOF. 1. Let $t_n \in J_n(d)$. We construct the path $p_n = p_n(t_n,d)$ in D of the form (4.9) satisfying (5.6) for $k := n$ and $r = 0,1,\ldots,n-1$. There exist indices $0 \leq r < m \leq n$ such that $t_r = t_m$. Hence

$$p_{rm} = \{(s_{r+1},t_r),(s_{r+1},t_{r+1}),\ldots,(s_m,t_{m-1}),(s_m,t_m)\}$$

is a closed path in D. And (5.6) leads to

$$0 > Y_m(t_m,d) - Y_r(t_r,d) = L_d^*(p_{rm})$$

and thus by Theorem 4.3 to

$$d < \mathrm{dist}(f,NOM,p_{rm}) \leq \mathrm{dist}(f,NOM,D) .$$

On the other hand, if $J_k(d)$ is empty for some k, we apply Theorem 5.2.

2. The proof of Part 2 of Theorem 5.3 has already been established in [12 ; Theorem 2] and in [13 ; Theorem 7.1] . □

6. Existence of Minimal Solutions.

EXAMPLE 6.1. Let $D = \{(s,t) \mid s/2 \leq t \leq s \, , \, 0 \leq s \leq 1/e\}$,

$$\mathrm{NOM} := \{ w(s,t) = x(s)+y(t) \mid x \in \ell_\infty([0,1/e]),\ y \in \ell_\infty([0,1/e]) \ ) \ .$$

a) If $f \in C(D)$ is defined by $f(0,0) := 0$, $f(s,t) :=$ loglog $1/s$ - loglog $1/t$ in $D \setminus \{(0,0)\}$, then dist(f,NOM,D) = 0 and there does not exist a minimal bounded solution in NOM.

b) If $f \in C(D)$ is defined by $f(s,t) := \sin(\text{loglog } 1/s) - \sin(\text{loglog } 1/t)$ in $D \setminus \{(0,0)\}$, $f(0,0) := 0$, then dist(f,NOM,D) = 0 and $x^*(s) := -y^*(s) := \sin(\text{loglog } 1/s)$, $0 < s \leq 1/e$, $x^*(0) := y^*(0) := 0$ is a minimal solution in NOM, but there does not exist a minimal solution $x(s) + y(t) \in$ NOM with continuous functions x and y .

We shall now establish conditions on the domain D under which each $f \in C(D)$ has a minimal solution in NOM (Property P1) or even a minimal solution w in NOM with continuous x and y . (Property P2). More detailed, we apply our Main Algorithm for $d^* :=$ dist(f,NOM,D) and check if the limits

$$(6.1) \qquad Y^*(t) := \lim_{k \to \infty} Y_k(t,d\ ) \quad , \quad X^*(s) := \lim_{k \to \infty} X_k(s,d\ )$$

exist and are bounded or even continuous on T and S .

DEFINITION. <u>We say that the domain</u> $D \subseteq S \times T$ <u>has the property</u> P2 <u>if there exist a positive integer</u> m <u>and a continuous function</u> $\omega \in C[0,\infty)$ <u>with</u> $\omega(0) = 0$ <u>such that for any</u> $t \in T$ <u>and any path</u> p <u>in</u> D <u>of the form</u> (4.9) <u>we can find a path</u> p' <u>in</u> D ,

$$(6.2) \qquad p' = \{(s_1',t_0'),(s_1',t_1'),(s_2',t_1'),\ldots,(s_k',t_{k-1}'),(s_k',t_k')\}$$

<u>satisfying</u>

$$(6.3) \qquad t_k' = t \quad , \quad |s_i' - s_i| \leq \omega(|t - t_k|) \ , \quad |t_j' - t_j| \leq \omega(|t - t_k|)$$

<u>for</u> i=1,2,...,k <u>and</u> j=0,1,...,k , <u>and</u>, <u>additionally</u>,

(6.4) $t'_r = t_r$, $s'_r = s_r$ <u>for</u> $r=0,1,\ldots,k-m$

<u>if</u> $k \geq m$ <u>and</u> t_{k-m} , t_{k-m+1} ,..., t_k <u>are pairwise distinct</u>.

REMARK 6.1. We notice that P2 implies P1. But P2 also implies that, for any $(\sigma,\tau) \in D$ and $t \in T$, there exists an $(s,t) \in D$ so that

$$(6.5) \qquad |s - \sigma| \leq \omega(|t - \tau|) \quad .$$

THEOREM 6.1. <u>Let</u> $f \in C(D)$ <u>and</u> $d^* := \mathrm{dist}(f,NOM,D)$.
1. <u>If the compact domain</u> $D \subseteq S \times T$ <u>has the property</u> P1 , <u>then the limits</u> $Y^*(t)$ <u>and</u> $X^*(s)$ <u>in</u> (6.1) <u>exist and are uniformly bounded on</u> T <u>and</u> S.

<u>The function</u> $w^*(s,t) = g(\, x^*(s)v(t) + u(s)y^*(t))$ <u>with</u> $x^* := u\, X^*$ <u>and</u> $y^* := v\, Y^*$ <u>is a minimal solution of</u> Problem A .

2. <u>If the compact domain</u> D <u>has the</u> property P2, <u>the limit function</u> Y^* <u>in</u> (6.1) <u>is continuous on</u> T , <u>and there exists a minimal solution</u> $w^{**}(s,t) = g(\, x^{**}(s)v(t) + u(s)y^*(t))$ <u>of</u> Problem A <u>with</u> $y^* := v\, Y^* \in C(T)$ <u>and</u> $x^{**} \in C(S)$.

PROOF. 1. We apply the Main Algorithm for d^* .
For any $k \geq 1$ and $t_k \in J_k(d\ )$ we construct the path $p_k = p_k(t_k,d\ )$ of the form (4.9) satisfying (5.6). Because of the property P1, we obtain a closed path p'_k in D by affixing at most 2m appropriate points at the last point of p_k. Applying (5.6) for r=0 and Theorem 4.3 ,

$$\begin{aligned} Y_k(t_k,d^*) &= Y_0(t_0,d^*) + L^*_{d^*}(p_k,f) \\ &\geq Y_0(t_0,d^*) + L^*_{d^*}(p'_k,f) - m(\|H_{d^*}\|_D + \|h_{d^*}\|_D) \\ &\geq -\|Y_0(\ ,d^*)\|_D - 2m\,\|H_{d^*}\|_D =: K_0 \ . \end{aligned}$$

For each $t \in T$ the sequence $Y_k(t,d^*)$, k=1,2,..., is non-increasing and uniformly bounded from below by K_0 . Hence the limit function Y^* exists on T and is bounded. We conclude the proof of Part 1 by using (5.2)-(5.3).

2. Because of (6.5), the function $Y_0(\ ,d^*)$ is continuous on T.

Let $k \geq 1$, $t \in T$, $t' \in T$. If $t' \in J_k(d^*)$ we put $t_k := t'$ and construct the path $p_k = p_k(t_k,d^*)$ of the form (4.9) obeying (5.6). Since $d^* = \mathrm{dist}(f,NOM,D)$, the numbers $\{t_0,t_1,\ldots,t_k\}$ are pairwise distinct. Hence, using the property P2 , we find a path p_k' in D of the form (6.2) satisfying (6.3) and (6.4). We observe that the relations (5.1)-(5.3) imply

$$(6.6)\qquad Y_k(t_k',d^*) \leq Y_r(t_r',d\ ) + \sum_{j=r+1}^{k} (\ H_{d^*}(s_j',t_j') - h_{d^*}(s_j',t_{j-1}')) ,$$

r=0,1,...,k-1. Combining (5.6), (6.3), (6.4), (6.6) we obtain

$$(6.7)\qquad Y_k(t,d^*) - Y_k(t',d^*) \leq \tilde{\omega}(|t - t'|)$$

with a function $\tilde{\omega} \in C[0,\infty)$, $\tilde{\omega}(0) = 0$, which is independent of k, t, t' . If $t' \notin J_k(d^*)$, (6.7) follows by induction. Exchanging t and t' in (6.7) we obtain

$$(6.8)\qquad |Y_k(t,d^*) - Y_k(t',d^*)| \leq \tilde{\omega}(|t - t'|)$$

for all $k \geq 1$ and all $t \in T$, $t' \in T$. Since P2 implies P1 , the limit function Y^* satisfies (6.8), hence $Y^* \in C(T)$, and X^* is bounded on S.

We define $f_0 \in C(D)$ and $f_1 \in C(D)$ by

$$f_0(s,t) := h_{d^*}(s,t) - Y^*(t) , \quad f_1(s,t) := H_{d^*}(s,t) - Y^*(t) .$$

The existence of a continuous solution w^{**} with $y^{**} := y^* := v\ Y^* \in C(T)$ and $x^{**} := u\ X^{**} \in C(S)$ follows from

LEMMA 6.2. <u>Suppose that</u> $D \subset \mathbb{R}^2$ <u>is compact</u> , $f_0 \in C(D)$, $f_1 \in C(D)$. <u>If</u>

$$(6.9)\qquad f_0(s,t) \leq X^*(s) \leq f_1(s,t) \quad , \quad (s,t) \in D$$

<u>has a solution</u> $X^* \in \ell_\infty(S)$, <u>then</u> (6.9) <u>also has a solution</u> $X^{**} \in C(S)$.

PROOF. We apply the Michael selection theorem ([24] , [19 ; p.183]):

Let $F : S \to 2^{\mathbb{R}}$ be the set-valued function defined by

$$F(s) := \left\{ z \in \mathbb{R} \;\middle|\; \max_{t \in T_s} f_0(s,t) \leq z \leq \min_{t \in T_s} f_1(s,t) \right\} \quad , \quad s \in S \,.$$

The compactness of D and the continuity of f_0 and f_1 imply that, for any open set $\mathcal{O} \subset \mathbb{R}$, the set $\{ s \in \mathbb{R} \mid F(s) \cap \mathcal{O} \neq \phi \}$ is also open: F is lower semicontinuous. Since $F(s)$ is a point or a non-empty closed interval, the Michael selection theorem guarantees the existence of a continuous selection $X^{**} : S \to \mathbb{R}$ for F. ◻ ◻

EXAMPLE 6.2. Let $T = [0,1]$, $\varphi_0 \in C(T)$, $\varphi_1 \in C(T)$, and $D = \{(s,t) \mid \varphi_0(t) \leq s \leq \varphi_1(t) \;, \; t \in T\}$. The domain D has the property P2 if

(6.10) $\quad \varphi_0(t) \leq \alpha \leq \varphi_1(t) \;, \quad t \in T \;$, for a real number α ,

and also if the functions φ_0 and φ_1 are strictly monotone and obey

(6.11) $\quad \varphi_0(t) < \varphi_1(t) \;, \quad t \in T \,.$

7. Appendix: Numerical Treatment of Problem A.

For computations, it is always reasonable and possible to replace the compact domain D by a finite (discrete) domain $D^*(\Delta)$, see Theorem 7.1. Let $D \subseteq S \times T$ be compact, $S \subseteq [a,b]$, $T \subseteq [\alpha,\beta]$, $D_0 := [a,b] \times [\alpha,\beta]$. Let $U \in C[a,b]$ and $V \in C[\alpha,\beta]$ be positive functions with $U_{|S} = u$, $V_{|T} = v$, and let $F \in C(D_0)$ be a function satisfying $F_{|D} = f$. Put

$$\begin{aligned} NOM := \{ w \in \ell_\infty(D_0) \mid\; & w(s,t) = g(\, x(s)V(t) + U(s)y(t)) \,, \\ & x \in \ell_\infty([a,b]) \text{ and } y \in \ell_\infty([\alpha,\beta]) \text{ are arbitrary} \} \end{aligned}$$

and $NOM^* := \{ w \in NOM \mid \; x \in C[a,b] \;,\; y \in C[\alpha,\beta] \}$.
Let

$$\Delta = \{ (\sigma_i,\tau_j) \quad a = \sigma_1 < \dots < \sigma_m = b \quad , \alpha = \tau_1 < \dots < \tau_n = \beta \},$$

$$|\Delta| := \max\{\sigma_{i+1} - \sigma_i \; ; \; \tau_{j+1} - \tau_j\} \text{ , and}$$

$$D(\Delta) := \bigcup \{K_{ij} = [\sigma_i, \sigma_{i+1}] \times [\tau_j, \tau_{j+1}] \mid K_{ij} \cap D \neq \phi\}$$

$$D^*(\Delta) := \Delta \cap D(\Delta) \; .$$

THEOREM 7.1. *For each* $f \in C(D)$,

$$\text{(7.1)} \qquad \operatorname{dist}(f,NOM^*,D) = \lim_{|\Delta| \to 0} \operatorname{dist}(F,NOM,D^*(\Delta))$$

$$= \lim_{|\Delta| \to 0} \operatorname{dist}(F,NOM,D(\Delta)) \; ;$$

$$\text{(7.2)} \qquad \operatorname{dist}(f,NOM,D) = \operatorname{dist}(f,NOM^*,D) \; .$$

PROOF. One has to use piecewise linear interpolation at the points σ_i and τ_j . The details are left to the reader. □

We now suppose that the domain D is a finite point set, i.e. that (4.5) holds. Applying the basic ideas of Lawler's monotony algorithm [22] (for finding optimal cycles in doubly weighted graphs) one obtains the following general algorithm for the discrete Problem A .

ALGORITHM 7.1. *Let* $D \subseteq S \times T$ *be of the form* (4.5).

Step 0. *Choose* d , $d < \operatorname{dist}(f,NOM,D)$.

Step 1. *Find a cycle* β_d *in* $G_d = [V,E,L_d]$ *of negative weight* $L_d(\beta_d)$ *and compute the number* d' , $d' > d$, *for which* $L_{d'}(\beta_d) = 0$.

Put $d := d'$ *and return to* Step 1.

If $L_d(\beta) \geq 0$ *for all cycles* β *in* G_d *then* $d = \operatorname{dist}(f,NOM,D)$; *go to* Step 2.

Step 2. *Solve the inequalities* (4.7) *for* $d = \operatorname{dist}(f,NOM,D)$. *By* (4.8) *we then obtain a minimal solution of* Problem A.

REMARKS. 1. Algorithm 7.1 terminates after finitely many iterations since D is finite.

2. For the computation of optimal cycles in doubly weighted graphs, versions

of Lawler's monotony algorithm have been suggested by Dantzig, Blattner, Rao [7] , Fox [9] , v.Golitschek [13] which differ only in the way how the cycles of negative weight are searched. Each of these methods for solving the negative cycle problem can be applied in Step 1 of Algorithm 7.1.

3. If g = identity, u = 1, v = 1, one only needs one iteration of Algorithm 7.1 if one uses our Main Algorithm and applies (5.9). See [11-13] .

4. For special cases of Problem A, Diliberto and Straus [8] and Aumann [1] have suggested an alternating direction method. See also [1-3; 17; 18; 25] .

REFERENCES

[1] Aumann, G. Über approximative Nomographie II. Bayer. Akad. Wiss. Math. Nat. Kl. S. B., (1959), 103-109.

[2] Buck, R.C. On approximation theory and functional equations. J. Approximation Theory 5 (1972), 228-237.

[3] Cheney, E.W. Approximating multivariate functions by combinations of univariate ones. Center for Numerical Analysis, The University of Texas, Report 148 (1979).

[4] Cheney, E.W. - v.Golitschek, M. On the algorithm of Diliberto and Straus for approximating bivariate functions by univariate ones. Numer. Funct. Anal. and Optimiz. 1 (1979), 341-363.

[5] Cheney, E.W. - Light, W.A. On the approximation of a bivariate function by the sum of univariate ones. J. Approximation Theory, to appear.

[6] Collatz, L. Approximation by functions of fewer variables. In: Lecture Notes in Mathematics 280 (1972), 16-31.

[7] Dantzig, G.B. - Blattner, W.O. - Rao, M.R. Finding a cycle in a graph with minimum cost to time ratio with applications to a ship routing problem. In: Theory of Graphs (P. Rosenstiehl, ed.),International Symposium, Rome 1966, pp. 77-83.

[8] Diliberto, S.P. - Straus, E.G. On the approximation of functions of several variables by the sum of functions of fewer variables. Pacific J. Math. 1 (1951), 195-210.

[9] Fox, B. Finding minimum cost-time ratio curcuits. Operations Research 17 (1969), 546-551.

[10] Fulkerson, D.R. - Wolfe, P. An algorithm for scaling matrices. SIAM Rev. 4 (1962), 142-146.

[11] v.Golitschek, M. Approximation of functions of two variables by the sum of two functions of one variable. In: Numerical Methods of Approximation Theory (L. Collatz, G. Meinardus, H. Werner, eds.), Birkhäuser Verlag, ISNM 52 (1980), pp. 117-124.

[12] v.Golitschek, M. An algorithm for scaling matrices and computing the minimum cycle mean in a digraph. Numer. Math. 35 (1980), 45-55.

[13] v.Golitschek, M. Optimal cycles in doubly weighted graphs and approximation of bivariate functions by univariate ones. Numer. Math. 39 (1982), 65-84.

[14] v.Golitschek, M. Approximating bivariate functions and matrices by nomographic functions. In: Quantitative Approximation (R. DeVore, K. Scherer, eds.), Academic Press, New York 1980, pp.143-151.

[15] v.Golitschek, M. - Rothblum, U.G. - Schneider, H. A conforming decomposition theorem, a piecewise linear theorem of the alternative, and scalings of matrices satisfying lower and upper bounds. Mathematical Programming, to appear.

[16] v.Golitschek, M. - Schneider, H. Applications of shortest path algorithms to matrix scalings. Numer. Math., to appear.

[17] Golomb, M. Approximation by functions of fewer variables.In: Numerical Approximation (R. Langer, ed.), Madison 1959, pp.275-327.

[18] Havinson, S.J. A Chebyshev theorem for the approximation of a function of two variables by the sum $\Phi(x) + \Psi(y)$. Izv. Akad. Nauk SSSR Ser. Mat. 33 (1969), 650-666.

[19] Holmes, R.B. Geometrical Functional Analysis and its Applications. Springer Verlag, New York 1975.

[20] Karp, R.M. A characterization of the minimum cycle mean in a digraph. Discrete Math. 23 (1978), 309-311.

[21] Lawler, E.L. Optimal cycles in doubly weighted directed linear graphs. In: Theory of Graphs (P. Rosenstiehl, ed.), International Symposium, Rome 1966, pp.209-213.

[22] Lawler, E.L. Optimal cycles in graphs and the minimal cost-to-time ratio problem.In: Periodic Optimization, Vol. I (A. Marzollo, ed.) Udine 1972, pp.37-60.

[23] Lawler, E.L. Combinatorial Optimization: Networks and Matroids. New York 1976.

[24] Michael, E. Continuous selections. Ann. Math. 63 (1956), 361-382.

[25] Ofman, J.P. Best approximation of functions of two variables by functions of the form $\Phi(x) + \Psi(y)$. Amer. Math. Soc. Transl. 44 (1965), 12-29.

[26] Rothblum, U.G. - Schneider, H. Characterizations of optimal scalings of matrices. Mathematical Programming 19 (1980), 121-136.

[27] Saunders, B.D. - Schneider, H. Flows on graphs applied to diagonal similarity and diagonal equivalence of matrices. Discrete Mathematics 24 (1978), 202-220.

[28] Yen, J.Y. Shortest Path Network Problems. Mathematical Systems in Economics 18, Meisenheim 1975.

International Series of
Numerical Mathematics, Vol. 65

A BIBLIOGRAPHY OF THE BERNSTEIN POWER SERIES OPERATORS OF MEYER - KÖNIG AND ZELLER AND THEIR GENERALIZATIONS

Eberhard L. Stark
Lehrstuhl A für Mathematik
RWTH
Aachen

Herrn Professor W. Meyer - König zum 70. Geburtstag am 26. Mai 1982 in Dankbarkeit gewidmet.

Concerning the approximation of nonperiodic functions by positive linear methods the best-known operators are the *Bernstein* polynomials *) and the singular integral of *Landau - Stieltjes* (Landau polynomials). Another positive linear approximation process which is attracting increasing attention was introduced by W. MEYER - KÖNIG and K. ZELLER [2]/[3] in 1959/60; for functions f on [0,1) it was originally defined by

$$(1) \qquad M_n(f;x) := (1-x)^n \sum_{k=o}^{\infty} f\left(\frac{k}{k+n}\right) \binom{k+n-1}{k} x^k , \qquad n \in \mathbb{N} .$$

A first generalization (1964) involving Laguerre polynomials is due to E.W. CHENEY and A. SHARMA in [7]; in particular, they showed as an extremal case of the general construction that the "slight" modification

$$(2) \qquad M_n^*(f;x) := (1-x)^{n+1} \sum_{k=o}^{\infty} f\left(\frac{k}{k+n}\right) \binom{k+n}{k} x^k$$

---

*) In this respect, the reader is referred to the author's *first* part of a corresponding bibliography, namely ≫ Bernstein - Polynome, 1912 - 1955 ≪ in "Functional Analysis and Approximation (Proc. Conf. Math. Res. Inst. Oberwolfach 1980; Eds. P.L. Butzer - B. Sz.-Nagy - E. Görlich; ISNM 60) Basel etc. 1981, 482 pp.; pp. 443 - 461." (There also exist "1. Nachtrag, March 1982", 3 pp., "2. Nachtrag, Aug. 1983", 3 pp.)

for $f \in C[0,1)$ (replacing $f(\frac{k}{k+n+1})$ by $f(\frac{k}{k+n})$ in $M_{n+1}(f;x)$ obviously leads to (2)) is more appropriate. This is due to the fact that, in contrast to (1), the modified version (2) preserves also linear functions, i.e., for the particular test functions $f_o(x) := 1$, $f_1(x) := x$, $0 \leqslant x \leqslant 1$, one has

$$M_n^*(f_o;x) \equiv 1 , \qquad M_n^*(f_1;x) \equiv x$$

for all $n \in \mathbb{N}$.

It should be mentioned that during the Baku conference on constructive theory of functions of 1962 (the proceedings of which appeared only in 1965) G.A. FOMIN [10],[11] and V.I. VOLKOV [12] introduced two further new classes of positive linear approximation processes. In [10],[11] the generalized Bernstein operators

$$(3) \qquad L_{m,n}(f;x) := \{1+(m-1)x\} \sum_{k=o}^{[\frac{n}{m}]} f(\tfrac{k}{n-(m-1)k})\binom{n-(m-1)k}{k} x^k (1-x)^{n-mk}$$

$m,n \in \mathbb{N}_o$, f bounded on $[0,1)$, were given. They reduce to the classical Bernstein polynomials for $m=1$, and to (2) for $m=0$; even the corresponding Voronovskaja - type expansion for general m is already contained in [11, p. 202]. In [12] a particular case of quite another general method is again given by (2); as a second specification there is, among others, the operator of Mirakjan - Szâsz. It is expressively mentioned ([12, p. 128]), without any reference however, that the operator (2) is a conjecture of A.G. RJATIN, a student of the Kalinin Pedagogical Institute; for a rather similar remark compare also [90, p. 22].

Strange to state that these three papers (obviously [3] does not appear in the references therein) have not (namely [11],[12]) or rarely been cited in the relevant literature; [10] is only mentioned in [26] (togehter with [3]),[39] (citing [7], however[12] does not appear), [43] (pointing out [3] but not [12]), and [30] (where, in turn, [3] and [7] are not referred to).

In this bibliography all papers - at least those known to the author - are collected which, with respect to (1) and/or (2),

(i) present results explicitly dealing with (1) or (2) ,
(ii) apply general theorems,
(iii) contain generalizations and corresponding results that

reduce to the above operators for certain parameter(s) involved. Not cited are reports (published later on as regular notes) and those numerous papers which only point out [3] and/or [7] as examples of further positive linear operators, without any concrete applications.

First occurances of these operators in the textbook literature on the subject are [46, p. 164 - 167] (1972) and [77, p. 24, 46,108] (1978); but as early as 1966 these operators were mentioned in the Notes - Section in [13], where [3] and [7] are cited; see also [90] - if possible at all!

Moreover it should be mentioned that it is known for some time that there exist some further general positive linear operators (e.g. those due to V.A. BASKAKOV, P. KESAVA MENON et al.; see, for instance, [1],[4]) from which the operators (1) may be derived by a simple transformation (consult [20],[55],[68],[75], among others); however, since this is a *rational* transformation the theory of the Meyer - König and Zeller operators remains self - contained for the essential parts.

Though completeness is of course aspired, there obviously will remain gaps. The author would like to thank all readers who will attract his attention to supplements and missing papers (in particular, [6]* and [31]* which have not been at the author's disposal - apart from the reviews) and provide him with (p)reprints. Finally, the author would like to thank all colleagues who helped him during the past years to write up just this version of the bibliography by sending relevant material and/or reading earlier manuscripts (including useful comments), in particular V.A. BASKAKOV (Moscow), D. LEVIATAN (Tel Aviv), M.W. MÜLLER (Dortmund) and, last not least, his student Dipl.-Math. ELFRIEDE KLUBERT who worked out most of the material (up to 1980) in her Diplomarbeit ("Die Bernsteinschen Potenzreihenoperatoren von M. Meyer - König und K. Zeller: eine integrierende Darstellung", RWTH Aachen, Jan. 1981, iii + 275 pp.).

---

[1] BASKAKOV, V.A., An instance of a sequence of positive linear operators in the space of continuous functions (Russ.). Dokl. Akad. Nauk SSSR 113(1957)249 - 251. MR 20 # 1153; Zbl 80, 52.

[2] MEYER - KÖNIG, W.-K. ZELLER, Pascal - Verteilung und Approximation durch Bernsteinsche Potenzreihen. Z. Angew. Math. Mech. 39(1959)380. Zbl 95, 128.

[3] MEYER - KÖNIG, W.-K. ZELLER, Bernsteinsche Potenzreihen. Studia Math. 19(1960)89 - 94. MR 22 # 2823; Zbl 91, 145.

[4] KESAVA MENON, P., A class of linear positive operators. J. Indian Math. Soc. (N.S.) 26(1962)77 - 80. MR 26 # 5431; Zbl 109, 289.

[5] RAMANUJAN, M.S., The moment problem. Math. Student 31(1963)201 - 206 (1964). MR 31 # 3803; Zbl 136, 107.

[6]* VOLKOV, V.I., On a sequence of linear positive operators in the space of continuous functions (Russ.). Kalinin. Gos. Ped. Inst. Učen. Zap. 29(1963)19 - 38. MR 28 # 4350; RŽM 1964,6B98.

[7] CHENEY, E.W. - A. SHARMA, Bernstein power series. Canad. J. Math. 16 (1964)241 - 252. MR 31 # 3770; Zbl 128, 290.

[8] JAKIMOVSKI, A. - M.S. RAMANUJAN, A uniform approximation theorem and its application to moment problems. Math. Z. 84(1964)143 - 153. MR 30 # 2272; Zbl 196, 318.

[9] RAMANUJAN, S.M., The moment problem in a certain function space of G.G. Lorentz. Arch. Math. (Basel) 15(1964)71 - 75. MR 28 # 4310; Zbl 136, 106.

[10] FOMIN, G.A., On convergence of certain constructions of linear operators (Russ.). In: Studies of Contemporary Problems of Constructive Theory of Functions (Russ.)(Proc. Second All - Union Conf., 8. - 13.10.1962; Ed. I.I. Ibragimov) Baku 1965, 638 pp.; pp. 193 - 199. MR 33 # 7770; Zbl 215, 178.

[11] FOMIN, G.A., The order of approximation of two times differentiable functions by certain linear operators (Russ.). In: Studies of Contemporary Problems of Constructive Theory of Functions (Russ.) (Proc. Second All - Union Conf., 8. - 13.10.1962; Ed. I.I. Ibragimov) Baku 1965, 638 pp.; pp. 200 - 206. MR 33 # 7771; Zbl 215, 177.

[12] VOLKOV, V.I., On a certain uniformly convergent sequence of linear positive operators in the space of continuous functions (Russ.). In: Studies of Contemporary Problems of Constructive Theory of Functions (Russ.)(Proc. Second All - Union Conf., 8. - 13.10.1962; Ed. I.I. Ibragimov) Baku 1965, 638 pp.; pp. 122 - 128. MR 34 # 548; Zbl 219, 248.

[13] CHENEY, E.W.: Introduction to Approximation Theory. McGraw - Hill, New York etc. 1966, xii + 259 pp.. MR 36 # 5568; Zbl 161, 252.

[14] LUPAŞ, A., On Bernstein power series. Mathematica (Cluj) 8(31)(1966) 287 - 296. MR 35 # 3332; Zbl 171, 309.

[15] CIMOCA, G. - A. LUPAŞ, Two generalizations of the Meyer - König and Zeller operator. Mathematica (Cluj) 9(32)(1967)233 - 240. MR 37 # 6652; Zbl 184, 92.

[16] JAKIMOVSKI, A. - D. LEVIATAN, Generalized Bernstein power - series. Math. Z. 96(1967)333 - 342. MR 35 # 7050; Zbl 146, 294.

[17] JAKIMOVSKI, A. - D. LEVIATAN, A property of approximation operators and applications to Tauberian constants. Math. Z. 102(1967)177 - 204. MR 36 #3015; Zbl 176, 348.

[18] LUPAŞ, A., Some properties of the linear positive operators (I). Mathematica (Cluj) 9(32)(1967)77 - 83. MR 35 # 7052; Zbl 155, 392.

[19] Lupaş, A. - M. Müller, Approximationseigenschaften der Gammaoperatoren. Math. Z. 98(1967)208 - 226. MR 35 # 7053; Zbl 171, 23.

[20] Müller, M.W., Die Folge der Gammaoperatoren. Dissertation, TH Stuttgart 1967, 87 pp.. MR 38 # 3669.

[21] BOEHME, T.K. - R.E. POWELL, Positive linear operators generated by analytic functions. SIAM J. Appl. Math. 16(1968)510 - 519. MR 37 # 1856; Zbl 172, 341.

[22] LEVIATAN, D., On approximation operators of the Bernstein type. J. Approx. Theory 1(1968)275 - 278. MR 38 # 1440; Zbl 165, 384.

[23] MÜLLER, M.W., Von links schräge Approximation durch Bernsteinsche Potenzreihen (Rum. sum.). Bul. Inst. Politehn. Iaşi 14(18), no. 1 - 2(1968)89 - 92. MR 39 # 3195; Zbl 175, 349.

[24] MÜLLER, M.W., Über die Ordnung der Approximation durch die Folge der Operatoren von Meyer - König und Zeller und durch die Folge deren erster Ableitungen (Rum. sum.). Bul. Inst. Politehn. Iaşi 14(18), no. 3 - 4(1968)83 - 90. MR 41 # 8892; Zbl 184, 92.

[25] MÜLLER, M., Gleichmäßige Approximation durch die Folge der ersten Ableitungen der Operatoren von Meyer - König und Zeller. Math. Z. 106 (1968)402 - 406. MR 38 # 1451; Zbl 159, 356.

[26] BASKAKOV, V.A., On certain linear positive operators (Russ.). Izv. Vysš. Učebn. Zaved. Matematika 1969, no. 10(89)(1969)11 - 20. MR 41 # 2267; Zbl 209, 448.

[27] EISENBERG, S.M., Moment sequences and the Bernstein polynomials. Canad. Math. Bull. 12(1969)401 - 411. MR 40 # 6123; Zbl 186, 112.

[28] JAKIMOVSKI, A. - D. LEVIATAN, Completeness and approximation operators. Publ. Ramanujan Inst. 1(1968/69)123 - 129. MR 42 # 6474; Zbl 198, 88.

[29] LEVIATAN, D., On the remainder in the approximation of functions by Bernstein - type operators. J. Approx. Theory 2(1969)400 - 409. MR 40 # 6131; Zbl 183, 328.

[30] MÜHLBACH, G., Über das Approximationsverhalten gewisser positiver linearer Operatoren. Dissertation, TU Hannover 1969, 93 pp..

[31]* RJATIN, A.G., The convergence of certain constructions of linear operators at points of discontinuity of the first kind (Russ.). Kalinin. Gos. Ped. Inst. Učen. Zap. 69(1969)124 - 134. MR 43 # 7821; RŽM 1969, 12B141.

[32] STANCU, D.D., Use of probabilistic methods in the theory of uniform approximation of continuous functions. Rev. Roumaine Math. Pures Appl. 14(1969)673 - 691. MR 40 # 606; Zbl 187, 325.

[33] SUZUKI, Y. - S. WATANABE, Some remarks on saturation problem in the local approximation, II. Tôhoku Math. J. (2) 21(1969)65 - 83. MR 37 # 6657; Zbl 215, 464.

[34] WALK, H., Approximation durch Folgen linearer positiver Operatoren. Arch. Math. (Basel) 20(1969)398 - 404. MR 42 # 2229; Zbl 191, 70.

[35] WATANABE, S. - Y. SUZUKI, Approximation of functions by generalized Meyer - Köning(!) and Zeller operator (Jap. sum.). Bull. Yamagata Univ. Nat. Sci. 7(1969)123 - 128. MR 41 # 8893.

[36] EISENBERG, S. - B. WOOD, Approximating unbounded functions with linear operators generated by moment sequences. Studia Math. 35(1970) 299 - 304. MR 42 # 6468; Zbl 199, 116.

[37] KING, J.P. - J.J. SWETITS, Positive linear operators and summability. J. Austral. Math. Soc. 11(1970)281 - 290. MR 42#2213; Zbl 199, 451.

[38] LUPAŞ, A. - M.W. MÜLLER, Approximation properties of the M_n-operators. Aequationes Math. 5(1970)19 - 37. MR # 5217; Zbl 205, 124.

[39] RJATIN, A.G., On the problem of the convergence of positive linear operators (Russ.). In: Interuniversity Scientific Conference on the Problem "Application of Functional Analysis in Approximation Theory". Proceedings of the Conference (Russ.). (Ed. V.N. Nikol'skiĭ) Kalinin. Gos. Ped. Inst., Kalinin 1970, 204 pp.; pp. 143 - 148.

[40] SIKKEMA, P.C., On the asymptotic approximation with operators of Meyer - König and Zeller. Indag. Math. 32(1970)428-440. MR 43 # 2402; Zbl 205, 81.

[41] SIKKEMA, P.C., On some research in linear positive operators in approximation theory. Nieuw Arch. Wisk. (3) 18(1970) 36 - 60. MR 41 # 5851; Zbl 189, 66.

[42] STANCU, D.D., Two classes of positive linear operators (Rum. sum.). An. Univ. Timişoara Ser. Şti. Mat.-Fiz. 8(1970)213 - 220. MR 48 # 11863; Zbl 276 # 41009.

[43] VOLKOV, V.I., Certain regular methods of the summation of series (Russ.). In: Proceedings of the Central Regional Union of Mathematical Departments; Functional Analysis and Theory of Functions, I (Russ.)(Eds. A.V. Efimov - A.L. Garkavi. - V.N. Nikol'skiĭ) Kalinin. Gos. Ped. Inst., Kalinin 1970, 148 pp.; pp. 38 - 75. MR 44 # 5669; RŽM 1971, 9B76.

[44] WALK, H., Approximation unbeschränkter Funktionen durch lineare positive Operatoren. Habilitationsschrift, Univ. Stuttgart 1970, 110 pp..

[45] BERENS, H., Pointwise saturation of positive operators. J. Approximation Theory 6(1972)135 - 146. MR 50 # 846; Zbl 262 # 41017.

[46] DEVORE, R.A., The Approximation of Continuous Functions by Positive Linear Operators. Lecture Notes Math. 293(1972), viii + 289 pp.. MR 54 # 8100; Zbl 276 # 41011.

[47] FELDMANN, D., Operatorenfolgen vom verallgemeinerten Voronovskaja Typ und deren Saturation. Dissertation, TU Hannover 1972, 79 pp..

[48] LORENTZ, G.G. - L.L. SCHUMAKER, Saturation of positive operators. J. Approx. Theory 5(1972)413 - 424. MR 49 # 9495; Zbl 233 # 41007.

[49] LUPAŞ, A., Die Folge der Betaoperatoren. Dissertation, Univ. Stuttgart 1972, 75 pp..

[50] MÜLLER, M.W. - H. WALK, Konvergenz- und Güteaussagen für die Approximation durch Folgen linearer positiver Operatoren. In: Constructive Function Theory (Proc. Internat. Conf., Golden Sands/Varna, 19. - 25.5.1970; Eds. B. Penkov - D. Vačov) Sofia 1972, 363 pp.; pp. 221 - 233. MR 51 # 3761; Zbl 243 # 41019.

[51] PETHE, S.P. - G.C. JAIN, Approximation of functions by a Bernstein-type operator. Canad. Math. Bull. 15(1972) 551 - 557. MR 47 # 687; Zbl 265 # 41013.

[52] SCHMID, G., Approximation unbeschränkter Funktionen. Dissertation, Univ. Stuttgart 1972, 79 pp..

[53] STANCU, D.D., A new generalization of the Meyer - König and Zeller operators (Rum. sum.). An. Univ. Timişoara Ser. Şti. Mat.-Fiz. 10(1972)207 - 214. MR 49 # 11113; Zbl 314 # 41010.

[54] BERENS, H., Pointwise saturation. In: Spline Functions and Approximation Theory (Proc. Symp. Univ. of Alberta, Edmonton, 29.5.-1.6. 1972; Eds. A. Meir - A. Sharma; ISNM 21) Basel - Stuttgart 1973, 386 pp.; pp. 11 - 30. MR 51 # 8682; Zbl 266 # 41015.

[55] RATHORE, R.K.S., Linear combinations of linear positive operators and generating relations in special functions. Dissertation, Indian Institute of Technology, Hauz Khas, New Delhi 1973, iv + 230 pp..

[56] STANCU, F., On the remainder in the approximation of functions by means of the Meyer - König and Zeller operator (Rum.; Engl. sum.). Stud. Cerc. Mat. 25(1973)619 - 627. MR 52 # 14769; Zbl 265 # 41014.

[57] SWETITS, J. - B. WOOD, Generalized Bernstein power series. Rev. Roumaine Math. Pures Appl. 18(1973)461 - 471. MR 47 # 7278; Zbl 253 # 41014.

[58] WALK, H., Lokale Approximation unbeschränkter Funktionen und ihrer Ableitungen durch eine Klasse von Folgen linearer positiver Operatoren. Mathematica (Cluj) 15(38)(1973)129 - 142. MR 52 # 6266; Zbl 294 # 41022.

[59] LUPAŞ, A., Mean value theorems for positive linear transformations (Rum.; Engl. sum.). Rev. Anal. Numer. Teoria Approximaţiei 3(1974) 121 - 140 (1975). MR 52 # 11426.

[60] STANCU, D.D., Evaluation of the remainders in certain approximation procedures by Meyer - König and Zeller - type operators. In: Numerische Methoden der Approximationstheorie, II. (Proc. Conf. Math. Res. Inst. Oberwolfach, Black Forest, 3. - 9.6.1973; Eds. L. Collatz - G. Meinardus; ISNM 26) Basel - Stuttgart 1975, 199 pp.; pp. 139 - 150. MR 52 # 9568; Zbl 331 # 41017.

[61] TIMMERMANS, C.A.: A generalization of a theorem of Mamedov. Rev. Anal. Numér. Théor. Approx. 4(1975)79 - 86. MR 58 # 29660; Zbl 363 # 41020.

[62] WALK, H., Über die Approximation unbeschränkter Funktionen durch lineare positive Operatoren. J. Reine Angew. Math. 276(1975)83 - 94. MR 53 # 13944; Zbl 308 # 41016.

[63] ESSER, H., On pointwise convergence estimates for positive linear operators on C[a,b]. Indag. Math. 38(1976)189 - 194. MR 53 # 8739; Zbl 327 # 41004.

[64] KNOOP, H.B. - P. POTTINGER, Ein Satz vom Korovkin - Typ für C^k - Räume. Math. Z. 148(1976)23 - 32. MR 54 # 3259; Zbl 322 # 41014.

[65] LEVIKSON, B., A new approximation operator generalizing Meyer - König and Zeller's power series. Canad. J. Math. 28(1976)301 - 311. MR 54 # 8101; Zbl 356 # 41012.

[66] POTTINGER, P., Zur linearen Approximation im Raum $C^k(I)$. Habilitationsschrift, Gesamthochschule Duisburg 1976, 116 pp..

[67] GÖTZ, B., Approximation durch lineare positive Operatoren und ihre Linearkombinationen. Dissertation, Univ. Stuttgart, 1977, 111 pp..

[68] HERMANN, T., Approximation of unbounded functions on unbounded intervals. Acta Math. Acad. Sci. Hungar. 29(1977)393 - 398. MR 56 # 16216; Zbl 371 # 41012.

[69] HÖLZLE, G.E., Quantitative Untersuchungen zur Approximation durch lineare positive Operatoren. Dissertation, Univ. Stuttgart 1977, 133 pp..

[70] JAIN, G.C. - S. PETHE, On the generalizations of Bernstein and Szász-Mirakyan operators. Nanta Math. 10(1977)185 - 193. MR 58 # 29654; Zbl 392 # 41012.

[71] KURC, W., Linear positive operators generated by Lagrange series and approximation of functions. In: Theory of Approximation of Functions (Russ.)(Proc. Conf. Kaluga, USSR; 24. - 28.7.1975; Eds. S.B. Stečkin - S.A. Teljakovskiĭ) Moscow 1977, 439 pp.; pp. 250 - 257. MR 80i : 41021; RŽM 1977,10B102.

[72] WOLFF, M.: On the theory of approximation by positive operators in vector lattices. In: Functional Analysis: Surveys and Recent Results (Proc. Conf. Paderborn, 17. - 21.11.1976; Eds. K.-D. Bierstedt - B. Fuchssteiner) North-Holland, Amsterdam - New York 1977, xii + 290 pp.; pp. 73 - 87. MR 57 # 6979; Zbl 371 # 41017.

[73] BECKER, M. - R.J. NESSEL, A global approximation theorem for Meyer-König and Zeller operators. Math. Z. 160(1978)195 - 206. MR 58 # 23273; Zbl 376 # 41007.

[74] ISMAIL, M.E.H., Polynomials of binomial type and approximation theory. J. Approximation Theory 23(1978)177 - 186. MR 81a : 41033; Zbl 385 # 41014.

[75] ISMAIL, M.E.H. - C.P. MAY, On a familiy of approximation operators. J. Math. Anal. Appl. 63(1978)446 - 462. MR 80a : 41017; Zbl 375 # 41011.

[76] MÜLLER, M.W., L_p - approximation by the method of integral Meyer - König and Zeller operators. Studia Math. 63(1978)81 - 88. MR 80a : 41021; Zbl 389 # 41009.

[77] MÜLLER, M.W., Approximationstheorie. Akad. Verl.-Ges., Wiesbaden 1978, 247 pp.. Zbl 382 # 41001.

[78] MÜLLER, M.W. - V. MAIER, Die lokale L_p - Saturationsklasse des Verfahrens der integralen Meyer - König und Zeller Operatoren. In: Linear Spaces and Approximation (Proc. Conf. Math. Res. Inst. Oberwolfach, Black Forest, 20. - 27.8.1977; Eds. P.L. Butzer - B. Sz.-Nagy; ISNM 40) Basel - Stuttgart 1978, 685 pp.; pp. 305 - 317. MR 58 # 23279; Zbl 412 # 41016.

[79] POTTINGER, P., On the C^k - approximation by Baskakov - operators. In: Fourier Analysis and Approximation Theory I, II (Colloquia Math. Soc. Jânos Bolyai, 19; Proc. Conf., Budapest, 16. - 21.8.1976; Eds. G. Alexits - P. Turān) Amsterdam - Oxford - New York 1978, 926 pp.; pp. 649 - 657. MR 80i : 41017; Zbl 447 # 41013.

[80] RATHORE, R.K.S., Lipschitz - Nikolskii constants and asymptotic simultaneous approximation of the M_n - operators. Aequationes Math. 17 (1978)391 - 393.

[81] RATHORE, R.K.S., Lipschitz - Nikolskii constants and asymptotic simultaneous approximation of the M_n - operators. Aequationes Math. 18 (1978)206 - 217. MR 80a : 41022; Zbl 379 # 41012.

[82] SCHURER, F. - F.W. STEUTEL, On the degree of approximation of functions in $C_1[0,1]$ by the operators of Meyer - König and Zeller. J. Math. Anal. Appl. 63(1978)719 - 728. MR 58 # 12120; Zbl 382 # 41012.

[83] VAN DER MEER, P.J.C., On the degree of approximation by certain linear positive operators. Indag. Math. 40(1978)467 - 478. MR 80c : 41013; Zbl 447 # 41014.

[84] VESELINOV, V.M., A general method for determining best constants in linear methods of Hausdorff approximation (Russ.). C.R. Acad. Bulgare Sci. 31(1978)1385 - 1388. MR 81c : 41051; Zbl 434 # 41017.

[85] GONSKA, H.H., Quantitative Aussagen zur Approximation durch lineare Operatoren. Dissertation, Gesamthochschule Duisburg,1979, ix +190 pp..

[86] KUDRJAVCEV, G.I., On the convergence of the derivatives of linear convex and smooth operators (Russ.). In: Application of Functional Analysis in Approximation Theory (Russ.). (Eds. A.L. Garkavi - A.V. Efimov - L.A. Markova - V.N. Nikol'skiĭ) Kalinin. Gos. Univ., Kalinin 1979, 163 pp.; pp. 61 - 65. MR 82c : 41023; RŽM 1979, 11B717.

[87] LEHNHOFF, H.-G., Lokale Approximationsmaße und Nikolskiĭ - Konstanten für positive lineare Operatoren. Dissertation, Univ. Dortmund 1979, 140 pp.. Zbl 433 # 41008.

[88] LEVIKSON, B., On the behaviour of a certain class of approximation operators for discontinuous functions. Acta Math. Acad. Sci. Hungar. 33(1979)299 - 306. MR 81b : 41054; Zbl 425 # 41023.

[89] SIKKEMA, P.C. - P.J.C. VAN DER MEER, The exact degree of local approximation by linear positive operators involving the modulus of continuity of the p-th derivative. Indag. Math. 41(1979)63 - 76. MR 80h : 41010; Zbl 399 # 41023.

[90] TIHOMIROV, N.B. - A.G. RJATIN, Linear Positive Operators and Singular Integrals. Kalinin. Gos. Univ., Kalinin 1979, 76 pp. (200 copies !). MR 81m : 45024.

[91] BLEIMANN, G. - P.L. BUTZER - L. HAHN, A Bernstein-type operator approximating continuous functions on the semi-axis. Indag. Math. 42 (1980)255-262. MR 81m: 41023; Zbl 437 # 41021.

[92] HÖLZLE, G.E., On the degree of approximation of continuous functions by a class of sequences of linear positive operators. Indag. Math. 42(1980)171-181. MR 81h : 41029; Zbl 427 # 41013.

[93] KHAN, R.A., Some probabilistic methods in the theory of approximation operators. Acta Math. Acad. Sci. Hungar. 35(1980)193-203. MR 81m : 41024; Zbl 437 # 41020.

[94] KURC, W., On some examples of linear positive operators in the space of continuous functions, related to the classical orthogonal polynomials. In: Constructive Function Theory '77 (Proc. Internat. Conf., Blagoevgrad, 30.5.-6.6.1977; Eds. B. Sendov - D. Vačov) Sofia 1980, 560 pp.; pp. 353-364. Zbl 452#41019; RŽM 1981,4B68.

[95] SHAW, SEN-YEN, Approximation of unbounded functions and applications to representation of semigroups. J. Approx. Theory 28(1980)238-259. MR 81f : 41027; Zbl 452 # 41020.

[96] WALK, H., Probabilistic methods in the approximation by linear positive operators. Indag. Math. 42(1980)445-455. MR 82c : 41024; Zbl 486 # 41016.

[97] IL'IN, V.F., On the question of convergence of sequences of the linear positive operators of G.A. Fomin (Russ.). VINITI 675-81(1981), 14 pp. [TIB Hannover S nat R 1/ZZ 3389(675-81)].

[98] MAIER, V. - M.W. MÜLLER - J. SWETITS, The local L_1-saturation class of the method of integral Meyer-König and Zeller operators. J. Approximation Theory 32(1981)27-31. Zbl 489#41022; RŽM 1982, 2B83.

[99] SIKKEMA, P.C. - P.J.C. VAN DER MEER - MARIA ROOS, Determination of the exact degree of local approximation by some linear positive operators involving the modulus of continuity of the p-th derivative. Indag. Math. 43(1981)117-128. MR 82g : 41024; RŽM 1982, 8B98.

[100] SINGH, S.P., An estimate on the M_n-operators. Boll. Un. Mat. Ital. (6) 1-A(1982)109-113.

[101] TOTIK, V., Approximation by Meyer-König and Zeller type operators. Math. Z. 182(1983)425-446.

[102] ALKEMADE, J.A.H., The second moment for the Meyer-König and Zeller operators. J. Approx. Theory, in print.

[103] TOTIK, V., Uniform approximation by Baskakov and Meyer - König and Zeller operators. Period. Math. Hungar., in print.

[104] TOTIK, V., Uniform approximation by positive operators on infinite intervals. Anal. Math., in print.

V Functions of a Complex Variable and Approximation

International Series of
Numerical Mathematics, Vol. 65

THE BEST HARMONIC APPROXIMANT TO A CONTINUOUS FUNCTION

Walter K. Hayman and Donald Kershaw and Terry J. Lyons

Walter K. Hayman, Dept. of Maths., Imperial College, London SW7 2BZ.

Donald Kershaw, Dept. of Maths., University of Lancaster, Lancaster LA1 4YL.

Terry J. Lyons, Dept. of Maths., Imperial College, London SW7 2BZ.

Suppose that f is bounded and continuous in a domain D in R^k. Then there exists a best harmonic approximant h to f in the uniform norm. If D is a Jordan domain, f is continuous in $\overline{D}$, and h is continuous in $\overline{D}$, then h is unique and can be characterised in terms of the sets in $\overline{D}$ where h - f assumes the extreme values $\mp m$. Examples are given to show that if these hypotheses are relaxed in various ways the conclusion may fail. For instance h need not be continuous in $\overline{D}$, even if f is continuous in $\overline{D}$, and if f is only bounded and continuous in D, h need not be unique.

1. Introduction. Suppose that D is a domain in R^k, $k \geq 1$, and that f(x) is continuous and bounded in D. We write

$$||f|| = \sup_{x \in D} |f(x)|,$$

and ask for a function h(x) harmonic in D, such that

$$(1.1) \qquad m = ||f-h||$$

is as small as possible, Such a function h will be called a best (harmonic) approximant to f.

We note first that a best approximant exists. In fact since f is bounded we may choose $h = 0$ to make m finite. Let m_0 be the lower bound of all possible values of m and choose h_n so that

$$m_n = ||f-h_n|| < m_0 + \frac{1}{n} .$$

The functions h_n are uniformly bounded and so equicontinuous in compact subsets of D and so, choosing a subsequence if necessary, we may assume that h_n converges, locally uniformly in D, to a harmonic limit h. Also for $x \in D$

$$|f(x) - h(x)| = \lim_{n\to\infty} |f(x) - h_n(x)| \leq \lim_{n\to\infty} (m_o + \frac{1}{n}) = m_o.$$

Thus

$$||f-h|| \leq m_o \quad \text{and so} \quad ||f-h|| = m_o.$$

We deduce that $m_o = 0$ if and only if f is harmonic in D. We shall assume from now on that this is not the case, so that $m_o > 0$. We assume that h is a best approximant to f, and ask whether h is unique and can be characterised.

2. Statement of Results. We can only obtain conclusions under rather strong restrictions. Accordingly we shall assume from now on that
(i) D is a Jordan domain with frontier Γ in R^k. Thus there exists a homeomorphism from the closed unit ball $|y| \leq 1$ in R^k onto $\bar{D}$ in which $|y| < 1$ corresponds to D and $|y| = 1$ to Γ. We use only the following consequence of this.
(i.a) If e is any compact subset of $\bar{D}$ and $\delta(e)$ is the union of those complementary domains of e which lie entirely in $\bar{D}$, then $\delta(e) \subset D$. We shall call

$$\hat{e} = e \cup \delta(e)$$

the hull of e. Thus $\hat{e}$ is also a compact subset of $\bar{D}$.

If e_1, e_2 are disjoint compact subsets of $\bar{D}$, we shall say that e_1, e_2 are linked if $\hat{e}_1 \cap \hat{e}_2 \neq \phi$.

We assume further that
(ii) f has a continuous extension to $\bar{D}$.
We can now state our result

THEOREM 1. *Suppose that there exists a best approximant h to f which has a continuous extension to $\bar{D}$. Then h is the unique best approximant* *to f. We now write f,h also for the extended functions on $\overline{D}$, define m by (1.1) and let e_+, e_- be the sets of all x on $\overline{D}$ where $h(x) - f(x) = m$, $h(x) - f(x) = -m$ respectively. Then e_+, e_- are linked.*

Conversely suppose that there exists h harmonic in D and continuous in $\overline{D}$, such that the sets e_+ and e_- as defined above are linked, where m is given by (1.1). Then h is the unique best approximant to f.

3. Proof of Theorem 1. We suppose that D is a Jordan domain, that f,h, are continuous in $\overline{D}$, that h is harmonic in D, but f is not, so that m, defined by (1.1) is positive and finite. We write

$$\mu = \inf_{x\in\overline{D}} (h(x) - f(x)), \qquad M = \sup_{x\in\overline{D}}(h(x) - f(x)).$$

Then

$$-m \leq \mu \leq M \leq m.$$

Suppose e.g. that $M = m-2\varepsilon < m$. Then $h_1(x) = h(x) + \varepsilon$ satisfies

$$-m + \varepsilon \leq \mu + \varepsilon \leq h_1(x) - f(x) \leq M + \varepsilon = m - \varepsilon, \qquad x \in \overline{D}.$$

Thus

$$||h_1 - f|| \leq m - \varepsilon < m,$$

so that h is not a best approximant to f. Hence if h is a best approximant to f we must have $M = m$ and similarly $\mu = -m$. Also, since $h - f$ is continuous in $\overline{D}$ and $\overline{D}$ is compact, the sets

$$e_+ = \{x|x\in\overline{D} \text{ and } h(x) - f(x) = m\}, \; e_- = \{x|x\in\overline{D} \text{ and } h(x) - f(x) = -m\}$$

are not empty. We now suppose that e_+ and e_- are linked.

With this hypothesis we show that h is the unique best approximant to f. Suppose contrary to this that $h_1(x)$ is any other function harmonic in D and such that

$$||h_1 - f|| \leq m.$$

We proceed to show that $h_1 = h$ in D. To see this let x_0 be a common point of $\hat{e}_+$ and $\hat{e}_-$. Since $m > 0$, e_+ and e_- are disjoint. Thus x_0 cannot lie in both e_+ and e_-. Suppose e.g. that x_0 is not in e_+. Then x_0 lies in a bounded complementary domain D_+ of e_+. It follows from the consequence (i.a) of hypothesis (i) that $D_+ \subset D$. Thus $u = h_1 - h$ is harmonic in D_+.

We now apply the maximum principle to u in D_+. Suppose that ξ is a frontier point of D_+. Then $\xi \in e_+$. Hence we have as $x \to \xi$ from inside D_+, since $h(x)$ is continuous in $\overline{D}_+$,

$$\limsup u(x) = \limsup \{h_1(x) - f(x)\} - \lim \{h(x) - f(x)\}$$
$$\leq m - m = 0.$$

Thus

(3.1) $\quad u(x) \leq 0$ in D_+.

We note next that $u(x_0) \geq 0$. In fact if $x_0 \in e_-$, we have

$$u(x_0) = h_1(x_0) - f(x_0) - (h(x_0)-f(x_0)) \geq -m + m = 0.$$

If x_0 is not in e_- then x_0 lies in a bounded complementary domain D_- of e_-. We apply the maximum principle to

$-u(x) = h(x) - h_1(x)$ in D_- and note that

$$\limsup -u(x) = \lim(h(x)-f(x)) - \liminf\{h_1(x)-f(x)\} \leq -m+m = 0$$

as x approaches any frontier point ξ of D_-, since ξ must belong to e_-. Hence $-u(x) \leq 0$ in D_- and so $u(x_0) \geq 0$. On combining this with (3.1) we deduce that $u(x_0) = 0$ and so $u(x) = 0$ throughout D_+. Since $u(x)$ is harmonic in the domain D, we deduce that $u(x) = 0$ throughout D. This completes the sufficiency part of Theorem 1.

Suppose next that h is a best approximant to f in D and that h is continuous in $\overline{D}$. Then, as was shown above, the sets e_+ and e_- cannot be empty. We assume that e_+ and e_- are not linked and obtain a contradiction. We write

$$E_+ = \hat{e}_+ , \quad E_- = \hat{e}_- , \quad E = E_+ \cup E_-$$

so that E_+, E_- are disjoint compact subsets of R^k. We suppose first that $k \geq 2$.

By our construction the complement $D_+ = R^k \setminus E_+$ of E_+ is the unbounded complementary domain of e_+ and so is an unbounded domain. The same is true of $D_- = R^k \setminus E_-$. Since $E_+ \cap E_- = \phi$, we deduce that the complement

$$D = D_+ \cap D_-$$

of E is also connected.

We now quote the following approximation Theorem of Brelot-Deny [1,2]

LEMMA 1. *If E is a compact set in R^k, $k \geq 2$, whose complement is connected, and $u(x)$ is harmonic in a neighbourhood of E, then given $\varepsilon > 0$ there exists $v(x)$ harmonic in R^k, such that $|v(x) - u(x)| < \varepsilon$ on E.*

We may surround E_+, E_- by disjoint open subsets G_+ and G_- and define $u(x) = 1$ on G_+, $u(x) = -1$ on G_-. Thus $u(x)$ is harmonic in the neighbourhood $G = G_+ \cup G_-$ of E and so we may approximate $u(x)$ on E by a function v harmonic in R^k. We deduce that there exist compact sets F_+ and F_- in $\overline{D}$ which include neighbourhoods of e_+, e_- w.r.t. $\overline{D}$ and are such that

(3.2) $\quad v(x) > 0,\quad$ and $\quad h(x) - f(x) > 0\quad$ in F_+,

(3.3) $\quad v(x) < 0\quad$ and $\quad h(x) - f(x) < 0\quad$ in F_-.

Also if

$$F = \overline{D} \setminus (F_+ \cup F_-)$$

then we deduce that the closure $\overline{F}$ of F does not meet e_+ or e_- so that

$$|h(x) - f(x)| < m \text{ in } \overline{F}.$$

Hence, since $\overline{F}$ is compact and $h(x) - f(x)$ is continuous on $\overline{F}$ we have

(3.4) $\quad |h(x) - f(x)| < m - 2\varepsilon\quad$ in $\overline{F}$,

where ε is a positive number.

We now choose α positive but so small that

(3.5) $\quad |\alpha\, v(x)| < \varepsilon\quad$ in $\overline{D}$.

Then $h_1(x) = h(x) - \alpha v(x)$ is a better harmonic approximant to f than $h(x)$ and this gives the required contradiction. In fact we have for $x \in F_+$ from (3.2), (3.4) and (3.5)

$$h_1(x) - f(x) = h(x) - f(x) - \alpha v(x) \leq m - \alpha v(x) < m$$

and

$$h_1(x) - f(x) = h(x) - f(x) - \alpha v(x) > -\varepsilon > -\tfrac{1}{2}m.$$

Thus

(3.6) $\quad |h_1(x) - f(x)| < m$

in F_+. Similarly (3.6) holds in F_-. Finally in F we deduce from (3.4) and (3.5)

$$|h_1(x) - f(x)| \leq |h(x) - f(x)| + |\alpha v(x)| \leq m - 2\varepsilon + \varepsilon = m - \varepsilon.$$

Thus (3.6) holds through $\overline{D}$ and so $h(x)$ cannot be the best harmonic approximant to f. This contradiction proves that e_1 and e_2 must be linked and completes the proof of Theorem 1, when $k \geq 2$.

If $k = 1$ the above argument must be modified. It is no longer true that if E_+, E_- are compact disjoint sets with connected complement on the line. i.e. disjoint closed intervals, then the same is true of their union. However we can argue directly as follows. Suppose that e_+ and e_- are not linked. Then no point of one of these sets lies between two points of the other. Thus there exists a real x_1 separating these sets, so that either

$$x < x_1 \text{ in } e_+ \text{ and } x > x_1 \text{ in } e_-$$

or

$$x > x_1 \text{ in } e_+ \text{ and } x < x_1 \text{ in } e_-.$$

We set $v(x) = x_1 - x$ in the former case and $v(x) = x - x_1$ in the latter case. Then we can still find neighbourhoods F_+, F_- of e_+ and e_- which satisfy (3.2) and (3.3) and now the proof of Theorem 3 is completed as before.

4. Some Examples. We proceed to illustrate Theorem 1 with some examples.

EXAMPLE 4.1. Let e_1, e_2 be disjoint non empty compact sets in R^k, let $\delta_j(x)$ be the distance of the point x from e_j and write

$$f(x) = \frac{\delta_2(x) - \delta_1(x)}{\delta_2(x) + \delta_1(x)}.$$

Then $f(x)$ is continuous in R^k, $-1 \leq f(x) \leq 1$ and $f(x) = 1$ precisely on e_1, $f(x) = -1$ on e_2. Thus if D is a Jordan domain containing e_1 and e_2, Theorem 1 shows that 0 is the best harmonic approximant to f in D if and only if e_1 and e_2 are linked. The example shows that any two linked sets can occur in Theorem 1.

EXAMPLE 4.2. Suppose that D is the unit disk in R^2 cut along the negative axis from -1 to 0. Define

$$f(z) = 1 \text{ on } |z| = 1$$

$$f(z) = \tfrac{1}{2} \text{ on } |z| = \tfrac{1}{2}$$

$$f(z) = -z, \quad -1 < z \leq 0$$

and extend $f(z)$ into the domains $D_1: |z| < \frac{1}{2}$ and $D_2: \frac{1}{2} < |z| < 1$, cut along the negative real axis, as a harmonic function with the above boundary values. Evidently $f(z)$ is continuous in $\overline{D}$. Also

$$f(z) - \tfrac{1}{2} = \tfrac{1}{2} \quad \text{on} \quad e_+ : |z| = 1$$

$$f(z) - \tfrac{1}{2} = -\tfrac{1}{2} \text{ on } \quad e_- : z = 0.$$

The sets e_+ and e_- are linked, but $\frac{1}{2}$ is not the best harmonic approximant to f. To see this we define $h(z)$ to be the harmonic function in D with boundary values $|z|$ on the negative real axis and $|z| = 1$. Then we have $0 < h(z) < 1$ in D and so

$$|h(z) - f(z)| = |h(z) - \tfrac{1}{2}| < \tfrac{1}{2}, \qquad |z| = \tfrac{1}{2}$$

while $h(z) - f(z) = 0$ at all other boundary points of D_1 and D_2. Since $h(z) - f(z)$ is harmonic in D_1, D_2, we deduce that

$$|h(z) - f(z)| < \tfrac{1}{2}$$

in $\overline{D}$. Thus $\frac{1}{2}$ is not the best harmonic approximant to f.

In this example D, although simply connected, is not a Jordan domain and the hypothesis (i.a) breaks down. In the proof of Theorem 1 we needed (i.a) to prove that $h(x)$ is harmonic in the bounded complementrary domains of e_+, in this case $|z| < 1$.

EXAMPLE 4.3. The following example shows that even when D is a plane Jordan domain and f is continuous in $\overline{D}$ the best harmonic approximant to f need not be continuous on the boundary of D, so that Theorem 1 does not cover all cases.

We take for D the strip

(4.1) $\qquad S = -1 < x < 1, \quad z = x + iy$

and define

(4.2) $\qquad f(x) = x(1 - |x|), \ -1 \leq x \leq 1$

(4.3) $\qquad f(x + iy) = 0, \quad x = \mp 1$

and f is continuous in $\overline{S}$ including ∞, and harmonic in S outside the real axis. We assert

THEOREM 2. *If $f(z)$ is the function defined above then the unique best approximant h to f in S is given by*

(4.4) $\qquad h(x + iy) = mx,$

where $m = 3-2\sqrt{2}$. Also

(4.5) $\qquad ||h - f|| = m.$

We note that $h(z)$ is discontinuous at $z = \mp i\,\infty$. By mapping S conformally onto the unit disk we obtain a continuous function f in $|z| \leq 1$, whose unique best approximant is discontinuous at two points on $|z| = 1$.

We prove first (4.5) with h given by (4.4). Since $h - f$ is harmonic in the half strip

$$S_+ : -1 < x < 1, \quad y > 0$$

it is sufficient to study the behaviour of $h - f$ on the boundary of S_+. Evidently

$$h - f = m, \text{ when } x = 1, \quad h - f = -m, \text{ when } x = -1$$

and

$$\overline{\lim_{z\to\infty}} |h(z) - f(z)| = \overline{\lim_{y\to\infty}} |h(x+iy)| \leq m.$$

Also for $y = 0$, $0 \leq x \leq 1$

$$h(x) - f(x) = x^2 + (m-1)x = \left(x + \frac{m-1}{2}\right)^2 - \left(\frac{m-1}{2}\right)^2.$$

Thus

$$-\left(\frac{m-1}{2}\right)^2 \leq h(x) - f(x) \leq \left(\frac{m+1}{2}\right)^2 - \left(\frac{m-1}{2}\right)^2 = m.$$

Since $m = 3 - 2\sqrt{2}$, this reduces to

$$(4.6) \qquad -m \leq h(x) - f(x) \leq m$$

for $0 \leq x \leq 1$ and similarly for $-1 \leq x \leq 0$. Thus (4.6) holds in S_+ and so by symmetry in the whole of S and this proves (4.5).

We also see that the set e_+ where $h(x) = f(x) + m$ consists of the line $y = 1$ and the point $z = \frac{1}{2}(m-1) = 1-\sqrt{2}$, and e_- consists of of the line $y = -1$ and the point $z = \sqrt{2}-1$. The two points of discontinuity at $\mp i\,\infty$ are limit points of both e_+ and e_-.

To prove Theorem 2 we need

LEMMA 2. *If $h(z)$ is harmonic in S given by (4.1) and satisfies $|h(z)| \leq M$ there, where M is a positive constant, then*

$$(4.7) \qquad h(x+iy) - h(-x + iy) \leq 2Mx, \quad 0 < x < 1.$$

Equality holds if and only if $h(z) \equiv Mx$ in S.

Consider

$$\phi(z) = h(x + iy) - h(-x+iy) - 2Mx$$

in the strip $S_1 : 0 < x \leq 1$. Then $\phi(z)$ is bounded and harmonic in S_1 and vanishes continuously on the line $x = 0$. Also as z approaches a boundary point $1 + iy_0$ of S_1 we have

$$\limsup \phi(z) \leq M + M - 2M = 0.$$

Thus $\phi(z) \leq 0$ in S_1, which proves (4.7). Also equality can hold only if $\phi(z) \equiv 0$. In this case choose $x = X = 1-\delta$, where δ is a small positive number. We deduce that

$$h(X + iy) = 2MX + h(-X + iy) \geq M(2X - 1).$$

Thus as z approaches any point $z_0 = 1 + iy$ from inside S we have

$$\lim h(z) = M.$$

Similarly as z approaches $z_0 = -1 + iy$ from inside S we have

$$\lim h(z) = -M.$$

Hence $h(z) - Mx \to 0$ as z approaches any finite boundary point of S and so $h(z) \equiv Mx$ and Lemma 2 is proved.

We can now complete the proof of Theorem 2. Suppose that $f(z)$ is the function defined in Theorem 2 and that $h(z)$ is harmonic in S and satisfies

$$||h - f|| \leq m = 3 - 2\sqrt{2}.$$

Choose $z_0 = \sqrt{2}-1$. Then

$$h(z_0) \geq f(z_0) - m, \text{ and } h(-z_0) \leq f(-z_0) + m.$$

Thus

$$h(z_0) - h(-z_0) \geq f(z_0) - f(-z_0) - 2m$$

$$=2\{(\sqrt{2}-1)(2-\sqrt{2}) - 3 + 2\sqrt{2}\} = 2(5\sqrt{2}-7) = 2mz_0. \tag{4.8}$$

Also $h(z)$ is bounded in S and as z approaches any finite boundary point of S we have

$$\limsup|h(z)|= \limsup |h(z) - f(z)| \leq m.$$

Thus $|h(z)| \leq m$ in S and now Lemma 2 and (4.8) show that $h(z) \equiv mx$. This proves Theorem 2.

5. Cases where the Best Approximant is Continuous. We note the following easy consequence of Theorem 1.

THEOREM 3. *Suppose that $f(x)$ vanishes on the frontier Γ of D and that $f(x) \le 0$ in D. Then if m is the minimum of $f(x)$ in D the best approximant to f is the constant $\frac{1}{2}m$.*

In fact if e_- is the frontier Γ of D and e_+ is the set of points in D where $f(x) = m$, e_+ lies in the hull $\overline{D}$ of e_- so that e_+ and e_- are linked. Also if $h(x) = -\frac{1}{2}m$, we have

$$|f(x) - h(x)| \le \tfrac{1}{2}m$$

and

$$h(x) = f(x) + \tfrac{1}{2}m \text{ in } e_+, \quad h(x) = f(x) - \tfrac{1}{2}m \text{ in } e_- .$$

Thus Theorem 3 follows from Theorem 1.

COROLLARY. *If D is a Jordan domain, regular for the problem of Dirichlet, and f is continuous in $\overline{D}$ and subharmonic (or superharmonic) in D, then the best approximant h to f is continuous in $\overline{D}$.*

Let $F(x)$ be the solution to the problem of Dirichlet with boundary values f on Γ. Then $f_0(x) = f(x) - F(x)$ is continuous in $\overline{D}$ and subharmonic in D and $f_0(x) = 0$ on Γ. Thus $f_0(x) \le 0$ in D and so has a constant best approximant $\frac{1}{2}m$. Thus $F(x) + \frac{1}{2}m$ is the best approximant of F. If f is superharmonic we consider $-f$ instead if f.

If $k = 1$ the domain D reduces to an open interval and harmonic functions in D are linear and so automatically continuous in $\overline{D}$. Thus the conclusions of Theorem 1 hold in this case. The following example shows however that if (ii) fails, i.e. if f is merely continuous and bounded in D, the best approximant need not be unique.

EXAMPLE 5.1.If $f(x) = (1-x)\sin 1/x$, $0 < x < 1$, then ax is a best harmonic approximant to $f(x)$ in $(0,1)$ for $|a| \le 1$.

We note that if $h(x)$ is continuous in $[0,1]$

$$\limsup_{x \to 0+} f(x) - h(x) = 1 - h(0), \quad \liminf_{x \to 0+} f(x) - h(x) = -1 - h(0).$$

Thus

$$\sup_{0<x<1} |h(x) - f(x)| \geq 1.$$

On the other hand if $h(x) = ax$, where $|a| \leq 1$, we have

$$|f(x) - h(x)| \leq |1-x| + |ax| \leq 1 - x + x = 1 \qquad 0 < x < 1,$$

so that ax is a best approximant to $f(x)$.

By considering f as a function of $(x, x_2, \ldots, x_k)$ in the hypercube

$$D : 0 < x < 1, \quad 0 < x_j < 1, \quad j = 2 \text{ to } k$$

we obtain a corresponding example in D. A best approximant h to f is a bounded harmonic function in D and so the limit

$$\lim_{x\to 0+} h(x, \xi_2, \xi_3, \ldots, \xi_k) = h(0,\xi)$$

exists almost everywhere in ξ on the face $x = 0$ of D. We deduce again that

$$||f - h|| \geq 1.$$

Hence $h(x) = ax$ is a best approximant to f in D, when $|a| \leq 1$.

REFERENCES

[1] Brelot, M., Sur l'approximation et la convergence dans la théorie des fonctions harmoniques ou holomorphes. Bull. Soc. Math. France 73 (1945), 55-70.

[2] Deny, J., Systèmes totaux de fonctions harmoniques. Ann. Inst. Fourier (Grenoble) 1 (1949), 103-113.

International Series of
Numerical Mathematics, Vol. 65

OPTIMAL APPROXIMANTS AND SZEGÖ'S INFIMUM

James Rovnyak[1)]
Department of Mathematics
University of Virginia
Charlottesville

Szegö's infimum is studied relative to a side condition on norms. A method based on Mellin transforms yields closed form expressions for the extremal functions when the weight function is the characteristic function of an arc on the unit circle.

1. Introduction

Let $D = \{z: |z| < 1\}$, $\Gamma = \partial D$, and let σ be normalized Lebesgue measure on Γ. We are interested in the mean square approximation of a given function h on Γ by functions $k \in H^2(\Gamma)$ relative to a nonnegative weight $w \in L^\infty(\sigma)$. By $H^2(\Gamma)$ we mean the subspace of $L^2(\sigma)$ consisting of the boundary functions of functions in the Hardy class $H^2(D)$. See Duren [2].

DEFINITION. *Given* $h \in L^2(wd\sigma)$, *by an* optimal approximant *to* h *we mean any* $\tilde{k} \in H^2(\Gamma)$ *such that*

$$\int_\Gamma |h-\tilde{k}|^2 \, wd\sigma \leq \int_\Gamma |h-k|^2 \, wd\sigma$$

for all $k \in H^2(\Gamma)$ *such that* $||k||_2 \leq ||\tilde{k}||_2$.

The norm on $H^2(\Gamma)$ is that of $L^2(\sigma)$: $||k||_2^2 = \int_\Gamma |k|^2 \, d\sigma$. An optimal

1) Research supported by NSF Grant MCS 81-02518.

approximant does the best job of approximating h among all functions of the same or smaller norm.

In this paper we compute an example related to *Szegö's infimum* (Grenander and Szegö [3], p. 44):
$\inf \{\int_\Gamma |e^{-i\theta}-k(e^{i\theta})|^2 w(e^{i\theta})d\sigma: k \in H^2(\Gamma)\} = \exp(\int_\Gamma \log w\, d\sigma)$, where the right side is interpreted as 0 if $\log w \notin L^1(\sigma)$. We shall exhibit the optimal approximants to $e^{-i\theta}$ when $w = \chi_\Delta$ is the characteristic function of an arc Δ on Γ. The case of a general weight function will be treated in [1] by a somewhat different method. A similar problem is given in Kreĭn and Nudel'man [4].

I wish to thank W. Splettstösser for a private communication in which he posed a related question. See [7], where a connection is also given with the problem of predicting a band limited function through the sampling of values.

2. Main Theorem

Let $\Delta = \{e^{i\theta}: a < \theta < b\}$, $0 < b-a < 2\pi$, and set $\alpha = e^{ia}$, $\beta = e^{ib}$.

THEOREM. *A function $\tilde{k} \in H^2(\Gamma)$ is an optimal approximant to $e^{-i\theta}$ relative to the weight $w = \chi_\Delta$ if and only if $\tilde{k}(e^{i\theta}) = \tilde{k}_s(e^{i\theta})$ for some $s \in (0,\infty)$, where*

$$(1) \qquad \tilde{k}_s(z) = z^{-1}\left[1 - \left(\frac{1-z/\alpha}{1-z/\beta}\right)^{-is}\right], \quad z \in D.$$

For each $s \in (0,\infty)$, $1-z\tilde{k}_s(z)$ is an outer function with

$$(2) \qquad |1-e^{i\theta}\tilde{k}_s(e^{i\theta})| = \begin{cases} \exp(-\pi s\sigma(\Gamma\setminus\Delta)), & e^{i\theta} \in \Delta, \\ \exp(\pi s\sigma(\Delta)), & e^{i\theta} \in \Gamma\setminus\Delta . \end{cases}$$

See Duren [2] for the definition of an outer function. In (1) we take the branch of $k_s(z)$ determined by $\left|\arg \frac{1-z/\alpha}{1-z/\beta}\right| < \pi$.

The proof uses a known solution to a Hilbert space extremal problem (Rosenblum [5]), which we recast in a form convenient to our purpose. Let $A \in \mathcal{B}(\mathcal{K},\mathcal{H})$, where $\mathcal{H},\mathcal{K}$ are Hilbert spaces. Let h be a given vector in $\mathcal{H}$. Call a vector $\tilde{k} \in \mathcal{K}$ *optimal* if

(3) $$||h-A\tilde{k}||_{\mathcal{H}} \leq ||h-Ak||_{\mathcal{H}}$$

for all $k \in \mathcal{K}$ such that $||k||_{\mathcal{K}} \leq ||\tilde{k}||_{\mathcal{K}}$. The totality of optimal vectors consists of the family

(4) $$k_\lambda = (A^*A + \lambda I)^{-1}A^*h, \qquad \lambda > 0,$$

together with $k_0 = \lim_{\lambda \downarrow 0} k_\lambda$ whenever this limit exists strongly.

In our application, $\mathcal{H} = L^2(\chi_\Delta d\sigma)$, $\mathcal{K} = H^2(D)$, and $A \in \mathcal{B}(\mathcal{K},\mathcal{H})$ is the natural embedding operator. Choose $h(e^{i\theta}) = e^{-i\theta} \in \mathcal{H}$ for the fixed vector in (3). Then the optimal vectors are precisely the optimal approximants to $e^{-i\theta}$ relative to the weight $w = \chi_\Delta$. Our task therefore is to compute the functions $\{k_\lambda\}_{\lambda>0}$ defined by (4). We shall see that, after a change of parameter, these functions coincide with the functions $\{\tilde{k}_s\}_{s>0}$ given by (1).

3. Calculation via Mellin Transforms

To compute the functions $\{k_\lambda\}_{\lambda>0}$ defined by (4) we pass via natural Hilbert space isomorphisms to a representation for A in which calculations are simple.

(5) $$\begin{array}{ccccc} \mathcal{K} = H^2(D) & \xrightarrow{A} & L^2(\chi_\Delta d\sigma) & = \mathcal{H} \\ U \uparrow & & \uparrow V & \\ \mathcal{K}_1 = H^2(\Pi) & \xrightarrow{A_1} & L^2(R^+) & = \mathcal{H}_1 \\ M \uparrow & & \uparrow N & \\ \mathcal{K}_2 = L^2_\phi(R) & \xrightarrow{A_2} & L^2(R) & = \mathcal{H}_2 \end{array} .$$

The first row in the diagram has already been described. In the second row, $H^2(\Pi)$ is the Hardy class associated with the half-plane $\Pi = \{z: \text{Im } z > 0\}$. The isomorphisms U,V are induced by the conformal mapping

$$\zeta(z) = i\left[\frac{1+\bar{\beta}z}{1-\bar{\beta}z} - \frac{1+\bar{\beta}\alpha}{1-\bar{\beta}\alpha}\right]$$

of D onto Π that maps $\Delta = \{e^{i\theta}: a < \theta < b\}$ onto $R^+ = (0,\infty)$:

$$(6) \qquad U: K(z) \to k(z) = 2i\pi^{1/2}\,\bar{\beta}(1-\bar{\beta}z)^{-1}\,K(\zeta(z)),$$

$$(7) \qquad V: F(x) \to f(e^{i\theta}) = 2i\pi^{1/2}\,\bar{\beta}(1-\bar{\beta}e^{i\theta})^{-1}\,F(\zeta(e^{i\theta})).$$

We define A_1 so as to make the top square in (5) commute.

The isomorphism M is van Winter's Mellin representation of $H^2(\Pi)$. Set $\phi(x) = 1 + \exp(2\pi x)$, and let $L^2_\phi(R) = L^2(\phi dx)$ be the Hilbert space of functions $\mathfrak{F}(x)$ on R such that $\|\mathfrak{F}\|^2 = \int_R |\mathfrak{F}(x)|^2\phi(x)dx < \infty$. Then

$$(8) \qquad M: \mathfrak{F}(x) \to F(z) = (2\pi)^{-1/2}\int_{-\infty}^{\infty} z^{-\frac{1}{2}-it}\,\mathfrak{F}(t)dt$$

is a Hilbert space isomorphism from $L^2_\phi(R)$ onto $H^2(\Pi)$ with inverse

$$(9) \qquad M^{-1}: F(z) \to \mathfrak{F}(x) = (2\pi)^{-1/2}\int_0^{\infty} t^{-\frac{1}{2}+ix}\,F(t)dt.$$

In (8) we take $0 < \arg z < \pi$. The integral in (9) is computed as the limit as $\varepsilon \downarrow 0$ of $\int_\varepsilon^{1/\varepsilon}$ in the metric of $L^2(R)$. See van Winter [9] or Rosenblum and Rovnyak [6].

The isomorphism N is the ordinary Mellin transform:

$$(10) \qquad N: \mathfrak{F}(x) \to F(x) = (2\pi)^{-1/2}\int_{-\infty}^{\infty} x^{-\frac{1}{2}-it}\,\mathfrak{F}(t)dt,$$

$$(11) \qquad N^{-1}: F(x) \to \mathfrak{F}(x) = (2\pi)^{-1/2}\int_0^{\infty} t^{-\frac{1}{2}+ix}\,F(t)dt.$$

The integrals are taken in the mean square sense. See Titchmarsh [8].

The operator A_2 in (5) is defined to make the bottom square commute. Comparison of (9) and (11) shows that A_2 is the inclusion mapping. Hence A_2^* is division by ϕ.

Fix $\lambda > 0$. Using the isomorphisms of (5), we transform (4) to

$$(12) \qquad k_{j\lambda} = (A_j^*A_j + \lambda I)^{-1}\,A_j^*h_j, \qquad j = 1,2,$$

where

$$(13)\qquad \begin{cases} h_2 = N^{-1}h_1 = N^{-1}V^{-1}h, \\ k_{2\lambda} = M^{-1}k_{1\lambda} = M^{-1}U^{-1}k_\lambda . \end{cases}$$

Since A_2 is the inclusion mapping and A_2^* is division by ϕ, by (12),

$$k_{2\lambda} = (\phi^{-1}+\lambda)^{-1}\phi^{-1}h_2 = (1+\lambda\phi)^{-1}h_2 \;.$$

By (7) and (11),

$$\begin{aligned} h_1(x) &= \pi^{-1/2}/(x-\zeta(0)), \\ h_2(x) &= \pi^{-1}\, 2^{-1/2} \int_0^\infty t^{-\frac{1}{2}+ix} (t-\zeta(0))^{-1}\, dt \\ &= i2^{1/2}\, \zeta(0)^{-\frac{1}{2}+ix} \Big/ (1+e^{-2\pi x}). \end{aligned}$$

The last equality follows by a residue calculation. Thus

$$k_{2\lambda}(x) = i2^{1/2}\, \zeta(0)^{-\frac{1}{2}+ix} [1+\lambda(1+e^{2\pi x})]^{-1}[1+e^{-2\pi x}]^{-1}.$$

By (13), $k_\lambda = UMk_{2\lambda}$. Hence by (6) and (8),

$$\begin{aligned} k_\lambda(z) &= 2i\pi^{1/2}\, \bar\beta(1-\bar\beta z)^{-1}k_{1\lambda}(\zeta(z)) \\ &= 2i\pi^{1/2}\, \bar\beta(1-\bar\beta z)^{-1}(2\pi)^{-1/2} \int_{-\infty}^\infty \zeta(z)^{-\frac{1}{2}-it}\, k_{2\lambda}(t)dt \\ &= i\, \frac{\bar\beta-\bar\alpha}{1-\bar\beta z} \int_{-\infty}^\infty \left(\frac{1-z/\alpha}{1-z/\beta}\right)^{-\frac{1}{2}-it} [1+\lambda(1+e^{2\pi t})]^{-1}[1+e^{-2\pi t}]^{-1}dt, \end{aligned}$$

where $\left|\arg \frac{1-z/\alpha}{1-z/\beta}\right| < \pi$. To evaluate the integral, set $u = e^{2\pi t}$ and use the formula

$$\begin{aligned} &(2\pi i)^{-1}\int_0^\infty w^{-i(\log u)/(2\pi)}[1+\lambda(1+u)]^{-1}(1+u)^{-1}du \\ &\quad = w^{1/2}(1-w)^{-1}[1-w^{-i(\log(1+1/\lambda))/(2\pi)}], \end{aligned}$$

$|\arg w| < \pi$, which holds by a residue calculation. Thus

$$(14) \qquad k_\lambda(z) = z^{-1}\left[1 - \left(\frac{1-z/\alpha}{1-z/\beta}\right)^{-i(\log(1+1/\lambda))/(2\pi)}\right].$$

Proof of Theorem. By (14), the family $\{k_\lambda\}_{\lambda>0}$ defined by (4) coincides with the family $\{\tilde{k}_s\}_{s>0}$ described in the theorem. These are all of the optimal approximants, except for $k_0 = \lim_{\lambda \downarrow 0} k_\lambda$ if the limit exists strongly. However, $||k_\lambda||_2 \to \infty$ as $\lambda \downarrow 0$, so the limit does not exist. The first assertion of the theorem follows. The second assertion may be verified using the explicit formula for $1-z\tilde{k}_s(z)$. We omit the straightforward details.

REFERENCES

[1] Anderson, J.M. - Rosenblum, M. - Rovnyak, J., Hilbert space extremal problems with constraints, in preparation.

[2] Duren, P.L., Theory of H^p Spaces, Academic Press, New York 1970.

[3] Grenander, U. - Szegö, G., Toeplitz Forms and their Applications, University of California Press, Berkeley 1958.

[4] Kreǐn, M.G. - Nudel'man, P.Ya., On some new problems for Hardy class functions with continuous families of functions with double orthogonality, Dokl. Akad. Nauk SSSR 209 (1973), 537-540; Soviet Math. Dokl. 14 (1973), 435-439.

[5] Rosenblum, M., Some Hilbert space extremal problems, Proc. Amer. Math. Soc. 16 (1965), 687-691.

[6] Rosenblum, M. - Rovnyak, J., Restrictions of analytic functions.II, Proc. Amer. Math. Soc. 51 (1975), 335-343.

[7] Splettstösser, W., Bandbegrenzte und effectiv bandbegrenzte Funktionen und ihre Praediktion aus Abtastwerten, Habilitationsschrift, Rhein.-Westf. Techn. Hochschule Aachen 1981.

[8] Titchmarsh, E.C., Theory of Fourier Integrals, 2nd ed., Oxford University Press, London 1948.

[9] van Winter, C., Fredholm equations on a Hilbert space of analytic functions, Trans. Amer. Math. Soc. 162 (1971), 103-139.

International Series of
Numerical Mathematics, Vol. 65

DOMAINS ALLOWING EXACT QUADRATURE IDENTITIES FOR HARMONIC FUNCTIONS - AN APPROACH BASED ON P.D.E.

Harold S. Shapiro
Mathematics Institute
Royal Institute of Technology
Stockholm, Sweden

The integral of a harmonic function u over a ball in $\mathbb{R}^d$ centered at x^o equals the volume of the ball times $u(x^o)$; this is the simplest "quadrature identity" of the type here under discussion. For certain domains other than balls analogous quadrature identities exist whereby the integral is exactly expressible as a finite linear combination of point evaluations. Such domains can be studied by converting the "quadrature" property into an equivalent boundary value problem. A new and convenient method for effecting this conversion is presented. It is based on techniques borrowed from the theory of partial differential equations and Sobolev spaces, which others have already successfully used in such areas as complex analysis and potential theory, polynomial and rational approximation, etc. As applied to quadrature identities, the strength of this method is its flexibility: it is adaptable to multidimensional problems, unbounded domains, quadrature identities of more general types, etc.

1. Introduction and notation

By Ω we always denote an open set in $\mathbb{R}^d$, and $L^p(\Omega)$ denotes the usual Lebesgue space. By $L^p_h(\Omega)$ and $L^p_a(\Omega)$ we denote respectively the subspaces of $L^p(\Omega)$ consisting of harmonic and analytic functions (the latter only in the case $d=2$). Regarding Sobolev spaces, we use the notations of [1].

We are here concerned mainly with Ω for which a ("quadrature") identity of the type

$$(1.1) \qquad \int_\Omega u\,dx = \sum_{j=1}^{n} c_j u(x^j)$$

holds for all $u \in L^1_h(\Omega)$. Here the x^j are points of Ω, and the c_j complex numbers, independent of u. (Closely related are analogous problems in which, on the right side of (1.1), there appear functionals of the type $\int u\, d\mu_i$, where the μ_i are measures (or distributions) of simple structure compactly supported in Ω. The method developed below can be adapted to problems of this type.)

For reasons of space we shall say very little here about the (by now quite considerable) history of such problems. The earliest work on quadrature identities seems to be due to Philip Davis [6]. For orientation see also [4] and Sakai's book [19] which has a good bibliography. (Sakai has also elaborated a profound generalization of the concept of quadrature identity involving subharmonic test functions, which is outside the scope of this paper.)

2. The basic lemma

Our current approach is the use (novel in the present context) of

LEMMA 2.1. *Let Ω be a bounded open set in $\mathbf{R}^d$, $1 < p < \infty$ and μ a distribution in Ω of the form $\mu = F + \nu$ where $F \in L^{p'}(\Omega)$ (here $p' = p/(p-1)$) and* supp ν *is compact. Suppose*

$$\langle u,\mu\rangle = 0, \quad \text{for all } u \in L^p_h(\Omega) \; . \tag{2.1}$$

Then there exists a distribution v *in Ω such that*

$$\Delta v = \mu \tag{2.2}$$

$$v = G + \lambda \tag{2.3}$$

where λ is a distribution of compact support in Ω and $G \in W^{2,p'}_0(\Omega)$.

REMARKS. In the case $\nu = 0$ we can moreover assert that $\lambda = 0$. This form of the lemma is familiar in connection with approximation by harmonic functions in L^p norm. The analogous lemma for $\bar{\partial}$ in place of Δ was given by Havin [13] and (in the presence of additional regularity assumptions, and for $p = 2$) can be traced back at least as far as a 1937 paper of Friedrichs [9]. It is important for our purposes that Ω be a completely arbitrary (bounded) open set.

In the case F = 0 one can also assert that G = 0. This is a special case of well known theorems on elliptic p.d.e. with constant coefficients (see [14, Ch. III], especially Corollary 3.7.1).

Lemma 2.1 follows easily from the two special cases $\nu = 0$, $F = 0$ just mentioned, as follows. Assume namely that each of these special cases has been established. Let $\psi \in C^\infty(\mathbf{R}^d)$ be a radial function with support in the ball $|x| \leqq a$, where a is less than the distance of supp ν from $\partial\Omega$, and such that $\int \psi\, dx = 1$. Then the convolution $\tilde{\nu} = \nu * \psi$ satisfies

$$\langle u,\tilde{\nu}\rangle = \langle u,\nu\rangle, \quad \text{for all u harmonic in } \Omega .$$

Hence (2.1) becomes

$$\langle u, F + \tilde{\nu}\rangle = 0, \quad \text{all } u \in L_h^p(\Omega) .$$

Since $F + \tilde{\nu}$ is in $L^{p'}(\Omega)$ we get: there is some $G \in W_o^{2,p'}(\Omega)$ such that $\Delta G = F + \tilde{\nu}$. But now, since $\nu - \tilde{\nu}$ is a distribution of compact support that annihilates all harmonic functions in Ω, there exists a distribution λ of compact support in Ω with $\Delta\lambda = \nu - \tilde{\nu}$. Hence $\Delta(G + \lambda) = F + \nu$, as was to be shown.

We outline briefly the (very well known) arguments to establish each of the two mentioned special cases of which Lemma 2.1 is an amalgam since a modified form of this reasoning will be needed in Section 4. Consider first the case $\nu = 0$. Thus, we are given $F \in L^{p'}(\Omega)$ satisfying

$$\int_\Omega F\, u\, dx = 0, \quad \text{all } u \in L_h^p(\Omega) . \tag{2.4}$$

Now, $u \in L_h^p(\Omega)$ is equivalent to: $u \in L^p(\Omega)$ and

$$\int u(\Delta\varphi)dx = 0, \quad \text{all } \varphi \in C_o^\infty(\Omega)$$

by "Weyl's lemma"). Thus by standard functional analysis (2.4) is equivalent to: F is in the $L^{p'}$ closure of $\{\Delta\varphi : \varphi \in C_o^\infty(\Omega)\}$, i.e. $\exists\varphi_n \in C_o^\infty(\Omega)$ such that

$$\Delta\varphi_n \to F \quad \text{in } L^{p'}(\Omega) .$$

Since $\{\Delta\varphi_n\}$ are bounded in $L^{p'}(\mathbf{R}^d)$ it follows from Marcinkiewcz's multiplier theorem that for each pair i,j the sequence $\{(\partial^2/\partial x_i \partial x_j)\varphi_n\}$ is bounded in $L^{p'}(\mathbf{R}^d)$. Also, by Poincaré's inequality [1, p. 159] $\{\varphi_n\}$ is bounded in $L^{p'}(\mathbf{R}^d)$, and thus $\{\varphi_n\}$ is a bounded sequence in $W^{2,p'}(\mathbf{R}^d)$, hence w.l.o.g. we may (by passing to a subsequence) assume there exists $G \in W^{2,p'}(\mathbf{R}^d)$ such that $\varphi_n \to G$ in the weak topology of $W^{2,p'}(\mathbf{R}^d)$, and *a fortiori* in the sense of distributions. Then $\Delta\varphi_n \to \Delta G$ in the sense of distributions, whence $\Delta G = \tilde{F}$, where $\tilde{F}$ denotes the extension of F to $\mathbf{R}^d$ by defining it as 0 outside Ω. Since $G \in W_o^{2,p'}(\Omega)$ this concludes the proof.

For the other special case of Lemma 2.1 we required, i.e. that where $F = 0$, one observes that if the compactly supported distribution ν annihilates harmonic functions, its Fourier transform (as a function on $\mathbb{C}^d$) vanishes on the variety where $\xi_1^2 + \ldots + \xi_d^2 = 0$, hence is an entire multiple of $\xi_1^2 + \ldots + \xi_d^2$. Division by this polynomial gives in turn the Fourier transform of a compactly supported distribution λ (in view of known estimates and the Paley-Wiener theorem) with $\Delta\lambda = \nu$, and since λ is a harmonic function on the unbounded component E of $\mathbf{R}^d \setminus \operatorname{supp} \nu$ vanishing for large $|x|$ it vanishes on E. ⌘

For the study of quadrature identities for analytic functions on plane domains, we require the analogous

LEMMA 2.2. *Let* Ω *be a bounded open set in* $\mathbf{R}^2$, $1 < p < \infty$ *and* μ *a distribution in* Ω *of the form* $\mu = F + \nu$ *where* $F \in L^{p'}(\Omega)$ (*here* $p' = p/(p-1)$) *and* supp ν *is compact. Suppose*

$$\langle f,\mu \rangle = 0, \quad \text{for all } f \in L_a^p(\Omega) .$$

Then there exists a distribution v *in* Ω *such that*

$$\frac{\partial v}{\partial \bar{z}} = \mu \tag{2.5}$$

$$v = G + \lambda \tag{2.6}$$

where λ *is a distribution of compact support in* Ω *and* $G \in W_o^{1,p'}(\Omega)$.

The proof is analogous to that of Lemma 2.1 (indeed one could formulate a generalization for elliptic systems which includes both lemmas).

In applications we require

COROLLARY 2.1. *Under the hypotheses of Lemma 2.1, if* $p' > d$, *then* v *and its first-order partial derivatives extend continuously to* $\partial\Omega$ *and vanish on* $\partial\Omega$.

This follows at once by applying the Sobolev embedding theorem to G in Lemma 2.1, and analogously:

COROLLARY 2.2. *Under the hypotheses of Lemma 2.2, if* $p' > 2$, *then* v *extends continuously to* $\partial\Omega$ *and vanishes there.*

3. Applications to quadrature identities

We illustrate here how the results in Section 2 apply to q.i. Note that we have restricted our domains to be bounded, indeed the above results are false if this hypothesis is omitted. (The boundedness assumption enters at only one point, but crucially, because we have used the Poincaré inequality; this inequality is true in certain unbounded domains too, see [2, pp. 73-77] and [1, pp. 158-160], and for such domains the results of Section 2 remain valid; of greater interest, however, is how to modify those results so as to dispense with all restrictions on Ω, and this will be indicated in Section 4.)

THEOREM 3.1. *Let* $\Omega \subset \mathbf{R}^d$, *where* $d \geqq 3$, *be a bounded open set and suppose the q.i. (1.1) holds for all* $u \in L^1_h(\Omega)$. *Then there exists a function* h *harmonic in* Ω *with* $h \in C^1(\bar{\Omega})$ *such that, writing*

$$w(x) = h(x) - \sum_{j=1}^{n} c_j |x - x^j|^{2-d} \tag{3.1}$$

we have, for a positive constant $a = a_d$,

$$\operatorname{grad} w(x) = ax \ , \quad x \in \partial\Omega \ . \tag{3.2}$$

REMARK. The analogous proposition with $d = 2$ is valid, with $|x - x^j|^{2-d}$ replaced by $\log |x - x^j|$. The vector field grad w on $\bar{\Omega}$ can, when $d > 2$, be viewed as the higher-dimensional analog of the "Schwarz function" [6, 4].

PROOF. We only treat the case $d \geqq 3$. Fix p such that $1 < p < d/(d-1)$ (so that $p' > d$). Then (1.1) holds for all $u \in L_h^p(\Omega)$ and so the distribution in Ω

$$\mu := 1 - \sum_{j=1}^{n} c_j \delta(x^j)$$

(which in this case is a bounded measure on Ω) satisfies the hypothesis of Lemma 2.2. Here $F = 1$ on Ω and ν is a finite linear combination of Dirac functionals. We conclude that there is a function v on Ω, which extends continuously to $\partial\Omega$ together with its partial derivatives of first order, such that

$$\Delta v = 1 - \sum_{j=1}^{n} c_j \delta(x^j)$$

holds on Ω. Integrating this,

$$v(x) = \frac{|x|^2}{2d} + b_d \sum_{j=1}^{n} c_j |x - x^j|^{2-d} + g(x) \tag{3.3}$$

where b_d is a positive constant depending only on d and g is in $C^1(\bar{\Omega})$ and harmonic in Ω. Since grad $v(x) = 0$ for $x \in \partial\Omega$, it is clear from (3.3) that, writing $h = (b_d)^{-1} g$, the w defined by (3.1) satisfies (3.2).

REMARK. It can be shown, conversely, that with some regularity assumptions on $\partial\Omega$, the existence of w satisfying (3.1) and (3.2) implies a quadrature identity (1.1). This is the "boundary value problem" version of the quadrature property referred to in the Introduction, i.e. the quest of a function harmonic in Ω except for finitely many prescribed singularities, whose gradient on $\partial\Omega$ coincides with the identity map. Given the x^j and the c_j, this problem is, for a specified Ω with $\{x^1,\dots,x^n\} \subset \Omega$, in general unsolvable (overdetermined), and the q.i. problem is equivalent to the search for solvable cases of the b.v.p., i.e. a "free boundary problem"; problems of this character, which arise frequently in hydrodynamics and other branches of physics, tend to be difficult and are the object of much contemporary research.

COROLLARY 3.1. (*Epstein and Schiffer [8]*). *If Ω is a bounded open set and (1.1) holds with $n = 1$ then Ω is a ball centered at x^1.*

PROOF. We may assume $x^1 = 0$. Then, from Theorem 3.1 we have

$$\operatorname{grad} h(x) = ax + c \operatorname{grad} |x|^{2-d}$$
$$:= b(x)x$$

for $x \in \partial\Omega$, where b is a smooth real-valued function, and h is harmonic in Ω. From this it can be deduced by an elementary argument (see [3]) that h is constant, whence from (3.3) there are constants $c > 0$ and C such that

$$v(x) = (2d)^{-1} |x|^2 + c|x|^{2-d} + C$$

vanishes with its gradient on $\partial\Omega$. Since the function

$$\varphi(r) := (2d)^{-1} r^2 + cr^{2-d} + C$$

has (we may assume $d > 2$) a unique critical point on $(0,\infty)$, $|x|$ is constant on $\partial\Omega$, which easily implies Ω is a ball. ⌘

We could deduce, for the case when Ω is a bounded plane domain admitting a q.i. relative to the class $L^1_a(\Omega)$ a result analogous to Theorem 3.1. The result thus obtained is (the specialization to bounded domains of) one of the main results of [4], namely Lemma 2.3 of that paper. The essence is, the q.i. is equivalent to the existence of a ("Schwarz") function S, analytic in Ω except for prescribed polar singularities at the "nodal points" of the quadrature formula, continuously extendible to $\partial\Omega$ and satisfying $S(z) = \bar{z}$ for $z \in \partial\Omega$.

REMARK. The point of view towards q.i. developed above is of course closely related to the "potential theoretic" method used in [8], the Cauchy transforms in [4] and generalization thereof in [18], in that all these transforms are precisely <u>explicit</u> methods to solve $\Delta v = \mu$ or $\frac{\partial v}{\partial \bar{z}} = \mu$. However, the present method can be used in situations where definition of those transforms is problematic or impossible (especially e.g. when Ω is "badly" unbounded) and the powerful machinery of elliptic regularity theory gives us the essential features of v that must otherwise be deduced from laborious potential-theoretic estimates.

4. Unbounded domains

If Ω is subjected to no restrictions of size, the conclusions in the preceding sections need not be true, and genuinely new q.i. phenomena appear. Thus, as M. Sakai discovered several years ago (and kindly pointed out in a letter to the author) there exist unbounded plane domains Ω satisfying a one-point q.i. relative to the class $L_a^1(\Omega)$. Moreover, such domains must be "very badly unbounded" for, as Sakai also proved,

$$\iint_\Omega \frac{dx\,dy}{1+x^2+y^2} = \infty$$

for any such Ω (see [19, § 11] for full details of these matters).

In this Section we show how to modify our earlier results when no "size" assumption on Ω is made. The results obtained enable one to find, for instance, all plane domains Ω satisfying one-point q.i. relative to $L_a^1(\Omega)$; for reasons of space this analysis will not be carried out in full here, but we illustrate our methods by giving a new proof of a simpler result, a recent theorem of M. Sakai on "null quadrature domains". Just as in the preceding sections the analysis could be carried out for harmonic functions in $\mathbf{R}^d$ as well as analytic functions in $\mathbf{R}^2 \equiv \mathbb{C}$, but for brevity we restrict attention in this section to the latter setting. We use usual "complex variable notation" $z = x + iy$ for points of $\mathbf{R}^2$.

LEMMA 4.1. *Let Ω be any open set in $\mathbf{R}^2$ and assume the remaining hypotheses of Lemma 2.2. Let $\tilde{F}$ denote the function equal to* F *on Ω and zero on $\mathbf{R}^2 \setminus \Omega$, and $\tilde{\mu} = \tilde{F} + \nu$. Then there is a distribution* v *on $\mathbf{R}^2$ and a function $G \in W^{1,p'}_{loc}(\mathbf{R}^2)$ such that*

$$\frac{\partial v}{\partial \bar{z}} = \tilde{\mu} \quad \text{on } \mathbf{R}^2 , \tag{4.1}$$

and $\lambda := v - G$ has compact support in Ω. Moreover, if $p' > 2$, G *vanishes on all of $\partial\Omega$ and for every $a > 0$,* G *is continuous and bounded on the set $\{z \in \mathbf{R}^2 : \text{dist}\,(z,\partial\Omega) < a\}$, and there is a constant* A *not depending on* z, *such that*

$$|G(z)| \leqq A(|z|^b + 1), \quad \text{all } z \in \mathbf{R}^2 \ (b = 1 - (2/p')) . \tag{4.2}$$

PROOF. We require first two further lemmas.

LEMMA 4.2. *Let* $B = B(\xi;R)$ *denote the open ball in* $\mathbb{R}^d$ *centered at* ξ, *with radius* R, E *a subset of* B *with positive d-dimensional Lebesgue measure, and* $s > 1$. *Then there is a constant* $c = c(R,d,s,E)$ *such that*

$$\int_B |\varphi|^s \, dx \leqq c \int_B \left(\sum_{j=1}^{d} \left|\frac{\partial\varphi}{x_j}\right|^s \right) dx$$

for every $\varphi \in C^\infty(\bar{B})$ *vanishing on* E.

A proof of this lemma, which is due to Morrey, may be found in [15,pp.82-83].

LEMMA 4.3. *Let* B *be any closed ball in* $\mathbb{R}^d$, *and* $s > d$. *Then there is a constant* $K = K(d,s)$ *such that*

$$|\varphi(x) - \varphi(x')| \leqq K\left[\int_B \left(\sum_{j=1}^{d} \left|\frac{\partial\varphi}{\partial x_j}\right|^s \right) dx\right]^{1/s} \cdot |x - x'|^b$$

where $b = 1 - (d/s)$ *and* $\varphi \in C^\infty(B)$.

This lemma is also due to Morrey. See for example [1, p. 109] or [12, p. 156] for a proof. It is crucial here that K does not depend on the radius of B.

Now let us return to the proof of Lemma 4.1. We need only discuss the case $\nu = 0$ since general ν are dealt with as before. Then, reasoning as in Section 2, we get a sequence $\{\varphi_n\} \subset C_o^\infty(\Omega)$ satisfying, for some C independent of n,

$$\iint_{\mathbb{R}^2} \left(\left|\frac{\partial\varphi_n}{\partial x}\right| + \left|\frac{\partial\varphi_n}{\partial y}\right| \right)^{p'} dx\, dy \leqq C \tag{4.3}$$

and $\partial\varphi_n/\partial\bar{z} \to F$ in $L^{p'}(\mathbb{R}^2)$. When $p' \leqq 2$ we also assume $\mathbb{R}^2 \setminus \bar{\Omega}$ is not empty. Let $z_o \in \mathbb{R}^2 \setminus \bar{\Omega}$ and $B_m := B(z_o;m)$, $m = 1,2,\ldots$. By Lemma 4.2, for each m the restrictions of the φ_n to B_m are a bounded sequence in $L^{p'}(B_m)$ and hence $W^{1,p'}(B_m)$, thus there is a subsequence converging in the weak topology of $W^{1,p'}(B_m)$. It now follows easily that there exists $G \in W^{1,p'}_{loc}(\mathbb{R}^2)$ and a subsequence of $\{\varphi_n\}$ which converges to G in the distributional sense. Hence $\partial G/\partial\bar{z} = \tilde{F}$.

To complete the proof we now assume $p' > 2$. From Lemma 4.3 and (4.3) we infer that for all z,z' in $\mathbf{R}^2$

$$|\varphi_n(z) - \varphi_n(z')| \leqq C|z - z'|^{1-(2/p')} \tag{4.4}$$

with C independent of n, z and z'. Since $\{\varphi_n\}$ is bounded in $W^{1,p'}(B)$ for each ball $B \subset \mathbf{R}^2$, by the Sobolev embedding theorem the φ_n are continuous and uniformly bounded on every ball, and together with (4.4) we see there is a subsequence converging uniformly on compact subsets of $\mathbf{R}^2$. The limit function must be G, and now letting z' be a fixed point of $\partial\Omega$, we get from (4.4) letting $n \to \infty$

$$|G(z)| \leqq C|z - z'|^{1-(2/p')}$$

which implies the remaining assertions of Lemma 4.1. ⌘

As an application we deduce the following, which (together with its generalization to $\mathbf{R}^d$ to be presented elsewhere) is our main new result.

THEOREM 4.4. *Let Ω be any open set in $\mathbf{R}^2$ such that $\partial\Omega$ contains a continuum, and suppose*

$$\iint_\Omega f \, dx \, dy = \sum_{j=1}^{n} c_j f(z_j) \tag{4.5}$$

for all $f \in L^1_a(\Omega)$, where the z_j are distinct points of Ω and c_j non-zero complex numbers (z_j and c_j independent of f). Then there is a function S such that

(i) *S is analytic in Ω except for simple poles at the z_j*

(ii) *S extends continuously to $\partial\Omega$ and $S(\zeta) = \bar{\zeta}$ for all $\zeta \in \partial\Omega$*

(iii) *$|S(z) \leqq C(1 + |z|)$ for all $z \in \Omega$ at distance $\geqq 1$ from $\{z_1,\ldots,z_n\}$.*

PROOF. Let ζ_1,ζ_2 be fixed distinct points of $\partial\Omega$ joined by a continuum $\Gamma = \Gamma(\zeta_1,\zeta_2)$ and $k(z) = ((z-\zeta_1)(z-\zeta_2))^{1/2}$. For a fixed choice of square root k is analytic and single valued in Ω, nonvanishing and extends continuously to $\partial\Omega \setminus \Gamma$. Let now g be any function in $L_a^{3/2}(\Omega)$. Since $k^{-1} \in L_a^3(\Omega)$, $f := gk^{-1}$ is in $L_a^1(\Omega)$, hence from (4.5),

$$\iint_\Omega g\, k^{-1}\, dx\, dy = \sum_{j=1}^n c_j'\, g(z_j)$$

holds for all $g \in L_a^{3/2}(\Omega)$, where $c_j' = k(z_j)^{-1} c_j$. By Lemma 4.1 with $p = 3/2$, $p' = 3$, $F = k^{-1}$, we get

$$\frac{\partial v}{\partial \bar{z}} = \tilde{F} - \sum_{j=1}^n c_j'\, \delta_{(z_j)} \tag{4.6}$$

where v coincides outside a compact subset of Ω with a continuous function G on $\mathbb{R}^2$ having all the properties enunciated in Lemma 4.1. Integrating (4.6) gives for $z \in \Omega$,

$$v(z) = \frac{\bar{z}}{k(z)} - \sum_{j=1}^n \frac{a_j}{z - z_j} - h(z) \tag{4.7}$$

where $a_j = c_j'/\pi$ and h is some function analytic in Ω. For $\varepsilon > 0$ let

$$N = N(\varepsilon,\zeta_1,\zeta_2) := \{z \in \Omega : \text{dist}\,(z,\Gamma) < \varepsilon\} .$$

Then from (4.7) we see that h extends continuously to the boundary of $\Omega' := \Omega \setminus \bar{N}$. From Lemma 4.1, v is bounded in a neighborhood of $\partial\Omega$, and since the first and second terms on the right side of (4.7) are bounded near $\partial\Omega'$, h is bounded on a neighborhood of $\partial\Omega'$. Also, from (4.7) and Lemma 4.1, $h(z) = O(|z|^{1/3})$ for large $z \in \Omega'$. Hence by a theorem of Phragmén-Lindelöf type due to Fuchs [10] h is bounded in Ω'. Since v vanishes on $\partial\Omega$ we get from (4.7)

$$\bar{z} = k(z) \left(h(z) + \sum_{j=1}^n \frac{a_j}{z - z_j}\right) \tag{4.8}$$

for all $z \in (\partial\Omega) \setminus N$. Denoting the function on the right by S, we see that S satisfies assertions (i) and (iii) of the theorem. Also it is clear that as we vary the pair ζ_1,ζ_2 the right side of (4.8), nominally dependent on ζ_1 and ζ_2 must remain the same since all functions so obtained have identical boundary values on a continuum. By thus varying ζ_1 and ζ_2 and letting ε be arbitrarily small we see that S also satisfies (ii), and this concludes the proof. ⋇

REMARK 1. An alternative approach to Theorem 4.4 would be to map Ω conformally onto a bounded domain (which is always possible by an elementary function) and apply the results of Section 2. However, since in higher dimensions essentially only inversions are available for this purpose, thus necessitating an assumption that $\mathbb{R}^d \setminus \bar{\Omega}$ have non-empty interior, we prefer the direct approach made possible by Lemma 4.1; moreover this lemma is of independent interest and has other applications.

REMARK 2. If in (4.5) we allow also terms involving $f'(z_j)$ or higher derivatives (we omitted such terms only to keep notations simple) we can by the same reasoning deduce the assertions of Theorem 4.4 except that S will correspondly have poles of higher order at the z_j.

For our next applications we require a lemma which is implicit in Sakai [18].

LEMMA 4.5. *Let Ω be an open connected subset of $\mathbb{C}$ containing a neighborhood of ∞ and suppose there exists a function* S *analytic in Ω such that*

(i) $S(z) = O(|z|) \, , \quad z \to \infty$

(ii) S *extends continuously to* $\Gamma := (\partial\Omega) \setminus \{\infty\}$ *and* $S(z) = \bar{z}$ *for* $z \in \Gamma$.

Then Γ is a subset of an ellipse (which may degenerate to a circle or line segment).

REMARK. The converse is easy to verify: writing the equation of an ellipse as $Q(z,\bar{z}) = 0$ where Q is a quadratic polynomial and solving for $\bar{z} := S(z)$ one checks that S satisfies (i) and (ii). For completeness we sketch a new, very simple proof of Lemma 4.5.

PROOF. S is regular at ∞, or has a simple pole there. Hence for large $|z|$

$$S(z) = C\,z + c_o + \frac{c_1}{z} + \dots \, . \tag{4.9}$$

Suppose first $|C| = 1$; by rotation we may achieve $C = 1$. Then $\mathrm{Re}(S(z) - z)$

is harmonic and bounded on Ω and vanishes on Γ, hence is identically zero, so $S(z) = z - i\,a$ where $a \in \mathbb{R}$, and Γ is a subset of the line $y = a/2$. If $|C| \neq 1$ it is easy to check that by a translation we can achieve $c_o = 0$ in (4.9) and we henceforth assume this is the case. If $C = 0$, then $zS(z)$ is analytic and bounded on Ω, real and non-negative on Γ and hence $zS(z) \equiv \rho^2$ with $\rho \geqq 0$ whence Γ is a subset of the circle (or point) $|z| = \rho$. Finally, if $|C| \neq 0$ or 1, we can by a rotation achieve $C > 0$. Writing

$$f(z) := \frac{S(z)+z}{2} = \left(\frac{C+1}{2}\right)z + \frac{1}{2}\frac{c_1}{z} + \dots$$

$$g(z) := \frac{S(z)-z}{2i} = \left(\frac{C-1}{2i}\right)z + \frac{1}{2i}\frac{c_1}{z} + \dots ,$$

then $h := ((C-1)/2)^2 f^2 + ((C+1)/2)^2 g^2$ is analytic and bounded in Ω, and on Γ equal to the real quantity $((C-1)/2)^2 x^2 + ((C+1)/2)y^2$, whence h is constant. Thus, this latter polynomial is constant on Γ, which consequently is a subset of an ellipse, concluding the proof.

COROLLARY 4.6. (Sakai [18]). _Let $\Omega \subset \mathbb{C}$ be an open connected set containing a neighborhood of ∞ such that $\partial\Omega$ contains a continuum and_

$$\iint_\Omega f\,dx\,dy = 0, \quad \underline{\text{all}}\ f \in L^1_a(\Omega) . \tag{4.10}$$

Then either Ω is the exterior of an ellipse (or circle) or $\partial\Omega$ is a subset of a line, and these domains do indeed satisfy (4.10).

PROOF (sketch). That (4.10) holds if Ω is a half-plane, or the exterior of an ellipse, can be checked by calculation (see [18]), which implies the last assertion of the Corollary. Now, Theorem 4.4 obviously applies _a fortiori_ to the present situation, so we deduce the existence in Ω of an everywhere analytic function S satisfying the hypotheses of Lemma 4.5. If $\partial\Omega$ is a subset of a line we are done, so we may assume $\Gamma := (\partial\Omega) \setminus \{\infty\}$ is a subset of a non-degenerate ellipse (or circle) and denote by E the bounded domain enclosed by this ellipse. We have to show $\Omega = \mathbb{C} \setminus \bar{E}$, i.e. that $\Gamma = \partial E$. Suppose this is not the case. Then $E \subset \Omega$, and for all $f \in L^1_a(\Omega)$

$$\iint_E f\,dx\,dy = \iint_\Omega f\,dx\,dy - \iint_{\mathbb{C}\setminus E} f\,dx\,dy ,$$

and both integrals on the right side are zero. Now by standard approximation techniques, as in [18], we can construct a sequence $\{f_n\}$ of rational functions with no poles outside Γ, which are $O(|z|^{-3})$ at infinity and satisfy

$$\lim_{n\to\infty} \iint_E |f_n - 1| \, dx \, dy = 0 .$$

Hence

$$\iint_E dx \, dy = \lim_{n\to\infty} \iint_E f_n \, dx \, dy \neq 0$$

and this contradiction completes the proof.

REMARK. There are other domains Ω, not containing a neighborhood of ∞, for which (4.10) holds; Sakai [18] finds all such. To get the complete result by methods of the present paper requires a corresponding extension of Lemma 4.5 which is fairly straightforward but cannot for reasons of space be dealt with here.

An interesting relation to potential theory is shown by

COROLLARY 4.7. (Dive [7]. Nikliborc [17]). *Let $D \subset \mathbb{R}^2$ be a bounded open set with $\mathbb{R}^2 \setminus \bar{D}$ connected and suppose the logarithmic potential u of a uniform mass distribution on D satisfies*

$$u(z) = Q(x,y)(= Q_1(z,\bar{z})), \quad \text{for } z = x + iy \in D \tag{4.11}$$

where Q (and Q_1) are quadratic polynomials. Then D is the interior of an ellipse (or circle).

REMARK. Dive and Nikliborc actually prove the d-dimensional generalization of this, which is the converse of a well-known result going back to Newton. We have been unable thus far to obtain the higher dimensional result by our methods.

PROOF. It is well known that $u \in C^1(\mathbb{R}^2)$. Now $f := \frac{\partial u}{\partial z}$ is analytic on $\mathbb{C} \setminus \bar{\mathbb{D}}$ and bounded and, from (4.11), for certain complex constants a,b,c

$$f(z) = az + b\bar{z} + c, \quad z \in D .$$

By continuity this holds for $z \in \bar{D}$. Hence $S(z) := b^{-1}(f(z) - az - c)$ satisfies in $\Omega := \mathbb{R}^2 \setminus \bar{D}$ the hypotheses of Lemma 4.5, hence $(\partial\Omega) \setminus \{\infty\} = \partial D$ is a subset of an ellipse (or circle), which implies D is the interior of that ellipse (or circle).

5. An operator introduced by K. Friedrichs

Let P denote the orthogonal projection of $L^2(\Omega)$ on $L^2_a(\Omega)$, where Ω is an open subset of $\mathbb{C}$, and let $(Kf)(z) := \overline{f(z)}$. Also, we shall always tacitly suppose that $\mathbb{C} \setminus \bar{\Omega}$ has positive logarithmic capacity (e.g. it suffices if this set contains a continuum) so as to assure that the space $L^2_a(\Omega)$ is infinite-dimensional. Then the operator $T = T_\Omega := PK$ from $L^2_a(\Omega) \to L^2_a(\Omega)$ was (with different notations) introduced by K. Friedrichs [9]. It is not linear, since for $a \in \mathbb{C}$ we have $T(a\, f) = \bar{a}\, T\, f$, however in $L^2_{a,r}(\Omega)$, the Hilbert space $L^2_a(\Omega)$ considered as a vector space over $\mathbb{R}$, T is linear; moreover, as easily verified, it is bounded, self-adjoint and positive. It satisfies the identity

$$(5.1) \qquad \langle f, T\, g\rangle = \iint_\Omega f\, g\, dx\, dy, \quad \text{all } f,g \in L^2_a(\Omega)$$

where $\langle\ ,\ \rangle$ is the usual inner product in $L^2(\Omega)$. T_Ω is intimately related to the biharmonic boundary value problem in Ω (and hence to planar elasticity theory). This was the origin of Friedrichs' interest in T_Ω; it turns out namely that the Dirichlet problem for the biharmonic equation permits a very elegant reformulation in terms of T_Ω; this was developed only sketchily by Friedrichs in [9], and he seems never to have returned to the matter. We hope to present elsewhere a more complete account of his theory, which in light of later developments concerning elliptic systems can now be considerably improved. Our present concern with the subject is a remarkable connection of T_Ω with quadrature identities.

THEOREM 5.1. *Let Ω be a planar open set whose boundary consists of a finite number of continua. Then, the following assertions are equivalent:*

(i) Ω *is a quadrature domain, in the sense that*

(5.2) $$\iint_\Omega h\,dx\,dy = \sum_{j=1}^{n}\left(\sum_{i=0}^{r_j} c_{ij}h^{(i)}(z_j)\right)$$

for all $h \in L_a^1(\Omega)$. Here the non-negative integers r_j, complex numbers c_{ij} and points $z_j \in \Omega$ do not depend on h.

(ii) The "Friedrichs operator" T_Ω is of finite rank.

The proof given below is based on an idea kindly communicated to us by Björn Gustafsson. We require first the following lemma due to L.I. Hedberg.

LEMMA 5.2. Let Ω be a planar open set whose boundary consists of finitely many continua. Then $L_a^p(\Omega)$, for $p \geqq 1$, is the closure of the bounded functions it contains.

For an indication of the proof (which Hedberg did not publish) see Prop. 14 on p. 112 of [14].

COROLLARY 5.3. Let Ω be as in Lemma 5.2, and also bounded. Let $p \not\equiv 0$ be a polynomial with no zeroes in Ω. Then

$$p \cdot L_a^2(\Omega) := \{p\,f : f \in L_a^2(\Omega)\}$$

is dense in $L_a^2(\Omega)$.

PROOF. Choose $c = 1/m$ (m a positive integer) so small that p^{-c} is in $L_a^2(\Omega)$. Then, for any $\varphi \in L_a^\infty(\Omega)$, the set $p \cdot L_a^2(\Omega)$ contains $p \cdot p^{-c}\varphi = p^{1-c}\varphi$, hence its closure contains $p^{1-c} \cdot L_a^2(\Omega)$ because of Lemma 5.2. Repeating this argument m times yields the desired conclusion.

PROOF of Theorem. That (i) $\Rightarrow$ (ii) is elementary and requires no topological assumptions on Ω: if S is the subspace of $L_a^2(\Omega)$ consisting of f for which

$$f^{(i)}(z_j) = 0\ , \quad i = 0,1,\ldots r_j\ ; \quad j = 1,2,\ldots n$$

then S has finite codimension and

$$\iint_\Omega f\, g\, dx\, dy = 0 , \quad \text{all } g \in L_a^2(\Omega)$$

as we see from (5.2) with $h = f\,g$; hence (5.1) asserts S is orthogonal to the range of $T = T_\Omega$ so this range is finite dimensional.

We prove (ii) ⇒ (i) under the extra assumption that Ω is bounded, for convenience. The simple modifications needed for the general case are left to the reader. Thus, suppose the range of T has finite (complex) dimension k. Then some nontrivial linear combination of $1,z,\ldots,z^k$ is orthogonal to this range, i.e. there is a polynomial $p \not\equiv 0$ such that (observing 5.1)

$$\iint_\Omega p\, g\, dx\, dy = 0 , \quad \text{all } g \in L_a^2(\Omega) . \tag{5.3}$$

Let $p = p_1 p_2$ where p_1, p_2 are polynomials and p_1 consists of those linear factors of p which vanish in Ω. Using Corollary 5.3 we derive from (5.3)

$$\iint_\Omega p_1\, g\, dx\, dy = 0 , \quad \text{all } g \in L_a^2(\Omega) . \tag{5.4}$$

Let $z_1,\ldots,z_n$ denote the zeroes of p_1. Just to simplify notations suppose they are distinct (the general case involving a trivial modification of the argument). Then from (5.4), $\iint_\Omega f\, dx\, dy$ vanishes for every $f \in L_a^2(\Omega)$ vanishing at $z_1,\ldots z_n$ and so if $P_1,\ldots,P_n$ are the Lagrange fundamental polynomials for $\{z_1,\ldots,z_n\}$ we have for every $h \in L_a^2(\Omega)$,

$$\iint_\Omega \left(h - \sum_{j=1}^n h(z_j)P_j\right) dx\, dy = 0$$

which is an identity of the form (5.2). This completes the proof, except that we have established (5.2) only for h in $L_a^2(\Omega)$ rather than $L_a^1(\Omega)$. However, the former class is dense in the latter, in view of Lemma 5.2, and we are done. ⋇

REMARK 1. Jaak Peetre pointed out to us that the assertion "T_Ω is of finite rank" has an interesting elementary reformulation: there exist linear functionals α_i, β_i (i = 1,2,... n) on $L_a^2(\Omega)$ such that

$$\iint_\Omega f\, g\, dx\, dy = \sum_{i=1}^n \alpha_i(f)\beta_i(g) , \quad \text{all } f,g \in L_a^2(\Omega) .$$

Incidentally, we don't know whether this (or (ii)) implies (i) for all Ω such that $L_a^2(\Omega)$ is non-trivial.

REMARK 2. An operator which serves much the same purposes as T_Ω is gotten by projecting $f(\bar{z})$ orthogonally into $L_a^2(\Omega')$, where Ω' is the reflection in the real axis of Ω. This operator is complex-linear, but maps $L_a^2(\Omega) \to L_a^2(\Omega')$. Yet another way to avoid "the curse of skew-linearity" is to work with the fully linear operator T_Ω^2. One can also generalize T_Ω by projecting $\varphi\bar{f}$ rather than $\bar{f}$, for some $\varphi \in L_a^\infty(\Omega)$; or working with $L^2(\partial\Omega;ds)$ and the corresponding Hardy space in place of $L_a^2(\Omega)$.

REMARK 3. Theorem 5.1, together with the fact already alluded to that the fundamental boundary value problems of linear two-dimensional elasticity theory can be reformulated as $(T_\Omega - \lambda I)f = g$ in $L_a^2(\Omega)$, can be interpreted as "explaining" why, in certain domains, these b.v. problems are solvable "in closed form" (and hence are favorites of authors on elasticity theory). Among such domains are (i) conformal images of the unit disc by rational functions, (ii) a half-plane, and (iii) the exterior of an ellipse (see [16] for numerous examples). All these are quadrature domains, where T_Ω degenerates (indeed it is $\equiv 0$ in cases (ii) and (iii)).

6. Concluding remarks.

6.1. The technique embodied in Lemmas 2.1 and 2.2 has other applications. Three of these which have not been indicated above and shall be presented elsewhere concern

(i) doubly orthogonal functions in the sense of S. Bergman ([5, 20]).

(ii) closely related to this, Friedrichs' eigenvalue problem (see [9, 11]) $T_\Omega f = \lambda f$, where Ω is a smoothly bounded domain and T_Ω the operator of Section 5.

(iii) best uniform approximation of continuous functions on open sets by analytic and harmonic functions.

6.2. A promising topic for future research seems to be the "free boundary problem" of Theorem 3.1. The main obstacle to progress seems to be our very limited knowledge about the vector field arising as the gradient of a harmonic function in $d \geq 3$ dimensions; especially in view of the physical significance of such vector fields, more intensive study of them seems indicated.

BIBLIOGRAPHY

[1] Adams, R.A., Sobolev Spaces, Academic Press, 1975.

[2] Agmon, S., Lectures on Elliptic Boundary Value Problems, Van Nostrand Math. Studies # 2, Princeton, 1965.

[3] Aharonov, D. - Schiffer, M. - Zalcman, L., Potato Kugel, Israel J. of Math. 40 (1981) 331-339.

[4] Aharonov, D. - Shapiro, H.S., Domains on which analytic functions satisfy quadrature identities, J. d'Analyse Math. 30 (1976) 39-73.

[5] Bergman, S. The Kernel Function and Conformal Mapping, second (revised) ed., Amer. Math. Soc. Math. Surveys No. 5, 1970.

[6] Davis, P., Simple quadratures in the complex plane, Pacific J. Math. 15 (1965) 813-824.

——, The Schwarz Function and Its Applications, Carus Math. Monographs No. 17, Math. Assoc. Amer. 1974.

[7] Dive, P., Attraction des ellipsoides homogènes et réciproque d'un théorème de Newton, Bull. Soc. Math. de France 59 (1931) 128-140.

[8] Epstein, B., - Schiffer, M., On the mean-value property of harmonic functions, J. d'Analyse Math. 14 (1965) 109-111.

[9] Friedrichs, K., On certain inequalities and characteristic value problems for analytic functions and for functions of two variables, Trans. Amer. Math. Soc. 41 (1937) 321-364.

[10] Fuchs, W.H.J., A Phragmén - Lindelöf theorem conjectured by D.J.Newman, Trans. Amer. Math. Soc. 267 (1981) 285-293.

[11] Garabedian, P.R., A partial differential equation arising in conformal mapping, Pacific J. Math. 1 (1951) 485-524.

[12] Gilbarg, D. - Trudinger, N., Elliptic Partial Differential Equations of Second Order, Grundl. der Math. Wiss. 224, Springer-Verlag 1977.

[13] Havin, V.P., Approximation in the mean by analytic functions, Dokl. Akad. Nauk SSSR 178 (1968) 1025-1028 (Russian; Engl. transl. in Soviet Math. Dokl. 9 (1968) 245-248.)

[14] Hörmander, L., Linear Partial Differential Operators, Grundl. der Math. Wiss. 116, Springer-Verlag 1969.

[15] Morrey, C., Multiple Integrals in the Calculus of Variations, Springer-Verlag 1966.

[16] Mushkhelishvili, N.I., Some Basic Problems of the Mathematical Theory of Elasticity, 5 ed., Moscow, 1966 (Russian; Engl. transl. Noordhof, Groningen 1953.)

[17] Nikliborc, W., Eine Bemerkung über die Volumpotentiale, Math. Zeitschr. 35 (1932) 625-631.

[18] Sakai, M., Null quadrature domains, J. d'Analyse Math. 40 (1981) 144-154.

[19] ——, Quadrature Domains, Lecture Notes in Math. 934. Springer-Verlag, 1982.

[20] Shapiro, H.S., Stefan Bergman's theory of doubly-orthogonal functions - an operator-theoretic approach, Proc. Royal Irish Acad. 79 (1979) 49-58.

[21] Shields, A.L., Weighted shift operators and analytic function theory, in Topics in Operator Theory, ed. C. Pearcy, Math. Surveys no. 13, Amer. Math. Soc. 1974, pp. 50-128.

International Series of
Numerical Mathematics, Vol. 65

A BANDLIMITED FUNCTION SIMULATING A DURATION-LIMITED ONE

Robert Gervais and Qazi Ibadur Rahman and Gerhard Schmeisser

Dépt. d'informatique et d'ingénierie, Collège militaire royal, Saint-Jean (Québec)

Dépt. de Math. et de Statistique, Université de Montréal, Montréal

Math. Institut, Universität Erlangen-Nürnberg, Erlangen

We construct a bandlimited function $\phi \not\equiv o$ for which $|\phi(x)|$ decreases nearly as fast as possible when $x \to \pm\infty$. Such a function may be approximately considered as a duration-limited one. It is of interest in the reconstruction of entire and harmonic functions of exponential type from given values.

1. Introduction and Statement of the Result.

It is well-known to engineers in communication theory that the properties "bandlimited" and "duration-limited" of a signal function $f \not\equiv o$ exclude one another. In the language of mathematicicans this simply means (in view of the Paley-Wiener theorem) that the restriction of an entire function $f \not\equiv o$ of exponential type to the real line cannot have compact support. In order to compromise in this dilemma we are looking for an entire function $\phi \not\equiv o$ of given exponential type δ for which $|\phi(x)|$ decreases so rapidly as $x \to \pm\infty$ that up to a certain degree of precision ϕ can be identified numerically with a function of compact support on $\mathbb{R}$. Such a function ϕ is of interest in connection with the sampling theorem of Whittaker-Kotel'nikov-Shannon *). In fact, if we replace the classical cardinal series of Whittaker, namely

$$C_\tau[f](z) := \sum_{n=-\infty}^{\infty} f(\frac{n\pi}{\tau}) \operatorname{sinc}(\tau z - n\pi) ,$$

*) For a thorough mathematical investigation of this theorem and various refinements, extensions and interpretations we refer to the work of Prof. P.L. Butzer and his school, i.e. [3], [4], [5], [6].

where

$$\operatorname{sinc} z := \begin{cases} \frac{\sin z}{z} & \text{if } z \neq 0 \\ 1 & \text{if } z = 0 , \end{cases}$$

by the modified series

$$\mathcal{C}_\tau[f](z) = \sum_{n=-\infty}^{\infty} f\left(\frac{n\pi}{\tau}\right) \operatorname{sinc}(\tau z - n\pi)\, \phi(\tau z - n\pi)$$

we obtain an interpolation operator $\mathcal{C}_\tau$ with the following properties:

(α) The series $\mathcal{C}_\tau[f](z)$ converges rapidly and allows efficient numerical evaluation.

(β) $\mathcal{C}_\tau$ reproduces even functions of exponential type which are unbounded on the real line.

(γ) For a uniformly continuous and bounded function f the error

$$\|f - \mathcal{C}_\tau[f]\|_{\mathbb{R}}$$

is of the order of best approximation to f by entire functions of exponential type τ.

Similarly, ϕ may be used in the reconstruction of entire harmonic functions of exponential type from given values at two parallel lines thus answering questions of Boas [2]. Details of all these results will appear elsewhere. Here we restrict ourselves to the explicit construction of the desired function ϕ.

It is known (see [1], [7, p.81]) that ϕ can behave like

$$\phi(x) = \mathcal{O}(e^{-w(|x|)}) \qquad (w > 0)$$

as $x \to \pm\infty$ provided w, besides of certain regularity conditions, satisfies

$$\int_a^\infty \frac{w(x)}{x^2}\, dx < \infty$$

for some $a > 0$. This latter condition on w cannot be relaxed. As far as the proofs of this result are constructive at all, the function ϕ is obtained in a very indirect way so that it is not really available for numerical purposes.

The following theorem gives a rather simple construction of ϕ which can be performed numerically.

THEOREM. For an $a > 0$ let

$$w : [a,+\infty) \to (0,+\infty)$$

be a continuously differentiable function such that

$$\text{(i)} \quad \int_a^\infty \frac{w(x)}{x^2}\,dx < \infty ,$$

(ii) $\varphi : x \mapsto xw'(x)$ is strictly increasing on $[a,+\infty)$ and tending to infinity,

(iii) $w'(a) = 0$.

Define

$$(1) \qquad \phi : \begin{cases} \mathbb{C} \to \mathbb{C} \\ z \mapsto \prod_{n=0}^{\infty} \operatorname{sinc}(z/\varphi^{-1}(n)), \end{cases}$$

where φ^{-1} denotes the inverse function of φ. Then ϕ is an entire function of exponential type τ, say, with the following properties:

(a) $\phi(0) = 1$,

(b) $\phi(-x) = \phi(x) \in \mathbb{R}$ for all $x \in \mathbb{R}$,

(c) $|\phi(x)| = \mathcal{O}(e^{-w(|x|)})$ as $x \to \pm\infty$,

(d) $\tau \leq \dfrac{1-w(a)}{a} + \displaystyle\int_a^\infty \frac{w(x)}{x^2}\,dx$.

REMARK 1. Note that condition (iii) has no influence on the growth property (c). In fact, if for some sufficiently large $a > 0$ we have $w'(a) > 0$, while all the other assumptions of the theorem are satisfied we can extend w to an interval $[b,+\infty)$ where $0 < b < a$ such that our theorem holds with a replaced by b. Furthermore, starting with an appropriate function $\tilde{w}$ instead of w such that $|\tilde{w}(x)/w(x)| \to \infty$ for $x \to \infty$ we can achieve that ϕ satisfies (a) - (c) but has in addition a type τ not exceeding a given bound $\delta > 0$ (see the example below).

REMARK 2. An infinite product of sinc functions has also been used by H.S. Shapiro [8] to construct an entire function of order 1/2 with a certain decreasing majorant on the positive part of the real line.

2. Proof of the Theorem.

For any $N > 1$ and $K > 1$ we find with the help of the integral comparison criterion

$$S_{N,K} := \sum_{n=o}^{N} \sum_{k=1}^{K} (\pi k \varphi^{-1}(n))^{-2} \leq \frac{1}{6} \sum_{n=o}^{N} (\varphi^{-1}(n))^{-2}$$

$$\leq \frac{1}{6} (a^{-2} + \int_{0}^{N} (\varphi^{-1}(t))^{-2} dt).$$

Substituting $t = \varphi(x)$ and putting $J := \varphi^{-1}(N)$ we obtain, using integration by parts

$$S_{N,K} \leq \frac{1}{6} (a^{-2} + \int_{a}^{J} \frac{d\varphi(x)}{x^2})$$

$$= \frac{1}{6} (a^{-2} + \frac{w'(J)}{J} + 2\frac{w(J)}{J^2} - 2\frac{w(a)}{a^2} + 4 \int_{a}^{J} \frac{w(x)}{x^3} dx).$$

Now for $N \to \infty$ also $J \to \infty$ and the right hand side remains bounded as a consequence of (i). Thus we can conclude that for $c_{kn} := \pi k \varphi^{-1}(n-1)$ the sum

$$\sum_{k,n \in \mathbb{N}} |c_{kn}|^{-2}$$

converges for every order of summation. By the Weierstraß factorization theorem and well-known facts on infinite products

$$\phi(z) := \prod_{k,n \in \mathbb{N}} \left(1 - \frac{z^2}{c_{kn}^2}\right)$$

represents one and the same entire function ϕ howsoever the infinite product on the right hand side is carried out. In particular ϕ coincides with the one defined in (1), since

$$\prod_{k=1}^{\infty}\left(1-\frac{z^2}{c_{kn}^2}\right) = \operatorname{sinc}\left(\frac{z}{\varphi^{-1}(n-1)}\right) .$$

The properties (a) and (b) are obvious. It is also readily seen that ϕ is of exponential type, say τ, provided

$$\sigma := \sum_{n=o}^{\infty} (\varphi^{-1}(n))^{-1} < \infty ,$$

and in this case $\tau \leq \sigma$. Using again the integral comparison criterion, the substitution $t = \varphi(x)$ and integration by parts, we obtain

$$\sigma \leq a^{-1} + \int_0^{\infty} \frac{dt}{\varphi^{-1}(t)} = a^{-1} + \int_a^{\infty} \frac{d\varphi(x)}{x}$$

$$= a^{-1} + \left(w'(x) + \frac{w(x)}{x}\right)\Big|_a^{\infty} + \int_a^{\infty} \frac{w(x)}{x^2}\, dx.$$

Now (i) not only shows $\sigma < \infty$ but also yields (d).

Let us finally verify the growth property (c). Given a sufficiently large positive x we put

$$N := [\varphi(x)]$$

where [r] denotes the integer part of r for every real r. Since for $z = x \in \mathbb{R}$ all the terms of the infinite product in (1) are of modulus at most 1 we may estimate

$$|\phi(x)| \leq \prod_{n=o}^{N} \left|\operatorname{sinc}\left(\frac{x}{\varphi^{-1}(n)}\right)\right| \leq \prod_{n=o}^{N} \frac{\varphi^{-1}(n)}{x}$$

$$= \exp\left(\sum_{n=o}^{N} \log\left(\frac{\varphi^{-1}(n)}{x}\right)\right) .$$

By our choice of N all the terms of the sum are negative and increasing with n. With the help of the integral comparison criterion, the substitution $t = \varphi(\xi)$ and integration by parts we conclude

$$\sum_{n=o}^{N} \log\left(\frac{\varphi^{-1}(n)}{x}\right) \leq \int_0^{N} \log\left(\frac{\varphi^{-1}(t)}{x}\right) dt + \log\left(\frac{\varphi^{-1}(N)}{x}\right)$$

$$\leq \int_0^{\varphi(x)} \log\left(\frac{\varphi^{-1}(t)}{x}\right) dt = -\int_0^{\varphi(x)} t\cdot d\log(\varphi^{-1}(t))$$

$$= - \int_a^x \varphi(\xi)\ d\log\xi = - (w(x) - w(a)).$$

Thus (c) is also shown.

3. An Example.

Let α and δ be any two given numbers satisfying $\alpha > 1$ and $\delta > o$. For a $\beta \in (1,\alpha)$ we put

$$w(x) := \frac{x}{(\log x)^{\beta}}$$

for large x. Then, in view of Remark 1, there exists an entire function ϕ of exponential type with the properties (a) - (c). If its type is at most $\sigma > o$, we put

$$\psi(z) := \phi(\frac{\delta}{\sigma} z)$$

and obtain an entire function ψ of exponential type at most δ for which

(a') $\psi(o) = 1$

(b') $\psi(-x) = \psi(x) \in \mathbb{R}$ for all $x \in \mathbb{R}$

and

(c') $|\psi(x)| = \mathcal{O}\left(\exp\left(-\frac{|x|}{(\log|x|)^{\alpha}}\right)\right)$ as $x \to \pm\infty$.

Unfortunately, in this case the function φ^{-1} cannot be calculated explicitly although we can show that $\varphi^{-1}(t)$ behaves roughly like $t(\log t)^{\beta}$ for large t. For practical purposes it may be preferable to start with φ^{-1} and calculate a corresponding w via the relation

$$w(x) - w(a) = \int_0^{\varphi(x)} t\cdot d\log(\varphi^{-1}(t)).$$

Thus, setting

$$a := (\log 2)^{\beta}$$

and

$$\varphi^{-1}(t) := \begin{cases} a\cdot(\frac{t}{2}+1) & \text{if } o \leq t \leq 2 \\ t\cdot(\log t)^{\beta} & \text{if } t \geq 2 \end{cases}$$

we obtain with appropriate constants $c_1, c_2 \in \mathbb{R}$

$$w(x) = c_1 + \int_2^{\varphi(x)} \left(1 + \frac{\beta}{\log t}\right) dt$$

$$> c_2 + \varphi(x).$$

Furthermore

$$\varphi(x) > \frac{x}{(\log x)^{\beta}}$$

for large x. Hence certainly

$$w(x) > \frac{x}{(\log x)^{\alpha}} \quad \text{as } x \to +\infty$$

and even (c') holds, if we turn from ϕ to ψ as above.

REFERENCES

[1] Beurling, A. - Malliavin P., On Fourier transforms of measures with compact support. Acta Math. 107 (1962), 291-309.

[2] Boas, R.P., A uniqueness theorem for harmonic functions. J. Approximation Theory 5 (1972), 425-427.

[3] Butzer, P.L., The Shannon sampling theorem and some of its generalizations; an overview. In: D. Vacov (ed.), Proc. Internat. Conference on Constructive Function Theory (Varna, Bulgaria, June 1981), to appear.

[4] Butzer, P.L. - Splettstösser, W., Approximation und Interpolation durch verallgemeinerte Abtastsummen. Ber. NRW 2708, Westdeutscher Verlag, Opladen 1977, 57pp.

[5] Butzer, P.L. - Stens, R.L., The Euler-Maclaurin summation formula, the sampling theorem, and approximate integration over the real axis. Linear Algebra Appl. 52/53 (1983), 141-155.

[6] Butzer, P.L. - Stens, R.L., The Poisson summation formula, Whittaker's cardinal series and approximate integration. In: Proc. Second Edmonton Conference on Approximation Theory, June 1982; CMS Conference Proc. Vol. 3; Amer. Math. Soc., Providence, R.I., 1983; pp. 19-36.

[7] Levinson, N., Gap and Density Theorems. Amer. Math. Soc. Colloq. Publ. 26, New York 1940.

[8] Shapiro, H.S., Weighted polynomial approximation and boundary behaviour of analytic functions. In: Contemporary Problems in the Theory of Analytic Functions. Proc. Internat. Conference Erewan 1965, Izdat. Nauka, Moscow 1966, pp.326-335.

International Series of
Numerical Mathematics, Vol. 65

SHANNON'S SAMPLING THEOREM CAUCHY'S INTEGRAL FORMULA, AND RELATED RESULTS

Paul L. Butzer, Sigmar Ries*, and Rudolf L. Stens
Lehrstuhl A für Mathematik
Rheinisch-Westfälische Technische Hochschule
Aachen

Cauchy's integral formula is compared with Shannon's sampling theorem. It is shown that each can be deduced from the other by elementary means.

1. Introduction; Preliminaries

Whereas Cauchy's integral formula plays the central role in complex function theory, it is Shannon's sampling theorem and its generalizations that do so in communication theory. The aim of this paper is to elucidate the connections between these two basic results. In the first part, Cauchy's formula for so called bandlimited functions is considered; it is shown that the classical Shannon theorem can be deduced from it. Conversely, this version of Cauchy's formula will be infered from the sampling theorem. So it can be said that both results are in a certain sense equivalent. The crucial point is that when passing from the Cauchy to the Shannon theorem one has only a very particular case of the integral formula at one's disposal and, in the converse direction, one is only allowed to use function theoretic results that are fully independent of Cauchy's formula. To make this clear our arguments are based upon Hille's book [11] and only results preceding Cauchy's theorem or Cauchy's formula are used. In regard to the direct implication, the situation is different from that in [12; 9, p. 115; 17, p. 54] where the sampling theorem was established via the residue theorem (which can be regarded as a

* The second named author was supported by the Stiftung Volkswagenwerk.

general form of Cauchy's integral formula). Moreover, the proofs in [9; 17] are not correct since they make use of the inequality $|\sin \alpha\xi| > (1/2)\exp\{\alpha r|\sin \varphi|\}$, $\xi = re^{i\varphi} \in \mathbb{C}$, which fails, e.g., if $\varphi = \pi/2$, $\alpha > 0$.

In the second part, a more general version of the integral formula is shown to be equivalent - in the same sense as above - to a kind of generalized sampling theorem.

Concerning notations, let $\mathbb{R}$ be the real line, $\mathbb{C}$ the complex plane, $\mathbb{Z}$ the set of all integers, and $\mathbb{N}$ the naturals. As usual, $L(\mathbb{R})$ is the space of all functions $f: \mathbb{R} \to \mathbb{C}$ which are Lebesgue integrable over $\mathbb{R}$. The class of entire functions $f: \mathbb{C} \to \mathbb{C}$ satisfying

(1.1) $$|f(z)| \leqslant Ke^{\sigma|y|} \qquad (z = x + iy \in \mathbb{C})$$

for some $K, \sigma \geqslant 0$ is denoted by B_σ. These functions are so called entire functions of exponential type σ, bounded along the real axis. If $f \in B_\sigma \cap L(\mathbb{R})$, then by the Paley-Wiener theorem

(1.2) $$f(z) = \frac{1}{\sqrt{2\pi}} \int_{-\sigma}^{\sigma} f^\wedge(u) e^{izu}\, du \qquad (z \in \mathbb{C})$$

where $f^\wedge(v) = (1/\sqrt{2\pi}) \int_{-\infty}^{\infty} f(u)e^{-ivu}\, du$, $v \in \mathbb{R}$, is the Fourier transform of f. Since a function f satisfying (1.2) is usually called bandlimited to $[-\sigma, \sigma]$, the members of B_σ will also be called bandlimited functions. As a matter of fact, in the proofs of Thms. 1,2 only inequality (1.1) is used, not the representation (1.2).

If C is a Jordan curve it is always assumed to be positive oriented. Its interior will be denoted by int C, its exterior by ext C. Let $C^* := C \cup \text{int } C$. When speaking of a function f holomorphic on C* it is assumed that f is holomorphic on an open set containing C*.

2. Cauchy's Integral Formula for Bandlimited Functions and Shannon's Theorem

CAUCHY'S INTEGRAL FORMULA FOR B_σ. *Let $f \in B_\sigma$ for some $\sigma > 0$, and C be a rectifiable Jordan curve. Then*

(2.1)
$$\frac{1}{2\pi i}\int_C \frac{f(\xi)}{(\xi - z)}\, d\xi = \begin{cases} f(z), & z \in \operatorname{int} C \\ 0, & z \in \operatorname{ext} C. \end{cases}$$

This particular version of the integral formula will be compared with the

SAMPLING THEOREM. *If* $f \in B_{\pi\tau}$ *for some* $0 < \tau < W$, *then*

(2.2)
$$f(z) = \sum_{k=-\infty}^{\infty} f\left(\frac{k}{W}\right) \frac{\sin\pi(Wz-k)}{\pi(Wz-k)} \qquad (z \in \mathbb{C}),$$

where the series is uniformly convergent on each bounded domain of $\mathbb{C}$.

Note that the assumptions made in this sampling theorem are slightly different from those in [4, 8]. Indeed, here τ has to be strictly less than W as can be seen from the counterexample $f(z) = \cos \pi Wz \in B_{\pi W}$. If, however, f is required to belong to $L(\mathbb{R})$ rather than to be only bounded on $\mathbb{R}$, then (2.2) holds in the limitng case $\tau = W$, too. See e.g. [4] for a survey on the sampling theorem.

THEOREM 1. *Cauchy's integral formula for functions* $f \in B_\sigma$ *implies Shannon's sampling theorem, and conversely.*

PROOF. To establish the first implication one may assume that $W = 1$. Indeed, if $W > 0$ is arbitrary, and $f \in B_{\pi\tau}$ for some $0 < \tau < W$, one just need apply the particular case $W = 1$ to the function $f(\cdot/W) \in B_{\pi(\tau/W)}$ to deduce

(2.3)
$$f\left(\frac{z}{W}\right) = \sum_{k=-\infty}^{\infty} f\left(\frac{k}{W}\right) \frac{\sin\pi(z-k)}{\pi(z-k)}.$$

The general result then follows by replacing z by zW. Moreover, one can assume that $z \notin \mathbb{Z}$; otherwise the series in (2.2) with $W = 1$ reduces to the term $k = z$.

To prove (2.2) for $0 < \tau < W = 1$ consider the integral

$$I_m(z) := \frac{\sin \pi z}{2\pi i} \int_{C_m} \frac{f(\xi)}{(\xi - z)\sin \pi\xi}\, d\xi,$$

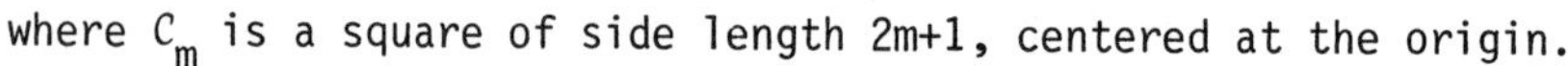
where C_m is a square of side length 2m+1, centered at the origin.

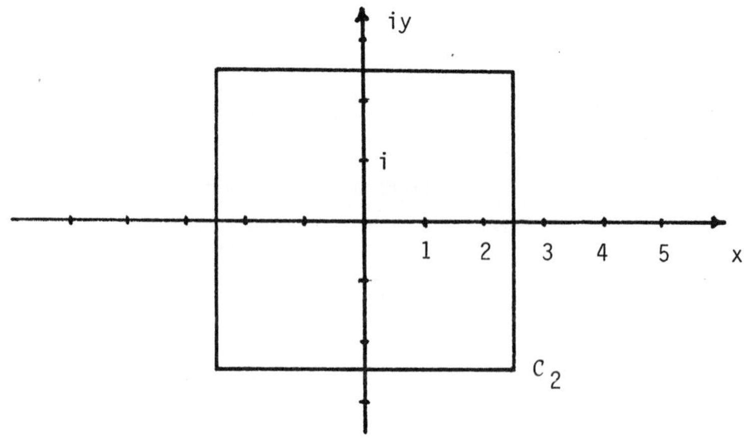

Here $m \in \mathbb{N}$ is chosen so large so that $z \in \operatorname{int} C_m$. Now the product representation of the sine function is needed, namely

$$(2.4)\qquad \begin{aligned} \sin \pi\xi &= \pi\, \xi \prod_{j=1}^{\infty} (1-(\xi/j)^2) \equiv \lim_{n\to\infty} A_n(\xi) \qquad (\xi \in \mathbb{C}), \\ A_n(\xi) &:= \pi\, \xi \prod_{j=1}^{n} (1-(\xi/j)^2), \end{aligned}$$

the convergence being uniform on bounded domains. (A proof of (2.4) avoiding the use of Cauchy's theorem or Cauchy's integral formula is in the appendix.) Since $\sin \pi\xi$ as well as $A_n(\xi)$, $n \in \mathbb{N}$, are bounded away from zero on C_m, it follows that $(\sin \pi\xi)^{-1} = \lim_{n\to\infty}(A_n(\xi))^{-1}$ uniformly on C_m, and so (cf. [3, p. 102])

$$(2.5)\qquad I_m(z) = \lim_{n\to\infty} \frac{\sin \pi z}{2\pi i} \int_{C_m} \frac{f(\xi)}{(\xi-z)A_n(\xi)}\, d\xi .$$

In order to calculate the integral on the right side, decompose the integrand into partial fractions. This yields

$$(2.6)\qquad \frac{f(\xi)}{(\xi-z)A_n(\xi)} = \frac{f(\xi)}{(\xi-z)A_n(z)} + \sum_{k=-n}^{n} \frac{f(\xi)}{(\xi-k)(k-z)A_n'(k)}$$

if one uses the formula

$$\frac{1}{p(\xi)} = \sum_{j=1}^{N} \frac{1}{(\xi - z_j)p'(z_j)} ,$$

valid for each polynomial p having simple zeros $z_1, z_2, \dots, z_N$ only. (Note that $z \notin \mathbb{Z}$ in (2.6)!)

Assuming now $n > m$, Cauchy's formula (2.1) gives

$$\frac{1}{2\pi i} \int_{C_m} \frac{f(\xi)}{(\xi - z)A_n(\xi)}\, d\xi = \frac{f(z)}{A_n(z)} + \sum_{k=-m}^{m} \frac{f(k)}{(k-z)A_n'(k)} ,$$

and by (2.4),(2.5) and the fact that $\lim_{n\to\infty} A_n'(k) = \pi(-1)^k$ (cf. La. 3 below) it follows that

$$I_m(z) = f(z) - \sin \pi z \sum_{k=-m}^{m} f(k)\frac{(-1)^k}{\pi(z-k)} = f(z) - \sum_{k=-m}^{m} f(k)\, \frac{\sin \pi(z-k)}{\pi(z-k)} .$$

The proof will be completed by showing that $\lim_{n\to\infty} I_m(z) = 0$, uniformly on bounded sets. If E is such a set, choose $M > 0$ such that $|\mathrm{Re}\, z| \leqslant M$ and $|\mathrm{Im}\, z| \leqslant M$ for all $z \in E$. Then one has for $m \geqslant M$ and ξ on C_m that $|\xi - z| \geqslant m + 1/2 - M$, i.e.,

$$|I_m(z)| \leqslant \frac{|\sin \pi z|}{2\pi(m+1/2-M)} \int_{C_m} \frac{|f(\xi)|}{|\sin \pi\xi|}\, d\xi$$

$$\equiv \frac{|\sin \pi z|}{2\pi(m+1/2-M)} \{I_m^{(1)} + I_m^{(2)} + I_m^{(3)} + I_m^{(4)}\} ,$$

where $I_m^{(1)}$ and $I_m^{(3)}$ denote the respective integrals along the vertical sides of C_m, and $I_m^{(2)}$ and $I_m^{(4)}$ along the horizontal sides. Hence it suffices to show that each of these integrals is of order $o(m)$, $m \to \infty$. Concerning the right vertical side, given by the parametrization $\xi(t) := (m + 1/2)(1+i(2t-1)$, $0 \leqslant t < 1$, one has by the identity

$$(2.7) \qquad |\sin \pi\xi|^2 = \sin^2 \pi x + \sinh^2 \pi y \qquad (\xi = x + iy \in \mathbb{C})$$

that

$$|\sin\pi\xi(t)| = \cosh[\pi(m+1/2)(2t-1)] \geqslant (1/2)\exp\{\pi(m+1/2)|2t-1|\} .$$

This implies by (1.1) that

$$|I_m^{(1)}| \leqslant (2m+1) \int_0^1 2K \exp\{\pi(m+1/2)(\sigma-1)|2t-1| \, dt$$

$$= \frac{4K}{\pi(1-\sigma)} \{1 - \exp\{\pi(m+1/2)(\sigma-1)\} = O(1) \qquad (m \to \infty) .$$

The same holds for $I_m^{(3)}$. As to $I_m^{(2)}$ and $I_m^{(4)}$, one has for ξ on the two horizontal sides that

$$\frac{f(\xi)}{|\sin \pi\xi|} \leqslant \frac{K\, e^{\sigma\pi(m+1/2)}}{\sinh \pi(m+1/2)} = o(1) \qquad (m \to \infty) ,$$

$(0 < \sigma < 1\,!)$ giving $|I_m^{(j)}| \leqslant (2m+2)o(1)$, $j = 2,4$. This completes the direct part.

Let us now show how to deduce Cauchy's formula from the sampling theorem. If C is a rectifiable Jordan curve, and $z \in \operatorname{int} C$, then by the uniform convergence of the sampling series

$$\text{(2.8)} \qquad \int_C \frac{f(\xi)}{(\xi-z)} d\xi = \sum_{k=-\infty}^{\infty} f(\tfrac{k}{W}) \int_C \frac{\sin\pi(W\xi-k)}{\pi(W\xi-k)(\xi-z)} d\xi .$$

Using the power series representation of the sine - function and applying the binomial formula yields

$$\frac{\sin\pi(W\xi-k)}{\pi(W\xi-k)} = \sum_{j=0}^{\infty} \frac{(-1)^j(\pi W)^{2j}}{(2j+1)!} (\xi-\tfrac{k}{W})^{2j}$$

$$= \sum_{j=0}^{\infty} \frac{(-1)^j(\pi W)^{2j}}{(2j+1)!} \sum_{\nu=0}^{2j} \binom{2j}{\nu}(z-\tfrac{k}{W})^{2j-\nu}(\xi-z)^{\nu} .$$

Hence it follows that

$$\text{(2.9)} \qquad \int_C \frac{\sin\pi(W\xi-k)}{\pi(W\xi-k)(\xi-z)} d\xi$$

$$= \sum_{j=0}^{\infty} \frac{(-1)^j(\pi W)^{2j}}{(2j+1)!} \sum_{\nu=0}^{2j} \binom{2j}{\nu}(z - \frac{k}{W})^{2j-\nu} \int_C (\xi-z)^{\nu-1}\, d\xi \ .$$

The latter integral can be computed to give $2\pi i$ for $\nu = 0$ (cf. [2, p. 239]) and 0 for $\nu \geqslant 0$ (cf. [11, p. 162]). So one obtains

$$\int_C \frac{\sin\pi(W\xi-k)}{\pi(W\xi-k)(\xi-z)}\, d\xi = 2\pi i \sum_{j=0}^{\infty} \frac{(-1)^j[\pi(zW-k)]^{2j}}{(2j+1)!}$$

$$= 2\pi i \frac{\sin\pi(Wz-k)}{\pi(Wz-k)} \ .$$

Inserting this identity into (2.8), and applying the sampling theorem once more, gives

$$\frac{1}{2\pi i} \int_C \frac{f(\xi)d\xi}{(\xi-z)} = \sum_{k=-\infty}^{\infty} f(\frac{k}{W}) \frac{\sin\pi(Wz-k)}{\pi(Wz-k)} = f(z) \ .$$

This proves the result for $z \in \text{int } C$. If $z \in \text{ext } C$, then one may proceed in exactly the same manner with the only exeption that the integral for $\nu = 0$ on the right of (2.9) vanishes too (cf. [2, p. 239]). The proof of Thm. 1 is complete.

3. Cauchy's Formula and a Generalized Sampling Theorem for Functions Holomorphic in a Rectangle

The considerations in Sec. 2 were restricted to bandlimited functions because Shannon's theorem cannot hold for arbitrary analytic functions since the series in (2.2) may diverge in this case. To investigate Cauchy's integral formula, however, one has to include functions which need only be holomorphic in a bounded region, and hence cannot be bandlimited. In order to keep the topological considerations simple, the matter is restricted to the case where the contour is the boundary of a rectangle. This particular case differs only in its topological but not in its analytical content from the general case (cf. [1, p. 109]).

CAUCHY'S INTEGRAL FORMULA FOR RECTANGLES. *Let C be the boundary of a rectangle. If the function f is holomorphic on C^*, then* (2.1) *holds.*

This version of the integral formula will turn out to be equivalent to the following

GENERALIZED SAMPLING THEOREM. *If C is the boundary of the rectangle with corners* $a+ib$, $-a+ib$, $-a-ib$, $a-ib$ *for any* $a,b>0$, *and if f is holomorphic in* C^*, *then there exists* $W_0>0$ *such that for all* $W>W_0$

$$(3.1) \qquad f(z) = \sum_{|k/W|\leq a} f\left(\frac{k}{W}\right) \frac{\sin\pi(Wz-k)}{\pi(Wz-k)} + R_W(z) \qquad (z\in \operatorname{int} C_W),$$

where C_W is the boundary of the rectangle with corners $\alpha(W)+ib$, $-\alpha(W)+ib$, $-\alpha(W)-ib$, $\alpha(W)-ib$, $\alpha(W)$ *being given by* $\alpha(W) := ([aW]+1/2)/W$, *and* $R_W(z)$ *is defined by*

$$R_W(z) := \frac{\sin\pi Wz}{2\pi i} \int_{C_W} \frac{f(\xi)}{(\xi-z)\sin\pi W\xi}\, d\xi .$$

REMARK. The symbol $\sum_{|k/W|\leq a}$ means that the sum is to be extended over all integers satisfying $|k/W|\leq a$ or, equivalently, $|k|\leq [aW]$, where $[x]$ denotes the largest integer $\leq x$. Note that the horizontal lines of C_W coincide (apart from the length) with those of C, whereas the vertical sides of C_W swing around the vertical sides of C when $W\to\infty$.

THEOREM 2. *Cauchy's integral formula for rectangles implies the generalized sampling theorem, and conversely.*

PROOF. Since f is holomorphic on an open set D containing C^*, there exists $W_0>0$ such that $C_W^*\subset D$ for all $W>W_0$. Now the integral $R_W(z)$ can be calculated in just the same manner as $I_m(z)$ above to give (3.1). This proves the direct implication.

Before showing the converse let us prove the following

COROLLARY. *Under the assumptions of the preceeding theorem there holds for* $t\in(-a,a)$

$$f(t) = \lim_{W\to\infty} \sum_{|k/W|\leqslant a} f(\tfrac{k}{W}) \frac{\sin\pi(Wt-k)}{\pi(Wt-k)} , \tag{3.2}$$

the convergence being uniform on each compact subinterval of $(-a,a)$.

PROOF. If $[c,d]$ is such a subinterval of $(-a,a)$, then there exists $W_1 \geqslant W_0$ such that the distance between $[c,d]$ and C_W is greater than some $\eta > 0$ (i.e. $|\xi - t| > \eta$ for $\xi \in C_W$, $t \in [c,d]$) for all $W > W_1$. Moreover, since C_W is contained in some compact subset of D for all $W \geqslant W_1$, it follows that there exists $M > 0$ such that $|f(\xi)| \leqslant M$ for all $\xi \in C_W$ and all $W \geqslant W_1$. Hence

$$|R_W(t)| \leqslant \frac{M|\sin \pi Wt|}{2\pi\eta} \int_C \frac{d\xi}{|\sin \pi W\xi|} \qquad (W \geqslant W_1) ,$$

and this integral can be shown to vanish for $W \to \infty$ by basically the same method as used in Sec. 2.

This result is in fact a version of the Shannon theorem for time limited functions already treated under various sets of conditions in [6; 7; 15; 16]. Here we have an entirely different proof, based on Cauchy's integral formula. It does give insight on the behaviour of the sampling series in the complex domain. Indeed, one might expect that (3.2) holds not only for real t but for all complex z in the interior of C_W. However, the series in (3.2) may diverge if $W \to \infty$ for complex values of the variable, as the following simple example shows.

LEMMA 1. *Let* Q *be the rectangle with corners* $1+2i$, $-1+2i$, $-1-2i$, $1-2i$, *and* $f(\xi) := \xi$. *Then*

$$\sum_{|k/W|\leqslant 1} f(\tfrac{k}{W}) \frac{\sin \pi(Wz-k)}{\pi(Wz-k)} \equiv \sum_{|k/W|\leqslant 1} \frac{k}{W} \frac{\sin \pi(Wz-k)}{\pi(Wz-k)}$$

diverges for $z = \gamma i$, $\gamma > 0$, *if* $W \to \infty$.

PROOF. Let $W = n$, where $n \in \mathbb{N}$ is odd. Then

$$\sum_{|k/n|\leq 1} \frac{k}{n} \frac{\sin \pi(\gamma ni-k)}{\pi(\gamma ni-k)} = \frac{i\, 2\sinh \gamma n\pi}{n\pi} \sum_{k=1}^{n} \frac{(-1)^{k+1} k^2}{(\gamma n)^2 + k^2} .$$

Now, since $k^2/((\gamma n)^2 + k^2)$ is nondecreasing in k and n is odd, the series on the right is seen to be greater than its first term, thus

$$\left| \sum_{|k/n|\leq 1} \frac{k}{n} \frac{\sin \pi(\gamma ni-k)}{\pi(\gamma ni-k)} \right| \geq \frac{2 \sinh \gamma n\pi}{\pi n((\gamma n)^2 + 1)} \to \infty \qquad (n \to \infty) .$$

This completes the proof.

This example shows that the corollary fails if t is replaced by a complex z. It is interesting to note that the same phenomenon arises in connection with Poisson's summation formula. If $f \in L(\mathbb{R})$ is of bounded variation over $\mathbb{R}$, then this formula states that (cf. [5, p. 202])

$$(3.3) \qquad \sqrt{2\pi} \sum_{k=-\infty}^{\infty} f(t+2k\pi) = \sum_{k=-\infty}^{\infty} f^{\wedge}(k)e^{ikt} \qquad (t \in \mathbb{R}) ,$$

Now, if $f \in B_\sigma \cap L(\mathbb{R})$, then (3.3) can be rewritten as

$$(3.4) \qquad \sqrt{2\pi} \sum_{k=-\infty}^{\infty} f(z+2k\pi) = \sum_{|k|\leq\sigma} f^{\wedge}(k)e^{ikz} \qquad (z \in \mathbb{C}) .$$

Indeed, the left hand side of (3.4) converges for all $z \in \mathbb{C}$ (see [14, p. 121 (4), p. 123] and represents an entire function [14, p. 127]. On the other hand, the series on the right of (3.4) coincides for $z = t \in \mathbb{R}$ with the right side of (3.3) since $f^{\wedge}$ vanishes for $|v| \geq \sigma$. So it follows that both sides of (3.4) are entire functions, coinciding on the real axis by (3.3), and hence for all $z \in \mathbb{C}$.

However, similarly to (3.2), there is no counterpart to (3.4) valid for functions which are just analytic on a rectangle. For example, choosing the rectangle C* as in the generalized sampling theorem, f being holomorphic on C*, then, redefining f to be zero outside of C*, it follows by (3.3) that

$$(3.5) \quad \sqrt{2\pi} \sum_{|t+2k\pi|\leq a} f(t+2k\pi) = \sum_{k=-\infty}^{\infty} \left\{\frac{1}{\sqrt{2\pi}} \int_{-a}^{a} f(u)e^{iku}\, du\right\}e^{ikt} \qquad (t \in (-a,a)) .$$

But this formula cannot be extended to the complex plane. In fact, take $f(z) = e^{-iz/4}$, $a = 2\pi$. Then the integrals in (3.5) are equal to $8/\sqrt{2\pi}(4k+1)$, and the series on the rigth would diverge if t would be replaced by any z with $z = t + iy$, $y \neq 0$.

Our final aim is to complete the proof of Thm. 2, i.e., to show how Cauchy's formula (2.1) for a rectangle can be recovered from (3.1). Without loss of generality the corners can be given as in the generalized sampling theorem, the general case then following by a linear transformation. Since f is holomorphic in a neighbourhood of C^*, there exists a rectangle with corners $a' + ib'$, $-a' + ib'$, $-a' - ib'$, $a' - ib'$, where $a' > a$, $b' > b$, such that f is holomorphic in a neighbourhood of this larger rectangle, too. Denoting its boundary by C', and applying (3.1) to C' instead of C, the integration of $f(\xi)/(\xi - z)$ gives

$$\frac{1}{2\pi i} \int_C \frac{f(\xi)}{(\xi-z)} d\xi = \sum_{|k/W| \leq a'} f(\tfrac{k}{W}) \frac{1}{2\pi i} \int_C \frac{\sin\pi(W\xi-k)}{\pi(W\xi-k)(\xi-z)} d\xi$$

$$+ \frac{1}{2\pi i} \int_C \frac{1}{2\pi i} \int_{C_W'} \frac{f(\zeta) \sin \pi W\xi}{(\xi-z)(\zeta-\xi)\sin \pi W\zeta} d\zeta \, d\xi .$$

Here C_W' is defined as C_W with a,b replaced by a', b'. The W has to be chosen so large that C_W' does not intersect C, and $(C_W')^*$ is in the domain of analyticity of f. Now, if $z \in \operatorname{int} C$, then the first integral on the right side can be calculated as in (2.9), and the order of integration in the double integral may be interchanged (cf. [3, p. 105]). Hence

$$\frac{1}{2\pi i} \int_C \frac{f(\xi)}{(\xi-z)} d\xi = \sum_{|k/W| \leq a'} f(\tfrac{k}{W}) \frac{\sin \pi(Wz-k)}{\pi(Wz-k)}$$

$$+ \frac{1}{2\pi i} \int_{C_W'} \frac{f(\zeta)}{\sin \pi W\zeta} \left\{\frac{1}{2\pi i} \int_C \frac{\sin \pi W\xi}{(\xi-z)(\zeta-\xi)} d\xi\right\} d\zeta$$

Since the inner integral is equal to $\sin Wz/(\zeta - z)$, which can be computed similarly to (2.9), the desired result follows by applying (3.1) once more. The case $z \in \operatorname{ext} C$ follows along the same lines.

4. Appendix

The aim of this section is to give elementary proofs of the product formula (2.4) as well as of the identity $\lim_{n\to\infty} A_n'(k) = \pi(-1)^k$.

LEMMA 2. *There holds*

$$\sin \pi\xi = \pi\xi \prod_{j=1}^{\infty} (1-(\xi/j)^2) \equiv \lim_{k\to\infty} A_k(\xi)$$

uniformly on each bounded subset of $\mathbb{C}$.

PROOF. For $k,n \in \mathbb{N}$ with $k<n$ one has the representation (cf. [10, p. 388])

$$\sin \pi\xi = U_k^{(n)}(\xi)\, V_k^{(n)}(\xi) \qquad (\xi \in \mathbb{C}), \tag{4.1}$$

where the functions $U_k^{(n)}$, $V_k^{(n)}$ are given by

$$U_k^{(n)}(\xi) := (2n+1) \sin \frac{\pi\xi}{2n+1} \prod_{j=1}^{k} (1-\xi_{j,n}^2), V_k^{(n)}(\xi) := \prod_{j=k+1}^{n} (1-\xi_{j,n}^2)$$

with $\xi_{j,n} := \sin(\pi\xi/(2n+1))/\sin(j\pi/(2n+1))$. Moreover, since $\lim_{n\to\infty} U_k^{(n)}(\xi) = A_k(\xi)$, it follows from (4.1) that $V_k(\xi) := \lim_{n\to\infty} V_k^{(n)}(\xi)$ exists, i.e.,

$$\sin \pi\xi = A_k(\xi)\, V_k(\xi) \qquad (\xi \in \mathbb{C};\ k \in \mathbb{N}).$$

Hence it remains to show that

$$\lim_{k\to\infty} V_k(\xi) = 1 \tag{4.2}$$

uniformly on bounded sets.

To this end, let E be a bounded set and let $M>0$ be such that $(\pi x)^2 + (2\pi y)^2 \leqslant M^2$ for all $\xi = x+iy \in E$. Then one has for $n \geqslant M$ by (2.7) that

$$\left|\sin^2 \frac{\pi\xi}{2n+1}\right| \leqslant \left(\frac{\pi x}{2n+1}\right)^2 + \left(\frac{2\pi y}{2n+1}\right)^2$$

and $\sin^2(j\pi/(2n+1)) \geq (2j/(2n+1))^2$ for $j \leq n$, noting that $\sinh\alpha \leq \alpha \sum_{\nu=0}^{\infty}[(2\nu+1)!]^{-1} \leq 2\alpha$ for $0 \leq \alpha \leq 1$ and $\sin\beta \geq 2\beta/\pi$, $0 \leq \beta \leq \pi/2$. This yields

$$|\xi_{j,n}^2| \leq \frac{(\pi x)^2 + (2\pi y)^2}{(2j)^2} \leq \frac{M^2}{4j^2} ,$$

and by the inequality $|\log(1+\xi)| \leq 2|\xi|$ for $|\xi| < 1/2$ (cf. [13, p. 435; 11, p. 145]) it follows that

$$|\log(1-\xi_{j,n}^2)| \leq (M/j)^2 \qquad (M \leq j \leq n)\ ^{1)} .$$

So one obtains in view of $|\exp(\xi) - 1| \leq \exp(|\xi|) - 1$, $\xi \in \mathbb{C}$, that

$$|V_k^{(n)}(\xi) - 1| = |\exp\{\sum_{j=k+1} \log(1 - \xi_{j,n}^2)\} - 1|$$

$$\leq \exp\{\sum_{j=k+1}^{n} (M/j)^2\} - 1 \qquad (M \leq k \leq n) .$$

Letting $n \to \infty$ this implies (4.2) since

$$|V_k(\xi) - 1| \leq \exp\{\sum_{j=k+1}^{\infty} (M/j)^2\} - 1 \qquad (k \geq M) .$$

LEMMA 3. <u>For</u> $k \in \mathbb{Z}$, $n \in \mathbb{N}$ <u>with</u> $|k| \leq n$ <u>there holds</u>

$$\lim_{n\to\infty} A_n'(k) = \pi(-1)^k$$

PROOF. The case $k = 0$ is obvious. For $1 \leq |k| \leq n$ let

$$B_n(\xi) := \pi\xi \prod_{\substack{j=1 \\ j\neq|k|}}^{n} (1-(\xi/j)^2) \qquad (\xi \in \mathbb{C}) ,$$

$$B(\xi) := \lim_{n\to\infty} B_n(\xi)$$

1) log is to be understood as the principal value

where the limit exists uniformly on bounded sets by La.2. It follows immediately from the definition of the derivative that $A_n'(k) = -2B_n(k)/k$, and since B is continuous on $\mathbb{C}$ one has

$$\lim_{n\to\infty} A_n'(k) = -\frac{2}{k} \lim_{\xi\to k} B(\xi) .$$

The assertion now follows, since, by La. 2,

$$B(\xi) = \frac{\sin \pi\xi}{1-(\xi/k)^2} \qquad (\xi \neq |k|) .$$

ACKNOWLEDGEMENT. The authors would like to thank Professor D.H. Mugler, University of Santa Clara, fellow of the Alexander von Humboldt-Foundation at the Lehrstuhl A für Mathematik, Aachen, for his critical reading of the manuscript.

REFERENCES

[1] Ahlfors, L.V., Complex Analysis (3rd Ed.). McGraw-Hill Book Company, New York, 1979.

[2] Apostol, T.M., Mathematical Analysis. Addison - Wesley Publishing Company, Reading, 1957.

[3] Behnke, H. - Sommer, F., Theorie der analytischen Funktionen einer komplexen Veränderlichen (Studienausgabe, 3rd Ed.). Springer Verlag, Berlin, 1976.

[4] Butzer, P.L., A survey of the Whittaker - Shannon sampling theorem and some of its extensions. J.Math. Res. Exposition 3(1983), 185-212.

[5] Butzer, P.L. - Nessel, R.J., Fourier Analysis and Approximation. Birkhäuser Verlag, Basel and Academic Press, New York, 1971.

[6] Butzer, P.L. - Ries, S. - Stens, R.L., The Whittaker - Shannon sampling theorem, related theorems and extensions. In: Proc. Jordan - IEEE Conf. (Amman, 25.-28.4.1984).

[7] Butzer, P.L. - Splettstösser, W., A sampling theorem for duration-limited functions with error estimates. Inform. and Control 34(1977), 55-65.

[8] Butzer, P.L. - Stens, R.L., The Poisson summation formula, Whittaker's cardinal series and approximate Integration. In: Proc. Second Edmonton Conf. on Approximation Theory (Edmonton, 7.-11.6.1983; Eds. Z. Ditzian - A. Meir - S. Riemenschneider - A. Sharma). Amer. Math Soc., Providence, 1983; pp. 19 - 36.

[9] Churkin, J.I. - Jakowlew, C.P. - Wunsch, G., Theorie und Anwendung der Signalabtastung. Verlag Technik, Berlin, 1966.

[10] Fichtenholz, G.M., Differential- und Integralrechnung II (6th Ed.). VEB Deutscher Verlag der Wissenschaften, Berlin, 1974.

[11] Hille, E., Analytic Function Theory I. Ginn and Company, Boston,1959.

[12] Jagerman, D.L. - Fogel, L., Some general aspects of the sampling theorem. IRE Trans. Inform. Theory IT -2, (1956), 139 - 146.

[13] Knopp, K., Theory and Applications of Infinite Series. Hafner Publishing Company, New York, 1971.

[14] Nikol'skiĭ, S.M., Approximation of Functions of Several Variables and and Imbedding Theorems. Springer - Verlag, Berlin, 1975.

[15] Ries, S. - Stens, R.L., Pointwise convergence of sampling series. In: Proc. Second European Signal Processing Conf. (Erlangen, 12. - 16.9.1983; Ed. H.W. Schüssler). North Holland Publishing Company, Amsterdam, 1983, pp. 5 - 7.

[16] de la Vallée - Poussin, Ch.J., Sur la convergence des formules d'interpolation entre ordonnées équidistantes. Bull. Acad. Roy. Belg. 1908, 319 - 410.

[17] Wunsch, G., Systemtheorie der Informationstechnik. Akademische Verlagsgesellschaft Geest & Portig, Leipzig, 1971.

VI Interpolation

International Series of
Numerical Mathematics, Vol. 65

RECENT RESULTS ON THE ALMOST EVERYWHERE DIVERGENCE OF LAGRANGE INTERPOLATION

Péter Vértesi
Mathematical Institute
of the Hungarian Academy of Sciences
Budapest

In this paper I try to write down the most significant results dealing with the almost everywhere divergence of Lagrange interpolation. Beginning with the classical results of Faber, Bernstein, Grünwald and Marcinkiewicz, our main interest are those, surprisingly many, important results which were developed in the last years. We consider theorems of almost everywhere divergence type (1) on finite interval,(2) on the complex unit circle and for periodic functions,(3) on infinite intervals. Finally, we mention some open problems, too. We refer 29 papers of 19 authors.

1. Introduction. Classical Results

1.1. One of the simplest approximating tools is the Lagrange interpolatory polynomial. As P. Turán wrote: "After ... the approximation theorem of Weierstrass, it was hoped that there exists a (non-equidistant) system of nodes for which Lagrange's interpolation polynomials converge uniformly for every function continuous in [-1,1]. The mathematical world was awakened from this dream by Faber who showed that there is no such system" ([28] p. 24, Part 4).

This result can be considered as the starting point of the divergence theory for Lagrange interpolation. To formulate the statement let us introduce some notations.

Let $X = \{x_{kn}\}$, $k = 1,2,\dots,n$, $n = 1,2,\dots$, be an interpolatory matrix in [-1,1], that means

$$(1.1)\qquad -1 \leq x_{nn} < x_{n-1,n} < \dots < x_{1n} \leq 1, \quad n = 1,2,\dots .$$

The corresponding Lagrange interpolatory polynomials will be denoted by

$L_n(f,X,x)$, $L_n(f,x)$, $L_n(f)$ or L_n. We have

(1.2) $$L_n(f,X,x) = \sum_{k=1}^{n} f(x_{kn})\ell_{kn}(X,x),$$

where

(1.3) $$\ell_{kn}(X,x) = \frac{\omega_n(X,x)}{\omega_n'(X,x_{kn})(x-x_{kn})}, \quad k=1,2,\dots,n,$$

(1.4) $$\omega(X,x) = c_n \prod_{k=1}^{n} (x-x_{kn}), \qquad c_n \neq 0 .$$

Here, sometimes omitting the superfluous notations, ℓ_k are the fundamental polynomials of Lagrange interpolation of degree exactly n-1 having the well-known property $\ell_k(x_j) = \delta_{kj}$.

As we know the expressions

(1.5) $$\lambda_n(X,x) = \sum_{k=1}^{n} |\ell_{kn}(X,x)|, \quad \lambda_n(X) = \|\lambda_n(X,x)\|$$

are of fundamental importance in the investigation of the Lagrange interpolation. They are the Lebesgue functions and the Lebesgue constants of the Lagrange interpolation, respectively ($\|g(x)\| := \max_{-1\leq x\leq 1}|g(x)|$). If we consider functions continuous on [-1,1] (shortly $f \in C$), Faber's result says as follows.

THEOREM 1.1 (G. Faber [4]). _For arbitrary_ $X \subset [-1,1]$ _there exists an_ $f_1 \in C$ _for which_

$$\overline{\lim_{n\to\infty}} \|L_n(f_1,X,x)\| = \infty .$$

_1.2._ But, of course, this result does not exclude a pointwise convergence result at least at a single point. This question was (negatively) answered almost twenty years later by

THEOREM 1.2 (S. Bernstein [2]). _For any_ $X \subset [-1,1]$ _there exist a point_ $x_0 \in [-1,1]$ _and_ $f_2 \in C$ _such that_

$$\overline{\lim_{n\to\infty}} |L_n(f_2,X,x_o)| = \infty .$$

1.3. The next natural question is: divergence result on a set of postive measure. For the (at least in interpolatory problems) "bad" equidistant matrix $E=\{-1+2(k-1)/(n-1)\}$ and the function $|x|$ Bernstein [4] proved that

$$\overline{\lim_{n\to\infty}} |L_n(|t|,E,x)| = \infty \quad \text{if} \quad x\in(-1,1) , \quad x\neq 0 .$$

The next result which was proved by G. Grünwald [13] states a similar theorem considering the Chebyshev matrix

$$T = \{x_{kn} = \cos\frac{2k-1}{2n}\pi , \ k=1,2,\dots,n , \ n=1,2,\dots\} .$$

He proved in 1935 that there exists an $f_3\in C$ such that

$$\overline{\lim_{n\to\infty}} |L_n(f_3,T,x)| = \infty \ \text{a.e. in} \ [-1,1] .$$

Later he (and independently) J. Marcinkiewicz obtained the next improvement.

THEOREM 1.3 (G. Grünwald [14], J. Marcinkiewicz [17]). One can find a function $f_4\in C$ for which

$$\overline{\lim_{n\to\infty}} |L_n(f_4,T,x)| = \infty \quad \text{for arbitrary} \ x\in[-1,1] . \tag{1.6}$$

2. Recent Results

A. Finite Interval

2.1. The importance of the previous Grünwald -Marcinkiewicz result is that their negative result holds true for the "very good" Chebyshev matrix T. And, as it turned out in several cases, if a negative interpolatory statement can be verified for T, then probably it holds for other matrices X. Or, with another word, the next conjecture of P. Erdös [6] was very reasonable: Prove the Grünwald - Marcinkiewicz result for arbitrary $X\subset[-1,1]$, at least a.e.

(and _not_ everywhere because it is fairly easy to construct a matrix X for which $\lim_{n\to\infty} L_n(f,X,x) = f(x)$, $f \in C$, on a dense subset of [-1,1]: we can take

$$X = \begin{bmatrix} x_1 & & & \\ x_1 & x_2 & & \\ x_1 & x_2 & x_3 & \\ . & . & . & . \end{bmatrix}$$

where $\{x_k\}$ is a dense subset of [-1,1] (see, say, [28], p. 30).

(Another result in this direction was obtained by P.P. Korovkin [15] and S.N. Mergeljan [18]: There exists a matrix $X_0 \subset [-1,1]$ and a perfect set $F \subset [-1,1]$ such that the Lagrange polynomials uniformly tend on F to any function continuous on F).

The first achievement is due to A.A. Privalov [21] and [22]. He considered the matrix $X^{(\alpha,\beta)} = \{x_{k,n}^{(\alpha,\beta)}\}$ where $x_{kn}^{(\alpha,\beta)}$ is the k-th root of the n-th Jacobi polynomial $P_n^{(\alpha,\beta)}(x)$, $\alpha,\beta > -1$ (see e.g. [25], Ch. IV.), _and essentially proved that for a suitable_ $f_5 \in C$

$$\overline{\lim_{n\to\infty}} |L_n(f_5, X^{(\alpha,\beta)}, x)| = \infty \text{ for any } x \in [-1,1] . \tag{2.1}$$

_2.2._ In 1977 N.B. Tihomirov [26] was able to settle the case when X has only different nodes (i.e. when $x_{kn} = x_{jN}$ iff $k = j$ and $n = N$).

THEOREM 2.1 (N.B. Tihomirov). _If_ $X \subset [-1,1]$ _has only different nodes then with a suitable_ $f_6 \in C$

$$\overline{\lim_{n\to\infty}} L_n(f_6, X, x) = \infty \text{ a.e. in } [-1,1] . \tag{2.2}$$

Later he slightly improved his statement ([27]) but was not able to handle matrices having repeating nodes. (Actually, in the proof of (2.1), Privalov first obtained another interpolatory matrix $Y^{(\alpha,\beta)} \subset X^{(\alpha,\beta)}$ which has only different nodes. The construction of $Y^{(\alpha,\beta)}$ uses some deep number theoretical results.)

One may think that to prove the Erdös' conjecture, it is enough to properly shift X such that the resulting matrix Y would have only different nodes; for Y we have Theorem 2.1 from where somehow we can obtain the statement for the matrix X. But to work out this plan seems to be very tedious. One of the possible difficulties can be demonstrated by the next statement of N.B. Tihomirov [27].

For any interpolatory matrix $X = \{x_{kn}\} \subset [-1,1]$ and any sequence $\{\varepsilon_n\}$ $\varepsilon_n > 0$, one can find another interpolatory matrix $Y = \{y_{kn}\} \subset [-1,1]$, a function $f_7 \in C$ and a subsequence $\{n_i\}$ such that

(1) $$\max_{1\leq k\leq n} |x_{kn}-y_{kn}| < \varepsilon_n ,$$

(2) $$\lim_{i\to\infty} \| L_{n_i}(f_7,X,x) - f_7(x)\| = 0 ,$$

(3) $$\overline{\lim_{i\to\infty}}\ L_{n_i}(f_7,Y,x) = \infty \text{ a.e. in } [-1,1] .$$

That means the same function can be "good" or "bad" for very close matrices.

2.3. Finally, in our paper with P. Erdös we chose a different way and succeeded to prove the Erdös' conjecture. Namely, in 1980 we got

THEOREM 2.2 (P. Erdös, P. Vértesi, [10], [11]). For any interpolatory matrix $X \subset [-1,1]$ one can find a function $F \in C$ such that

(2.3) $$\overline{\lim_{n\to\infty}}\ |L_n(F,X,x)| = \infty \text{ a.e. in } [-1,1] .$$

REMARK 1. It is easy to see that the divergence set can be dense and of second category (see [29]).

REMARK 2. Using a theorem of A. del Junco and J. Rosenblatt [23] p. 186, we can see that there exists a dense G_δ subset R from B (= the Banach-space of the functions $f \in C$) such that for arbitrary $F \in R$ (2.3) holds true.

REMARK 3. In (2.3) "$\overline{\lim}$" cannot be replaced by "lim" or "$\underline{\lim}$". Indeed, as P. Erdös [6], p. 384 showed one can construct a point group such that for every $f \in C$ and every $x_0 \in [-1,1]$ there exists a sequence n_k (depending on f and x_0) with $\lim_{k\to\infty} L_{n_k}(f,x_0) = f(x_0)$.

B. Complex and trigonometric cases

2.4. In this part we investigate the previous problem on a special complex domain, on the unit circle $\Gamma = \{z;\ z = \exp(i\theta),\ 0 \leq \theta < 2\pi\}$.

Let the interpolatory matrix Z be defined on Γ, that means $Z = \{z_{kn} = \exp(i\theta_{kn})\}$, $k = 1,2,\ldots,n$; $n = 1,2,\ldots$, with

$$(2.4) \qquad 0 \leq \theta_{1n} < \theta_{2n} < \ldots < \theta_{nn} < 2\pi\ , n = 1,2,\ldots,$$

$L_n(f,Z,z)$, $\ell_{kn}(Z,z)$, $\omega_n(Z,z)$, $\lambda_n(Z,z)$ and $\lambda_n(Z)$ (or their short forms) are the notations corresponding to 1.1. Now $|z| \leq 1$ and the norm $\|\cdot\|$ is taken on Γ , $f \in AC$ (= f is analytic on $U = \{z;\ |z| < 1\}$ and continuous on Γ) .

The "complex Faber's result" is due to S.Y. Alper [1]:

Proving that for any matrix Z, $\lambda_n(Z) > \ln n/(8\sqrt{\pi})$ (the corresponding result for $\lambda_n(x)$ was verified by Faber [4] and Bernstein [2]), he obtained that for any Z there exists an $f_1 \in AC$ for which

$$\overline{\lim_{n\to\infty}}\ \|L_n(f_1,Z,z)\| = \infty\ .$$

In a very recent paper A.H. German [12] essentially proved as follows.

THEOREM 2.3 (A.H. German). If Z has only different nodes then for a suitable $f_2 \in AC$

$$\overline{\lim_{n\to\infty}}\ |L_n(f_2,Z,z)| = \infty \quad \text{a.e. on } \Gamma\ .$$

2.5. Finally in 1982, using a proper representation of $\ell_{kn}(z)$ and extensions for complex domain results previously proved only on [-1,1], I was able to obtain the next

THEOREM 2.4 (P. Vértesi [29]). For arbitrary interpolatory $Z \subset \Gamma$ there exists an $F \in AC$ such that

$$\overline{\lim_{n\to\infty}}\ |L_n(F,Z,z)| = \infty \quad \text{for almost all } z \in \Gamma\ .$$

2.6. The trigonometric case can be easily settled using the complex one. Namely, if $f \in \tilde{C}$ (= f is 2π - periodic and continuous), n is odd, $L_n(f,\Theta,\theta)$ are the trigonometric interpolatory polynomials of degree (n-1)/2 based on the nodes (2.4) then we can prove

THEOREM 2.5 (P. Vêrtesi [29]). For any matrix $\{\theta_{kn}\}$, $k = 1,2,\ldots,n$, $n = 1,3,5,\ldots$, one can find a function $f(\theta) \in \tilde{C}$ for which

$$\overline{\lim_{n\to\infty}} |L_n(f,\Theta,\theta)| = \infty \text{ a.e. on the real line.}$$

C. Infinite interval

2.7. Contrary to the case of the finite interval we know very little on the infinite ones,i.e.,when instead of (1.1) we have $X \subset (-\infty,\infty)$ or

$$(2.5) \qquad -\infty < x_{nn} < x_{n-1,n} < \ldots < x_{1n} < \infty .$$

Even the analogy of Faber's result has not been proved yet.

Generally, the corresponding theorem deal with

(1) Special nodes (Laguerre or Hermite ones see. e.g. L.P. Povchun [19]).

(2) Nodes where the maximal distance of the consecutive ones tends to zero with $n \to \infty$, see e.g. [16].

(3) A finite interval $[a,b] \subset (-\infty,\infty)$(see e.g. [20]).

As an illustration we quote the next statement.

Let $X^{(\alpha)} = x_{kn}^{(\alpha)}$, $1 \leq k \leq n$, $n = 1,2,\ldots$, where $x_{kn}^{(\alpha)}$ is the k-th root of the n-th Laguerre polynomials $L_n^{(\alpha)}(x)$, $\alpha > -1$ (see [25], Ch.V.).

THEOREM 2.6 (L.P. Povchun [19]). Let $X = X^{(\alpha)}$, $\alpha > -1$. Then there exists a function f(x) being continuous on $[0,\infty)$ such that

$$\overline{\lim_{n\to\infty}} |L_n(f,X^{(\alpha)},x)| = \infty \quad \text{a.e. on } [0,\infty) .$$

2.8. Dealing with infinite intervals, the main problems may be formulated in the following way.

The counterexample functions F(x) generally have the form $\sum_{k=1}^{\infty} c_k \varphi_k(x)$, where $\sum |c_k| < \infty$ and $\varphi_k(x)$ are uniformly bounded polynomials, say, in [-1,1].

On the other hand, if $x \to \infty$, even the continuity of F(x) is questionable. Of course, one can "cut" somehow $\varphi_k(x)$, but then we could not use the very useful relation $L_N(\varphi_k, x) \equiv \varphi_k(x)$ if $N > \deg \varphi_k$.

2.9. In our recent paper with J. Szabados we were able to overcome these problems and generalize Theorem 2.2 to infinite intervals.

THEOREM 2.7 (J. Szabados, P. Vértesi [24]). For any interpolatory matrix $X \subset (-\infty,\infty)$ one can find a function F(x) being continuous on $(-\infty,\infty)$ such that

$$\overline{\lim_{n\to\infty}} |L_n(F,X,x)| = \infty \text{ for almost all } x \in (-\infty,\infty).$$

Let us remark that the proof is based on the Erdös - Vértesi theorem.

3. Some Open Problems

There are many problems on the divergence of Lagrange interpolation. Some are easy but most of them seem to be difficult.

Prove (or disprove)

1. There is an $X \subset [-1,1]$ so that for every $f \in C$ $L_n(f,x_0) \to f(x_0)$ holds for at least one x_0 for which $\overline{\lim}_{n\to\infty} \lambda_n(x_0) = \infty$ (P. Erdös [6] and P. Erdös, P. Vértesi [11]).

2. If $\overline{\lim}_{n\to\infty} \lambda_n(X,x) = \infty$ for all $x \in [-1,1]$, there exists an $f \in C$ such that

$$\overline{\lim_{n\to\infty}} |L_n(f,X,x)| = \infty \quad \text{whenever } x \in [-1,1]$$

(P. Erdös [6], p. 384; this is another form of Problem 1).

3. There exists an $f \in C$ such that

$$\overline{\lim_{n\to\infty}} \left\{ \frac{1}{n} \sum_{k=1}^{n} L_n(f,T,x) \right\} = \infty \text{ if } x \in [-1,1]$$

(P. Erdös [5], p. 12; the proof of P. Erdös and G. Grünwald [9] unfortunately is incorrect (last formula of p. 92). It proves the weaker result where the summands are replaced by their moduli. In [5] P. Erdös proved that $\{\ldots\} = o(\ln \ln n)$.)

4. Formulate Problems 1 - 3 for the complex case and for $X \subset (-\infty,\infty)$.
5. Prove Theorem 2.4 for a general complex domain.
6. What can one say on the behaviour of $L_n(f,Z,z)$ if $f \in AC$ and $|z| < 1$? (V. Totik and E. Saff).
7. Find the smallest possible function class for which Theorem 2.7 still holds true.

\- * -

Further problems can be found e.g. in P. Turán [28], P. Erdös [5],[6], [7],[8].

REFERENCES

[1] Alper, S.Y., On the convergence of Lagrange interpolation on complex domains. Usp. Mat. Nauk 11(5) (1956), 44 - 50 (Russian).

[2] Bernstein, S.N., Quelques remarques sur l'interpolation. Math. Ann. 79 (1918), 1 - 12.

[3] Bernstein, S.N., Sur la limitation des valeurs d'un polynome. Bull. Acad. Sci. de l'URSS 8(1931), 1025 - 1050.

[4] Faber, G., Über die interpolatorische Darstellung stetiger Funktionen. Jahresber. der Deutschen Math. Ver. 23 (1914), 190 - 210.

[5] Erdös, P., Some theorems and remarks on interpolation. Acta Sci. Math. (Szeged) 12(1950), 11 - 17.

[6] Erdös, P., Problems and results on the theory of interpolation. Acta Math. Acad. Sci. Hungar. 9 (1958), 381 - 388.

[7] Erdös, P., Problems and results on the convergence and divergence properties of the Lagrange interpolation polynomials and some extremal problems. Mathematica (Cluj), 10(33)(1)(1968), 65 - 73.

[8] Erdös, P., Problems and Results on Polynomials and Interpolation. Aspects of Contemporary Complex Analysis, Academic Press, 1980. p. 383 - 393.

[9] Erdös, P. - Grünwald, G., Über die arithmetischen Mittelwerte der Lagrangeschen Interpolationspolynome. Studia Math. 7(1938), 82 - 95.

[10] Erdös, P. - Vértesi, P., On the almost everywhere divergence of Lagrange interpolation. Proc. of the Conference on Approximation and Function Spaces held in Gdansk, August 1979, 270 - 278.

[11] Erdös, P. - Vértesi, P., On the almost everywhere divergence of Lagrange interpolatory polynomials for arbitrary system of nodes. Acta Math. Acad. Sci. Hungar. 36 (1980), 71 - 89 and 38 (1981), 263.

[12] German, A.H., Interpolation on complex domain. Anal. Math. 6 (1980), 121 - 135 (Russian).

[13] Grünwald, G., Über die Divergenzerscheinungen der Lagrangeschen Interpolationspolynome. Acta Sci. Math. (Szeged) 7 (1935), 207 - 221.

[14] Grünwald, G., Über die Divergenzerscheinungen der Lagrangeschen Interpolationspolynome stetiger Funktionen. Annals of Math. 37 (1936), 908 - 918.

[15] Korovkin, P.P., On the closure of Chebyshev systems. Dokl. Akad. Nauk SSSR. 78 (1951), 853 - 855 (Russian).

[16] Kutepov, V.A., Lagrange interpolation on unbounded sets. International Conference on the Theory of Approximation of Functions, USSR, Kiev, May 30 - June 6, 1983. Abstracts, p. 109 (Russian).

[17] Marcinkiewicz, J., Sur la divergence des polynomes d'interpolation. Acta Sci. Math. (Szeged), 8 (1937), 131 - 135.

[18] Mergeljan, S.N., Certain questions of the constructive theory of functions. Trudy Mat. Inst. Steklov. 37 (1951) (Russian).

[19] Povchun, L.P., On the almost everywhere divergence of Lagrange interpolation on the Laguerre nodes. Izv. Vyss. Ucebn. Zaved. Matematika 7 (1976), 114 - 116 (Russian).

[20] Povchun, L.P., Divergence of interpolation processes at a fixed point. Izv. Vyss. Ucebn. Zaved. Matematika. 3 (1980), 56 - 60 (Russian).

[21] Privalov, A.A., Divergence of Lagrange interpolation based on the Jacobi abscissas on sets of positive measure. Sibirsk. Mat. Z. 18 (1976), 837 - 859 (Russian).

[22] Privalov, A.A., Approximation of functions by interpolation polynomials. In "Fourier Analysis and Approximation Theory" I - II North-Holland Publ. Co., (Amsterdam-Oxford-New York), 1978, p. 659 - 671.

[23] del Junco, A. - Rosenblatt, J., Counterexamples in Ergodic Theory and Number Theory. Math. Ann. 245 (1979), 185 - 197.

[24] Szabados, J. - Vértesi, P., On the almost everywhere divergence of Lagrange interpolation on infinite interval. Acta Sci. Math. (Szeged) (to appear).

[25] Szegö, G., Orthogonal Polynomials. AMS Coll. Publ. 23, 1974.

[26] Tihomirov, N.B., On the divergence of interpolatory processes. Coll. papers on diff. geom. Kalinin, 1977, p. 104 - 111 (Russian).

[27] Tihomirov, N.B., On the divergence of interpolatory processes on sets of total measure. Appl. of funct. anal. to approx. theory. Kalinin Gos. Univ., Kalinin. 1980 p. 138 - 150 (Russian).

[28] Turán, P., On Some Open Problems of Approximation Theory. J. Approximation Theory. 29 (1980), 23 - 85.

[29] Vértesi, P., On the almost everyhwere divergence of Lagrange interpolation. Acta Math. Acad. Sci. Hungar. 39 (1982), 367 - 377.

International Series of
Numerical Mathematics, Vol. 65

UNIFORM CONVERGENCE OF SOME POISED PROBLEMS OF HERMITE - BIRKHOFF INTERPOLATION

R.B. Saxena and H.C. Tripathi
Department of Mathematics and Astronomy
Lucknow University
Lucknow, India

In this paper we study a special Hermite-Birkhoff interpolation problem which is similar to (0,2,3) interpolation. It consists of finding a polynomial $p_n(X,x)$ of degree at most 2n+1 such that $p_n^{(j)}(X,x) = f_i{}^j$ $(i=1,\ldots,n;j=0,2,3)$ for an arbitrary given set of real nodes $X: -1 \leq x_n \leq \ldots \leq x_1 \leq 1$ and arbitrary real numbers $f_i{}^j$. We call it quasi-(0,2,3) interpolation. We prove a theorem which gives a sufficient condition under which the quasi-(0,2,3) interpolation for every $f \in C^2[-1,1]$ converges uniformly on [-1,1]. The error estimates for Chebyshev nodes are derived. Further we generalise the quasi-(0,2) interpolation to quasi-(0,2,...,2r-2,2r), $r \geq 1$ interpolation and carry out the same scheme of investigations as in the case of quasi-(0,2,3) interpolation.

1. Introduction

Let n and k be natural numbers, and let $E = (e_{ij})$, $(i = 1,2,\ldots,k; j = 0,1,\ldots,n-1)$, be an incidence matrix of k rows and n columns, where $e_{ij} = 0$ or 1 and $\sum_{i,j} e_{ij} = n$. Let X denote the ordered k-tuple of reals $-1 \leq x_k < x_{k-1} < \ldots < x_1 \leq 1$, and let $e = \{(i,j) : e_{ij} = 1\}$. Then the pair (E,X) describes the Hermite-Birkhoff (H-B) interpolation problem of finding a polynomial p(x) of degree $\leq n-1$ such that

$$(1.1) \qquad p^{(j)}(x_i) = f_i^j \;, \quad (i,j) \in e \;,$$

where f_i^j are prescribed reals. If, for an arbitrary X and prescribed E, (1.1) has a unique solution, then the problem is said to be poised. On the other hand, if for a given E, (1.1) has a unique solution only for a special choice of X, then it is said to be conditionally poised.

J. Surânyi and P. Turán [6] were the first to study a special problem what they call (0,2)-interpolation which corresponds to the matrix E of order $n \times 2n$ with prescribed $e = \{(i,j) : i = 1, \ldots, n; \ j = 0,2\}$. This study was followed by R.B. Saxena and A. Sharma [4] for the matrix E of order $n \times 3n$ with $e = \{(i,j) : i = 1, \ldots, n; j = 0,1,3\}$. They called it (0,1,3) interpolation. These problems are not poised but only conditionally poised (cf. Sharma [5]), and have been studied for the special set of nodes X consisting of the zeros of the polynomial $\Pi_n(x) = (1-x^2)P'_{n-1}(x)$, where $P_{n-1}(x)$ stands for the (n-1)th Legendre polynomial with $P_{n-1}(1) = 1$. Similar studies for different pairs (E,X) have been made by various authors. There is now an abundance of available literature on H-B interpolation.Almost all problems studied so far are non-poised.

Very recently Suzuki [7] has investigated a poised problem what he calls quasi-(0,2) interpolation. This problem is similar to (0,2) interpolation and corresponds to the incidence matrix E of order $n \times (n+2)$ with $e = \{(1,0), (n,0)$ and $(i,2) : i = 1, \ldots, n)\}$. He has obtained sufficient conditions for the uniform convergence of quasi-(0,2) polynomials to a function $f \in C^2[-1,1]$.

The problem of (0,2,3) interpolation corresponds to the pair (E,X), where E is a matrix of order $n \times 3n$ with $e = \{(i,j) : i = 1, \ldots, n; \ j = 0,2,3\}$. It is known (cf. Sharma [5]) that the (0,2,3) problem is poised. But it is a fact that till now the explicit form of the fundamental polynomials for (0,2,3) problem are not known for real nodes. Even for the Π_n-nodes, the explicit form of these polynomials is not known. An attempt in this direction was made by Saxena [3] as early as in 1959 when a solution was found to the problem, (including uniform convergence) corresponding to the pair (E,X), where $X = \Pi_n(x)$ and E is of order $n \times (2n+2)$ with $e = \{(1,3), (n,3)$ and $(i,j) : i = 1, \ldots, n; j = 0,2\}$. This problem is similar to the (0,2,3) problem but is non-poised.

Following Suzuki [7] we consider now the problem corresponding to the incidence matrix E of order $n \times (2n+2)$ with $e = \{(1,0), (n,0)$ and $(i,j) : i = 1, \ldots \ldots, n; j = 2,3\}$. This problem is also similar to the (0,2,3) problem. The essential difference between the two problems is that the first one is non-poised whereas our problem is poised. In fact, our E can be decomposed into horizontal sums of two incidence matrices of Lagrange and Hermite interpolations which are known to be poised. We call it quasi-(0,2,3) interpolation.

In section 2, we shall find the explicit polynomials of quasi-(0,2,3) interpolation corresponding to an arbitrary set X and in section 3 we investigate the convergence behaviour of quasi-(0,2,3) polynomials for functions $f \in C^2[-1,1]$. Here we give error estimates for Chebyshev nodes. Further in section 4 and 5 we generalise the quasi-(0,2) interpolation to quasi-$(0,2,\dots,2r-2,2r)$, $r \geq 1$ interpolation and carry out the same scheme of investigations as in the case of quasi-(0,2,3) interpolation.

2. quasi-(0,2,3) interpolation- explicit representation.

As mentioned above quasi-(0,2,3) interpolation is poised and, hence, for an arbitrary set of nodes X and arbitrary real numbers $f_1^0, f_n^0; f_1^2, f_i^3$, $i=1,2,\dots,n$, there exists a unique polynomials $p_n(X,x)$ of degree at most $2n+1$ such that

$$(2.1)\qquad \begin{aligned} p_n(X,x_i) &= f_i^0 , \; i=1,\dots,n , \\ p_n^{(j)}(X,x_i) &= f_i^j , \; i=1,\dots,n ; j=2,3 . \end{aligned}$$

It is easy to see that all solutions of (2.1) form a linear space of dimension $2n+2$, the basis elements of which are polynomials each of degree at most $2n+1$. If we denote them by $u_1(X,x)$, $u_n(X,x)$, $v_i(X,x)$ and $w_i(X,x)$, $i=1,\dots,n$, then evidently we can write

$$(2.2)\qquad p_n(X,x) = f_1^0 \, u_1(X,x) + f_n^0 \, u_n(X,x) + \sum_{i=1}^{n} [f_i^2 v_i(X,x) + f_i^3 w_i(X,x)] .$$

In view of the interpolatory requirements we find that the expressions

$$(2.3)\qquad u_1(X,x) = \frac{x-x_n}{x_1-x_n} , \; u_n(X,x) = \frac{x_1-x}{x_1-x_n} .$$

$$(2.4)\qquad v_i(X,x) = \int_{x_n}^{x_1} G(x,t) \, h_i(X,t) \, dt ,$$

$$(2.5)\qquad w_i(X,x) = \int_{x_n}^{x_1} G(x,t) \, \bar{h}_i(X,t) \, dt , \; i=1,\dots,n ,$$

each of which is, in fact, a polynomial of degree $\leq 2n+1$, are the required basis polynomials. Here $h_i(X,t)$ and $\bar{h}_i(X,t)$ are fundamental polynomials (each of degree at most 2n-1) of Hermite-Fêjer (H-F) interpolation

$$(2.6)\qquad H_n(f,X,t) = \sum_{i=1}^{n} f(x_i)h_i(X,t) + \sum_{i=1}^{n} f'(x_i)\bar{h}_i(X,t)$$

associated with the function f(x) and

$$(2.7)\qquad G(x,t) = \begin{cases} \dfrac{(t-x_n)(x-x_1)}{x_1-x_n}, & t \leqslant x \\ \dfrac{(x-x_n)(t-x_1)}{x_1-x_n}, & x \leqslant t, \end{cases}$$

is the well-known Green's function. This solves the problem of explicit representation of quasi-(0,2,3) interpolation.

3. quasi-(0,2,3) interpolation- Uniform Convergence.

Let f(x) be a function given on $[x_n,x_1]$ such that $f_1^0 = f(x_1), f_n^0 = f(x_n)$; $f_i^2 = f''(x_i)$ and $f_i^3 = f'''(x_i)$, $i = 1, \ldots, n$. Then, on making this substitution in (2.2), the explicit expression for the quasi-(0,2,3) polynomial for the function f(x) associated with the nodes X is given by

$$(3.1)\qquad p_n(f,X,x) = f(x_1)\frac{x-x_n}{x_1-x_n} + f(x_n)\frac{x_1-x}{x_1-x_n} + \\ + \sum_{i=1}^{n} f''(x_i)\int_{x_n}^{x_1} G(x,t)h_i(X,t)\,dt + \sum_{i=1}^{n} f'''(x_i)\int_{x_n}^{x_1} G(X,t)\bar{h}_i(X,t)dt.$$

We shall now prove the following

THEOREM 1. _Let_ $f(x) \in C^2[-1,1]$ _and_ $X : -1 \leqslant x_n < \ldots < x_1 \leqslant 1$ _be the set of nodes. Then_

$$\|p_n(f,X,x) - f(x)\| \to 0$$

whenever

$$\|H_n(f'',X,x) - f''(x)\| \to 0 \quad \text{as} \quad n \to \infty .$$

Here $\|\cdot\| = \max_{-1\leq x\leq 1} |\cdot|$.

PROOF. Let

$$R_n(f,X,x) = p_n(f,X,x) - f(x) .$$

Then

$$(3.2) \qquad R_n(f,X,x) = \int_{x_n}^{x_1} G(x,t)[p_n''(f,X,t) - f''(t)]\,dt .$$

On differentiating (3.1) twice and using (2.6) we have

$$p_n''(f,X,x) = \sum_{i=1}^{n} f''(x_i)h_i(X,x) + \sum_{i=1}^{n} f'''(x_i)\overline{h}_i(X,x) = H_n(f'',X,x) .$$

So that (3.2) can be written as

$$R_n(f,X,x) = \int_{x_n}^{x_1} G(x,t)[H_n(f'',X,t) - f''(t)]\,dt .$$

Since

$$\max_{-1\leq x\leq 1} \int_{x_n}^{x_1} |G(x,t)|\,dt \leq \frac{1}{2} ,$$

we have

$$|R_n(f,X,x)| \leq \frac{1}{2} \max_{-1\leq x\leq 1} |H_n(f'',X,x) - f''(x)|$$

from which the assertion of the theorem follows.

REMARK. A consequence of the above theorem is that, for functions $f(x) \in C^2[-1,1]$, the convergence behaviour of $p_n(f,X,x)$ is the same as that of $H_n(f'',X,x)$. Thus the whole theory of convergence of H-F interpolation is directly applicable to the study of the convergence of quasi-(0,2,3) interpolation.

Error estimates for Chebyshev nodes.

Let $X = T$ denote the Chebyshev nodes of first kind, i.e., the roots of $T_n(x)$, and $X = U$ the Chebyshev nodes of second kind, i.e., the roots of

$(1-x^2)U_{n-2}(x)$. If only the continuity of $f''(x)$ is known, we may take $f'''(x_i)=0$; then for any $f\in C^2[-1,1]$ the following error estimates hold:

$$|p_n(f,T,x)-f(x)| \leq \frac{C_1}{n} T_n^2(x) \sum_{k=1}^{n} [\omega(\frac{\sqrt{1-x^2}}{k}) + \omega(\frac{1}{k^2})] + C_2\, \omega(\frac{|T_n(x)|}{n^2})$$

(cf. [2]) and

$$|p_n(f,U,x)-f(x)| \leq \frac{C_3}{n} \sum_{k=1} [\omega(\frac{\sqrt{1-x^2}}{k}) + \omega(\frac{1}{k^2})] + \frac{C_4}{n^2}$$

(cf. [1])for $-1\leq x\leq 1$. Here $\omega(\cdot)$ denotes the modulus of continuity of the second derivative of f, and C_1, C_2, C_3 and C_4 are numerical constants.

4. quasi-(0,2,...,2r-2,2r), $r\geq 1$ interpolation - explicit representation.

Let

$$X: -1 \leq x_n < \dots < x_1 \leq 1$$

be an arbitrary set of nodes and

$$f_i^{2j},\ i=1,n\,;\ j=0,1,\dots,r-1\,;\quad f_i^{2r},\ 1=1,\dots,n$$

a set of 2r+n given reals . The problem of quasi-(0,2,...,2r-2,2r) interpolation is to seek a polynomial $q_n(f,X,x)$ of degree at most 2r+n-1 such that

$$q_n^{(2j)}(f,X,x_i) = f_i^{2j},\ i=1,n\,;\ j=0,1,\dots,r-1$$

(4.1)

$$q_n^{(2r)}(f,X,x_i) = f_i^{(2r)},\ i=1,\dots,n\,.$$

We can easily see that this problem is a poised one. So any solution of (4.1) is the polynomial we are seeking. A solution of (4.1) can be written as

$$(4.2)\qquad q_n(f,X,x) = \sum_{j=0}^{r-1} [f_i^{2j}\Lambda_j(x) + f_n^{2j}\Lambda_j(x_1+x_n-x)] + \sum_{i=1}^{n} f_i^{2r}\lambda_i(X,x)\,,$$

where the basis functions $\Lambda_j(x)$ and $\lambda_i(X,x)$ are polynomials, each of degree at most 2r+n-1, determined by the following requirements:

$$\Lambda_0(x_1) = 1\,, \quad \Lambda_0(x_n) = 0$$

(4.3) $$\Lambda_j^{(2\nu)}(x_1) = \Lambda_j^{(2\nu)}(x_n) = 0\,, \quad \nu = 0,1,\ldots, j-1\,;$$

$$\Lambda_j^{(2j)}(x_1) = 1\,, \Lambda_j^{(2j)}(x_n) = 0\,.\; j = 1,2,\ldots,r-1\,; \Lambda_j^{(2r)}(x_i) = 0, i = 1,\ldots,n$$

and

(4.4) $$\lambda_i^{(2j)}(X,x_1) = \Lambda_i^{(2j)}(X,x_n) = 0$$

$$\lambda_i^{(2r)}(X,x_k) = \delta_{ik}\,, \; j = 0,1,\ldots,r-1;\; i,k = 1,\ldots,n\,.$$

From (4.3), we find that $\Lambda_j(x)$ is a polynomial of degree 2j+1 and satisfies the following differential relation

$$\Lambda_0(X,x) = \frac{x-x_n}{x_1-x_n}$$

(4.5) $$\Lambda_j''(X,x) = \Lambda_{j-1}(X,x)$$

$$\Lambda_j(X,x_1) = \Lambda_j(X,x_n) = 0\,, j = 1,2,\ldots, r-1\,.$$

The polynomial $\Lambda_j(x)$ is the jth Lidstone polynomial (Widder [8]) over the interval $[x_n,x_1]$. The explicit expression of $\Lambda_j(x)$ can be obtained in terms of the Green's function (2.7) as a solution of the boundary value problem (4.5). Thus we have

$$\Lambda_j(x) = \int_{x_n}^{x_1} G(x,t)\,\Lambda_{j-1}(t)\,dt\,,$$

and, by an iteration process [8],

(4.6) $$\Lambda_j(x) = \int_{x_n}^{x_1} G_j(x,t)\left(\frac{t-x_n}{x_1-x_n}\right) dt \quad ,$$

where

$$G_1(x,t) = G(x,t)\ ,$$

(4.7)

$$G_j(x,t) = \int_{x_n}^{x_1} G(x,y)G_{j-1}(y,t)\,dy\ ,\ j=2,3,\ldots,r\ .$$

Similarly, the polynomial $\lambda_i(X,x)$ can be obtained as a solution of the differential system

$$\lambda_i^{(2r)}(X,x) = l_i(X,x)$$

(4.8)

$$\lambda_i^{(2j)}(X,x_1) = \lambda_i^{(2j)}(X,x_n) = 0,\ j=0,1,\ldots,r-1\ ;\ i=1,2,\ldots,n\ ,$$

where $l_i(X,x)$ is a fundamental polynomial of Lagrange interpolation. The unique solution of (4.8) is

(4.9) $$\lambda_i(X,x) = \int_{x_n}^{x_1} G_r(x,t)\ l_i(X,t)\,dt\ ,\ i=1,\ldots,n\ .$$

Thus (4.2), (4.6) and (4.9) completely determine the polynomial $q_n(f,X,x)$.

5. quasi-(0,2,...,2r-2,2r) interpolation - uniform convergence.

Let $f(x)$ be a function given on $[x_n,x_1]$ such that

$$f_i^{2j} = f^{(2j)}(x_i)\ ,\ i=1,n\ ;\ j=0,1,\ldots,r-1\ .$$

$$f_i^{2r} = f^{(2r)}(x_i)\ ,\ r\geqslant 1,\ i=1,\ldots,n\ .$$

Then, on making these substitutions in (4.2) we have

(5.1)

$$q_n(f,X,x) = \sum_{j=o}^{r-1}[f^{(2j)}(x_1)\int_{x_n}^{x_1} G_j(x,t)\,\frac{t-x_n}{x_1-x_n}\,dt + f^{(2j)}(x_n)\int_{x_n}^{x_1} G_j(x,t)\,\frac{x_1-t}{x_1-x_n}\,dt] + \sum_{i=1}^{n} f^{(2r)}(x_i)\int_{x_n}^{x_1} G_r(x,t)l_i(X,t)\,dt\ .$$

THEOREM 2. Let $f(x) \in C^{2r}[-1,1]$, $r \geq 1$ and $X: -1 = x_n < x_{n-1} < \dots < x_1 = 1$ be the set of nodes. Then the polynomial $q_n(f,X,x)$ in (5.1) satisfies the inequality

$$(5.2) \qquad |f(x) - q_n(f,X,x)| \leq C \max_{-1\leq x\leq 1} |L_n(f^{(2r)}, X,x) - f^{(2r)}(x)|, \quad -1 \leq x \leq 1 ,$$

where C is a constant.

PROOF. Let $R_n(f,X,x) = q_n(f,X,x) - f(x)$. Owing to the interpolation properties of $q_n(f,X,x)$, $R_n(f,X,x)$ can be taken as a particular solution of the differential system

$$R_n^{(2r)}(f,X,x) = q_n^{(2r)}(f,X,x) - f^{(2r)}(x)$$

$$R_n^{(2j)}(f,X,x) = R_n^{(2j)}(f,X,-1) = 0, \quad j = 0,1, \dots, r-1 ,$$

which can be written as

$$R_n(f,X,x) = \int_{-1}^{1} G_r(x,t)[q_n^{(2r)}(f,X,t) - f^{(2r)}(t)]\, dt .$$

On differentiating (5.1) 2r times and using (4.8) and (4.9), we have

$$R_n(f,X,x) = \int_{-1}^{1} G_r(x,t)[L_n(f^{(2r)},X,t) - f^{(2r)}(t)]\, dt ,$$

where

$$L_n(f^{(2r)}, X,t) = \sum_{i=1}^{n} f^{(2r)}(x_i) l_i(X,t)$$

is the Lagrange interpolation polynomial. Thus,

$$(5.3) \qquad |R_n(f,X,x)| \leq C \max_{-1\leq x\leq 1} |L_n(f^{(2r)},X,x) - f^{(2r)}(x)|$$

because

$$\max_{-1\leq x\leq 1} \left| \int_{-1}^{1} G_r(x,t)\, dt \right| < C \text{ - a constant} .$$

Concluding remarks.

From this theorem we find that quasi-(0,2,...,2r-2,2r), $r \geq 1$, polynomials for functions $f \in C^{2r}[-1,1]$ have a convergence behaviour similar to that of Lagrange interpolation. Because of the irregular behaviour of Lagrange interpolstion, the quasi-(0,2,...,2r-2,2r) interpolation is not a "good" convergence sequence for functions whose 2r th derivative $(r \geq 1)$ is continuous on [-1,1]. To obtain a convergent sequence, we should generalise quasi-(0,2,3) interpolation to quasi-(0,2,...,2r-2,2r,2r+1) interpolation in the same manner as quasi-(0,2) was generalised to (0,2,...,2r-2,2r) interpolation. To avoid repitition, we do not give the details here but simply point out that the resulting process will hehave like Hermite interpolation.

REFERENCES

[1] Bojanic, R. - Prasad, J. - Saxena, R.B., An upper bound for the rate of Hermite-Fējer process on the extended Chebyshev nodes of second kind, J. Approximation Theory 26(1979), 195 - 203.

[2] Goodenough, S.J. - Mills, T.M., A new estimate for the approximation of functions by Hermite-Fējer interpolation polynomials, Ibid. 31(1981), 253 - 260.

[3] Saxena, R.B., On modified (0,2) interpolation, Acta Math. Acad. Sci. Hung. 10(1959), 177 - 192.

[4] Saxena, R.B. - Sharma, A., On some interpolatory properties of Legendre polynomials, Ibid. 9(1958), 345 - 358.

[5] Sharma, A., Some poised and non-poised problems of interpolation, SIAM Review 14, 1(1972), 129 - 155.

[6] Surānyi, J. - Turān, P., Notes on interpolation I, Acta Math. Acad. Sci. Hung. 6(1955), 67 - 69.

[7] Suzuki, C., Some poised Lacunary interpolation polynomials, J. Information Proc. 5, 1(1982), 38 - 44.

[8] Widder, D.V., Completely convex functions and Lidstone series, Trans. Amer. Math. Soc. 51(1942), 387 - 398.

International Series of
Numerical Mathematics, Vol. 65

SPLINE INTERPOLATION OF POWER-DOMINATED DATA

A.Jakimovski and D.C.Russell and M.Stieglitz

School of Math.Sciences, Tel-Aviv University, Tel-Aviv, Israel

Department of Mathematics, York University, Toronto M3J 1P3, Canada

Mathematisches Institut I, Universität Karlsruhe, 7500 Karlsruhe 1, Germany

Let (x_k) be a bi-infinite knot sequence for which the mesh ratio is smaller than some exponential order (in particular, the local mesh ratio must be finite, but the global mesh ratio may be infinite). Let ρ be a non-negative real number, and (y_k) a data sequence for which $y_k = O(|x_k|^\rho)$ as $k \to \pm\infty$. We prove the existence and uniqueness of a spline function of any previously specified odd degree, with knots (x_k), which interpolates (y_k) (that is, $S(x_k) = y_k$ for all k) and which is dominated by exactly the same power, namely $S(t) = O(|t|^\rho)$ as $t \to \pm\infty$. We may replace O by o throughout. We also obtain a series representation for $S(\cdot)$ in terms of (y_k) and of a sequence of "fundamental splines" $L_k(\cdot)$ which decay exponentially near $\pm\infty$.

0. Introduction

Throughout this paper, $x := (x_i)_{i\in\mathbb{Z}}$ will be a fixed strictly increasing real sequence. For $m\in\mathbb{Z}_{++} := \{1,2,\ldots\}$, π_{m-1} denotes the set of all polynomials (in a real variable, with complex coefficients) of degree not exceeding m-1, and $\$_{m,x}$ is the space of spline functions of degree m-1 (or order m) with simple knots x, defined by

$$(0.1)\qquad \$_{m,x} := \{ S(\cdot) \mid S\in C^{m-2}(\mathbb{R}) \text{ and } S|_{(x_i,x_{i+1}]} \in \pi_{m-1}\ (\forall\, i\in\mathbb{Z}) \},$$

the continuity condition $S\in C^{m-2}(\mathbb{R})$ being omitted if $m = 1$.

We now define the class of "power-dominated" functions $f:\mathbb{R}\to\mathbb{C}$ and sequences $y := (y_i)_{i\in\mathbb{Z}} \in \omega$ (ω denotes the space of all bi-infinite complex-valued sequences). Let $\rho\in\mathbb{R}_+ := \{t \mid t \geq 0\}$ and write

$$(0.2)\qquad F_\rho := \{ f(\cdot) \mid f(t) = O(|t|^\rho) \text{ as } |t|\to\infty \},$$

$$(0.3)\qquad Y_{\rho,x} := \{ y \mid y_i = O(|x_i|^\rho) \text{ as } |i|\to\infty \},$$

where the constants implied by the O may depend on ρ.

Given a power-dominated *data-sequence* $y \in Y_{\rho,x}$, it is our purpose here to investigate the existence, uniqueness, and representation of a power-dominated spline function which will *interpolate* y *at the knots* x, namely a spline $S(\cdot) \in \$_{m,x} \cap F_\rho$ such that $S(x_i) = y_i$ $(\forall i \in \mathbb{Z})$.

Schoenberg [6] raised and solved this problem for *splines of odd degree* in the *cardinal case* $x_i := i$, for any $\rho \in \mathbb{R}_+$. De Boor [1] solved the problem for splines of odd degree, for $\rho = 0$ (namely for a *bounded* data sequence) and for knots with a *finite global mesh ratio*, namely

$$(0.4) \qquad M_x := \sup_{i,j} \frac{x_{i+1} - x_i}{x_{j+1} - x_j} < +\infty ;$$

this is equivalent to $0 < \underline{c} \leqslant x_{i+1} - x_i \leqslant \overline{c} < \infty$ ($\underline{c}, \overline{c}$ fixed), $\forall i \in \mathbb{Z}$. Our main Theorem 1 generalizes both of these results, by relaxing the condition on the knots ((0.4) is replaced by the much less restrictive condition (1.2) below), as well as allowing $\rho \in \mathbb{R}_+$. Recently Rong-qing Jia [4] has examined in more detail the interpolation of *bounded* data by *cubic* splines, and in this case he is also able to relax condition (0.4); see Remark 1 below.

There is also an analogous theorem in which the classes (0.2) and (0.3) are re-defined, for $\rho \in \mathbb{R}_+$, with o in place of 0, namely

$$(0.5) \qquad F_\rho^0 := \{ f(\cdot) \mid f(t) = o(|t|^\rho) \text{ as } |t| \to \infty \},$$

$$(0.6) \qquad Y_{\rho,x}^0 := \{ y \mid y_i = o(|x_i|^\rho) \text{ as } |i| \to \infty \}.$$

Schoenberg's methods in [6] depend in several places on the translation-invariance of the sequence of integers, a property which fails for more general knot-sequences, and accordingly a different treatment must be sought. We are, however, able to utilize several significant ideas from the paper of de Boor [1]. We remark that Schoenberg [6] also considers *splines of even degree* — then his knots are at $x_i := i + \frac{1}{2}$ but the interpolation remains on the integers, $S(i) = y_i$; this case consequently generalizes to quite a different problem, requiring separate treatment.

1. Statement of the Main Results

The growth and smoothness conditions to be imposed on our knot-sequence will be as follows.

(1.1) $\quad x_i < x_{i+1}\ (\forall i)$ and $\lim_{i\to-\infty} x_i = -\infty, \quad \lim_{i\to+\infty} x_i = +\infty;$

(1.2) $\quad x_i < x_{i+1}\ (\forall i)$ and, given some $\varepsilon > 0$, $\exists\ \kappa_\varepsilon > 0$ such that

$$\forall\ i,j \in \mathbb{Z}, \quad \frac{x_{i+1} - x_i}{x_{j+1} - x_j} \leq \kappa_\varepsilon\ e^{\varepsilon|i-j|}.$$

We shall denote by E_ε the set of sequences x such that

(1.3) $$E_\varepsilon := \{\ x = (x_i)\ |\ x \text{ satisfies (1.1) and (1.2)}\ \}.$$

In all of our theorems we require $x \in E_\varepsilon$, but in some of the lemmas a weaker hypothesis on x suffices.

If $y[x_i,\ldots,x_{i+m}]$ denotes the usual divided difference of the sequence y at the points $x_i,\ldots,x_{i+m}$, we define the sequence space

(1.4) $$\ell^m_{2,x} := \{\ y \in \omega\ |\ \|y\|_{\ell^m_{2,x}} := (m^{-1} \sum_i (x_{i+m} - x_i)\,|y[x_i,\ldots,x_{i+m}]|^2)^{\frac{1}{2}} < \infty\ \}.$$

Also $\mathcal{L}_2 := \mathcal{L}_2(\mathbb{R})$ denotes the Lebesgue function space with the usual norm, and the unit sequences are

(1.5) $$e^{(k)} \in \omega\ (\forall\ k \in \mathbb{Z}), \text{ where } \quad e^{(k)}_k = 1,\ e^{(k)}_j = 0\ (j \neq k).$$

LEMMA A. _Let_ x _satisfy_ (1.1), _and_ $m \in \mathbb{Z}_{++}$. _If_ $y \in \ell^m_{2,x}$ _then there is a unique spline function_ $S_y := S_y(\cdot)$ _with the properties_

$$S_y \in \$_{2m,x}, \quad S^{(m)}_y \in \mathcal{L}_2, \quad S_y(x_i) = y_i\ (\forall\ i \in \mathbb{Z}).$$

Moreover, $\exists\ c_m$ _such that_

$$\|S^{(m)}_y\|_{\mathcal{L}_2} \leq c_m \|y\|_{\ell^m_{2,x}}.$$

PROOF. See Jakimovski and Russell [3, Theorems 8(a) and 9]. #

Now for each k, $e^{(k)} \in \ell^m_{2,x}$ (the series in (1.4), which defines its norm, is finite). Thus, given $m \in \mathbb{Z}_{++}$ and x satisfying (1.1), we may take $y = e^{(k)}$ in Lemma A and obtain, for each $k \in \mathbb{Z}$, the existence of a unique spline $L_k(\cdot) := L_{k,2m}(\cdot) := L_{k,2m}(\cdot\,;x)$ such that

(1.6) $$L_k \in \$_{2m,x}, \quad L^{(m)}_k \in \mathcal{L}_2, \quad L_k(x_i) = e^{(k)}_i\ (\forall\ i \in \mathbb{Z}),$$

(1.7) $$\|L^{(m)}_k\|_{\mathcal{L}_2} \leq c_m \|e^{(k)}\|_{\ell^m_{2,x}}.$$

Following the nomenclature of Schoenberg [6,p.408] and de Boor [1,Lemma 1], we call the L_k fundamental splines. Our main theorem is now the following (but see particularly Remark 1 at the end of this paper).

THEOREM 1. *Let* $m \in \mathbb{Z}_{++}$, $x \in E_\varepsilon$ *for every* $\varepsilon > 0$, $\rho \in \mathbb{R}_+$, $y \in \omega$. *In order that* $S(\cdot)$ *should exist, with the properties*

$$S \in \$_{2m,x} \cap F_\rho \tag{1.8}$$

$$S(x_i) = y_i \quad (\forall i \in \mathbb{Z}) \tag{1.9}$$

it is necessary and sufficient that

$$y \in Y_{\rho,x} \tag{1.10}$$

and then the spline $S(\cdot)$ *is unique and has the representation*

$$S(t) = \sum_{k\in\mathbb{Z}} y_k L_{k,2m}(t\,;x) \quad (\forall t \in \mathbb{R}), \tag{1.11}$$

where the series converges uniformly on any compact subset of $\mathbb{R}$.

We also have the following supplementary results.

THEOREM 2. *Theorem 1 holds with* F_ρ, $Y_{\rho,x}$ *replaced respectively by* F^0_ρ, $Y^0_{\rho,x}$.

THEOREM 3. *Let* $m \in \mathbb{Z}_{++}$, $x \in E_\varepsilon$ *for every* $\varepsilon > 0$, $\rho \in \mathbb{R}_+$. *If we provide* F_ρ *and* $Y_{\rho,x}$ *with the respective norms*

$$\|f\|_{F_\rho} := \sup_{t\in\mathbb{R}} \{\,|f(t)|\,/\,(1+|t|)^\rho\}, \tag{1.12}$$

$$\|y\|_{Y_{\rho,x}} := \sup_{i\in\mathbb{Z}} \{\,|y_i|\,/\,(1+|x_i|)^\rho\}, \tag{1.13}$$

then $\$_{2m,x} \cap F_\rho$ *and* $Y_{\rho,x}$ *are Banach spaces; and* (1.11) *defines a bijective linear map* $T : Y_{\rho,x} \to \$_{2m,x} \cap F_\rho$ *which is an isomorphism with* $\|T^{-1}\| \le 1$.

THEOREM 4. *Let* $m \in \mathbb{Z}_{++}$, $x \in E_\varepsilon$ *for every* $\varepsilon > 0$, $\rho \in \mathbb{R}_+$. *If we provide* F^0_ρ, $Y^0_{\rho,x}$ *with the same norms as* F_ρ, $Y_{\rho,x}$ *respectively, then* $\$_{2m,x} \cap F^0_\rho$ *and* $Y^0_{\rho,x}$ *are Banach spaces; and* (1.11) *defines an isomorphism* $T_0 : Y^0_{\rho,x} \to \$_{2m,x} \cap F^0_\rho$ *with* $\|T_0^{-1}\| \le 1$. *Also* $\$_{2m,x} \cap F^0_\rho$ *has Schauder basis* $\{L_{k,2m}(\cdot\,;x)\}_{k\in\mathbb{Z}}$, *with representation* $S(\cdot) = \sum_{k\in\mathbb{Z}} S(x_k) L_k(\cdot)$ *in the* F^0_ρ*-norm.*

2. Preliminaries: Null Splines and Fundamental Splines

We first require some elementary consequences of the mesh condition (1.2) and remark that here, and throughout the paper, c and c_ε represent constants (which may depend on x, and later also on m and ρ) which need not be the same at each occurrence.

LEMMA 1. Let x satisfy (1.2) for some given $\varepsilon > 0$. Then

$$1 \leq \mu := \mu_x := \sup_{|i-j|=1} \frac{x_{i+1} - x_i}{x_{j+1} - x_j} < +\infty ; \tag{2.1}$$

$$\exists\, c_\varepsilon > 0 \text{ such that } \frac{1+|x_i|}{1+|x_j|} \leq c_\varepsilon\, e^{\varepsilon|i-j|} \quad (\forall\, i,j). \tag{2.2}$$

PROOF. The deduction of (2.1) (the existence of a finite local mesh ratio μ) is immediate: take $|i-j| = 1$ in (1.2).

In proving (2.2) we may assume, without loss of generality, that $x_{-1} < 0 \leq x_0$ (or $x_{-1} \leq 0 < x_0$). First note that, for $r > k$, $x_r - x_k = (x_r - x_{r-1}) + \ldots + (x_{k+1} - x_k)$, and then (1.2) gives

$$x_r - x_k \leq \kappa_\varepsilon[e^{\varepsilon(r-k-1)} + \ldots + e^\varepsilon + 1](x_{k+1} - x_k) \leq c'_\varepsilon e^{\varepsilon(r-k)} (x_{k+1} - x_k), \tag{2.3}$$

$$x_r - x_k \leq \kappa_\varepsilon[1 + e^\varepsilon + \ldots + e^{\varepsilon(r-k-1)}](x_r - x_{r-1}) \leq c'_\varepsilon e^{\varepsilon(r-k)} (x_r - x_{r-1}). \tag{2.4}$$

If $i \geq 0$, (2.3) gives $|x_i| = x_i \leq x_i - x_{-1} \leq c'_\varepsilon\, e^{\varepsilon(i+1)}(x_0 - x_{-1})$, while if $i < 0$, (2.4) gives $|x_i| = -x_i \leq x_0 - x_i \leq c'_\varepsilon\, e^{\varepsilon(-i)} (x_0 - x_{-1})$, so that in any case we have

$$\exists\, c_\varepsilon \text{ such that } |x_i| \leq c_\varepsilon\, e^{\varepsilon|i|} \quad (\forall\, i). \tag{2.5}$$

It follows from (2.5) that (2.2) holds if $j = 0$, for any i.

Suppose that $1 \leq j \leq i$. Then $x_{j-1} \geq 0$ and, by (2.3),

$$\frac{1+|x_i|}{1+|x_j|} = \frac{x_i + 1}{x_j + 1} \leq \frac{x_i - x_{j-1}}{x_j - x_{j-1}} \leq c'_\varepsilon\, e^{\varepsilon(i-j+1)} \leq c''_\varepsilon\, e^{\varepsilon|i-j|} .$$

If $0 \leq i < j$ we have trivially that $\dfrac{1+|x_i|}{1+|x_j|} \leq 1 \leq e^{\varepsilon|i-j|}$.

If $i < 0 < j$, (2.5) gives $\dfrac{1+|x_i|}{1+|x_j|} \leq 1 + |x_i| \leq c'''_\varepsilon\, e^{\varepsilon(-i)} \leq c'''_\varepsilon\, e^{\varepsilon(j-i)}$.

The required result (2.2) thus holds for $j \geq 0$ and for all i.

If $j \leq -1$ we apply the results already found to $t_i := -x_{-i-1}$. #

(2.6) Examples of Knot-sequences. It may be verified that the following sequences $x = (x_i)_{i\in\mathbb{Z}}$ belong to E_ε (i.e., satisfy (1.3)) for every $\varepsilon > 0$:

(a) $x_i := \log^\sigma(1 + |i|)\cdot\operatorname{sgn} i$, $\sigma > 0$;

(b) $x_i := |i|^\sigma \operatorname{sgn} i$, $\sigma > 0$;

(c) $x_i := e^{|i|^\sigma} \operatorname{sgn} i$, $0 < \sigma < 1$.

Of course, the local mesh ratio μ_x is then finite in each case, by Lemma 1, but in all of these examples with the exception of (b) with $\sigma = 1$ (the cardinal case), the global mesh ratio M_x defined in (0.4), is infinite.

A null-spline S is one which vanishes at every knot: $S(x_i) = 0$ ($\forall i$); it is non-trivial if $S(t) \neq 0$ for at least one t. It is important for uniquness of spline interpolation that there should be no non-trivial null-splines. In the case $\rho = 0$ of Lemma 2 below, de Boor [1,Lemma 4] required only a finite local mesh ratio (property (2.1) above), but for higher powers of ρ we need to use the full force of (1.2).

We shall need to use the normalized B-splines defined as usual, for $p\in\mathbb{Z}_{++}$, $i\in\mathbb{Z}$, $t\in\mathbb{R}$, by the divided differences

$$N_{i,p}(t) := N_{i,p}(t\,;x) := (x_{i+p} - x_i)\,[x_i,\dots,x_{i+p}](\cdot - t)_+^{p-1} ;$$

(for $p = 1$ we define $0_+^0 = 1$). For a survey of their properties see, for example, de Boor [2]. We remark that they are non-negative, of bounded support, and belong to $\$_{p,x}$. Moreover, the map $(a_i) \mapsto \sum_i a_i N_{i,p}(\cdot)$ (the series being finite, since $N_{i,p}(t) = 0$ outside $(x_i, x_{i+p}]$) defines a bijection from ω to $\$_{p,x}$.

LEMMA 2. _If_ $m\in\mathbb{Z}_{++}$, $x\in E_\varepsilon$ _for every_ $\varepsilon > 0$, $\rho\in\mathbb{R}_+$, _then_ $\$_{2m,x}\cap F_\rho$ _contains no non-trivial null-spline._

PROOF. If $m = 1$ the result is obvious, so we assume $m \geqslant 2$.

Now assume that $\$_{2m,x}\cap F_\rho$ does contain a non-trivial null-spline S, and that x satisfies (1.2); then (following de Boor [1,Lemma 4]) we shall show that (1.1) is contradicted.

Because of the unique representation of S, and of its derivatives $S^{(j)}$, in terms of B-splines, the sequences (α_i), $(\alpha_i^{(j)})$ are uniquely defined by

$$S =: \sum_i \alpha_i N_{i,2m}, \quad S^{(j)} =: \sum_i \alpha_i^{(j)} N_{i,2m-j} \quad (j = 0,1,\dots,m).$$

Since $(\alpha_i^{(j)})$ can be expressed in terms of $(\alpha_i^{(j-1)})$, it is possible by induction to estimate $|\alpha_i^{(m)}|$ in terms of values of $|\alpha_k|$. This has already been done by de Boor [1,(4.5)] to give

$$(2.7) \qquad |\alpha_i^{(m)}| \leq \frac{(2m-1)!\,2^m}{(m-1)!}(x_{i+m}-x_i)^{-m} \max_{i-m\leq k\leq i} |\alpha_k| .$$

Since $S \in \$_{2m,x}$, a lemma of Jakimovski and Russell [3,Lemma 5] shows that for each k there are numbers θ_{kj}, a_{kj} $(j = 0,1,\ldots,2m-1)$, with $\theta_{kj} \in [x_k, x_{k+2m}]$ and $|a_{kj}| \leq c = c_m$, such that

$$\alpha_k = \sum_{j=0}^{2m-1} a_{kj}\, S(\theta_{kj}) .$$

Using this, with $S \in F_\rho$ and Lemma 1 (2.5), we obtain

$$(2.8) \qquad |\alpha_k| \leq c \sum_{j=0}^{2m-1} |S(\theta_{kj})| \leq c' \max(|x_k|^\rho, |x_{k+2m}|^\rho) \leq c'_\varepsilon e^{\varepsilon\rho|k|} .$$

Now define $\beta_i := \{\frac{1}{m}(x_{i+m}-x_i)\}^{\frac{1}{2}} \alpha_i^{(m)}$; then (2.7) and (2.8) give

$$(2.9) \qquad |\beta_i|^2 \leq c''_\varepsilon e^{2\varepsilon\rho|i|} / (x_{i+m}-x_i)^{2m-1} .$$

However, since S is assumed non-trivial, it follows from [1,Corollary p.44, and Remark pp.45-46] that there are positive constants c,λ such that, either for all $j \geq 2$ or for all $j \leq -2$,

$$(2.10) \qquad \sum_{i\in I_j} |\beta_i|^2 \geq c\, e^{\lambda|j|}, \quad I_j := (\,2(m-1)j,\ 2(m-1)(j+1)] .$$

By combining (2.9), (2.10) and (1.2), and choosing ε so small that, in the combination of the exponential factors, the λ term remains dominant ($4\varepsilon\rho(m-1) < \lambda$ turns out to be enough) we can follow through the final part of the proof of [1,Lemma 4] to deduce that we must then have either $\lim_{i\to+\infty} x_i < +\infty$ or $\lim_{i\to-\infty} x_i > -\infty$. Since this contradicts (1.1), which is part of the hypothesis $x \in E_\varepsilon$, the lemma is established. #

Our other main preliminary requirement is to show that the fundamental splines L_k decay exponentially. Both Schoenberg [6,Theorem 2] and de Boor [1, § 4] investigated this in the cases they considered. In view of (1.7), we note first the following inequality.

LEMMA 3. *If* $m \in \mathbb{Z}_{++}$ *and* x *satisfies* (2.1), *then* $\exists\, c = c(m,x)$ *such that*

$$\|e^{(k)}\|_{\ell^m_{2,x}} \leq c\,(x_{k+1} - x_k)^{-m+\frac{1}{2}} \quad (\forall\, k)\,.$$

PROOF. Define $z_i^{(k)} := e^{(k)}[x_i,\dots,x_{i+m}]$ $(\forall\, i,k)$ so that, by (1.4),

$$\|e^{(k)}\|^2_{\ell^m_{2,x}} = \frac{1}{m}\sum_i (x_{i+m} - x_i)\,|z_i^{(k)}|^2\,. \tag{2.11}$$

Now, by (1.5), $z_i^{(k)} = 0$ for $i > k$ and for $i < k-m$. Otherwise we have, for $k-m \leq i \leq k$, and using (2.1),

$$|z_i^{(k)}| = \{(x_{i+m}-x_k)\dots(x_{k+1}-x_k)\cdot(x_k-x_{k-1})\dots(x_k-x_i)\}^{-1}$$

$$\leq \{(x_{k+1}-x_k)^{i+m-k}\,(x_k - x_{k-1})^{k-i}\}^{-1}$$

$$\leq (x_{k+1} - x_k)^{k-i-m}\,\mu^{k-i}\,(x_{k+1} - x_k)^{i-k} \leq \mu^m\,(x_{k+1} - x_k)^{-m}. \tag{2.12}$$

Also, by (2.1),

$$x_{i+m}-x_i = (x_{i+m}-x_{i+m-1})+\dots+(x_{i+1}-x_i) \leq (\mu^{m-1}+\dots+\mu+1)(x_{i+1} - x_i)$$

$$\leq m\,\mu^{m-1}\cdot\mu^m\,(x_{k+1} - x_k) \quad \text{for} \quad 0 \leq k-i \leq m\,, \tag{2.13}$$

and we combine (2.11),(2.12),(2.13) to give

$$\|e^{(k)}\|^2_{\ell^m_{2,x}} \leq \frac{1}{m}\,\mu^{2m}\,(x_{k+1} - x_k)^{-2m}\sum_{i=k-m}^{k}(x_{i+m}-x_i)$$

$$\leq (m+1)\,\mu^{4m-1}\,(x_{k+1} - x_k)^{-2m+1}\,. \qquad \#$$

LEMMA 4. *Let* $m \in \mathbb{Z}_{++}$, $x \in E_\varepsilon$ *for some given* $\varepsilon > 0$. *Then* $\exists\, \lambda = \lambda(m,x) > 0$ *and* $\exists\, c_\varepsilon = c_\varepsilon(m,x) > 0$ *such that*

$$|L_{k,2m}(t)| \leq c_\varepsilon\, e^{-[\lambda-\varepsilon(m-\frac{1}{2})]\,|k-j|} \quad (\forall\, t \in [x_j,x_{j+1}],\ \forall\, k,j \in \mathbb{Z})\,.$$

PROOF. Let $k \in \mathbb{Z}$ and $t \in [x_j,x_{j+1}]$. Then since x satisfies (1.1) we have, for $k \leq j$, as shown by de Boor [1,p.50,last 6 lines],

$$|L_k(t)| \leq (x_{j+1}-t)\dots(x_{j+m}-t)\,\frac{1}{m!}\,[m\,/\,(x_{j+m}-t)]^{\frac{1}{2}}\,\|L_k^{(m)}\|_{\mathcal{L}_2[t,x_{j+m}]}\,.$$

Since x satisfies (2.1) we also have, as in (2.13),

$$(x_{j+1}-t)\dots(x_{j+m}-t)^{\frac{1}{2}} \leq (x_{j+m}-t)^{m-\frac{1}{2}} \leq [m\,\mu^{m-1}\,(x_{j+1} - x_j)]^{m-\frac{1}{2}}$$

and hence

$$|L_k(t)| \leq c\,(x_{j+1}-x_j)^{m-\frac{1}{2}}\, \|L_k^{(m)}\|_{\mathcal{L}_2[x_j,+\infty)} \,.$$

Similarly, for $k \geq j+1$, we get

$$|L_k(t)| \leq c\,(x_{j+1}-x_j)^{m-\frac{1}{2}}\, \|L_k^{(m)}\|_{\mathcal{L}_2(-\infty,x_{j+1}]} \,.$$

Now, according to [1,Corollary p.39, n:=k-j-1, n:=j-k respectively] there are constants $c > 0$, $\lambda > 0$ such that

$$\|L_k^{(m)}\|_{\mathcal{L}_2(-\infty,x_{j+1}]} + \|L_k^{(m)}\|_{\mathcal{L}_2[x_j,+\infty)} \leq c\,\|L_k^{(m)}\|_{\mathcal{L}_2(\mathbb{R})}\, e^{-\lambda|k-j|}$$

and so we obtain

$$|L_k(t)| \leq c\,(x_{j+1}-x_j)^{m-\frac{1}{2}}\, e^{-\lambda|k-j|}\, \|L_k^{(m)}\|_{\mathcal{L}_2} \quad \text{for any } j,k$$

$$\text{(2.14)} \qquad \leq c \left(\frac{x_{j+1}-x_j}{x_{k+1}-x_k}\right)^{m-\frac{1}{2}} e^{-\lambda|k-j|} \,, \quad \text{by (1.7) and Lemma 3.}$$

If we now apply the full force of (1.2), our result follows. #

Finally, we need the elementary result that the limit of a pointwise convergent sequence of splines, is a spline.

LEMMA 5. _Let_ $m \in \mathbb{Z}_{++}$ _and_ x _be a strictly increasing knot-sequence. If_ $S_n \in \$_{m,x}$ $(n = 1,2,\ldots)$ _and_ $\lim_n S_n(t) =: S(t)$ _(pointwise) then_ $S \in \$_{m,x}$.

PROOF. Compare, for example, the proof of [3,Theorem 1(a)] . #

3. Proofs of the Theorems

PROOF OF THEOREM 1. The necessity part is trivial, since obviously from (0.2) and (0.3), $S \in F_\rho$ and $S(x_i) = y_i$ $(\forall\, i)$ together imply $y \in Y_{\rho,x}$.

We suppose, therefore, that $y \in Y_{\rho,x}$, and note first that there is a t m o s t o n e spline function with the required properties. For if there were splines S_1, S_2 both satisfying (1.8) and (1.9), then $S := S_1 - S_2$ would be a null-spline in $\$_{2m,x} \cap F_\rho$. But since $x \in E_\varepsilon$ for every $\varepsilon > 0$, Lemma 2 shows that S must be trivial, namely $S_1 = S_2$.

We now assume (1.10) and show that (1.11) does in fact define a spline function with the required properties (1.8) and (1.9). Clearly (1.11) implies (1.9), so it remains to establish (1.8). By (1.11) and (2.2), we have

$$
\begin{aligned}
(1+|t|)^{-\rho}\ |S(t)| &\leq c\,(1+|x_j|)^{-\rho} \sum_k |y_k|\,|L_k(t)| && \text{for } t\in[x_j,x_{j+1}] \\
(3.1)\qquad &\leq c' \sum_k \left(\frac{1+|x_k|}{1+|x_j|}\right)^{\rho} |L_k(t)| && \text{since } y\in Y_{\rho,x} \\
&\leq c_\varepsilon \sum_k e^{\varepsilon\rho|k-j|}\, e^{-[\lambda-\varepsilon(m-\frac12)]|k-j|} && \text{by Lemma 4} \\
&= c_\varepsilon \sum_{r\in\mathbb{Z}} \theta^{|r|} = O_\varepsilon(1) && \text{independently of } j,
\end{aligned}
$$

where we choose ε so small that $0 < e^{-[\lambda-\varepsilon(m-\frac12+\rho)]} =: \theta < 1$. It follows that $S\in F_\rho$ and that the series (1.11) converges locally uniformly.

Since $L_k\in \$_{2m,x}$ for each k, $S_{nr} := \sum_{n\leq k\leq r} y_k L_k$ now defines a double sequence of splines in $\$_{2m,x}$ which converges to S and hence, by Lemma 5, $S\in \$_{2m,x}$. #

PROOF OF THEOREM 2. If $y\in Y^{o}_{\rho,x}$ we require only routine modification of the proof of Theorem 1, in which (3.1) is replaced by

$$(1+|t|)^{-\rho}\ |S(t)| \leq \sum_{|k|\leq k_0} |y_k|\,|L_k(t)| + \varepsilon \sum_{|k|>k_0}\left(\frac{1+|x_k|}{1+|x_j|}\right)^{\rho}|L_k(t)|$$

and the finite sum on the right tends to zero as $|t|\to\infty$, by Lemma 4. #

PROOFS OF THEOREMS 3 AND 4. It is a routine matter to show that F_ρ and $Y_{\rho,x}$, with the norms defined in (1.12) and (1.13), are complete normed linear spaces. If now $S_n\in \$_{2m,x}\cap F_\rho$, $S\in F_\rho$, $\|S_n-S\|_{F_\rho}\to 0$ then, for any $t\in\mathbb{R}$, $|S_n(t)-S(t)| \leq (1+|t|)^{\rho}\,\|S_n-S\|_{F_\rho}\to 0$ as $n\to\infty$; thus $S_n\to S$ pointwise and so, by Lemma 5, $S\in \$_{2m,x}$. Hence $\$_{2m,x}\cap F_\rho$ is closed in F_ρ, and $\$_{2m,x}\cap F_\rho$ is therefore also a Banach space.

Theorem 1 shows that $T: Y_{\rho,x}\to \$_{2m,x}\cap F_\rho$, defined by (1.11), is bijective; moreover, T^{-1} is bounded, with $\|T^{-1}\|\leq 1$, because

$$\|T^{-1}(S)\|_{Y_{\rho,x}} = \sup_i \frac{|S(x_i)|}{(1+|x_i|)^{\rho}} \leq \sup_t \frac{|S(t)|}{(1+|t|)^{\rho}} = \|S\|_{F_\rho}.$$

Finally, the open mapping theorem shows that T is bounded.

A similar proof yields the first part of Theorem 4; the final part, concerning a basis for $\$_{2m,x}\cap F^{o}_{\rho}$, is already implicit in the proof of Theorem 2 above. #

REMARK 1. In our theorems we have stated our hypothesis on the mesh ratio in the convenient form that x should satisfy (1.2) for *every* $\varepsilon > 0$. However, in the two places in our proofs where we use this fact, we only require to be able to pick *some suitable fixed* $\varepsilon > 0$: in the proof of Lemma 2, we picked ε so that $4\varepsilon\rho(m-1) < \lambda$, and in the proof of Theorem 1 we picked ε so that $\varepsilon(m-\frac{1}{2}+\rho) < \lambda$. Since the constant λ is somewhat difficult to determine, we were unable to specify our ε precisely, but it should be noted that in Theorems 1-4, Lemma 2, and Remark 2 below, it suffices to take $x \in E_{\varepsilon_0}$ for a suitably small fixed $\varepsilon_0 > 0$ (that is, we do not need the full hypothesis that $x \in E_\varepsilon$ for every $\varepsilon > 0$). For example, the computations of Rong-qing Jia [4] show that in the case $\rho = 0$, $m = 4$ we may take any ε_0 in $0 < \varepsilon_0 < \log[\frac{1}{2}(3+\sqrt{5})]$.

REMARK 2 (Representation of the monomials). An important example of Theorem 1 occurs when we take $y_i = x_i^s$ ($\forall i \in \mathbb{Z}$), where $x \in E_\varepsilon$ for a suitable $\varepsilon > 0$ (see Remark 1) and $s \in \{0,1,\ldots,2m-1\}$. Then $S(t) := t^s$ is a spline in $\$_{2m,x} \cap F_s$ which satisfies $S(x_i) = y_i$; hence by Theorem 1,

$$(3.2) \qquad t^s = \sum_{k\in\mathbb{Z}} x_k^s L_{k,2m}(t;x), \quad \forall t \in \mathbb{R}, \quad s = 0,1,\ldots,2m-1.$$

This is the analogue for fundamental splines of Marsden's formula for B-splines [5,Theorem 1], namely

$$(3.3) \qquad t^s = \sum_{k\in\mathbb{Z}} \xi_{k,m}^{(s)} N_{k,m+1}(t;x), \quad \forall t \in \mathbb{R}, \ s = 0,1,\ldots,m,$$

where the ξ are symmetric polynomials defined by

$$\binom{m}{s} \xi_{k,m}^{(s)} := \sum_{k+1 \le i_1 < \ldots < i_s \le k+m} x_{i_1} x_{i_2} \ldots x_{i_s} \ (s = 1,\ldots,m), \quad \xi_{k,m}^{(0)} := 1.$$

REFERENCES

[1] de Boor, C., Odd-degree spline interpolation at a bi-infinite knot sequence, Approximation Theory, Bonn 1976, pp.30-53. Springer Lecture Notes No. 556.

[2] de Boor, C., Splines as linear combinations of B-splines:a survey, Approximation Theory II (Sympos.Proc.,Texas 1976, ed. G.G.Lorentz, C.K.Chui, L.L.Schumaker), pp.1-47. Academic Press, New York/London 1976.

[3] Jakimovski, A., - Russell, D.C., On an interpolation problem and spline functions, General Inequalities 2 (Conf.Proc., Oberwolfach 1978, ed. E.F.Beckenbach), pp.205-231. Birkhäuser Verlag, Basel/Stuttgart 1980.

[4] Jia, Rong-qing, On a conjecture of C.A.Micchelli concerning cubic spline interpolation at a bi-infinite knot sequence, J.Approx.Theory 38 (1983), 284-292.

[5] Marsden, M.J., An identity for spline functions with applications to variation-diminishing spline approximation, J.Approx.Theory 3 (1970), 7-49.

[6] Schoenberg, I.J., Cardinal interpolation and spline functions:II. Interpolation of data of power growth, J.Approx.Theory 6 (1972), 404-420.

VII Orthogonal Functions and Harmonic Analysis

International Series of
Numerical Mathematics, Vol. 65

TWO OF MY FAVORITE WAYS OF OBTAINING ASYMPTOTICS FOR ORTHOGONAL POLYNOMIALS*

Paul Nevai
Department of Mathematics
The Ohio State University
Columbus, OH 43210

Improvements of the continuous and discrete Liouville-Steklov method for proving asymptotic formulas for orthogonal polynomials are discussed, and a short survey of recent asymptotic results is given.

1. Introduction

Let $d\alpha$ be a nonnegative measure on the real line such that all the moments of $d\alpha$ are finite and the support of $d\alpha$ is an infinite set. Then there exists a unique system of polynomials $\{p_n(d\alpha)\}_{n=0}^{\infty}$, $p_n(d\alpha,x) = \gamma_n(d\alpha)x^n + \dots$, $\gamma_n(d\alpha) > 0$ which satisfies

$$\int_{\mathbb{R}} p_n(d\alpha)\, p_m(d\alpha)d\alpha = \delta_{nm} .$$

These polynomials are called the orthogonal (orthonormal) polynomials with respect to $d\alpha$. If $d\alpha$ is absolutely continuous then we use the notation $p_n(w)$ where $w = \alpha'$.

Finding asymptotic expressions as $n \to \infty$ for orthogonal polynomials occupies a major part of the theory of orthogonal polynomials, and frequently it turns out to be a formidable challenge, especially when the orthogonal polynomials do not possess explicit differential equations, generating func-

*This paper is based upon research supported by the National Science Foundation under grant No. MCS-83-00882.

tions or integral representations. The purpose of this paper is to provide the reader with some ideas as to how Liouville-Steklov's highly successful method can be adapted to the case when the orthogonal polynomials either satisfy complicated differential equations or are given by a recursion formula. Many problems mentioned in this paper are only partially resolved at the present time and we sincerely hope that this work will contribute to raising new efforts by researchers who otherwise might remain unaware of these problems.

2. The Continuous Liouville-Steklov Method

All the classical orthogonal polynomials (Jacobi, Hermite, Laguerre) satisfy a simple second order linear homogeneous differential equation with known initial conditions and all these equations can successfully be attacked by Liouville-Steklov's method [18], [63] which immediately yields asymptotics for these polynomials. There exist other orthogonal polynomial systems as well which satisfy differential equations. However, these equations are not so simple anymore, and their solutions may require great efforts. For example, Shohat [60] considered weight functions w of the form

$$w(x) = A^{-1} \exp(\int BA^{-1}\, dx) \tag{2.1}$$

where A and B are certain fixed polynomials and, using an idea going back to Laguerre, he proved that the corresponding orthogonal polynomials $p_n(w)$ satisfy

$$A_n p_n'' + B_n p_n' + C_n p_n = 0$$

where A_n, B_n and C_n are polynomials in x, each of fixed degree independent of n, with coefficients eventually depending on n. In a subsequent paper [62] Shohat simplified his arguments. Not being aware of Shohat's works which were recently brought to my attention by Dick Askey, I rediscovered his methods in [48] where I applied it to polynomials orthogonal with respect to $\exp(-x^4)$.

Let us illustrate Shohat's method [62] on the example of the Hermite polynomials h_n which we define to be orthonormalized with respect to $\exp(-x^2)$. Let the recurrence formula for h_n be

$$x\, h_n = a_{n+1} h_{n+1} + a_n h_{n-1}$$

where $h_n(x) = \gamma_n x^n + \dots$ and $a_n = \gamma_{n-1}/\gamma_n$. By orthogonality

$$n/a_n = \int_{-\infty}^{\infty} [h_n(x)h_{n-1}(x)]' \exp(-x^2)dx = 2\int_{-\infty}^{\infty} h_n(x)h_{n-1}(x)x \exp(-x^2)dx = 2a_n$$

so that $a_n = \sqrt{n/2}$. If R is an arbitrary polynomial of degree less than $n - 1$ then again by orthogonality relations

$$\int_{-\infty}^{\infty} h_n'(x)\; R(x) \exp(-x^2)dx = \int_{-\infty}^{\infty} [h_n(x)\; R(x)]' \exp(-x^2)dx =$$

$$= 2\int_{-\infty}^{\infty} h_n(x)\; R(x)x \exp(-x^2)dx = 0 .$$

Thus

$$h_n'(x) = \text{const}\; h_{n-1}(x)$$

and by comparing leading coefficients we obtain

$$h_n'(x) = \sqrt{2n}\; h_{n-1}(x) . \tag{2.2}$$

Differentiating this identity we get

$$h_n''(x) = 2\sqrt{n(n-1)}\; h_{n-2}(x)$$

and applying the recurrence formula to h_{n-2} we end up with

$$h_n''(x) = 2\sqrt{2n}\left(x\; h_n(x) - \sqrt{n/2}\;\; h_{n-1}(x)\right) . \tag{2.3}$$

Now we can eliminate h_{n-1} from (2.2) and (2.3), and we obtain the differential equation for the Hermite polynomials

$$h_n'' - 2x\; h_n' + 2n\; h_n = 0$$

which is convenient to rewrite as

$$z'' + (2n + 1 - x^2)z = 0 \quad , \quad z = \exp(-x^2/2)\; h_n(x) . \tag{2.4}$$

The very same idea works for the weight function w defined by (2.1), in particular for $w(x) = |x|^{\varepsilon} \exp(-Q(x))$, $x \in \mathbb{R}$, where Q is a polynomial, though the formulas become much more complicated to the extent that at this time we see no particular pattern emerging and we have to repeat the arguments for each individual weight function. If $w(x) = \exp(-x^4)$, $x \in \mathbb{R}$, then the orthogonal polynomials $p_n(w)$ satisfy [48], [62]

$$(2.5) \qquad z'' + f_n z = 0$$

where

$$z = p_n(x)\ \phi_n(x)^{-1/2} \exp(-x^4/2) ,$$

$$\phi_n(x) = a_{n+1}^2 + a_n^2 + x^2 ,$$

and

$$f_n(x) = 4a_n^2[4\ \phi_n(x)\ \phi_{n-1}(x) + 1 - 4a_n^2\ x^2 - 4x^4 - 2x^2\ \phi_n(x)^{-1}] -$$

$$- 4x^6 - 4x^4\ \phi_n(x)^{-1} - 3x^2\ \phi_n(x)^{-2} + 6x^2 + \phi_n(x)^{-1} .$$

Here the coefficients a_n are the (unique) positive solutions of

$$n = 4a_n^2(a_{n+1}^2 + a_n^2 + a_{n-1}^2) , \quad n = 1, 2, \ldots, a_0 = 0 .$$

The sequence $\{a_n\}$ has been investigated in [21], [37] and [48]. Lew and Quarles [37] proved

$$a_n^2 = (n/12)^{1/2}\ (1 + \frac{1}{24n^2} + O(n^{-4})) , \quad n = 1, 2, \ldots .$$

A student of mine, Rong Sheen, is writing his Ph.D. dissertation about orthogonal polynomials corresponding to $w(x) = \exp(-x^6/6)$, $x \in \mathbb{R}$ (the normalizing factor $1/6$ in the weight is introduced for reason of convenience). Sheen [58] proved that in this case the orthogonal polynomial also satisfy (2.5) with

$$z = p_n(x)\ \phi_n(x)^{-1/2} \exp(-x^6/12) ,$$

where

$$\phi_n(x) = a_{n+1}^2(a_{n+2}^2 + a_{n+1}^2 + a_n^2) + a_n^2(a_{n+1}^2 + a_n^2 + a_{n-1}^2) + x^2(a_{n+1}^2 + a_n^2 + x^2) ,$$

$$\delta_n(x) = a_n^2\ x(a_{n+1}^2 + a_n^2 + a_{n-1}^2 + x^2) ,$$

and

$$f_n(x) = - x^{10}/4 - x^5 \phi_n(x)^{-1} \phi_n'(x)/2 + 5x^4/2 - 3 \phi_n'(x)^2 \phi_n(x)^{-2}/4 +$$
$$+ \phi_n(x)^{-1} \phi_n''(x)/2 + a_n^2 \phi_n(x) \phi_{n-1}(x) + \delta_n'(x) - \delta_n(x)^2 - \delta_n(x)x^5 -$$
$$- 4 \delta_n(x)x^3 \phi_n(x)^{-1} - 2x\delta_n(x) \phi_n(x)^{-1}(a_{n+1}^2 + a_n^2) .$$

In the above formulas the coefficients a_n are the (again unique) positive solutions of the Freud type recurrence [21]

$$n = a_n^2[a_{n+2}^2 a_{n+1}^2 + a_{n+1}^4 + 2 a_{n+1}^2 a_n^2 + a_{n+1}^2 a_{n-1}^2 + a_n^4 + 2 a_n^2 a_{n-1}^2 +$$
(2.6)
$$+ a_{n-1}^4 + a_{n-1}^2 a_{n-2}^2] , \quad n = 1 , 2 ,\ldots, a_0 = 0 , a_{-1} = 0 .$$

The (so far unpublished) asymptotics for a_n in (2.6) is

$$a_n^2 = (n/10)^{1/3} \left(1 + \frac{1}{18n^2} + O(n^{-4})\right) , \quad n = 1 , 2 ,\ldots$$

(Máté-Sheen-Nevai).

Having digested (?) the last two differential equations, the reader may well develop the false impression that the situation and the asymptotics are out of control. However, this is indeed a false impression, though I have to acknowledge that handling other weight functions and the corresponding differential equations without new ideas will become quite an accomplishment. Let us illustrate my favorite method of solving equations of the type (2.5) asymptotically on the example of the Hermite polynomials which satisfy (2.4). This method is an improvement of Liouville-Steklov's method. First of all, using (2.2) and the recurrence formula, it is easy to find asymptotics for $h_n(0)$ and $h_n'(0)$ as $n \to \infty$. The original method of Liouville and Steklov would consist of writing (2.4) as

$$z'' + (2n + 1)z = x^2 z \tag{2.7}$$

and then solving this equation as a non-homogeneous second order equation with constant coefficients. The resulting formula would provide asymptotics for $h_n(x)$ as $n \to \infty$ for fixed values of x. The improvement that I suggest is the following. We form the characteristic equation

$$t^2 + (2n + 1 - x^2) = 0$$

whose roots

$$t_{1,2}(x) = \pm i\sqrt{2n + 1 - x^2}$$

are uniformly purely imaginary as long as $x^2 \leq \sigma(2n + 1)$ where $0 < \sigma < 1$ is fixed. Now we introduce a new function y defined by

$$y = z' + t_1 z .$$

Then y satisfies the equation

$$(2.8) \qquad y' + t_2 y = -i \frac{x \operatorname{Im} y}{2n + 1 - x^2} , \quad x^2 \leq \sigma(2n + 1) .$$

The main advantage of (2.8) as opposed to (2.7) is that $x(2n + 1 - x^2)^{-1}$ is much smaller than $|t_2|$ for $x^2 \leq \sigma(2n + 1)$. We can solve (2.8) and we obtain

$$(2.9) \qquad y(x) = \exp(- \int_0^x t_2) \; [y(o) - i \int_0^x \frac{u \operatorname{Im} y(u)}{2n+1 - u^2} \exp(\int_0^u t_2) du] ,$$

for $x^2 \leq \sigma(2n+1)$. Knowing how $y(0)$ behaves as $n \to \infty$ it follows from (2.9) and some simple estimation that

$$(2.10) \qquad |h_n(x)| \leq \text{const } n^{-1/4} \exp(x^2/2)$$

uniformly for $n = 1 , 2 ,...$ and $x^2 \leq \sigma(2n + 1)$. Note that (2.10) is a consequence of Plancherel-Rotach's deep asymptotic formula for the Hermite polynomials [63, p. 201], whereas we obtained it by quite simple arguments. Once we have (2.10), we can take the imaginary parts in (2.9) and we arrive at

$$(2.11) \qquad \sqrt{2n + 1 - x^2} \; z(x) = \operatorname{Im}[y(o) \exp(- \int_0^x t_2)] + O(|x| n^{-1/4})$$

uniformly for $n = 1 , 2 ,...$ and $x^2 \leq \sigma(2n + 1)$. If x is chosen so that $x\, n^{-1/2} \to 0$ $(n \to \infty)$ then (2.11) is the asymptotic formula for the Hermite polynomials which is better than (8.22.8) of [63] though not as good as Plancherel-Rotach's asymptotics. This method can be further improved [49] and one can actually prove the Plancherel-Rotach type asymptotics.

Finding asymptotics for the orthogonal polynomials corresponding to $w(x) = \exp(-x^k)$, $k = 4$, 6 is based on the above described approach applied to (2.5).

THEOREM [49]. *Let* $w(x) = \exp(-x^4)$, $x \in \mathbb{R}$ *and let* p_n *denote the corresponding orthonormal polynomials. Let* $0 < \varepsilon < \pi/2$ *be fixed and let* $x = (4n/3)^{1/4} \cos\theta$. *Then the asymptotic formula*

$$(2.12)\qquad \begin{aligned} p_n(x)\exp(-x^4/2) &= 12^{1/8}\,\pi^{-1/2}\,n^{-1/8}\,(\sin\theta)^{-1/2}\cdot \\ &\cdot \cos\Big[\frac{n}{12}(12\theta - 4\sin 2\theta - \sin 4\theta) + \frac{\theta}{2} - \frac{\pi}{4}\Big] + O(n^{-9/8}) \end{aligned}$$

holds uniformly for $n = 1, 2, \ldots$ *and* $\varepsilon \le \theta \le \pi - \varepsilon$. *For a given interval* Δ

$$(2.13)\qquad \begin{aligned} p_n(x)\exp\Big(-\frac{x^4}{2}\Big) &= 12^{1/8}\,\pi^{-1/2}\,n^{-1/8}\cdot \\ &\cdot\Big[\big(1 + B_1(x)\,n^{-1/2}\big)\cos\big((64/27)^{1/4}\,x\,n^{3/4} + (1/12)^{1/4}\,x^3\,n^{1/4} - n\pi/2\big) + \\ &+ \big(B_2(x)n^{-1/4} + B_3(x)n^{-3/4}\big)\sin\big((64/27)^{1/4}\,x\,n^{3/4} + (1/12)^{1/4}\,x^3\,n^{1/4} - \\ &- n\pi/2\big)\Big] + O(n^{-9/8}) \end{aligned}$$

uniformly for $n = 1, 2, \ldots$ *and* $x \in \Delta$, *where*

$$B_1(x) = \frac{1}{8}\Big(\frac{3}{4}\Big)^{1/2}\Big(x^2 + \frac{9}{10}x^6 - \frac{81}{400}x^{10}\Big),$$

$$B_2(x) = \frac{1}{2}\Big(\frac{3}{4}\Big)^{1/4}\Big(-x + \frac{9}{20}x^5\Big)$$

and

$$B_3(x) = \frac{1}{4}\Big(\frac{3}{4}\Big)^{3/4}\Big(-\frac{3}{4}x^3 + \frac{163}{560}x^7 + \frac{81}{1600}x^{11} - \frac{243}{32000}x^{15}\Big).$$

THEOREM [58]. *If* $w(x) = \exp(-x^6/6)$, $x \in \mathbb{R}$, $0 < \varepsilon < \pi/2$ *and* $x = (32n/5)^{1/6} \cos\theta$ *then*

$$p_n(w,x)\exp(-x^6/12) = 10^{1/12}\,\pi^{-1/2}\,n^{-1/12}(\sin\theta)^{-1/2}\,\cdot \tag{2.14}$$
$$\cdot\cos\left[\frac{n}{60}(60\theta - 15\sin 2\theta - 6\sin 4\theta - \sin 6\theta) + \frac{\theta}{2} - \frac{\pi}{4}\right] + O(n^{-13/12})$$

holds uniformly for $n = 1, 2, \ldots$ *and* $\varepsilon \leq \theta \leq \pi - \varepsilon$. *For* x *belonging to a fixed interval an asymtotic formula analogous to* (2.13) *is valid.*

The reason for $x = (4n/3)^{1/4}\cos\theta$ in (2.12) and $x = (32n/5)^{1/6}\cos\theta$ in (2.14) is that the largest zeros of the corresponding orthogonal polynomials asymptotically behave [22] (and [48]) as $(4n/3)^{1/4}$ and $(32n/5)^{1/6}$ respectively, and therefore, (2.12) and (2.14) are indeed Plancherel-Rotach type asymptotics. For $w(x) = \exp(-|x|^{\lambda})$, Freud made a conjecture [21], [23] that was recently proved by Rahmanov [56] without being aware of Freud's conjecture.

THEOREM [56]. *Let* $w(x) = \exp(-|x|^{\lambda})$, $x \in \mathbb{R}$, $\lambda > 1$ *and let* X_n *be the largest zero of* $p_n(w)$. *Then*

$$\lim X_n\, n^{-1/\lambda} = \left[\sqrt{\pi}\,\frac{\Gamma(\frac{\lambda}{2})}{\Gamma(\frac{\lambda+1}{2})}\right]^{1/\lambda} . \tag{2.15}$$

On the basis of [21], [22], [40], [47], [48], [49], [55], [56], [57], [58], [64], [65] and [66] we are ready to state the following

CONJECTURE. *Let* $W(x) = \exp(-|x|^{\lambda})$, $x \in \mathbb{R}$, $\lambda > 1$, *and let* $p_n(w)$ *denote the orthonormal polynomials with respect to* w . *Let* $0 < \varepsilon < \pi/2$ *be fixed and let* $x = [\sqrt{\pi}\; n\; \Gamma(\lambda/2)\; \Gamma((\lambda+1)/2)^{-1}]^{1/\lambda} \cos\theta$. *Then the asymptotic formula*

$$p_n(w,x)\, w(x)^{1/2} = [\Gamma(\lambda+1)\,\Gamma(\lambda/2)^{-1}\,\Gamma(\lambda/2+1)^{-1}]^{1/2\lambda}\,\pi^{-1/2}\,n^{-1/2\lambda}\,(\sin\theta)^{-1/2}\,\cdot$$

$$(2.16) \qquad \cdot \cos\left[n\left(\theta - \operatorname{sign}(\cos\theta) \cdot |\cos\theta|^{\lambda} \int_{|\cos\theta|}^{1} t^{-\lambda} (1-t^2)^{-1/2}\, dt\right) + \frac{\theta}{2} - \frac{\pi}{4}\right] + O\left(n^{-1-1/2\lambda}\right)$$

holds uniformly for $n = 1, 2, \ldots$ *and* $\varepsilon \leq \theta \leq \pi - \varepsilon$.

Proving (2.16) would advance the theory of orthogonal polynomials on infinite intervals to such an extent that I hereby offer $100.00 to anybody who succeeds in proving (2.16). There is another conjecture (by Freud [22]) which has generated great interest and which has not been resolved yet.

CONJECTURE [22]. *Let* w *be defined as in the previous conjecture. Let* a_n *be the recursion coefficients for the corresponding orthogonal polynomials and* X_n *be the largest zero of* $p_n(w)$. *Then*

$$\lim_{n \to \infty} X_n a_n^{-1} = 2 .$$

Important steps have been taken towards the solution of this conjecture by Mhaskar-Saff [40] and Rahmanov [56].

3. The Discrete Liouville-Steklov Method

If the orthonormal polynomials are given by a recurrence formula

$$(3.1) \qquad a_{n+1} p_{n+1} + (b_n - x)p_n + a_n p_{n-1} = 0 \quad , \quad n = 0, 1, \ldots,$$

where $p_n(x) = \gamma_n x^n + \ldots$, $a_n = \gamma_{n-1}/\gamma_n$, $b_n = \int x p_n^2 \, d\alpha$ and $p_{-1} = 0$, which is a second order linear homogeneous difference equation, then my favorite method to obtain asymptotics for the orthogonal polynomials is a perturbation tecnique which has its roots in the perturbation theory of differential equations and which can be traced back to the works of Poincaré [50] on the behavior of the ratios of solutions of difference equations and which was further developed by Blumenthal [11], Ford [19], Shohat [59], Geronimus [30], Hukuwara [41], Baxter [10], Case [13], [14], [15], and which was finally crystallized in Guseinov [33], Geronimo-Case [25], [26], Geronimo [24],

Nevai [42], [43], [46], Geronimo-Nevai [27] and Máté-Nevai [39]. This technique is essentially a discrete version of Liouville-Steklov's method [18], [63] (Case-Geronimo would prefer calling it a discrete scattering theory method), and it can be described as follows. Let us assume that a_n and b_n in (3.1) converges to finite limits a and b as $n \to \infty$ and consider the solutions of the limiting equation

$$a\, p_{n+1} + (b - x)p_n + a\, p_{n-1} = 0 \,. \tag{3.2}$$

One can expect similarity between the behavior of solutions of (3.1) and (3.2). However, when a in (3.2) equals 0 we are obviously in trouble and in order to avoid this we will assume $a > 0$. If this is the case then without loss of generality we may suppose $a = 1/2$ and $b = 0$ and then (3.2) becomes

$$\frac{1}{2}\, p_{n+1} - x\, p_n + \frac{1}{2}\, p_{n-1} = 0 \qquad (n = 0\,,\,1\,,\ldots) \tag{3.3}$$

whose solutions are the second kind Chebyshev polynomials

$$U_n(x) = \frac{\sin(n+1)\theta}{\sin\theta} \quad , \quad x = \cos\theta \,.$$

If we require nothing but

$$\lim_{n\to\infty} a_n = \frac{1}{2} \quad \text{and} \quad \lim_{n\to\infty} b_n = 0 \tag{3.4}$$

then the only common behavior (as far as pointwise asymptotics goes) between solutions of (3.1) and (3.3) is that

$$\lim_{n\to\infty} \frac{p_{n+1}(x)}{p_n(x)} = \lim_{n\to\infty} \frac{U_{n+1}(x)}{U_n(x)}$$

(p_n satisfies (3.1)) holds for every $x \in \mathbb{C} \setminus \operatorname{supp}(d\alpha)$ [42, p. 33], and this is basically Poincaré's theorem adapted for orthogonal polynomials. However, if we take the Pollaczek polynomials [51], [63] which satisfy (3.1) with

$$a_n = 1/2 + c_1 n^{-1} + O(n^{-2}) \quad , \quad b_n = c_2 n^{-1} + O(n^{-2})$$

(c_1 and c_2 are nonzero constants), then we can see that they are essentially different from the Chebyshev polynomials on both $[-1,1]$ and $\mathbb{C}\setminus[-1,1]$. Those who are familiar with Szegö's theory [20], [28], [32], [63] know that this happens because the Pollaczek weight is not in Szegö's class. If more than (3.4) is known, for example if

$$\sum_{n=1}^{\infty} \{|a_n - \frac{1}{2}| + |b_n|\} < \infty \tag{3.5}$$

holds, then one can indeed successfully compare (3.1) and (3.3).

Let (3.5) be satisfied and let Δ be a proper subinterval of $(-1,1)$. Let p_n be defined by (3.1) and let $x = \cos\theta \in \Delta$. Define Φ_n by

$$\Phi_n(\theta) = p_n(x) - e^{i\theta} p_{n-1}(x) .$$

Then

$$\Phi_{n+1} - e^{-i\theta} \Phi_n = p_{n+1} - 2xp_n + p_{n-1}$$

that is by (3.1)

$$\Phi_{n+1} - e^{-i\theta} \Phi_n = (1 - 2a_{n+1})p_{n+1} - 2b_n p_n + (1 - 2a_n)p_{n-1} . \tag{3.6}$$

It is easy to see that

$$|p_j(x)| \leq \text{const}|\Phi_n(\theta)| \quad , \quad j = n-1, n, n+1 \tag{3.7}$$

for $x \in \Delta$ and $n = 1, 2, \ldots$. Thus by (3.5)

$$|\Phi_{n+1}| \leq |\Phi_n| (1 + \varepsilon_n) \quad , \quad x \in \Delta \quad , \quad n = 1, 2, \ldots \tag{3.8}$$

where $\varepsilon_n \geq 0$,

$$\sum_{n=1}^{\infty} \varepsilon_n < \infty .$$

Repeated applications of (3.8) leads to

$$|\Phi_n| \leq \text{const} \exp(\sum_{j=1}^{\infty} \varepsilon_j) \leq \text{const} .$$

Thus by (3.7)

(3.9) $$|p_n(x)| \leq \text{const}$$

uniformly for $x \in \Delta$ and $n = 1, 2, \ldots$. Now from (3.6) we obtain

$$e^{in\theta}\,\Phi_n = \sum_{k=0}^{n} \{(1 - 2a_k)p_k - 2b_{k-1}\,p_{k-1} + (1 - 2a_{k-1})p_{k-2}\}e^{ik\theta}$$

and applying (3.5) and (3.9)

(3.10) $$\Psi(\theta) = \lim_{n \to \infty} e^{in\theta}\,\Phi_n(\theta)$$

follows uniformly for $x \in \Delta$. Rewriting (3.10) as

$$e^{-i\theta}\,\Phi_n(\theta) = e^{-i(n+1)\theta}\,\Psi(\theta) + o(1)$$

and taking the imaginary parts we finally arrive at the asymptotic formula

$$\sin\theta\, p_n(\cos\theta) = -\mathrm{Im}\, e^{-i(n+1)\theta}\,\Psi(\theta) + o(1)$$

uniformly for $x \in \Delta$ and $n \to \infty$. The actual determination of the function Ψ can be achieved by employing techniques developed in [42]. We say that $d\alpha$ (or w) belongs to the Szegö class S if $\mathrm{supp}(d\alpha) = [-1,1]$ (or $\mathrm{supp}(w) = [-1,1]$) and $\log \alpha'(\cos\theta)$ (or $\log w(\cos\theta)$) is integrable. The Szegö function $D(w)$ is defined by

$$D(w,z) = \exp\{\frac{1}{4\pi} \int_{-\pi}^{\pi} \log(w(\cos\theta)|\sin\theta|)\,\frac{1 + z\,e^{-i\theta}}{1 - z\,e^{-i\theta}}\,d\theta\} , \quad |z| < 1$$

[63, (10.2.10)] and $D(d\alpha)$ is defined by $D(d\alpha) = D(\alpha')$. The function Ψ in (3.10) is closely related to the Szegö function of $d\alpha$ whenever $d\alpha$ belongs to the Szegö class. However, it is not true that $d\alpha \in S$ if (3.5) holds. What is known is that if $\mathrm{supp}(d\alpha) \subset [-1,1]$ and (3.5) is satisfied then $d\alpha \in S$ [42, p. 124]. Another result of this type is that if

$$\sum_{n=1}^{\infty} \log n\{|a_n - \frac{1}{2}| + |b_n|\} < \infty$$

then $\alpha' \in S$ [46]. I conjectured a long time ago that (3.5) is sufficient for α' to belong to the Szegö class, but this has not been proved yet.

Using various improvements and modifications of this perturbation method one can get asymptotics for the orthogonal polynomials satisfying (3.1) under different conditions on the recurrence coefficients. These asymptotics may hold either inside the support of $d\alpha$ or in the complex plane cut along $\mathrm{supp}(d\alpha)$. Here are a few examples.

THEOREM [42, p. 143]. _Let_ (3.5) _be satisfied. Then_ $d\alpha$ _can be written in the form_

$$d\alpha(t) = \alpha'(t)dt + d\alpha_j(t) \tag{3.11}$$

where α' _is supported in_ $[-1,1]$, α' _is positive and continuous in_ $(-1,1)$ _and_ α_j _is a pure jump function such that_ α_j _is constant in_ $(-1,1)$ _and outside a finite interval as well. Moreover,_

$$\sqrt{\alpha'(x)\sqrt{1-x^2}}\; p_n(d\alpha,x) = \sqrt{\frac{2}{\pi}}\, \cos(n\theta + \phi(\theta)) + o(1) \quad , \quad x = \cos\theta \tag{3.12}$$

uniformly for $n = 1, 2, \ldots$ _and_ x _belonging to any fixed proper subinterval of_ $(-1,1)$ _where_ ϕ _is a continuous function of_ θ _for_ $0 < \theta < \pi$. _If_ K _is any set of the complex plane such that_ $\overline{K} \cap [-1,1] = \emptyset$ _then_

$$\lim_{n \to \infty} p_n(d\alpha,z)\, (z + \sqrt{z^2 - 1})^{-n} = g(z)$$

holds uniformly for $z \in K$ _where that branch of_ $\sqrt{z^2 - 1}$ _is chosen for which_ $|z + \sqrt{z^2 - 1}| > 1$ _if_ $z \notin [-1,1]$. _Here_ g _is analytic in_ $\mathbb{C}$ _cut along_ $[-1,1]$ _and vanishes precisely at the points of_ $\mathrm{supp}(d\alpha)\setminus[-1,1]$.

This result has also been obtained by Geronimo-Case [26].

THEOREM [46, p. 380]. _Let_

$$\sum_{n=1}^{\infty} \{|a_n a_{n+1} - \frac{1}{4}| + |b_n + b_{n+1}| + |a_n - \frac{1}{2}|^2 + b_n^2\} < \infty . \tag{3.13}$$

Then $d\alpha$ _is of the form_ (3.11) _where_ α' _is positive and continuous on_ $(-1,0) \cup (0,1)$ _with_ $\mathrm{supp}(\alpha') \subset [-1,1]$ _and_ α_j _is a jump function which is constant on_ $(-1,0) \cup (0,1)$ _and outside a finite interval. The asymptotic_ formula (3.12) holds uniformly inside $(-1,0) \cup (0,1)$ and ϕ is a continuous function in $(0,\pi/2) \cup (\pi/2,\pi)$.

The Szegö polynomials [63, (4.1.6)] p_n which are orthogonal with respect to $w(x) = |x|^\varepsilon$ $(-1 \leq x \leq 1)$, satisfy (3.1) with

$$a_n = \frac{1}{2} - (-1)^n \frac{\varepsilon}{4n} + O(\frac{1}{n^2}) \quad , \quad b_n = 0$$

[42, p. 127], and therefore this theorem's conclusions are the best. If we form orthogonal polynomials with respect to $d\alpha$ which is the sum of the Chebyshev weight and a mass point at zero then again it is easy to check [42, p. 131] that (3.13) holds. When $d\alpha$ is in the Szegö class then the function ϕ in (3.12) is exactly the argument of Szegö's function $D(d\alpha,e^{i\theta})$ [63, Ch. 12].

The essence of the previous two theorems is that if either (3.5) or (3.13) holds then the corresponding orthogonal polynomials (at least locally) behave like the Chebyshev polynomials. As soon as we exit the Szegö class we cannot expect similarity between the orthogonal polynomials and the Chebyshev polynomials anymore. This observation may be illustrated by the Pollaczek polynomials whose weight functions is essentially $\exp(-(1 - x^2)^{-1/2})$ $(-1 \leq x \leq 1)$ and whose asymptotics has the form

$$\sqrt{\alpha'(x)\sqrt{1-x^2}}\ p_n(x) = \sqrt{\frac{2}{\pi}}\ \cos(n\theta + \phi_1(\theta)\log n + \phi_2(\theta)) + o(1) \quad , \quad x = \cos\theta \ ,$$

for $-1 < x < 1$ ([63, p. 393]). The recursion coefficients for the Pollaczek polynomials form sequences of bounded variation. Therefore it is natural to consider the condition

$$(3.14) \qquad \sum_{n=1}^{\infty} \{|a_{n+1} - a_n| + |b_{n+1} - b_n|\} < \infty \ .$$

The way to handle orthogonal polynomials satisfying (3.4) and (3.14) has been worked out in [39] and what comes out is the following (yet unpublished) result.

THEOREM. Let (3.4) and (3.14) be satisfied. Then $d\alpha$ can be expressed in the form (3.11) where α' is positive and continuous in $(-1,1)$, α' vanishes outside $[-1,1]$, α_j is a jump function and α_j is constant in $(-1,1)$ and outside a finite interval. For the orthogonal polynomials the asymptotic formula

$$(3.15)\quad \sqrt{\alpha'(x)\sqrt{1-x^2}}\; p_n(d\alpha,x) = \sqrt{\frac{2}{\pi}}\, \cos\Big(\sum_{k=1}^{n} \arg t_k(x) + \phi(\theta)\Big) + o(1)\,,$$

$x = \cos\theta$, holds uniformly for x inside $(-1,1)$ and $n = 1, 2, \ldots$ where ϕ is a continuous function in $(0,\pi)$ and

$$t_k(x) = x - b_k + \sqrt{(x - b_k)^2 - 4\, a_k a_{k+1}}\ .$$

Note that $\lim t_k(x) = e^{i\theta}$ if (3.4) holds. The strength of this theorem can be highlighted by the fact that at the present time every known orthogonal polynomial system has the property that if the conclusions of the theorem hold then (3.4) and (3.14) are satisfied as well.

All these theorems must have suitable generalizations when the absolutely continuous portion of the distribution function lives on several disjoint intervals. One would be delighted to see such extensions. Examples of orthogonal polynomials on disjoint intervals are given in [1], [7], [9], [31], [35] and [38].

Another interesting problem is recovering the recursion coefficients in (3.1) from properties of $d\alpha$ [42], [38] (and the references therein).

Finally, we mention [2] which is a good starting point to generalize (3.13).

4. Concluding Remarks

The alert reader is probably trying to find out whether the words "two of my favorite ways" in the title mean that I have other favorite methods of finding asymptotics. Well, I do have another favorite which I call "asymptotics via weak asymptotics". This method is perhaps the most powerful, and it uses Szegö's theory [63] and techniques developed in [42], [44], [45] and Frey [23]. In an upcoming paper with Máté we will apply this approach to obtain asymptotics for orthogonal polynomials whose weighted function w behaves like $\exp(-(1 - x^2)^{-1/2})$ on $[-1,1]$, that is like the Pollaczek weight.

When there exists an integral representation for the orthogonal polynomials then the method of steepest descent [18], [63] seems to be to the best way of getting asymptotic expansions, but I am not aware of any recent applications of this method with the exception of one asymptotics for the Pollaczek polynomials on the infinite intervals [63, p. 393] by Bill Goh (unpublished and unwritten).

If one can find a reasonable generating function for the orthogonal polynomials then Darboux's method may produce the right asymptotics. Darboux's method was applied by Pollaczek [51], [52], [53] in investigating the orthogonal polynomials which are named after him. Recently R. Askey, M. Ismail, J. Wilson and their coauthors discovered several new sets of orthogonal polynomials and used Darboux's method to determine their asymptotic behavior. The interested reader may consult [2], [3], [4], [5], [6], [7], [8], [12], [16], [17], [34], [35], [36], [67], [68] and [69].

REFERENCES

[1] Ahiezer, N. I., Elements of the Theory of Elliptic Functions, Nauka, Moscow, 1970.

[2] Al-Salam, W. A. - Allaway, W. R. - Askey, R. A., Sieved ultraspherical polynomials, manuscript.

[3] Al-Salam, W. A. - Ismail, M., Orthogonal polynomials associated with the Rogers-Ramanujan continued fraction, Pacific J. Math. 104 (1983), 269-283.

[4] Askey, R. A., Orthogonal polynomials old and new, and some combinatorial connections, in "Proceedings of the 1982 Waterloo Conference in Combinatorics" (to appear).

[5] Askey, R. A. - Ismail, M., A generalization of ultraspherical polynomials, in "Studies in Pure Mathematics", P. Turán memorial volume, ed. by P. Erdős, Birkhäuser, Boston, 1983.

[6] Askey, R. A. - Ismail, M., The Rogers q-ultraspherical polynomials, in "Approximation Theory III"., ed. by W. Cheney, Academic Press, New York, 1980, 175-182.

[7] Askey, R. A. - Ismail, M., Recurrence relations, continued fractions and orthogonal polynomials, manuscript.

[8] Askey, R. A. - Wilson, J. A., Some basic hypergeometric orthogonal polynomials that generalize the Jacobi polynomials, AMS Memoirs (to appear).

[9] Barkov, G. I., Systems of polynomials orthogonal on two symmetric intervals (in Russian), Izv. Vysh. Uceb. Zaved. Matematika 4(17) (1960), 3-16.

[10] Baxter, G., A convergence equivalence related to polynomials orthogonal on the unit circle, TAMS 99 (1961), 471-487.

[11] Blumenthal, O., Über die Entwicklung einer willkürlichen Funktion nach den Nennern des Kettenbruches für $\int_{-\infty}^{\infty} [\phi(\xi)/(z-\xi)]d\xi$, Dissertation, Göttingen, 1898.

[12] Bustoz, J. - Ismail, M., The associated ultraspherical polynomials and their q-analogues, Can. J. Math. 34 (1982), 718-736.

[13] Case, K. M., Orthogonal polynomials from the viewpoint of scattering theory, J. Math. Physics 15 (1974), 2166-2174.

[14] Case, K. M., Orthogonal polynomials revisited, in "Theory and Application of Special Functions", ed. by R. A. Askey, Academic Press, New York, 1975, 289-304.

[15] Case, K. M., Orthogonal polynomials, II, J. Math. Physics 16 (1975), 1435-1440.

[16] Chihara, T. S., An Introduction to Orthogonal Polynomials, Gordon and Breach, New York, 1978.

[17] Chihara, T. S. - Ismail, M., Orthogonal polynomials suggested by a queueing model, Advances in Appl. Math. 3 (1982), 441-462.

[18] Erdélyi, A., Asymptotic Expansions, Dover Publ., Inc., New York, 1956.

[19] Ford, W. R., On the integration of the homogeneous linear difference equation of second order, TAMS 10 (1909), 319-336.

[20] Freud, G., Orthogonal Polynomials, Pergamon Press, New York, 1971.

[21] Freud, G., On the coefficients in the recursion formulae of orthogonal polynomials, Proc. Royal Irish Acad. 76 (1976), 1-6.

[22] Freud, G., On the greatest zero of an orthogonal polynomial, manuscript, 1978.

[23] Frey, T., On asymptotic behavior of orthogonal polynomials (in Russian), Matem. Sbornik 49 (1959), 133-180.

[24] Geronimo, J. S., A relation between the coefficients in the recurrence formula and the spectral function for orthogonal polynomials, TAMS 260 (1980), 65-82.

[25] Geronimo, J. S. - Case, K. M., Scattering theory and polynomials orthogonal on the unit circle, J. Math. Physics 20 (1979), 299-310.

[26] Geronimo, J. S. - Case, K. M., Scattering theory and polynomials orthogonal on the real line, TAMS 258 (1980), 467-494.

[27] Geronimo, J. S. - Nevai, P., Necessary and sufficient conditions relating the coefficients in the recurrence formula to the spectral function for orthgonal polynomials, SIAM J. Math. Analysis 14 (1983), 622-637.

[28] Geronimus, Ya. L., Polynomials Orthogonal on a Circle and Interval, Pergamon Press, New York, 1960.

[29] Geronimus, Ya. L., Polynomials orthogonal on a circle and their applications, AMS Translations, Ser. 1, 3 (1962), 1-78.

[30] Geronimus, Ya. L., On asymptotic properties of polynomials which are orthogonal on the unit circle, and on certain properties of positive harmonic functions, AMS Translations, Ser. 1, 3 (1962), 79-106.

[31] Geronimus, Ya. L., Orthogonal polynomials, AMS Translations, Ser. 2, 108 (1977), 37-130.

[32] Grenander, U. - Szegö, G., Toeplitz Forms and Their Applications, University of California Press, Berkeley, 1958.

[33] Guseinov, G. S., The determination of an infinite Jacobi matrix from the scattering data, Soviet Math. Dokl. 17 (1976), 596-600.

[34] Ismail, M., The zeros of basic Bessel functions, the functions $I_{\nu+ax}(x)$, and associated orthogonal polynomials, J. Math. Analysis and Appl. 86 (1982), 1-19.

[35] Ismail, M. - Mulla, F. S., On the generalized Chebyshev polynomials, SIAM J. Math. Analysis (to appear).

[36] Ismail, M. - Wilson, J. A., Asymptotic and generating relations for the q-Jacobi and ${}_4\Phi_3$ polynomials, J. Approximation Th. 36 (1982), 43-54.

[37] Lew, J. S. - Quarles, D. A., Jr., Nonnegative solutions of a nonlinear recurrence, J. Approximation Th. 38 (1983), 357-379.

[38] Magnus, A., Recurrence coefficients for orthogonal polynomials on connected and nonconnected sets, in "Padé Approximation and its Applications", Lecture Notes in Mathematics, No. 765, Springer-Verlag, New York, 1979, 150-171.

[39] Máté, A. - Nevai, P., Absolute continuity of measures associated with orthogonal polynomials, in "Approximation Theory, IV", ed. by L. Schumaker, Academic Press, 1983.

[40] Mhaskar, H. N. - Saff, E. B., Extremal problems for polynomials with exponential weights, TAMS (to appear).

[41] Miller, K. S., Linear Difference Equations, W. A. Benjamin, Inc., New York, 1968.

[42] Nevai, P., Orthogonal Polynomials, AMS Memoirs, vol. 213, 1979.

[43] Nevai, P., On orthogonal polynomials, J. Approximation Th. 25 (1979), 34-37.

[44] Nevai, P., An asymptotic formula for the derivatives of orthogonal polynomials, SIAM J. Math. Analysis 10 (1979), 472-477.

[45] Nevai, P., Distribution of zeros of orthogonal polynomials, TAMS 249 (1979), 341-361.

[46] Nevai, P., Orthogonal polynomials defined by a recurrence relation, TAMS 250 (1979), 369-384.

[47] Nevai, P. - Dehesa, J. S., On asymptotic average properties of zeros of orthogonal polynomials, SIAM J. Math. Analysis 10 (1979), 1184-1192.

[48] Nevai, P., Orthogonal polynomials associated with $\exp(-x^4)$, Canadian Math. Soc. Conference Proc. 3 (1983).

[49] Nevai, P., Asymptotics for orthogonal polynomials associated with $\exp(-x^4)$, SIAM J. Math. Analysis (to appear).

[50] Poincaré, H., On linear ordinary differential and finite difference equations (in French), Amer. J. Math. 7 (1885), 203-258.

[51] Pollaczek, F., On a generalization of the Legendre polynomials (in French), C. R. Acad. Sci. Paris 228 (1949), 1363-1365.

[52] Pollaczek, F., On a four parameter family of orthogonal polynomials (in French), C. R. Acad. Sci. Paris 230 (1950), 2254-2256.

[53] Pollaczek, F., On a Generalization of the Jacobi Polynomials (in French), Mémorial des Sc. Math. 121 (1956), Paris.

[54] Rahmanov, E. A., On the asymptotics of the ratio of orthogonal polynomials, I, II (in Russian), Matem. Sbornik 103 (145) (1977), 237-252 and 118 (160) (1982), 104-117.

[55] Rahmanov, E. A., On asymptotic properties of polynomials orthogonal on the real line (in Russian), Doklady Akad. Nauk USSR 261 (1981), 282-284.

[56] Rahmanov, E. A., On asymptotic properties of polynomials orthogonal on the real line (in Russian), Matem. Sbornik 119 (161) (1982), 163-203.

[57] Saff, E. B., Incomplete and orthogonal polynomials, manuscript.

[58] Sheen, R., Orthogonal Polynomials Associated with $\exp(-x^6/6)$, Ph.D. Dissertation (in preparation).

[59] Shohat, J., On the asymptotic expressions for Jacobi and Legendre polynomials derived from finite difference equations, Amer. Math. Monthly 33 (1926), 354-361.

[60] Shohat, J., On a wide class of algebraic continued fractions and the corresponding Chebyshev polynomials (in French), Comptes Rendus Acad. Sci. Paris 191 (1930), 989-990.

[61] Shohat, J., General Theory of Chebyshev's Orthogonal Polynomials (in French), Mémorial des Sc. Math 66 (1934), Paris.

[62] Shohat, J., A differential equation for orthogonal polynomials, Duke Math. J. 5 (1939), 401-417.

[63] Szegö, G., Orthogonal Polynomials, Amer. Math. Soc., New York, 1967.

[64] Ullman, J. L., Orthogonal polynomials associated with an infinite interval, Mich. Math. J. 27 (1980), 353-363.

[65] Ullman, J. L., On orthogonal polynomials associated with the infinite interval, in "Approximation Theory III", ed. by E. Cheney, Academic Press, New York, 1980, 889-895.

[66] Ullman, J. L., A generalization of Blumenthal's theorem on the three term recursion relationship for orthogonal polynomials, manuscript.

[67] Wilson, J. A., Hypergeometric series, recurrence relations and some new orthogonal functions, Ph.D. Dissertation, University of Wisconsin, Madison, 1978.

[68] Wilson, J. A., Some hypergeometric orthogonal polynomials, SIAM J. Math. Analysis 11 (1980), 690-701.

[69] Wilson, J. A., Asymptotics for the ${}_4F_3$ polynomials, manuscript.

International Series of
Numerical Mathematics, Vol. 65

CONVOLUTION STRUCTURES FOR EIGENFUNCTION EXPANSIONS ARISING FROM REGULAR STURM-LIOUVILLE PROBLEMS

W.C. Connett and A.L. Schwartz
Department of Mathematical Sciences
University of Missouri-St. Louis
St. Louis, Missouri

The eigenfunctions associated with a regular Sturm-Liouville problem behave "like" trigonometric expansions in many ways - for example there are various asymptotic estimates and equiconvergence theorems. In order to utilize the full machinery of harmonic analysis, however, it is necessary to have some substitute for the group structure so useful in arguments concerning trigonometric expansions. That substitute is a positive convolution which is then utilized to prove various maximal function inequalities.

1. Introduction

Much of the machinery of harmonic analysis can only be utilized in an environment where there is a positive convolution. If not absolutely necessary, it is at least extremely helpful to be able to make estimates at a convenient point and then "carry" these estimates around to other points using the convolution when proving maximal function inequalities or multiplier theorems for various orthogonal expansions ([3],[4],[6],[7]). Unfortunately, it is difficult to tell if a given orthogonal expansion has a bounded convolution and extremely difficult to tell if it has a positive convolution by just looking at the basis functions themselves (see for example [1],[5], and the references given there).

Much more structure is needed, and in fact that structure is found in the class of orthogonal expansions that arise as eigenfunction expressions associated with a regular Sturm-Liouville operator and a typical set of unmixed boundary conditions.

2. Eigenfunction Expansions

Let $I = [0,pi]$ and consider the Sturm-Liouville problem in I given by

(1) $$y'' + (l-q)y = 0$$

(2) $$Ay(0) - By'(0) = 0$$

(3) $$Ay(pi) + By'(pi) = 0 .$$

If q is a real continuous function on I and if not both A and B are zero there are countably many eigenvalues $l_0, l_1, l_2, \ldots$ forming a monotone sequence with $l_n \to \infty$ as $n \to \infty$. The eigenfunction belonging to l_n has exactly n zeros in (0,pi) [2, p. 212].

It is no loss to assume that $l_0 = 0$ since this can be achieved by adding a constant to q. The numbers $m_k = l_k^{1/2}$ will also be called eigenvalues. The Sturm comparison theorem may be used to obtain

(4) $$\lim_{k\to\infty} m_k/k = 1 .$$

For this discussion it will be assumed that $B = 1$, $A \geqslant 0$,

(5) $$q(t) \geqslant q(s), \qquad 0 \leqslant t < s \leqslant pi/2 ,$$

and

(6) $$q(pi-s) = q(s) .$$

Let u_k denote the eigenfunction belonging to m_k which satisfies $u_k(0) = 1$ and $u_k'(0) = A$ and let

$$h_k = \{\int u_k^2(s)ds\}^{-1}$$

(all integrals extend over I unless otherwise indicated). Some useful properties of the u_k's are contained in the following.

Lemma 1. (i) $u_k(pi-s) = (-1)^k\, u_k(s)$

(ii) There is $M > 0$ such that

$$\|u_k\|_\infty \leqslant M \qquad (k = 0,1,2,\dots)$$

(iii) $$\lim_{k\to\infty} h_k = 2/pi$$

(iv) There are constants $b > 0$ and $K > 0$ independent of k such that

$$u_k(s) \geqslant b \text{ if } 0 \leqslant ks \leqslant K .$$

(v) There is a constant m such that

$$0 < m \leqslant u_o(s) .$$

PROOF. The parity property (i) follows from (2),(3), and (6).

Now observe that $y = u_k$ solves the inhomogeneous equation

$$y'' + m_k^2 y = qu_k$$

so we obtain by variation of parameters

$$u_k(t) = \cos m_k t + \frac{A}{m_k} \sin m_k t + \frac{1}{m_k} \int_0^t \sin m_k(t-s)q(s)u_k(s)\,ds . \tag{7}$$

Thus $\|u_k\|_\infty \leqslant 1 + (|A| + \|q\|_\infty \|u_k\|_\infty)/m_k$ so (ii) follows from (4), and (7) yields $u_k(t) = \cos m_k t + O(m_k^{-1})$ from which (iii) and (iv) are immediately obtained.

The eigenfunction u_o has no zeros in $(0,pi)$ and $u_o(0) = 1$, part (i) of the lemma implies $u_o(pi) = 1$ so (v) follows.

The eigenfunction expansion of any f in L^1 is obtained by setting

$$f^0(k) = \int f(t)\, u_k(t)\,dt \tag{8}$$

and writing

$$f(t) \sim \sum f^0(k)h_k u_k(t) .$$

Sums are taken over all non-negative integers unless otherwise stated.

If f in L^2 the series converges in L^2 to f. We write $AC = \{f : \sum|f^0(k)| < \infty\}$ so that for f in AC "~" may be replaced by "=" .

LEMMA 2. *If f is twice continuously differentiable on I then f in AC, so that AC is dense in the continuous functions on I.*

PROOF. Use (1) to rewrite (8) as

$$f^0(k) = m_k^{-2}\{\int f(t)q(t) - \int f(t)u_k''(t)\}$$

and integrate by parts twice to obtain $f^0(k) = O(m_k^{-2}) = O(k^{-2})$.

3. Convolution

We can now define generalized translation and convolution at least for functions in AC by

$$f_s(t) = \sum f^0(k)h_k u_k(s)u_k(t)$$

$$f*g(x) = \sum f^0(k)g^0(k)h_k u_k(t) = \int f_s(t)g(t)\,dt .$$

The proof of the following is inspired by ideas in [8].

LEMMA 3. *If f in AC then*

(i) *if $f \geq 0$, then $f_s \geq 0$ for all s in I ,*

(ii) there is a positive constant C such that $\|f_s\|_\infty \leq C\|f\|_\infty$.

PROOF. We first assume $f > 0$ on I and let n be so large that

$$g(t) = \sum_{k=o}^{n} f^0(k)\ h_k u_k(t) \geqslant 0 \quad (t \text{ in } I).$$

Let

$$w(s,t) = \sum_{k=o}^{n} f^0(k)\ h_k u_k(s) u_k(t)$$

then w satisfies

$$w_{ss} - w_{tt} + [q(t) - q(s)]w = 0,\ w_t(s,0) \geqslant 0, \text{ and } w(s,0) \geqslant 0\ .$$

The second relation follows since

$$w_t(s,0) = \sum_{k=o}^{n} f^0(k)\ h_k u_k'(0) u_k(s) = Ag(s)\ .$$

It now follows from the weak maximum principle for hyperbolic equations [9, pp. 197-99] that $w(s,t) \geqslant 0$ if $0 \leqslant t \leqslant s \leqslant pi - t$. Finally $w \geqslant 0$ in all of $I \times I$ since $w(t,s) = w(s,t)$ and by Lemma 1 (i) $w(pi-s, pi-t) = w(s,t)$. Now let n increase without bound and (i) follows.

If $f \geqslant 0$, let $c > 0$ and apply the argument to $f + cu_o$ to obtain $f_s(t) + cu_o(s)\ u_o(t) \geqslant 0$ for each $c > 0$.

From Lemma 1 (v) we have

$$m^{-1}\ \|f\|_\infty\ \ u_o \pm f \geqslant 0$$

so by part (i) of this lemma

$$m^{-1}\ \|f\|_\infty\ u_o(s) u_o(t) \pm f_t(s) \geqslant 0\ .$$

We are now in a position to build a Banach algebra with multiplication *. Let L^1 consist of those f for which

$$\|f\|_1 = \int |f(t)| u_o(t)\, dt\ .$$

This is equivalent to the usual L^1-norm, but we can obtain a sharper inequality with this norm.

THEOREM A. _The operation * can be extended to all of_ L^1 _so that if_ f, g _in_ L^1 _then_

(i) If $f \geqslant 0$ and $g \geqslant 0$(p.p.) then $f*g \geqslant 0$ (p.p.)

(ii) $$(f*g)^0(n) = f^0(n)g^0(n)$$

(iii) $$\|f*g\|_1 \leqslant \|f\|_1 \|g\|_1 .$$

PROOF. Assume that f and g are in AC and consider the functional defined on continuous functions h by $A(h) = \int f*g(s)h(s)\,ds$. Then if h in AC,

$$A(h) = \int \int f(t)g(s)h_t(s)\,ds\,dt \tag{9}$$

so by Lemma 3, there is a constant C_1 such that

$$|A(h)| \leqslant C_1 \|f\|_1 \|g\|_1 \|h\|_\infty ,$$

and since AC is dense in the continuous functions on I, the Riesz Representation Theorem implies

$$\int |f*g(s)|\,ds \leqslant C_1 \|f\|_1 \|g\|_1$$

or

$$\|f*g\|_1 \leqslant \|u_0\|_\infty C_1 \|f\|_1 \|g\|_1 . \tag{10}$$

Now, thanks to (10) the convolution can be extended to all of L^1 because AC is dense in L^1, so from (9)

$$\int f*g(s)h(s)\,ds = \int\int f(s)g(t)h_t(s)\,ds\,dt \tag{11}$$

for f,g in L^1 and h in AC.

Now supose $f,g \geqslant 0$, then if h in AC is non-negative, the right side of (11) is non-negative so (i) follows, and (ii) is obtained by letting $h = u_k$.

To derive (iii) first assume f and g are non-negative in L^1, then

$$\|f*g\|_1 = (f*g)^0(0) = f^0(0)g^0(0) = \|f\|_1 \|g\|_1 .$$

Finally if f and g are arbitrary functions in L^1 let $F = |f|$ and $G = |g|$ so that

$$\|f*g\|_1 \leqslant \|F*G\|_1 = \|F\|_1 \|G\|_1 = \|f\|_1 \|g\|_1 .$$

4. Maximal Functions

The classical maximal function of Hardy and Littlewood is defined by

$$Hf(t) = \sup_{r>0} \left|\frac{1}{2r} \int f(s)\, ds\right| = \sup_{r>0} \left| \int f(t-s) h_r(s)\, ds\right|$$

where $h_r(s)=2r^{-1}$ if $|s| \leqslant r$ and $h_r(s) = 0$ otherwise. These two formulations are distinct in the present context. An analogue to the second formula is

$$Mf(t) = \sup_{r>0} |f*k_r(t)|;$$

$k_r(s) = (\int_0^r u_0(t)\, dt)^{-1}$ if $0 \leqslant s \leqslant r$, $k_r(s) = 0$ if $s > r$.

The function Mf is more useful as a tool for harmonic analysis, but it is somewhat more difficult to obtain inequalities for it than for Hf. Nevertheless, we prove

THEOREM B. *There are constants* A_p *such that*

$$\|Mf\|_p \leqslant A_p \|f\|_p \qquad 1 < p \leqslant \infty .$$

The proof of Theorem B will follow from Stein's Maximal Theorem [10, p. 73]. Define the semigroup T^r as follows: suppose f in AC, let

$$v(r,s) = \sum \exp(-l_n r) h_n f^0(n) u_n(s) = \int f(t) \{\sum \exp(-l_n r) h_n u_n(s) u_n(t)\}\, dt ,$$

then v solves the problem

$$v_{ss} - q(s)\, v - v_t = 0, \qquad v(0,s) = f(s) .$$

If $f \geqslant 0$ then $v(r,s) \geqslant 0$ by the maximum principle for parabolic differential equations [9, p. 172], so the quantity in the braces is non-negative. We shall make use of the related kernel

$$W_r(s,t) = [u_o(s)u_o(t)]^{-1} \sum \exp(-l_n r) h_n u_n(s) u_n(t)$$

which evidently satisfies

(12) $$W_r(s,t) \geqslant 0$$

and

(13) $$\int W_r(s,t)\, u_o^2(t)\, dt = 1 .$$

We now define

$$T^o f(s) = f(s)$$

$$T^r f(s) = \int f(t) W_r(s,t) u_o^2(t)\, dt$$

so that we have

LEMMA 4. $\{T^r : 0 \leqslant r < \infty\}$ satisfies the hypotheses of Stein's Maximal Theorem.

PROOF. Let $F = u_o f$, so that

$$T^r f(s) = \int u_o^{-1}(t) F(t)\, W_r(s,t)\, u_o^2(t)\, dt = u_o^{-1}(s) T^r F(s)$$

which tends to $f(s)$ as $r \to 0$.

Now if f and g are measurable and $p^{-1} + q^{-1} = 1$, (12) implies

$$\int |T^r f(s) g(s)|\, u_o^2(s)\, ds \leqslant \iint |f(t)|\, |g(s)| \{W_r(s,t)\, u_o^2(s) u_o^2(t)\}\, ds\, dt$$

$$\leqslant \{ \int |f(t)|^p\, u_o^2(t) dt\}^{1/p} \{\int |g(s)|^q\, u_o^2(s)\, ds\}^{1/q} .$$

The last inequality is obtained by Hölder's inequality with the weight in the braces followed by Fubini's theorem and an application of (12). Thus T is a contraction on $L^p(u_o^2(s)\, ds)$ for $1 \leqslant p \leqslant \infty$.

The other hypotheses of Stein's Maximal theorem are all obviously true by direct computation.

PROOF OF THEOREM B. Let

$$P_r(t) = u_o(t)W_{r^2}(0,t) = \sum \exp(-l_n r^2)h_n u_n(t)$$

so that if we can find positive numbers Q and E such that

$$(14) \qquad k_r(t) \leq E\, P_{Qr}(t) \qquad (r>0,\ t \text{ in } I)$$

then $|k_r*f(t)| \leq k_r*|f|(t) \leq EP_{Qr}*|f|(t) = E\, u_o(t)\, T^r(f/u_o)$ so that

$$Mf(t) \leq E\, \|u_o\|_\infty \sup_{r>o} T^r(f/u_o)(t)$$

and Theorem B will follow from Stein's Maximal Theorem and Lemma 1 (v).

The various parts of Lemma 1 ensure the existence of positive constants J, C_1, C_2, C_3, C_4 with the following properties for n = 0, 1, 2, ... and s in I

$$C_1 \leq h_n \leq C_2\ ,$$

$$C_3 n \leq m_n \leq C_4 n$$

$$u_n(s) \geq b \quad \text{if} \quad 0 \leq ns \leq J^2\ .$$

The constants b, m, and M will have the same meaning as in Lemma 1.

Assume $t < Jr < pi$ and define E_1, E_2 and E_3 to be the sets of integers n such that $n \leq J/r$, $J/r < n \leq J^2/s$, and $J^2/s < n$ respectively; and let

$$S_j = \sum_{E_j} \exp(-l_n r^2)h_n u_n(t) \quad (j=1,2,3)\ .$$

Then

$$S_1 \geq \exp(-C_4^2 J^2)\ C_1 bJ/r$$

since for n in E_1, $nt < nJr < J^2$. If n in E_2, $u_n(t) \geq 0$ so $S_2 \geq 0$.

We now make the additional assumption that $t < r/Q$ for some $Q > 1$ so that if n in E_3, $n > J^2Q/r$ and so

$$S_3 \leqslant C_2 M \sum_{n>J^2Q/r} \exp(-C_3^2 n^2 r^2).$$

The series is dominated by

$$\int_{J^2Q/r}^{\infty} \exp(-C_3^2\, x^2 r^2)\, dx \; = \; (C_3 r)^{-1} \int_{C_3 J^2 Q} \exp(-y^2)\, dy\,,$$

so that if Q is sufficiently large $|S_3| \leqslant 1/2\ S_1$ thus there is a constant C_5 such that

$$P_r(t) \geqslant C_5/r \qquad (0 < t < r/Q),$$

but

$$k(t) \leqslant (mr)^{-1} \qquad (0 < t < r)$$

so for some constant E

$$k_{r/Q}(t) \leqslant EP_r(t)$$

and (14) is proved.

REFERENCES

[1] Askey, R. and Wainger, S., A convolution structure for Jacobi Series. Amer. J. Math. 91 (1969), 463 - 485.

[2] Coddington, E.A. and Levinson, N., Theory of Ordinary Differential Equations, McGraw Hill Book Co, New York, (1955)

[3] Connett, W.C., and Schwartz, A.L., The Theory of Ultraspherical Multipliers, Memoirs of The Amer. Math. Soc. 183, Providence, (1977).

[4] Connett, W.C. and Schwartz, A.L., The Littlewood-Paley theory for Jacobi expansions, Trans. Am. Math. Soc. 251 (1979), 219 - 234.

[5] Gasper, G. Positivity and Special Functions, Theory and Application of Special Functions, R. Askey Editor, Academic Press, (1975).

[6] Gilbert, J.E., Maximal theorems for some orthogonal series I, Trans. Am. Math. Soc., 145 (1969), 495 - 515.

[7] Gilbert, J.E., Maximal theorems for some orthogonal series II, J. Math. Anal. and Appl. 31(1970), 349 - 368.

[8] Koornwinder, T.H., New proof of the positivity of generalized translation for Jacobi series, Mathematisch Centrum, Amsterdam, (1975), preprint.

[9] Protter, M.H. and Weinberger, H.F., Maximum Principles in Differential Equations, Prentice-Hall, Englewood Cliffs, N.J. (1967).

[10] Stein, E., Topics in Harmonic Analysis Related to the Littlewood-Paley Theory, Ann. of Math. Studies, No. 63, Princeton University Press, Princeton, (1970).

International Series of
Numerical Mathematics, Vol. 65

PRODUCT FORMULAS FOR BESSEL, WHITTAKER, AND JACOBI FUNCTIONS VIA THE SOLUTION OF AN ASSOCIATED CAUCHY PROBLEM

Clemens Markett *)
Lehrstuhl A für Mathematik
Rheinisch-Westfälische Technische Hochschule
Aachen

Delsarte's approach to generalized translation operators via the solution of an associated Cauchy problem is used to derive the product formulas for the Bessel, Whittaker, and Jacobi functions in kernel form. As essential prerequisites, explicit representations of the corresponding Riemann functions are given for three cases. The main part of the paper deals with the Jacobi case, for which the derivation of the translation kernel is carried out explicitly. In the Bessel case, the results of Delsarte are covered, and in the Whittaker case, a generalization of previous results of the author on the Laguerre polynomial product formula is obtained.

1. Introduction

For $\alpha \geqslant \beta \geqslant -\frac{1}{2}$, $\lambda \in \mathbb{C}$, and $0 < x < \infty$ let the Jacobi functions $\varphi_\lambda^{\alpha,\beta}(x)$ be defined by

$$\varphi_\lambda^{\alpha,\beta}(x) = R^{\alpha,\beta}_{(i\lambda-\alpha-\beta-1)/2}(\operatorname{ch} 2x) ,$$

$$R_\mu^{\alpha,\beta}(z) = F(-\mu,\mu+\alpha+\beta+1;\alpha+1;\frac{1-z}{2}) ,$$

where F denotes the hypergeometric function. If $\mu = n$, for some non-negative integer n, the latter function reduces to the Jacobi polynomial $R_n^{\alpha,\beta}(z) = P_n^{\alpha,\beta}(z)/P_n^{\alpha,\beta}(1)$. Using this relation together with an analytic continuation argument, Koornwinder [8] obtained, as a corollary to the addition formula for Jacobi polynomials, the product formula for Jacobi functions

*) Supported by the Deutsche Forschungsgemeinschaft under grant No. Go 261/5-2.

$$(1.1)\quad R_\mu^{\alpha,\beta}(x)\, R_\mu^{\alpha,\beta}(y) = \frac{2\,\Gamma(\alpha+1)}{\Gamma(\alpha-\beta)\Gamma(\beta+1/2)\Gamma(1/2)} \cdot \int_{\psi=0}^{\pi} \int_{r=0}^{1} R_\mu^{\alpha,\beta}\Big(\tfrac{1}{2}(x+1)(y+1) + \tfrac{1}{2}(x-1)y-1)r^2 + \sqrt{(x^2-1)(y^2-1)}\, r\cos\psi - 1\Big) \cdot (1-r^2)^{\alpha-\beta-1}\, r^{2\beta+1} \sin^{2\beta}\psi \, dr\, d\psi ,$$

where $x \geq 1$, $y \geq 1$, $\mu \in \mathbb{C}$, $\alpha > \beta > -\frac{1}{2}$. A direct analytic proof of (1.1) as well as its kernel form were given by Flensted - Jensen and Koornwinder [5].

Principally, the product formula for a given function system admits an extension to a certain function space, which then defines the generalized translation operator associated with the given system. The main purpose of this paper is to show that the kernel form of (1.1) can also be obtained by using the approach to generalized translations via the solution of an associated Cauchy problem. This method was initiated by Delsarte [3] in the particular case of the Hankel translation, which immediately leads to Sonine's product formula for the Bessel functions. Recently the author [10], [11] has shown that a proof of Watson's product formula for the Laguerre polynomials is possible by this method and that it can be used as a starting point for estimating generalized translation operators, which arise from perturbations of the Laguerre differential operator. It will be shown here that this way of proof of the Laguerre product formula can be extended to yield the product formula for the Whittaker functions, which was obtained by Glaeske [6] via a generalization of Watson's proof. The generalized translations associated with the Bessel, Whittaker, and Jacobi system have been used as the basic tools in defining the convolution structures associated with the Hankel transform, the so-called Laguerre - Pinney transform, and the Jacobi transform, respectively.

Another aspect of this approach is that the three product formulas can be derived simultaneously. This is due to the basic property of the Bessel, Whittaker, and Jacobi functions to be eigenfunctions of Sturm - Liouville differential equations which can be transformed in such a way that they differ by the potential function only. In § 2, the differential equations as well as the corresponding product formulas (Theorem 1) will be stated. In § 3, we briefly repeat the main steps in Delsarte's approach, i.e., the underlying

Cauchy problem, the characteristic boundary value problem for the associated Riemann function, as well as the formula relating the Riemann function with the resulting translation kernel. Explicit representations of the Riemann functions associated with the Bessel, Whittaker, and Jacobi differential operators are given in § 4, from which the product formulas can be derived. This is carried out for the Jacobi case in § 5. Concluding remarks follow in § 6.

2. Sturm - Liouville Equations and Product Formulas

For $\alpha \geqslant -\frac{1}{2}$, we consider the second order singular differential operator

$$D^{\alpha}_{q,x} - \frac{d^2}{dx^2} + \frac{2\alpha+1}{x}\frac{d}{dx} - q(x), \quad 0 < x < \infty ,$$

where in the Bessel, Whittaker, and Jacobi cases, the potential function q is chosen to be

$$q_1(x) \equiv 0 , \qquad q_2(x) = x^2 ,$$

$$q_3(x) = (\tfrac{1}{4} - \alpha^2)(x^{-2} - \mathrm{sh}^{-2}x) + (\tfrac{1}{4} - \beta^2)\,\mathrm{ch}^{-2}x \qquad (\alpha \geqslant \beta \geqslant -\tfrac{1}{2}) ,$$

respectively. The unified form of the Sturm - Liouville equations then reads

$$(2.1) \qquad D^{\alpha}_{q,x}\, u_\lambda(x) + \lambda^2\, u_\lambda(x) = 0 \qquad (0 < x < \infty,\ \lambda \in \mathbb{C}) ,$$

where the eigenfunctions u_λ are supposed to satisfy the initial conditions $u_\lambda(0) = 1$, $u'_\lambda(0) = 0$. For $q = q_i, i = 1,2,3$, the unique solutions are the Bessel, Whittaker, and Jacobi functions normalized by appropriate factors which are independent of λ, namely

$$u^{(1)}_\lambda(x) = j_\alpha(\lambda x) = \begin{cases} 2^\alpha\,\Gamma(\alpha+1)(\lambda x)^{-\alpha}\, J_\alpha(\lambda x) , & \lambda x \neq 0 \\ 1 , & \lambda x = 0 , \end{cases}$$

$$u^{(2)}_\lambda(x) = x^{-\alpha-1}\, M_{\lambda^2/4,\, \alpha/2}(x^2) =$$

$$= \exp(-x^2/2)\ {}_1F_1((\alpha+1)/2 - \lambda^2/4;\ \alpha+1; x^2)\ ,$$

$$u_\lambda^{(3)}(x) = (\mathrm{sh}\ x/x)^{\alpha+1/2}(\mathrm{ch}\ x)^{\beta+1/2}\ \varphi_\lambda^{\alpha,\beta}(x)\ .$$

We are looking for the product formulas of these eigenfunctions in kernel form, i.e.,

$$u_\lambda(x)\ u_\lambda(y) =$$

(2.2)
$$\begin{cases} \int_0^\infty u_\lambda(z) K_q^\alpha(x,y,z) z^{2\alpha+1}\, dz\ , & \alpha > -\frac{1}{2} \\ \frac{1}{2}\{u_\lambda(|x-y|) + u_\lambda(x+y)\} + \int_0^\infty u_\lambda(z) K_q^{-1/2}(x,y,z)\, dz\ , & \alpha = -\frac{1}{2} \end{cases}$$

for $x,y > 0$, where the weight functions $w(z) = z^{2\alpha+1}$ are chosen according to the selfadjoint form of DE(2.1). So the task is to determine the kernels $K_q^\alpha(x,y,z)$ explicitly in the three cases. The results are listed in the following theorem.

THEOREM 1. <u>For</u> $x,y > 0$, $|x-y| < z < x+y$, <u>the Bessel, Whittaker, and Jacobi product formulas in the form</u> (2.2) <u>have the following kernels.</u>

(i) $K_{q_1}^\alpha(x,y,z) =$

$$\begin{cases} \frac{\Gamma(\alpha+1)2^{1-2\alpha}}{\Gamma(\alpha+1/2)\Gamma(1/2)} (xyz)^{-2\alpha} ([(x+y)^2 - z^2][z^2-(x-y)^2])^{\alpha-1/2} & , \alpha > -\frac{1}{2} \\ 0\ , \quad \alpha = -\frac{1}{2}\ , \end{cases}$$

(ii) $K_{q_2}^\alpha(x,y,z) =$

$$\begin{cases} K_{q_1}^\alpha(x,y,z)\ j_{\alpha-1/2}(\frac{1}{2}\sqrt{[(x+y)^2 - z^2][z^2-(x-y)^2]})\ , & \alpha > -\frac{1}{2} \\ -\frac{1}{4} xyz\ j_1(\frac{1}{2}\sqrt{[(x+y)^2 - z^2][z^2-(x-y)^2]})\ , & \alpha = -\frac{1}{2}\ , \end{cases}$$

(iii) $K^{\alpha}_{q_3}(x,y,z) =$

$$\begin{cases} \frac{\Gamma(\alpha+1)2^{1-2\alpha}}{\Gamma(\alpha+1/2)\Gamma(1/2)}(xyz)^{-\alpha-1/2}\left(\frac{[\mathrm{ch}^2(x+y)-\mathrm{ch}^2 z][\mathrm{ch}^2 z-\mathrm{ch}^2(x-y)]}{\mathrm{sh}\,x\,\mathrm{ch}\,x\,\mathrm{sh}\,y\,\mathrm{ch}\,y\,\mathrm{sh}\,z\,\mathrm{ch}\,z}\right)^{\alpha-1/2} \cdot \\ \cdot F\left(\alpha+\beta,\alpha-\beta;\alpha+\frac{1}{2};\frac{[\mathrm{ch}(x+y)-\mathrm{ch}\,z][\mathrm{ch}\,z-\mathrm{ch}(x-y)]}{4\,\mathrm{ch}\,x\;\mathrm{ch}\,y\,\mathrm{ch}\,z}\right), \quad \alpha>-\frac{1}{2} \\ 0, \quad \alpha=-\frac{1}{2}. \end{cases}$$

For $z \leqslant |x-y|$ _and for_ $z \geqslant x+y$, _the kernels vanish._

REMARK. If $\alpha = -\frac{1}{2}$, Theorem 1 states that there are no integral terms in the Bessel and Jacobi product formulas. This is due to the fact that in both cases the eigenfunctions of (2.1) reduce just to $u_\lambda(x) = \cos\lambda x$, so that (2.2) reduces to the cosine product formula. In the Whittaker case for $\alpha = -\frac{1}{2}$, the integral term does not vanish. It can be considered as a generalization of the particular Laguerre product formula due to Boersma (cf. [9]).

3. Translation and Riemann Function

The main idea of Delsarte was to consider the generalized translation as the solution of a Cauchy problem. Concerning the product formula, this amounts to consider $u_\lambda(x)\, u_\lambda(y)$, i.e., the left hand side of (2.2), as the solution $u(x,y)$ of

$$(3.1) \qquad \begin{cases} (D^{\alpha}_{q,x} - D^{\alpha}_{q,y})u(x,y) = 0 & (0 < y \leqslant x), \\ u(x,0) = u_\lambda(x), \quad u_y(x,0) = 0 & (x > 0). \end{cases}$$

For $0 < x < y$ one uses the symmetry of the product and interchanges the roles of x and y. In order to solve (3.1) via Riemann's method one has to determine the corresponding Riemann function $A^{\alpha}_{q}(\xi,\eta;x,y)$, say. It is defined as the solution of the characteristic boundary value problem

$$
(3.2) \qquad \begin{cases} (D^{\alpha}_{q,\xi} - D^{\alpha}_{q,\eta})^{*}\, v(\xi,\eta) = 0 & \text{if } (\xi,\eta) \in \Delta_{xy} \\[2ex] v(\xi,\eta) = \left(\frac{\xi\eta}{xy}\right)^{\alpha+1/2} & \text{if } \xi - \eta = x - y \text{ or } \xi + \eta = x + y , \end{cases}
$$

where the asterisk denotes the adjoint operator and Δ_{xy} stands for the triangular region in the (ξ,η)-plane with vertices (x,y), $(x-y,0)$, and $(x+y,0)$, $0 < y \leqslant x$. If A^{α}_{q} as well as

$$
(3.3) \qquad B^{\alpha}_{q}(\xi,\eta;x,y) = -\frac{1}{2}\,\eta^{2\alpha+1}\,\frac{\partial}{\partial\eta}\,[(\xi\eta)^{-2\alpha-1}\,A^{\alpha}_{q}(\xi,\eta;x,y)]
$$

are continuous on Δ_{xy}, and if A^{α}_{q} is symmetric with respect to x and y, the kernel of the product formula can immediately be obtained from the Riemann function by (cf. [1],[10])

$$
(3.4) \qquad K^{\alpha}_{q}(x,y,\xi) = \begin{cases} \lim_{\eta\to 0+} B^{\alpha}_{q}(\xi,\eta;x,y) & \text{if } |x-y| < \xi < x + y , \\[2ex] 0 & \text{elsewhere} . \end{cases}
$$

This will be the basic relation in the proof of Theorem 1.

4. Riemann Functions Associated with the Bessel, Whittaker, and Jacobi Differential Operators

It is convenient to transform (3.2) into a normal form. Setting

$$
X = X(\xi,\eta) = \tfrac{1}{4}(\xi+\eta)^{2} , \quad X_{0} = X(x,y) ,
$$

$$
Y = Y(\xi,\eta) = \tfrac{1}{4}(\xi-\eta)^{2} , \quad Y_{0} = Y(x,y) ,
$$

and

$$
V(X,Y,X_{0},Y_{0}) = \left(\frac{X_{0}-Y_{0}}{X-Y}\right)^{\alpha+1/2} v\,(\xi,\eta;x,y) ,
$$

problem (3.2) turns into

$$(4.1)\qquad \begin{cases} V_{XY} + Q(X,Y)\,V = 0 & \text{if } (X,Y)\in\Delta_{X_oY_o} \\ V(X,Y;X_o,Y_o) = 1 & \text{if } X = X_o \text{ or } Y = Y_o\,, \end{cases}$$

where $\Delta_{X_oY_o}$ is the image of Δ_{xy} under the transformation and

$$Q(X,Y) = \frac{\alpha^2 - 1/4}{(X-Y)^2} - \frac{q(\sqrt{X}+\sqrt{Y}) - q(\sqrt{X}-\sqrt{Y})}{4\sqrt{XY}}\,.$$

For the Bessel potential $q_1 \equiv 0$, the second term on the right hand side vanishes, while it equals 1 for the Whittaker potential $q_2(x) = x^2$. The corresponding solutions of (4.1) on a certain subset of $\Delta_{X_oY_o}$ were given by Riemann [12] and Henrici [7] in terms of hypergeometric functions of one and two auxiliary variables, respectively. The complete solutions then follow by analytic continuation using Euler's transformations

$$(4.2)\qquad \begin{aligned} F(a,b;c,z) &= (1-z)^{-b}\,F(c-a,b;c;\tfrac{z}{z-1}) \\ F(a,b;c;z) &= (1-z)^{-a}\,F(a,c-b;c;\tfrac{z}{z-1}) \end{aligned} \qquad (|z|<1, |\tfrac{z}{z-1}|<1)\,.$$

(As a general reference for properties of the hypergeometric function we use [4]). Returning to the variables ξ,η again, and setting

$$\chi(\xi,\eta;x,y) = \tfrac{1}{16}[(x+y)^2 - (\xi+\eta)^2][(\xi-\eta)^2 - (x-y)^2]\,,$$

$$\psi(\xi,\eta;x,y) = \tfrac{1}{16}[(x+y)^2 - (\xi-\eta)^2][(\xi+\eta)^2 - (x-y)^2]\,,$$

$$\Phi(\xi,\eta;x,y) = \chi/\psi\,,$$

so that $1-\Phi = \xi\eta\,xy/\psi > 0$, the Riemann functions associated with the Bessel and Whittaker differential operators, respectively, are given by

$$(4.3)\qquad A^{\alpha}_{q_i}(\xi,\eta;x,y) = \left(\frac{\xi\eta}{xy}\right)^{\alpha+1/2}(\xi\eta xy)^{1/2-|\alpha|}\,a^{\alpha}_{q_i}(\xi,\eta;x,y)\,,$$

for $i = 1,2$ and $\alpha \geq -1/2$, where

$$a^{\alpha}_{q_1} = \psi^{|\alpha|-1/2}\, F(\tfrac{1}{2} - |\alpha|, \tfrac{1}{2} - |\alpha|; 1; \Phi) ,$$

$$a^{\alpha}_{q_2} = \psi^{|\alpha|-1/2} \sum_{k=0}^{\infty} F(k + \tfrac{1}{2} - |\alpha|, \tfrac{1}{2} - |\alpha|; k+1; \Phi) \frac{(-\chi)^k}{k!k!}$$

(cf.[10],[11]). Notice that the representation of $A^{\alpha}_{q_2}$ differs from that in [10] and can be obtained by interchanging the order of summations in Henrici's double sum.

In case of the Jacobi potential $q_3 = q_3(\alpha,\beta)$, another substitution is needed to transform the partial differential equation of (4.1) into an appropriate form. Indeed, setting $r(X) = 2\mathrm{ch}^2\sqrt{X} - 1$ one obtains, after some calculations,

$$Q(X,Y) = r'(X)r'(Y) \left\{ \frac{\alpha^2 - 1/4}{[r(X)-r(Y)]^2} - \frac{\beta^2 - 1/4}{[r(X)+r(Y)]^2} \right\}. \tag{4.4}$$

With an argument due to Henrici [7] the corresponding solution of (4.1) can then be reduced to a solution of a similar problem with $r(X)$, $r(Y)$ replaced by X,Y. The latter problem was solved by Chaundy [2] and leads to

$$V_{q_3}(X,Y;X_o,Y_o) = \sum_{n=o}^{\infty} \left\{ \sum_{k=o}^{\infty} \frac{(1/2+\beta)_k (1/2-\beta)_k}{k!(k+n)!} w_1^{\,k} \right\} \frac{(1/2+\alpha)_n (1/2-\alpha)_n}{n!} w_2^{\,n} , \tag{4.5}$$

where

$$w_1 = \frac{(\tilde{X}_o - \tilde{X})(\tilde{Y} - \tilde{Y}_o)}{(\tilde{X}+\tilde{Y})(\tilde{X}_o+\tilde{Y}_o)} , \qquad w_2 = \frac{(\tilde{X}-\tilde{X}_o)(\tilde{Y}-\tilde{Y}_o)}{(\tilde{X}-\tilde{Y})(\tilde{X}_o-\tilde{Y}_o)}$$

and $\tilde{X} = r(X)$ etc., as long as w_1 and w_2 remain inside the unit disk. The latter condition is obviously satisfied for w_1, whereas w_2 ranges over the negative semi-axis when (X,Y) ranges over $\Delta_{X_oY_o}$. To establish the complete solution via analytic continuation we first interchange the order of summations in (4.5) which is justified for small values of $|w_2|$, and then apply Euler's transformations (4.2) according to $\alpha>0$ or $\alpha\leq 0$. Setting now

$$\chi(\xi,\eta;x,y) = \frac{1}{4}[\mathrm{ch}(x+y)-\mathrm{ch}(\xi+\eta)][\mathrm{ch}(\xi-\eta)-\mathrm{ch}(x-y)]\ ,$$

$$\psi(\xi,\eta;x,y) = \frac{1}{4}[\mathrm{ch}(x+y)-\mathrm{ch}(\xi-\eta)][\mathrm{ch}(\xi+\eta)-\mathrm{ch}(x-y)]\ ,$$

$$\omega(\xi,\eta;x,y) = \mathrm{ch}\xi\ \mathrm{ch}\eta\ \mathrm{ch}x\ \mathrm{ch}y\ ,\quad \Phi=\chi/\psi\ ,\quad \Omega=\chi/\omega\quad ,$$

so that $1-\Phi=\mathrm{sh}\xi\ \mathrm{sh}\eta\ \mathrm{sh}x\ \mathrm{sh}y/\psi>0$, one finally arrives at the Riemann function associated with the Jacobi differential operator

$$\text{(4.6)}\quad \begin{aligned} A^{\alpha}_{q_3}(\xi,\eta;x,y) &= \left(\frac{\xi\eta}{xy}\right)^{\alpha+1/2}(\mathrm{sh}\xi\ \mathrm{sh}\eta\ \mathrm{sh}x\ \mathrm{sh}y)^{1/2-|\alpha|}\ a^{\alpha}_{q_3}(\xi,\eta;x,y)\ ,\\ a^{\alpha}_{q_3} &= \psi^{|\alpha|-1/2}\sum_{k=0}^{\infty}F(k+\tfrac{1}{2}-|\alpha|,\tfrac{1}{2}-|\alpha|;k+1;\Phi)\ \frac{(1/2+\beta)_k(1/2-\beta)_k}{k!\ k!}\ \Omega^k\ .\end{aligned}$$

5. Derivation of the Jacobi kernel

The proof of the three parts of Theorem 1 follows by applying (3.3), (3.4) to the Riemann functions (4.3),(4.6), which are continuous on Δ_{xy} and symmetric with respect to x and y. For the Bessel and Whittaker cases compare [10]. Obviously, the analysis in the Whittaker case is the same as in the particular Laguerre case. In the following we confine ourselves to the Jacobi case.

Here we have to calculate the limit as $\eta\to 0+$ of

$$\text{(5.1)}\quad B^{\alpha}_{q_3}(\xi,\eta;x,y) = -\frac{1}{2}(xy\xi)^{-\alpha-1/2}(\mathrm{sh}x\ \mathrm{sh}y\ \mathrm{sh}\xi)^{1/2-|\alpha|}\cdot\begin{cases}\eta^{2\alpha+1}\frac{\partial}{\partial\eta}[\eta^{-\alpha-1/2}(\mathrm{sh}\eta)^{1/2-\alpha}]a^{\alpha}_{q_3}(\eta)+\eta^{\alpha+1/2}(\mathrm{sh}\eta)^{1/2-\alpha}\frac{\partial}{\partial\eta}a^{\alpha}_{q_3}(\eta), & \alpha>0\\ \eta^{2\alpha+1}\frac{\partial}{\partial\eta}[(\mathrm{sh}\eta/\eta)^{\alpha+1/2}]\ a^{-\alpha}_{q_3}(\eta)+(\eta\,\mathrm{sh}\eta)^{\alpha+1/2}\frac{\partial}{\partial\eta}\,a^{-\alpha}_{q_3}(\eta), & -\frac{1}{2}\leq\alpha\leq 0\ .\end{cases}$$

For $\alpha=\beta=-\frac{1}{2}$ and $\alpha=\frac{1}{2}\geq\beta\geq-\frac{1}{2}$, one immediately obtains

$$a_{q_3}^{-1/2}(\eta) = 1\,, \quad a_{q_3}^{1/2}(\eta) = F(\tfrac{1}{2}+\beta,\ \tfrac{1}{2}-\beta;\ 1;\Omega)\,,$$

so that

$$(5.2) \qquad B_{q_3}^{-1/2}(\eta) \equiv 0\,, \quad B_{q_3}^{1/2}(\eta) \to \frac{1}{2xy\xi}\, a_{q_3}^{1/2}(0) \qquad (\eta\to 0+)\,.$$

For general α, it is easy to see that

$$(5.3) \qquad \begin{aligned} &\eta^{2\alpha+1}\frac{\partial}{\partial\eta}[\eta^{-\alpha-1/2}(\mathrm{sh}\eta)^{1/2-\alpha}] \to -2\alpha && (\eta\to 0+,\ \alpha>0)\,,\\ &\eta^{2\alpha+1}\frac{\partial}{\partial\eta}[(\mathrm{sh}\eta/\eta)^{\alpha+1/2}] = O(\eta^{2\alpha+2}) && (\eta\to 0+,\ -\tfrac{1}{2}<\alpha\leq 0),\end{aligned}$$

and so it remains to investigate the behavior of $a_{q_3}^{\gamma}$ and its derivative at zero for each $\gamma\geq 0$. The result is

LEMMA 1. <u>Let</u> $\gamma\geq 0$ <u>and let</u> $\psi_0=\psi(\xi,0;x,y)$, $\Omega_0=\psi_0/(\mathrm{ch}x\ \mathrm{ch}y\ \mathrm{ch}\xi)$. <u>Then</u>

(i) $$\lim_{\eta\to 0+} a_{q_3}^{\gamma}(\eta) = \frac{\Gamma(2\gamma)}{\Gamma^2(\gamma+1/2)}\,\psi_0^{\gamma-1/2}\, F(\tfrac{1}{2}+\beta,\ \tfrac{1}{2}-\beta;\gamma+\tfrac{1}{2};\Omega_0) \qquad (\gamma>0)\,,$$

(ii) $$a_{q_3}^{0}(\eta) = O(\log\tfrac{1}{\eta}) \qquad (\eta\to 0+),$$

(iii) $$\lim_{\eta\to 0+} (\mathrm{sh}\eta)^{1-2\gamma}\frac{\partial}{\partial\eta}a_{q_3}^{\gamma}(\eta) =$$

$$-\frac{\Gamma(1-2\gamma)}{\Gamma^2(1/2-\gamma)}\,\psi_0^{-\gamma-1/2}(\mathrm{sh}x\ \mathrm{sh}y\ \mathrm{sh}\xi)^{2\gamma}F(\tfrac{1}{2}+\beta,\ \tfrac{1}{2}-\beta;\ \tfrac{1}{2}-\gamma;\Omega_0) \quad (0\leq\gamma<\tfrac{1}{2})$$

(iv) $$\frac{\partial}{\partial\eta}a_{q_3}^{\gamma}(\eta) = O(1) \qquad (\eta\to 0+;\ \gamma>\tfrac{1}{2})\,.$$

PROOF. The behavior of $a_{q_3}^{\gamma}$ is governed by the hypergeometric function of Φ, since Φ tends to 1- when η tends to $0+$. Part (i) follows by interchanging the limit and summation and by using Gauss' summation theorem

$$(5.4)\qquad F(a,b;c;1) = \frac{\Gamma(c)\Gamma(c-a-b)}{\Gamma(c-a)\Gamma(c-b)} \qquad (c \neq 0,-1,\ldots;\mathrm{Re}\,c > \mathrm{Re}(a+b))$$

Concerning (ii), i.e., in the limiting case $\gamma = 0$, we use the asymptotic behavior

$$F(a,b;a+b;z) = \frac{\Gamma(a+b)}{\Gamma(a)\Gamma(b)} \log(\frac{1}{1-z}) + O(1) \qquad (z \to 1-)$$

instead of (5.4). In the remaining two parts, term by term differentiation of $a^{\gamma}_{q_3}$ is allowed for $\eta > 0$, so that

$$\frac{\partial}{\partial\eta} a^{\gamma}_{q_3}(\eta) = (\gamma - \frac{1}{2})\, \psi^{\gamma-3/2}\, \psi'(\eta) \sum_{k=0}^{\infty} \ldots \Omega^k$$

$$+ \psi^{\gamma-1/2} \sum_{k=1}^{\infty} F(k+\frac{1}{2}-\gamma,\ \frac{1}{2}-\gamma;\ k+1;\Phi)\frac{(1/2+\beta)_k(1/2-\beta)_k}{k!(k-1)!}\, \Omega^{k-1}\, \Omega'(\eta)$$

$$+ \psi^{\gamma-1/2} \sum_{k=0}^{\infty} \frac{(k+1/2-\gamma)(1/2-\gamma)}{k+1}\, F(k+\frac{3}{2}-\gamma,\ \frac{3}{2}-\gamma;k+2;\Phi)\Phi'(\eta).$$

$$\cdot \frac{(1/2+\beta)_k(1/2-\beta)_k}{k!\ k!}\, \Omega^k$$

$$= b^{\gamma}_1(\eta) + b^{\gamma}_2(\eta) + b^{\gamma}_3(\eta),$$

say. An easy calculation shows that

$$\chi'(\eta) \to -\frac{1}{2}\,\mathrm{shx\ xhy\ sh}\xi, \quad \psi'(\eta) \to \frac{1}{2}\,\mathrm{shx\ shy\ sh}\xi,$$

$$\Phi'(\eta) \to -\,\mathrm{shx\ shy\ sh}\xi/\psi_0, \quad \Omega'(\eta) \to -\frac{1}{2}\frac{\mathrm{shx\ shy\ sh}\xi}{\mathrm{chx\ chy\ ch}\xi}$$

as $\eta \to 0+$, and, with similar arguments as in part (i), one obtains $b^{\gamma}_i(\eta) = O(1)$ if $\gamma > 0$, and $b^{\gamma}_i(\eta) = O(\log\frac{1}{\eta})$ if $\gamma = 0$, for $i = 1,2$ and $\eta \to 0+$. If $\gamma > \frac{1}{2}$, $b^{\gamma}_3(\eta)$ is uniformly bounded, too, so that part (iv) follows. For $0 \leqslant \gamma < \frac{1}{2}$, one first applies Euler's transformation

$$(5.5)\qquad F(a,b;c;z) = (1-z)^{c-a-b}\, F(c-a,c-b;c;z) \qquad (|z| < 1),$$

and then (5.4) to deduce (iii). So Lemma 1 is proved.

The proof of Theorem 1, (iii), now follows by inserting (5.3) and Lemma 1 into (5.1). For $\alpha > 0$, the first term on the right-hand side of (5.1) is the main part, while for $-\frac{1}{2} < \alpha \leq 0$, it is the second term. The two other terms vanish as $\eta \to 0+$. In both cases, the limits are of the same form. Thus for each $\alpha > -\frac{1}{2}$, one obtains

$$K^{\alpha}_{q_3}(x,y,\xi) = \lim_{\eta\to 0+} B^{\alpha}_{q_3}(\xi,\eta;x,y)$$

$$= \frac{1}{2}\frac{\Gamma(2\alpha+1)}{\Gamma^2(\alpha+1/2)}(xy\xi)^{-\alpha-1/2}(\mathrm{sh}x\ \mathrm{sh}y\ \mathrm{sh}\xi/\psi_0)^{1/2-\alpha}\cdot$$

$$\cdot F(1/2+\beta,1/2-\beta,\ \alpha+1/2;\ \Omega_0)$$

$$= \frac{\Gamma(\alpha+1)2^{1-2\alpha}}{\Gamma(\alpha+1/2)\Gamma(1/2)}(xy\xi)^{-\alpha-1/2}\left(\frac{\psi_0[1-\Omega_0]}{\mathrm{sh}x\ \mathrm{sh}y\ \mathrm{sh}\xi}\right)^{\alpha-1/2}\cdot$$

$$\cdot F(\alpha-\beta,\ \alpha+\beta;\ \alpha+1/2;\ \Omega_0)$$

by Legendre's duplication formula for $\Gamma(x)$ and (5.5), where

$$1-\Omega_0 = \frac{[\mathrm{ch}(x+y)+\mathrm{ch}\xi][\mathrm{ch}\xi+\mathrm{ch}(x-y)]}{4\ \mathrm{ch}x\ \mathrm{ch}y\ \mathrm{ch}\xi}.$$

Together with (5.2) for $\alpha = -\frac{1}{2}$, this yields the assertion.

6. Concluding Remarks

In the ultrasperical cases $\beta = \alpha$ as well as in the cases $\beta = -\alpha$, the translation kernel has a simpler structure than in the general Jacobi case, since the hypergeometric function in Theorem 1, (iii), then reduces to 1. It may be noted that this simplification can also be achieved right from the beginning of the above derivation, namely in the Riemann function itself. Indeed, if $\beta^2 = \alpha^2$, $Q(X,Y)$ in (4.4) reduces to

$$Q(X,Y) = s'(X)s'(Y)\frac{\alpha^2-1/4}{[s(X)-s(Y)]^2},\quad s(X) = [r(X)]^2.$$

Hence, using Henrici's argument again, one arrives at the same characteristic boundary value problem as in the Bessel case. This may be considered as a reason for the similarity of the Bessel and the ultraspherical kernels.

As already mentioned, in the Bessel case and thus also in the ultraspherical case, the Riemann functions are constructed by means of one auxiliary variable, whereas in the Whittaker and the general Jacobi cases, two of them were used. A similar phenomenon can be found in other important formulas in the field as, for example, in integral representations for the corresponding eigenfunctions in terms of single or double integrals.

ACKNOWLEDGEMENT

The author would like to thank Professor E. Görlich for helpful comments and suggestions.

REFERENCES

[1] Braaksma, B.L.J. - de Snoo, H.S.V., Generalized translation operators associated with a singular differential operator. In: Theory of Ordinary and Partial Differential Equations (B.D. Sleeman, I.M. Michaels, eds.) Lecture Notes Math. 415, Springer, Berlin 1974, 62 - 77.

[2] Chaundy, T.W., Linear partial differential equations (I). Quart. J. Math. Oxford Ser. 9 (1938), 234 - 240.

[3] Delsarte, J., Sur une extension de la formule de Taylor. J. Math. Pures Appl. 17 (1936), 213 - 231.

[4] Erdelyi, A., et al., Higher Transcendental Functions, Vol. 1. McGraw Hill, New York 1953.

[5] Flensted - Jensen, M. - Koornwinder, T.H., The convolution structure for Jacobi function expansions. Ark. Mat. 11(1973), 245 - 262.

[6] Glaeske, H.-J., Die Laguerre - Pinney - Transformation. Aequationes Math. 22(1981), 73 - 85.

[7] Henrici, P., A survey of I.N. Vekua's theory of elliptic differential equations with analytic coefficients. Z. Angew. Math. Phys. 8(1957), 169 - 203.

[8] Koornwinder, T.H., The addition formula for Jacobi polynomials, I. Summary of results. Indag. Math. 34 (1972), 188 - 191.

[9] Markett, C., Mean Cesàro summability of Laguerre expansions and norm estimates with shifted parameter. Anal. Math. 8 (1982), 19 - 37.

[10] Markett, C., A new proof of Watson's product formula for Laguerre polynomials via a Cauchy problem associated with a singular differential operator. (to appear)

[11] Markett, C., Norm estimates for generalized translation operators associated with a singular differential operator. (to appear)

[12] Riemann, B., Über die Fortpflanzung ebener Luftwellen von endlicher Schwingungsweite. Gesammelte Math. Werke, Leipzig 1892, 156 - 181.

International Series of
Numerical Mathematics, Vol. 65

AN OUTLINE OF THE FOURIER TRANSFORMATION AS INTEGRATION WITH RESPECT TO A VECTOR MEASURE

P. Masani
Departments of Mathematics
University of Pittsburgh
Pittsburgh, PA 15260

We present a unified theory of the p,p' Fourier transformation $F_{p,p'}$ over any locally compact abelian group Γ for $p \in [1,2]$, $p' := p/(p-1)$, based on (proper) Lebesgue integration with respect to a vector measure. This generalizes the case $p = 2$ settled in 1969, [8]. It transpires that for $p \in (1,2)$, $L_p(\Gamma)$ is only a proper subset of the domain of $F_{p,p'}$.

1. Introduction

The theory of the Fourier transformation is a theory of improper integration. Only in case $f \in L_1(\mathbb{R})$ do we have

$$f^{\vee}(\lambda) = \int_{\mathbb{R}} e^{it\lambda} f(t)dt, \tag{1.1}$$

where the last is a (proper) Lebesgue integral. In all other cases, in order that the equation

$$f^{\vee}(\lambda) = ⨍_{\mathbb{R}} e^{it\lambda} f(t)dt \tag{1.2}$$

should make sense, we have to interpret the sign $⨍_{\mathbb{R}}$ as an improper integral. For instance for $f \in L_2(\mathbb{R})$, we must interpret $⨍_{\mathbb{R}}$ as $\underset{A \to \infty}{\text{l.i.m.}} \int_{-A}^{A}$.

The only way to get a theory based on proper Lebesgue integration is by using vector-valued measures, cf. [8,9] for details. For instance, for the group $\mathbb{R}$, we must first recall that for $f \in \ell_2 = L_2(\mathbb{N})$, the equation

(1.3) $$f^{\vee} = \sum_{n=-\infty}^{\infty} f(n)e(n) \text{ in } L_2[0,2\pi], \quad e(n)(\theta) := e^{ni\theta}, \quad \theta \in [0,2\pi]$$

prevails,and for $f \in L_2(\mathbb{R})$ write by analogy

(1.4) $$f^{\vee} = \int_{-\infty}^{\infty} f(t)\xi(dt) \text{ in } L_2(\mathbb{R}), \xi(B)(\lambda) := \int_B e^{i\lambda t}dt, \quad \lambda \in \mathbb{R}$$

replacing the o.n. basis $e(\cdot)$ of $L_2[0,2\pi]$ by its "continuous analogue", the the <u>orthogonally scattered</u> (<u>o.s.</u>) <u>basic</u> $L_2(\mathbb{R})$-valued measure $\xi(\cdot)$. Such o.s. Hilbert space valued measures and their integration is discussed in our paper [7]. In the later paper [8] we demonstrated that a theory of the Fourier-Plancherel transform for any l.c.a. group can be built on the vectorial pattern of equation (1.4).

The definition of integration with respect to o.s. measures adopted by us in [7] was a restatement in abstract Hilbert spacial terms of Doob's definition of stochastic integration [2:p.426-], the idea of which goes back to Wiener and Paley. In the early 1970s D.R. Lewis [5,6] and independently E.G.F. Thomas [12] studied countably additive (c.a.) measures ξ on a δ-ring $\mathcal{D}$ with values in a Banach or even locally convex topological vector space X, as well as their integrals $\int_{\Omega} f(\omega)\xi(d\omega)$ for scalar-valued f, extending thereby the theory in Dunford-Schwartz [3]. When X is a Hilbert space H and ξ is o.s., their theory reduces precisely to the one we developed in [7], (see 2.15 below).

Now the measure ξ defined in (1.4) takes values in all Banach spaces $L_q(\mathbb{R})$, for $q \in [2,\infty]$, and is indeed c.a. under their topologies. The question naturally arises whether for any $p \in [1,2]$, the equation (1.4) continues to hold in $L_{p'}(\mathbb{R})$ for $f \in L_p(\mathbb{R})$, and whether the $f^{\vee}$ so obtained is the Hausdorff-Young Fourier transform of f. In the present paper we answer this question in the affirmative for any l.c.a. group Γ replacing $\mathbb{R}$. To state the issue more fully, write $\xi_{p'}$ for ξ when ξ is treated as a measure with values in $L_{p'}(\Gamma^{\wedge})$, where $\Gamma^{\wedge}$ is the dual group of Γ, and let $L_{1,\xi_{p'}}(\Gamma)$ be the set of $\xi_{p'}$-integrable functions. We then show that

(1.5) $$\forall\, p \in [1,2], \quad L_p(\Gamma) \subseteq L_{1,\xi_{p'}}(\Gamma),$$

and moreover that $\forall\, f \in L_p(\Gamma)$, the $f^{\vee}$ defined by the analogue of (1.4) is the

Hausdorff-Young Fourier transform in $L_{p'}(\Gamma^\wedge)$ of f. Equality prevails in (1.5) for all Γ when $p=2$, but it fails for $\Gamma=\mathbb{N}$, the group of integers, when $p\in(1,2)$. The situation for $p=1$ is not yet clear. Examples of functions

$$f\in L_{1,\xi_{p'}}(\mathbb{N}) \setminus L_p(\mathbb{N}), \quad p\in(1,2),$$

based on a theorem of Hardy and Littlewood, [13:II,p.109,(3.19)], have been given by H. Helson.[1] This means that the p,p' Fourier transformation, conceived vectorally, has a larger domain than $L_p(\Gamma)$, for $p\in(1,2)$. Unfortunately, the "Riesz-Fischer Theorem" for the subject, to wit that $L_{1,\xi_{p'}}(\Gamma)$ is a Banach space, remains open. As of now we only know that it is a normed vector space. There are many such open questions.

Space limitations will compel us to omit many proofs.

2. Integration with respect to $\xi\in CA(\mathcal{D},X)$

2.1 Notation. In this section (i) $\mathcal{D}$ is a δ-ring over a space Ω; (ii) $B_{\mathcal{D}}$ and $A_{\mathcal{D}}$ are the σ-ring and σ-algebra generated by $\mathcal{D}$; (iii)

$$\mathcal{D}^{loc} := \{B:\ B\subseteq\Omega \ \ \&\ \ \forall\, D\in\mathcal{D},\ B\cap D\in\mathcal{D}\};$$

(iv) X is a Banach space over $\mathbb{F}$;[2] (v) $\xi\in CA(\mathcal{D},X)$, the set of countably additive measures on $\mathcal{D}$ to X; (vi) $s_\xi(\cdot)$ on 2^Ω is the semi-variation of ξ; (vii) the subspace spanned by ξ is[3]

$$\mathcal{S}_\xi := S\{\xi(D):\ D\in\mathcal{D}\}\subseteq X.$$

For $x^*\in X^*$, the adjoint of X,[4] $x^*\circ\xi\in CA(\mathcal{D},\mathbb{F})$, its total variation measure

1 Who attended the October, 1982,Maryland meeting of the AMS, where the writer raised this issue, after referring to an error in his Abstract [11]. Professor Helson, to whom the writer is much indebted, communicated his example in December, 1982.

2 $\mathbb{F} := \mathbb{R}$ or $\mathbb{C}$. $\mathbb{R}_{0+} := \{x:\ x\in\mathbb{R}\ \&\ x\geqslant 0\}$.

3 For $A\subseteq X$, $S(A)$ is the closure of the linear manifold spanned by A.

4 The adjoint of X is defined by $X^* := \{\overline{x'(\cdot)}:\ x'\in X'\}$, where X' is the dual of X, the bar being complex conjugation. The use of X^* instead of X' facilitates the subsumation of the Hilbert space case.

$|x^* \circ \xi|$, which is easily seen to be in $CA(\mathcal{D}, \mathbb{R}_{0+})$, has a unique Hahn-Caratheodory extension to $\mathcal{B}_{\mathcal{D}}$, and a further extension to $A_{\mathcal{D}}$ obtainable by assigning the value ∞ to sets in $A_{\mathcal{D}} \setminus \mathcal{B}_{\mathcal{D}}$. We shall denote this last extension by $|x^* \circ \xi|^{\cdot}$; thus

$$(2.2) \qquad \forall\, x^* \in X^* \qquad |x^* \circ \xi|^{\cdot} \in CA(A_{\mathcal{D}}, [0,\infty]),$$

The classes $L_{1,|x^* \circ \xi|^{\cdot}} := L_1(\Omega, A_{\mathcal{D}}, |x^* \circ \xi|^{\cdot}; \mathbb{F})$ will play an important role.

The class $L_{1,\xi}(\Omega) := L_1(\Omega, \mathcal{D}, \xi; \mathbb{F})$ of ξ integrable functions over sets in $\mathcal{D}^{loc}$ is defined by the condition:

$$(2.3) \left\{ \begin{array}{l} f \in L_{1,\xi}(\Omega), \text{ iff } f \in \bigcap_{x^* \in X^*} L_{1,|x^* \circ \xi|^{\cdot}}, \text{ and } \forall\, C \in \mathcal{D}^{loc}, \exists\, x_C \in X \text{ such that} \\ \qquad \forall\, x^* \in X^*, \quad x^*(x_C) = \int_C \overline{f(\omega)}(x^* \circ \xi)(d\omega). \end{array} \right.$$

For any $C \in \mathcal{D}^{loc}$, this x_C is unique, and we define the integral of f with respect to ξ over $C \in \mathcal{D}^{loc}$ by

$$(2.4) \left\{ \begin{array}{l} \forall\, f \in L_{1,\xi}(\Omega) \;\&\; \forall\, C \in \mathcal{D}^{loc}, \\ \int_C f(\omega)\xi(d\omega) := \text{the (unique) } x_C \text{ in } X \text{ in (2.3).} \end{array} \right.$$

In particular for $C = \Omega$, we shall follow the probabilists and write

$$(2.5) \qquad \forall\, f \in L_{1,\xi}(\Omega), \quad E_\xi(f) := \int_\Omega f(\omega)\xi(d\omega) \in X .$$

The existence of x_C for $C \in \mathcal{D}^{loc}$ is automatic for $f \in \bigcap_{x^* \in X^*} L_{1,|x^* \circ \xi|^{\cdot}}$ when X is <u>weakly</u> <u>Σ-complete</u> in the sense of E.G.F. Thomas [12].[5] Consequently

$$(2.6) \left\{ \begin{array}{l} \text{For weakly } \Sigma\text{-complete } X, \; L_{1,\xi}(\Gamma) = \bigcap_{x^* \in X^*} L_{1,|x^* \circ \xi|^{\cdot}} \\ \text{i.e. } \; f \in L_{1,\xi}(\Omega) \Leftrightarrow \forall\, x^* \in X^* \; \int_\Omega |f(\omega)|\,|x^* \circ \xi|^{\cdot}(d\omega) < \infty. \end{array} \right.$$

5 See Appendix for the definition.

The definition (2.3), from which it is clear that any f in $L_{1,\xi}(\Omega)$ is $A_{\mathcal{D}}$, Borel measurable, suggests the introduction of two norms for functions f which are $A_{\mathcal{D}}$, Borel measurable on Ω to $\mathbb{F}$:

$$(2.7)\quad \begin{cases} |f|_{\xi} := \sup\limits_{\substack{x^*\in X^* \\ |x^*|\leq 1}} |f|_{1,|x^*\circ\xi|} \in [0,\infty], \\[2ex] \|f\|_{\xi} := \sup\limits_{A\in\mathcal{D}^{loc}} \sup\limits_{\substack{x^*\in X^* \\ |x^*|\leq 1}} \left|\int_A f(\omega)(x^*\circ\xi)(d\omega)\right| \in [0,\infty]. \end{cases}$$

Some of the fundamental properties of the space $L_{1,\xi}(\Omega)$ and of ξ integration established by Lewis and Thomas are stated in the next two theorems.

THEOREM 2.8. (a) For any $f\in L_{1,\xi}(\Omega)$,

$$\|f\|_{\xi} \leq |f|_{\xi} \leq 4\|f\|_{\xi} < \infty,$$

and the following equalities are equivalent:

$$\|f\|_{\xi} = 0,\ |f|_{\xi} = 0,\ s_{\xi}(\text{supp } f) = 0,\ |\xi|(\text{supp } f) = 0.$$

(b) $L_{1,\xi}(\Omega)$ is a pre-Banach space under the equivalent norms $|\cdot|_{\xi}$, $\|\cdot\|_{\xi}$, when functions differing on sets of zero semivariation $s_{\xi}(\cdot)$ (or equivalently, zero total-variation $|\xi|(\cdot)$) are identified.

(c) The linear manifold $\mathcal{S}(\mathcal{D},\mathbb{F})$ of $\mathcal{D}$-simple, $\mathbb{F}$-valued functions on Ω is everywhere dense in $L_{1,\xi}(\Omega)$.

(d) $$f\in L_{1,\xi}(\Omega)\ \&\ C\in\mathcal{D}^{loc} \Rightarrow f\cdot\chi_C\in L_{1,\xi}(\Omega).$$

(e) For weakly Σ-complete X,

$$f\in L_{1,\xi}(\Omega) \Leftrightarrow f \text{ is } A_{\mathcal{D}} \text{ Borel measurable on } \Omega\ \&\ |f(\cdot)|\in L_{1,\xi}(\Omega).$$

THEOREM 2.9. (a) $\forall\, D\in\mathcal{D}$, $E_{\xi}(\chi_D) = \xi(D)$.

(b) $$f \in L_{1,\xi}(\Omega) \ \& \ C \in \mathcal{D}^{loc} \Rightarrow E_\xi(f\chi_C) = \int_C f(\omega)\xi(d\omega)\,.$$

(c) The integration E_ξ is a linear contraction of norm $|E|_\xi \leqslant 1$ on $L_{1,\xi}(\Omega)$ (with-either norm) into X:

$$|E_\xi(f)|_X \leqslant \|f\|_\xi \leqslant |f|_\xi, \quad f \in L_{1,\xi}(\Omega).$$

(d) The weak (or strong) closure of Range E_ξ is $\mathcal{S}_\xi \subseteq X$, cf. 2.1(vii) .

(e) Let $f \in L_{1,\xi}(\Omega)$ and $\forall\ C \in \mathcal{D}^{loc}$, $\eta_f(C) := \int_C f(\omega)\xi(d\Omega)$. Then [6] $\eta_f \in BCA(\mathcal{D}^{loc}, X)$.

(f) The correspondence $f \to \eta_f$ is a linear isometry on the space $L_{1,\xi}(\Omega)$ with norm $\|\cdot\|_\xi$ into the Banach space $BCA(\mathcal{D}^{loc}, X)$ with the sup norm.

For the application to the Fourier transformation (§4) we need in our picture a non-negative measure corresponding to Haar measure. We accordingly stipulate that

(2.10) $$\mu \in CA(\mathcal{B}_\mathcal{D}, [0,\infty]),\ \forall\ D \in \mathcal{D},\ \mu(D) < \infty \ \& \ \xi \ll \mu \text{ on } \mathcal{D}.$$

A moot question then arising is whether any of the spaces $L_{p,\mu}(\Omega)$ fall inside $L_{1,\xi}(\Omega)$. To address this question, first note that $\forall\ x^* \in X^*$, $x^* \circ \xi \ll \mu$ on $\mathcal{D}$, and by the classical Radon-Nikodym theorem,

(2.11) $$\forall\ x^* \in X^*,\ T(x^*) := \psi_{x^*} := \frac{d(x^* \circ \xi)}{d\mu} \in L_1^{loc}(\Omega).$$

The result (2.6) thus becomes

(2.12) $$\left\{ \begin{array}{l} \text{For weakly } \Sigma\text{-complete } X\,, \\ f \in L_{1,\xi}(\Omega) \Leftrightarrow \forall\ x^* \in X^*,\ \int_\Omega |f(\omega)|\,|\psi_{x^*}(\omega)|\,\mu(d\omega) < \infty. \end{array} \right.$$

But by the Hölder inequality, we have for any $p \in [1,\infty]$, any $A_\mathcal{D}$, Borel measurable f in $\mathbb{C}^\Omega$ and all $x^* \in X^*$

(2.13) $$|f|_{1,|x^* \circ \xi|} := |f \cdot \psi_{x^*}|_{1,\mu} \leqslant |f|_{p,\mu} |\psi_{x^*}|_{p',\mu} \leqslant \infty\ .$$

6 We write BCA for bounded and countably additive.

This immediately answers our question for weakly Σ-complete spaces X:

THEOREM 2.14. Let (i) the Banach space X be weakly Σ-complete, (ii) (2.10) hold, (iii) $p \in [1,\infty]$ be such that in (2.11) Range $T \subseteq L_{p',\mu}(\Omega)$, i.e.

$$\forall\, x^* \in X^*, \quad \psi_{x^*} := \frac{d(x^* \circ \xi)}{d\mu} \in L_{p',\mu}(\Omega).$$

Then (a) $L_{p,\mu}(\Omega) \subseteq L_{1,\xi}(\Omega)$, (b) Moreover,

$$\forall\, f \in L_{p,\mu}(\Omega), \quad |f|_\xi \leqslant |f|_{p,\mu} \sup_{x^* \in X^*, |x^*| \leqslant 1} |\psi_{x^*}|_{p',\mu} \leqslant \infty .$$

PROOF. Let $f \in L_{p,\mu}(\Omega)$. Then by (ii) and (2.13), the inequality on the RHS of (2.12) holds for all $x^* \in X^*$. Hence $f \in L_{1,\xi}(\Omega)$. The inequality between the norms is clear from (2.7) and (2.13). □

For X, which are not weakly Σ-complete, the answer has to be worked out individually.

A useful application of the last theorem is to the case in which H is a Hilbert space and ξ is o.s. We then have the following theorem.

THEOREM 2.15. Let (i) $X = H$ be a Hilbert space over $\mathbb{F}$, (ii) $\xi \in CAOS(\mathcal{D},H)$ and (iii) $\mu(\cdot)$ be the Hahn extension of its control measure $|\xi(\cdot)|_H^2$ to $A_{\mathcal{D}}$. Then (a) $L_{1,\xi}(\Omega) = L_{2,\mu}(\Omega)$, and $E_\xi = S$, where S is the integration with respect to ξ defined in [7: 5.4 - 5.6]; thus E_ξ is an isometry on $L_{2,\mu}(\Omega)$ onto $\mathcal{S}_\xi \subseteq H$; (b) $\forall\, f \in L_{2,\mu}(\Omega)$, $\| f \|_\xi = |f|_\xi = |f|_{2,\mu}$.

3. The $L_{p'}(\Gamma^\wedge)$-valued measure ξ, $p \in [1,2]$

We now turn to the specific vector measures encountered in Fourier transform theory. Our notation will be as follows:

3.1. Notation.

(i) Γ is an (additive, Hausdorff) l.c.a. group,

(ii) $\mathcal{B}$ is the σ-ring generated by the topology of Γ,

(iii) m is a Haar measure for Γ with domain $\mathcal{B}$, [7]

7 We follow [4: I,pp. 193-194,(i)-(viii)] in regard to the definition of Haar measure. Unless Γ is compact, $\mathcal{B}_m$ is not a σ-ring but a δ-ring.

(iv) $\mathcal{B}_m := \{B : B \in \mathcal{B}\ \&\ m(B) < \infty\}$,

(v) $\mathcal{B}_\sigma = \sigma\text{-ring}(\mathcal{B}_m)$, $A_\sigma = \sigma\text{-alg}(\mathcal{B}_m)$,

(vi) For $p \in [1,\infty]$, $L_p(\Gamma) := L_p(\Gamma,\mathcal{B},m;\mathbb{C})$, & $|\cdot|_{p,m}$ is the usual norm for $L_p(\Gamma)$,

(vii) $\mathcal{S}_m(\Gamma) :=$ the family of $\mathcal{B}_m$ simple functions on Γ to $\mathbb{C}$,

(viii) $\Gamma^\wedge$ is the (multiplicative) character group of Γ,

(xi) $\mathcal{B}^\wedge$, $m^\wedge$, $\mathcal{B}^\wedge_{m^\wedge}$, $\mathcal{B}^\wedge_\sigma$, $A^\wedge_\sigma$ $L_p(\Gamma^\wedge)$, $\mathcal{S}_{m^\wedge}(\Gamma^\wedge)$ are defined for $\Gamma^\wedge$ in the same way as $\mathcal{B}$, m, $\mathcal{B}_m$, etc. are defined for Γ. It is understood that $m^\wedge$ is chosen as the precise dual of m in the sense of [8: 3.17],

(x) The letters t and λ will stand for the members of Γ and $\Gamma^\wedge$ respectively, (these being regarded as the "time" and "frequency" domains).

DEFINITION 3.2. $\forall\ B \in \mathcal{B}_m$, $\xi(B) := \chi^\vee_B$ <u>on</u> $\Gamma^\wedge$, <u>where</u>

$$\chi^\vee_B(\lambda) := \int_B \lambda(t)m(dt),\ \lambda \in \Gamma^\wedge,$$

By the Riemann-Lebesgue theorem (maximal ideal theory of $L_1(\Gamma)$) we see that

$$\forall\ B \in \mathcal{B}_m,\ \xi(B) = \chi^\vee_B \in C_0(\Gamma^\wedge). \tag{3.3}$$

Also, cf. [8: 4.4],

$$\left.\begin{aligned} &\forall\ B \in \mathcal{B}_m,\ \xi(B) \in L_\infty(\Gamma^\wedge) \cap L_2(\Gamma^\wedge);\\ &\text{in fact } \forall\ A,B \in \mathcal{B}_m,\\ &\max_{\lambda\in\Gamma^\wedge} |\xi(B)(\lambda)| = |\xi(B)|_{\infty,m^\wedge} = m(B) < \infty,\\ &(\xi(A),\xi(B))_{2,m^\wedge} = m(A \cap B),\ |\xi(B)|_{2,m^\wedge} = \sqrt{m(B)} < \infty . \end{aligned}\right\} \tag{3.4}$$

From (3.4) we easily infer that

$$\left.\begin{aligned} &\forall\ B \in \mathcal{B}_m\ \&\ \forall\ p \in (1,2),\ |\xi(B)|_{p',m^\wedge} \leqslant m(A)^{1/p} < \infty\\ &\forall\ B \in \mathcal{B}_m,\ \xi(B) \in \bigcap_{1\leqslant p\leqslant 2} L_{p'}(\Gamma^\wedge). \end{aligned}\right\} \tag{3.5}$$

Now let X stand for $C_o(\Gamma^\frown)$ or $L_{p'}(\Gamma^\frown)$ for $p \in [1,2]$. Clearly the measure ξ is finitely additive on $\mathcal{B}_m$ to X, and by (3.4), (3.5), $\xi \ll m$. Hence ξ is countably additive. The attributes of ξ uncovered up to now may therefore be summed up in the two assertions:

$$(3.6)\quad \begin{cases} \xi \in CA(\mathcal{B}_m, C_o(\Gamma^\frown)) \cap \bigcap_{1\leqslant p\leqslant 2} CA(\mathcal{B}_m, L_{p'}(\Gamma^\frown)); \\ \xi \ll m, \text{ under each of the } C_o \text{ or } L_{p'} \text{ norms.} \end{cases}$$

NOTATION 3.7. *We shall write* $\xi_{p'}$ *instead of* ξ *when we treat* ξ *as a measure in the Banach space* $L_{p'}(\Gamma^\frown)$ *for* $p \in [1,2]$.

For each $p \in [1,2]$, $\xi_{p'}$ is a *stationary measure* in the sense of [10: 2.3], and for its total variation measure $|\xi_{p'}|(\cdot)$, it may be shown that either (i) $\forall$ non-void open $V \subseteq \Gamma$, $|\xi_{p'}|(V) = \infty$, or (ii) $|\xi_{p'}|(\cdot) = m(\cdot)$. Moreover, (ii) holds only when $p' = \infty$, or when $p \in (1,2]$ and Γ is discrete (and therefore m is the cardinality measure). Although $\xi_{p'} \ll m$, for all $p \in [1,2]$, only in the cases $p = 1$ or Γ is discrete do we have the Radon-Nikodym theorem for the measures $\xi_{p'}$ and m.

4. The p,p' Fourier Transformation

Let $p \in [1,2]$, and consider the measure $\xi_{p'}$ defined in 3.2. and 3.7. It follows from (3.6) that the conditions 2.1(v) and (2.10) are fulfilled when $\Omega = \Gamma$, $\mathcal{D} = \mathcal{B}_m$, (and therefore $A_{\mathcal{D}} = A_\sigma$), $X = L_{p'}(\Gamma^\frown)$, $\xi = \xi_{p'}$, and $\mu = m$. For $p \in (1,2]$, $X = L_{p'}(\Gamma^\frown)$ is reflexive and therefore weakly Σ-complete, and all the results of §2 can be applied. For $p = 1$, our X, viz. $L_\infty(\Gamma^\frown)$, is not weakly Σ-complete and some of the results in §2 (e.g. 2.14) are not applicable.

From (2.5) we see that if $p \in [1,2]$, then

$$\forall\, f \in L_{1,\xi_{p'}}(\Gamma),\quad E_{\xi_{p'}}(f) = \int_\Gamma f(t)\xi_{p'}(dt) \in L_{p'}(\Gamma^\frown).$$

In the light of this and of our initial remarks (§1), we adopt the following definition:

DEFINITION 4.1. *For* $p \in [1,2]$, *we call the operator* $E_{\xi_{p'}}$ *on* $L_{1,\xi_{p'}}(\Gamma)$ *to* $L_{p'}(\Gamma^\frown)$

<u>the</u> <u>(indirect)</u> p,p' <u>Fourier</u> <u>transformation</u>, <u>and</u> <u>for</u> <u>any</u> $f \in L_{1,\xi_{p'}}(\Gamma)$, <u>we</u> <u>call</u> $f^{\vee} := E_{\xi_{p'}}(f)$ <u>the</u> <u>(indirect)</u> p,p' <u>Fourier</u> <u>transform</u> (FT) <u>of</u> f.

From Def. 3.2. it is clear that for $B \in \mathcal{B}_m$, $\xi_{p'}(B)$ is indeed the indirect Hausdorff-Young FT of χ_B. Since, cf. 2.9(a),(c), $E_{\xi_{p'}}(\chi_B) = \xi_{p'}(B)$ and $E_{\xi_{p'}}$ is linear, it follows that for all f in $\mathcal{S}_m(\Gamma)$, our $f^{\vee}$ is the p,p' indirect Hausdorff-Young FT of f. We now assert a bit more:

LEMMA 4.2. <u>Let</u> (i) $p \in [1,2]$ <u>and</u> (ii) $\forall f \in L_p(\Gamma)$, $\tilde{f}$ <u>be</u> <u>the</u> <u>(indirect)</u> p,p' <u>Hausdorff-Young</u> FT <u>of</u> f. <u>Then</u>

$$\forall f \in L_p(\Gamma) \cap L_{1,\xi_{p'}}(\Gamma) \, , \; \tilde{f} = f^{\vee}.$$

It follows from the last lemma that for all $p \in [1,2]$ our p,p' FT will subsume the Hausdorff-Young p,p' FT iff $L_p(\Gamma) \subseteq L_{1,\xi_{p'}}(\Gamma)$. We are thus brought back to the inclusion issue raised in (2.11) - 2.14, and for the weakly Σ-complete cases $p \in (1,2]$ to the condition 2.14(ii). It is convenient to address the terminal cases $p = 2$ and $p = 1$ before considering $p \in (1,2)$:

MAIN THEOREM 4.3. (a) $L_2(\Gamma) = L_{1,\xi_2}(\Gamma)$; (b) $L_1(\Gamma) \subseteq L_{1,\xi_\infty}(\Gamma)$; (c) $\forall p \in (1,2)$, $L_p(\Gamma) \subseteq L_{1,\xi_{p'}}(\Gamma)$. <u>Thus</u> <u>our</u> 2,2 FT <u>is</u> <u>equal</u> <u>to</u> <u>the</u> <u>classical</u> 2,2 FT <u>and</u> <u>for</u> $p \in [1,2)$, <u>our</u> p,p' FT <u>extends</u> <u>the</u> <u>Hausdorff-Young</u> p,p' FT.

PROOF. (a) We know, cf. (3.8), that $\xi_2 \in CAOS(\mathcal{B}_m, L_2(\Gamma^\wedge))$ and has the control measure m. Hence by Thm. 2.15, we have (a). (b) The measure ξ_∞ takes values in $L_\infty(\Gamma^\wedge)$ which is not weakly Σ-complete, and the N.&S.C. that $f \in L_{1,\xi_\infty}(\Gamma)$ is, cf. (2.3),

$$(\alpha) \; \forall \Phi \in L_\infty(\Gamma^\wedge)^*, \; f \in L_{1,|\Phi \circ \xi_\infty|} \; ;$$

&

$$(\beta) \; \forall C \in \mathcal{B}_m^{loc}, \; \exists F_C \in L_\infty(\Gamma^\wedge) \text{ such that}$$

$$\forall \Phi \in L_\infty(\Gamma^\wedge)^*, \; \Phi(F_C) = \int_C \overline{f(t)} (\Phi \circ \xi_\infty)(dt).$$

But since, cf. (3.3), ξ takes values in $C_0(\Gamma^\wedge)$, this N.&S.C. can, by use of A.2, be rephrased:

$$(1)\quad \left\{ \begin{array}{ll} & (\alpha)\ \forall\ \phi \in L_1(\Gamma^\wedge),\ f\cdot\phi^\wedge \in L_1(\Gamma), \\ \& & \\ & (\beta)\ \forall\ C \in \mathcal{B}_m^{loc},\ \exists\, F_C \in L_\infty(\Gamma^\wedge) \text{ such that} \\ & \quad \forall\ \Phi \in L_\infty(\Gamma^\wedge)^*,\ \Phi(F_C) = \int_C \overline{f(t)}\phi_0^\wedge(t)m(dt), \\ & \quad \text{where } \phi_0 \text{ is related to } \Phi \text{ as in A.2(a).} \end{array} \right.$$

Now let $F \in L_1(\Gamma)$. Then since for any $\phi \in L_1(\Gamma^\wedge)$, $\phi^\wedge \in C_0(\Gamma)$, therefore $f\cdot\phi^\wedge \in L_1(\Gamma)$. Thus (1)($\alpha$) holds. But so does (1)(β), upon taking

$$F_C(\lambda) := \int_C f(t)\lambda(t)m(dt),\ C \in \mathcal{B}_m^{loc}\ ,$$

noting that $F_C \in C_0(\Gamma^\wedge)$, applying A.2(a) and then the Fubini theorem. Hence $f \in L_{1,\xi_\infty}(\Gamma)$. Thus (b) holds.

(c) Let $p \in (1,2)$. By Thm. 2.14 we have only to show that

$$\forall\ \Phi \in \{L_{p'}(\Gamma^\wedge)\}^*,\ \frac{d(\Phi\circ\xi_{p'})}{dm} \in L_{p'}(\Gamma)$$

By Lemma A.3 we may, using the Riesz representation, rephrase this condition as

$$(\mathrm{I})\qquad \forall\ \phi \in L_p(\Gamma^\wedge),\ \frac{d\nu_\phi}{dm} \in L_{p'}(\Gamma),$$

where

$$(2)\qquad \forall\ \Delta \in \mathcal{B}_m,\ \nu_\phi(\Delta) := \int_{\Gamma^\wedge} \phi(\lambda)\overline{\chi^\vee_\Delta(\lambda)}m^\wedge(d\lambda).$$

Proof of (I). Define, cf. (2.11),

$$(3)\qquad \forall\ \phi \in \mathcal{S}_{m^\wedge}(\Gamma^\wedge),\ T(\phi) := \frac{d\nu_\phi}{dm} \in L_1^{loc}(\Gamma),$$

and then define $T_{p,q}$ as in the Riesz-Thorin Thm. A.6. Grant momentarily that

$$(4)\qquad T \text{ is of type } (1,\infty)\ \&\ (2,2) \text{ with } |T_{1,\infty}| \leqslant 1 \geqslant |T_{2,2}|.$$

Then by A.6, T is of type (p,p') and $|T_{p,p'}| \leqslant 1$, i.e. we have (I), and the proof is finished.

To turn to (4), we know, cf. [8: 5.9 & 5.8], that

$$\forall\ \phi \in L_2(\Gamma^\frown),\ T_{2,2}(\phi) := \frac{d\nu_\phi}{dm} = \tilde{\phi} \in L_2(\Gamma),$$

and that $T_{2,2}$ is an isometry. Thus T is of type (2,2) and $|T_{2,2}| = 1$. Next, let $\phi \in L_1(\Gamma^\frown)$. Applying in (2), the Hölder inequality and then the first equality in (3.4), we get

$$|\nu_\phi(\Delta)| \leqslant |\phi|_{1,m^\frown} \cdot |\chi^\vee_\Delta|_{\infty,m^\frown} = |\phi|_{1,m^\frown} \cdot m(\Delta),$$

and therefore, cf.(3),

$$|T_{1,\infty}(\phi)|_{\infty,m} = \left|\frac{d\nu_\phi}{dm}\right|_{\infty,m} \leqslant |\phi|_{1,m^\frown}.$$

Thus T is of type $(1,\infty)$ and $|T_{1,\infty}| \leqslant 1$. □

Also, prevailing are the norm relations:

$$(4.4)\quad \left\{\begin{array}{l} \|f\|_{\xi_2} = |f|_{\xi_2} = |f|_{2,m}\ ; \\ \&\ \|f\|_{\xi_{p'}} \leqslant |f|_{\xi_{p'}} \leqslant |f|_{p,m} \text{ for } 1 \leqslant p < 2. \end{array}\right.$$

These follow from (2.15)(b); A.2(b),(a) for p=1, and 2.14(b), $|T|_{p,p'} \leqslant 1$ for $1<p<2$.

The last theorem leaves open the question of the equality of $L_p(\Gamma)$ and $L_{1,\xi_{p'}}(\Gamma)$ for $p \in (1,2)$. We shall now show that the answer is negative, at least for the group $\Gamma = \mathbb{N}$ of integers.

THEOREM 4.5. $\forall\ p \in (1,2),\ \exists\ f \in L_{p,\xi_{p'}}(\mathbb{N}) \setminus L_p(\mathbb{N})$.

PROOF. Given $p \in (1,2)$ define

$$(1)\qquad f(n) := \left\{\frac{1}{n \cdot \log n}\right\}^{1/p} \cdot \chi_{[2,\infty)}(n),\ n \in \mathbb{N}.$$

Then obviously $\sum_{-\infty}^{\infty}|f(n)|^p = \infty$ and $f \notin L_p(\mathbb{N})$. Now take any ϕ in $L_p(C)$, where C is the circle group. Then, importantly, by Thm. A.7,

$$(2) \qquad \sum_{-\infty}^{\infty} |\phi^{\wedge}(n)|^{p'} \nu_p\{n\} < \infty .$$

We shall complete the proof by showing that

$$(I) \qquad \sum_{-\infty}^{\infty} |f(n)\phi^{\wedge}(n)| < \infty ,$$

i.e., cf. (A.5), that f satisfies the condition for membership in $L_{1,\xi_{p'}}(\mathbb{N})$.

<u>Proof of (I)</u>. Since $|f(n)\phi^{\wedge}(n)| = |\{f(n)/\nu_p(n)\}\phi^{\wedge}(n)|\nu_p(n)$, it follows that $\text{LHS}(I) = |\{f/\nu_p\}\cdot\phi^{\wedge}|_{1,\nu_p}$. Hence by the Hölder inequality for the measure ν_p,

$$(3) \qquad \text{LHS}(I) \leqslant |f/\nu_p|_{p,\nu_p} \cdot |\phi^{\wedge}|_{p',\nu_p} .$$

By (2), the second factor on the right is finite. As for the first factor, we note that for $n \geqslant 8 > e^2$, $n \cdot \log n > 1 + n$; consequently

$$|f/\nu_p|_{p,\nu_p}^p = \sum_{n=2}^{\infty} |f(n)/\nu_p(n)|^p \cdot \nu_p(n)$$

$$\leqslant S_7 + \sum_{n=8}^{\infty} \frac{1}{(1+n)^{1+\varepsilon}} < \infty,$$

since $\varepsilon > 0$. Thus by (3), $\text{LHS}(I) < \infty$, i.e. we have (I). □

Say that a function f in $\mathbb{C}^\Gamma$ <u>balances</u> a function g in $\mathbb{C}^\Gamma$, iff $fg \in L_1(\Gamma)$. Then Thm. 4.5 answers negatively the following question[8] for $p \in (1,2)$:

QUESTION 4.6. <u>Let</u> $p \in [1,2]$. <u>If a</u> $\mathcal{B}$, <u>Borel measurable function</u>, f <u>in</u> $\mathbb{C}^\Gamma$ <u>balances the</u> (<u>classical</u>) FT $\phi^{\wedge}$ <u>of each</u> ϕ <u>in</u> $L_p(\Gamma^{\wedge})$, <u>then is</u> f <u>in</u> $L_p(\Gamma)$?

---

8 This was the precise question that the writer raised at the AMS meeting, to which Professor Helson responded. The f used in our proof of 4.5 is the simplest of a class of functions proposed by him. See footnote 1.

Appendix

In much of this paper, the Banach space X of interest is reflexive, and therefore weakly Σ-complete in the following sense, cf. Thomas [12],:

DEFINITION A.1. *We say that a Banach space X over $\mathbb{F}$ is "weakly Σ-complete" iff for any sequence $(x_n)_1^\infty$ in X, the following conditions are equivalent*

$$(\alpha) \qquad \forall\, x^* \in X^*, \quad \sum_{n=1}^{\infty} |x^*(x_n)| < \infty,$$

$$(\beta) \qquad \exists\, x \in X \ \text{such that}\ \forall\, x^* \in X^*, \quad \sum_{n=1}^{\infty} x^*(x_n) \ \text{converges}$$

unconditionally to $x^*(x)$ *in* $\mathbb{F}$.

Bessaga and Pelczynski [1] have shown that such X's are precisely those not containing copies of c_0.

For $p = 1$, $\xi_{p'} = \xi_\infty$ takes values in the space $X = L_\infty(\Gamma^\wedge)$, which is not weakly Σ-complete. The membership of $L_{1,\xi_\infty}(\Gamma)$ is therefore governed by (2.3), not (2.6). Thanks to the Riemann-Lebesgue theorem, however, we can avoid the adjoint of $L_\infty(\Gamma^\wedge)$, in favour of the more convenient adjoint of $C_0(\Gamma^\wedge)$:

LEMMA A.2. *Let* $\Phi \in \{L_\infty(\Gamma^\wedge)\}^*$, $\Psi := \mathrm{Rstr}_{C_0(\Gamma^\wedge)}\Phi$. *Then*
(a) $\exists_1\, \phi_0 \in L_1(\Gamma^\wedge)$ *such that* $\forall\, \psi \in C_0(\Gamma^\wedge)$,

$$\Psi(\psi) = \int_{\Gamma^\wedge} \overline{\psi(\lambda)}\phi_0(\lambda) m^\wedge(d\lambda), \quad |\Psi| = |\phi_0|_{1,m^\wedge};$$

(b) $\forall\, \Delta \in \mathcal{B}_m$, $(\Phi \circ \xi_\infty)(\Delta) = \int_\Delta \phi_0^\wedge(t) m(dt)$, *where $\phi_0^\wedge$ is the direct* FT *of ϕ_0 in* $L_1(\Gamma^\wedge)$:

$$\phi_0^\wedge(t) := \int_{\Gamma^\wedge} \overline{\lambda(t)}\phi_0(\lambda) m^\wedge(d\lambda), \quad t \in \Gamma;$$

(c) $\Phi \circ \xi_\infty \ll m$ *and*

$$\frac{d(\Phi \circ \xi_\infty)}{dm}(\cdot) = \phi_0^\wedge(\cdot), \quad \frac{d|\Phi \circ \xi_\infty|}{dm} = |\phi_0^\wedge(\cdot)|, \quad \text{a.e.}(m) \text{ on } \Gamma.$$

Next, for $p \in (1,2]$, the Riesz theorem for $\{L_{p'}(\Gamma^\wedge)\}^*$, which is applicable since $\xi_{p'}$ takes values in $L_{p'}(\Gamma^\wedge)$, cf. (3.6), 3.7, readily gives us :

LEMMA A.3. Let $p\in(1,2]$ and $\Phi\in\{L_{p'}(\Gamma^\wedge)\}^*$. Then (a) $\exists_1\ \phi\in L_p(\Gamma^\wedge)$ such that $\Phi\circ\xi_{p'}=\nu_\phi$, where

$$\nu_\phi(\Delta) := \int_{\Gamma^\wedge} \phi(\lambda)\overline{\chi_\Delta^\vee(\lambda)}m^\wedge(d\lambda), \quad \Delta\in\mathcal{B}_m;$$

(b) $\Phi\circ\xi_{p'}\in CA(\mathcal{B}_m,\mathbb{C})$, $\Phi\circ\xi_{p'}\ll m$ and

$$\frac{d(\Phi\circ\xi_{p'})}{dm}(\cdot)=\frac{d\nu_\phi}{dm}(\cdot),\quad \frac{d|\Phi\circ\xi_{p'}|}{dm}(\cdot)=\left|\frac{d\nu_\phi}{dm}(\cdot)\right|, \text{ a.e.}(m) \text{ on } \Gamma .$$

Now for $\Gamma=\mathbb{N}$, and therefore $\Gamma^\wedge=C$, the measure $\nu_\phi(\cdot)$ reduces to

$$\nu_\phi(\Delta) = \int_C \phi(\theta)\overline{\{\sum_{n\in\Delta} e^{ni\theta}\}}\frac{d\theta}{\sqrt{(2\pi)}}=\frac{1}{\sqrt{(2\pi)}}\sum_{n\in\Delta}\phi^\wedge(n),$$

where $\Delta\subseteq\mathbb{N}$ is finite. Consequently, the last lemma yields the following:

LEMMA A.4. If in A.3, $\Gamma=\mathbb{N}$ (and therefore $\Gamma^\wedge=C$), and $\Phi\in\{L_{p'}(C)\}^*$, then (a) $\exists_1\ \phi\in L_p(C)$ such that

$$(\Phi\circ\xi_{p'})(\Delta) = \frac{1}{\sqrt{(2\pi)}}\sum_{n\in\Delta}\phi^\wedge(n), \ \text{finite } \Delta\subseteq\mathbb{N},$$

(b)
$$|\Phi\circ\xi_{p'}|(n) = \frac{1}{\sqrt{(2\pi)}}|\phi^\wedge(n)|, \ n\in\mathbb{N}.$$

Combining this and (2.6), we get:

$$\text{(A.5)}\quad \left\{\begin{array}{l} \forall\ p\in(1,2], \\ f\in L_{1,\xi_{p'}}(\mathbb{N}) \Leftrightarrow \forall\ \phi\in L_p(C),\ \sum_{n\in\mathbb{N}}|f(n)||\phi^\wedge(n)|<\infty. \end{array}\right.$$

We also need the following version of the Riesz-Thorin theorem:

THEOREM A.6. Let (i) T be a linear operator on $\mathcal{S}_{m^\wedge}(\Gamma^\wedge)$ to $L_1^{loc}(\Gamma)$, (ii) $T_{p,q}$

be the $L_p(\Gamma^\wedge) \times L_q(\Gamma)$ -closure of the graph of T, where $p,q \in [1,\infty]$. Say that T is of type (p,q), iff $T_{p,q}$ is (single-valued and) continuous on $L_p(\Gamma^\wedge)$ to $L_q(\Gamma)$. Then

$$T \text{ is of types } (1,\infty) \ \& \ (2,2) \text{ and } |T_{1,\infty}| \leq 1 \geq |T_{2,2}|$$
$$\Rightarrow \forall\, p \in (1,2),\ T \text{ is of type } (p,p') \ \&\ |T|_{p,p'} \leq 1.$$

Finally, the following theorem of Hardy and Littlewood, cf. Zygmund [13: II,p.109,3.19(i)], is needed to prove Thm. 4.5.

THEOREM A.7. Let (i) $p \in [1,2]$, (ii) ν_p be the purely atomic measure over $\mathbb{N}$ such that

$$\nu_p\{n\} = 1/(1+|n|)^{2-p}, \quad n \in \mathbb{N}.$$

Then $\forall\, g \in L_p(C)$, $g^\wedge \in L_{p',\,\nu_p}(\mathbb{N})$, where

$$g^\wedge(n) := \frac{1}{\sqrt{(2\pi)}} \int_0^{2\pi} e^{-ni\theta} g(\theta)d\theta, \quad n \in \mathbb{N}.$$

REFERENCES

[1] Bessaga, C. - Pelczynski, A., On bases and unconditional convergence of series in Banach spaces. Studia Math. 17 (1958), 151-164.

[2] Doob, J.L., Stochastic processes, Wiley, New York, 1953.

[3] Dunford, N. - Schwartz, J., Linear operators, I, Interscience, New York, 1958, 1963.

[4] Hewitt, E. - Ross, K., Abstract Harmonic Analysis, Vols. I,II, Springer Verlag, New York, 1963, 1970.

[5] Lewis, D.R., Integration with respect to vector measures, Pacific J. Math. 23 (1970), 157-165.

[6] Lewis, D.R., On integrability and summability in vector spaces, Illinois J. Math. 16 (1972), 294 - 307.

[7] Masani, P., Orthogonally scattered measures, Adv. in Math. 2 (1968), 61-117.

[8] Masani, P., Explicit form for the Fourier-Plancherel transform over locally compact abelian groups, in Abstract spaces and Approximation, edited by P.L. Butzer and B.Sz. Nagy, Basel 1969, 162-182.

[9] Masani, P., Quasi-isometric measures and their applications, Bull. Amer. Math. Soc. 76 (1970), 427-528.

[10] Masani, P., The theory of stationary vector-valued measures over ℝ, Crelle's J. 339 (1983), 105-132.

[11] Masani, P., Fourier transformation as integration with respect to a vector measure, Abstracts Amer. Math. Soc. 3 (No. 6) (1982), p.444.

[12] Thomas, E.G.F., L'integration par rapport a une measure de Radon vectorielle, Ann. Inst. Fourier (Grenoble) 20: 2, (1970), 55-191.

[13] Zygmund, A., Trigonometric Series, Vols. I,II, English transl., 2nd Ed., Cambridge Univ. Press, London and New York, 1968.

International Series of
Numerical Mathematics, Vol. 65

RADAR AMBIGUITY FUNCTIONS AND THE LINEAR SCHRÖDINGER REPRESENTATION

Walter Schempp
Lehrstuhl für Mathematik I
Universität Siegen
Siegen

In radar analysis there exists an analogue of the Heisenberg uncertainty principle of quantum mechanics. Quantum mechanics stands here for the quantum-mechanical description, at a given instant of time, of a non-relativistic particle. These uncertainty principles suggest that there should exist a common mathematical structure behind both quantum mechanics and the theory of signals. It is the aim of the present article to establish that the concept of real Heisenberg nilpotent group $\tilde{A}(\mathbb{R})$ lies at the foundations of both of these fields. The crucial point is to endow the time-frequency plane $\mathbb{R} \oplus \mathbb{R}$ with the structure of a two dimensional real symplectic vector space so that it gets symplectomorphic to the tangent plane to the "Schrödinger coadjoint orbit" at the point 1 in the Kirillov orbit picture for the unitary dual of $\tilde{A}(\mathbb{R})$. Various applications of this geometric relationship between signal theory and nilpotent harmonic analysis are pointed out.

1. Radar Auto-Ambiguity Functions

Consider a radar pulse s of finite energy transmitted in the form

$$s(t) = f(t)e^{2\pi i\omega t}.$$

The parameter $t \in \mathbb{R}$ denotes time and the monochromatic high-frequency carrier $e^{2\pi i\omega t}$ is modulated in amplitude by the function f of time which varies much more slowly than the cycles of the carrier. The signals received in most communication systems are of this type. In what follows we will assume that the signal envelope function f of s belongs to the vector space $\mathscr{S}(\mathbb{R})$ of all complex-valued $\mathscr{C}^{\infty}$ functions on $\mathbb{R}$, rapidly decaying at infinity. Notice that

$\mathcal{S}(\mathbb{R})$ is an everywhere dense vector subspace of the complex Hilbert space $L^2(\mathbb{R})$. To measure the distance of a remote radar target it is necessary to estimate the time x at which the echo from it arrives at the receiver. If time is counted from the transmission of the radar pulse s, the range is $\frac{1}{2}cx$, where c denotes the velocity of electromagnetic radiation. If the target is not stationary but is moving at a certain range rate toward or away from the radar antenna, the carrier frequency of the echo signal differs from that of the transmitted pulse s because of the Doppler effect. If we pick the transmitted frequency ω as our basic reference frequency, the echo pulse will have the form

$$s_{echo}(t) = f_{echo}(t)e^{2\pi i\omega t}$$

where

$$f_{echo}(t) = f(t+x)e^{2\pi iyt}$$

and x is the time delay mentioned above and y is the change in carrier frequency. The Doppler frequency shift is given by $y = 2\frac{v}{c}\omega$ where v is the radial velocity, i.e., the component of target velocity in the direction of the radar antenna. Radar (=RAdio Detection And Ranging) systems are not merely a device for discovering distant objects. A radar system must find where the targets actually are and how they are moving. For this purpose it must distinguish or resolve the two narrowband signals s and s_{echo} with the same radar carrier in the presence of white Gaussian random noise. The time delay x and the Doppler frequency shift y which chiefly serve to the signal resolution are involved in the mathematical expectation $\langle f_{echo}|f\rangle$ upon which the structure of the receiver and its performance actually depend. The expectation value for the pair (x,y) of the time-frequency plane $\mathbb{R} \oplus \mathbb{R}$ is written

$$\int_{\mathbb{R}} f(t+x)\overline{f}(t)^{2\pi iyt}dt.$$

Omitting the time independent phase factor $e^{-\pi ixy}$ which is not essential in the present context, the correlation takes the symmetric form of the radar auto-ambiguity function (cf. Woodward [10])

$$H(f;x,y) = \int_{\mathbb{R}} f(t+\tfrac{1}{2}x)\overline{f}(t-\tfrac{1}{2}x)e^{2\pi iyt}dt$$

with respect to the complex envelope $f \in \mathscr{S}(\mathbb{R})$. Its range $H(f;\mathbb{R},\mathbb{R})$ is called radar auto-ambiguity surface over the time-frequency plane $\mathbb{R} \oplus \mathbb{R}$ (see [3] for computer plots of some typical examples).

The following properties (I) and (II) of the radar auto-ambiguity function $H(f;.,.) \in \mathscr{S}(\mathbb{R} \oplus \mathbb{R})$ are immediate. The property (III) follows by an application of the Cauchy-Schwarz inequality.

(I) $H(f;0,0) \geqq 0$

(II) $H(f;-x,-y) = \overline{H}(f;x,y)$ for all pairs $(x,y) \in \mathbb{R} \oplus \mathbb{R}$
("Hermitean central symmetry")

(III) $|H(f;x,y)| \leqq H(f;0,0)$ for all pairs $(x,y) \in \mathbb{R} \oplus \mathbb{R}$
("Central peak property")

Thus for every signal envelope function $f \in \mathscr{S}(\mathbb{R})$ the associated radar auto-ambiguity surface is peaked at the origin (0,0) of the time-frequency plane $\mathbb{R} \oplus \mathbb{R}$. A second signal arriving with time delay x and Doppler frequency shift y that lie under this central peak will be difficult to distinguish from the first signal. For most types of signals utilized in practice, for instance, if the transmitter sends out a train of coherent pulses of the same carrier frequency in order to make the measurement of the radial velocity more accurately, the radar auto-ambiguity surface exhibits additional peaks elsewhere over the time-frequency plane $\mathbb{R} \oplus \mathbb{R}$. These "sidelobes" may conceal weak signals with arrival times and carrier frequencies far from those of the first signal. In a measurement of the arrival time and frequency of a single signal, the random noise may cause one of the subsidiary peaks leading to errors in the result. The taller the sidelobes, the greater the probability of gross errors in time delay and Doppler frequency shift and hence in tracking the radar target. It is desirable, therefore, for the central peak of the radar auto-ambiguity surface to be as narrow as possible, and for there to be as few and as low sidelobes as possible. From this reasoning it follows that the geometry of the radar auto-ambiguity surface over the time-frequency plane is of fundamental importance for radar waveform design (cf. Section 4 infra).

2. Uncertainty Principles

One of the basic tenets of quantum mechanics is the Heisenberg uncertainty principle. According to this principle, not all physical quantities observed in any realizable experiment (even in principle only) can be determined simultaneously with an arbitrarily high degree of accuracy at a given instant of time. There are mutually exclusive ("conjugate") quantities the measurement errors of which are interrelated. Even under ideal experimental conditions, an increase in the measurement accuracy of one quantity can be achieved only at the expense of decreasing the measurement accuracy of another quantity. For instance, the position coordinate q of a non-relativistic quantum-mechanical particle and its corresponding momentum p in the q-direction is one example of two such mutually exclusive quantities. The selfadjoint operators with domains in $L^2(\mathbb{R})$ associated to q and p in the theory of quantum mechanics satisfy the Heisenberg canonical commutation relations. The group-theoretic embodiment of these canonical commutation relations is the real Heisenberg nilpotent group $\tilde{A}(\mathbb{R})$ (cf. Cartier [1]).

Although there is no analogue of Planck's constant in signal theory, there is a similar situation with radar measurements. The properties of the radar auto-ambiguity function $H(f;.,.)$ are such that the measurement accuracy of the distance and radial velocity are interrelated. Indeed, an application of the Plancherel theorem to the "uncertainty inequality"

$$\text{(IV)} \quad \|D_1 H(f;.,.)\|_2^2 + \|D_2 H(f;.,.)\|_2^2 \geq \frac{1}{4\pi} \|H(f;.,.)\|_2^2$$

$$\text{where } D_1 = \frac{1}{2\pi i}\frac{\partial}{\partial x}, \quad D_2 = \frac{1}{2\pi i}\frac{\partial}{\partial y}$$

shows the impossibility of concentrating the Wigner quasi-probability density function around a point (cf. Wigner [9]). Thus an improvement of the range accuracy results in a worsening of the range rate accuracy, and vice versa. By changing the signal parameters, it is only possible to redistribute the ambiguity in determining the distance of the radar target and its radial velocity simultaneously, but not to decrease it. Consequently the uncertainty inequality (IV) leads to the same radar uncertainty principle as the familiar Heisenberg inequality formulated in terms of the spread of the pulse envelope

$f \in \mathcal{S}(\mathbb{R})$.

The analogy between the uncertainty principle of quantum mechanics on the one hand and the radar uncertainty principle on the other hand suggests that the real Heisenberg nilpotent group $\tilde{A}(\mathbb{R})$ constitutes a mathematical structure that has significance also in signal theory. It is the purpose of the present article to support this idea.

3. The Symplectic Time-Frequency Plane

We look at the time-frequency plane $\mathbb{R} \oplus \mathbb{R}$ as the real vector space of all columns $v = \begin{bmatrix} x \\ y \end{bmatrix}$, $x \in \mathbb{R}$, $y \in \mathbb{R}$ and define the symplectic (=nondegenerate skew symmetric bilinear) form B on $\mathbb{R} \oplus \mathbb{R}$ according to

$$B(v_1,v_2) = B(\begin{bmatrix} x_1 \\ y_1 \end{bmatrix},\begin{bmatrix} x_2 \\ y_2 \end{bmatrix}) = x_1y_2-x_2y_1 = \det\begin{pmatrix} x_1 & x_2 \\ y_1 & y_2 \end{pmatrix}.$$

In the following we shall consider the time-frequency plane as the standard two dimensional real symplectic vector space $(\mathbb{R} \oplus \mathbb{R};B)$ similarly to the phase space of mechanics which also carries a natural symplectic structure. Denote by $\tilde{A}(\mathbb{R})$ the three dimensional real Heisenberg nilpotent group. We shall use the realization

$$\tilde{A}(\mathbb{R}) = (\mathbb{R} \oplus \mathbb{R}) \oplus \mathbb{R}$$

with the group law

$$(v_1,z_1).(v_2,z_2) = (v_1+v_2,z_1+z_2+\tfrac{1}{2}B(v_1,v_2)).$$

Clearly the subgroup $\tilde{Z} = \{(\begin{bmatrix} 0 \\ 0 \end{bmatrix},t) \mid t \in \mathbb{R}\}$ is the center of $\tilde{A}(\mathbb{R})$ isomorphic to $\mathbb{R}$ and obviously $\tilde{Z} = [\tilde{A}(\mathbb{R}),\tilde{A}(\mathbb{R})]$ where the bracket denotes the commutator subgroup. Hence $\tilde{A}(\mathbb{R})$ is a real two step nilpotent Lie group. Its Lie algebra is the three dimensional Heisenberg Lie algebra $\mathfrak{n}$. The center $\mathfrak{z}$ of $\mathfrak{n}$ is the Lie algebra of $\tilde{Z}$. We have

$$\mathfrak{n} = (\mathbb{R} \oplus \mathbb{R}) \oplus \mathfrak{z}$$

and the canonical basis $\{X,Y,T\}$ of $\mathfrak{n}$ with $T \in \mathfrak{z}$ satisfies the Heisenberg canonical commutation relations of quantum mechanics

$$[X,Y] = T, \; [X,T] = [Y,T] = 0.$$

Let now $\mathcal{H}$ be a complex Hilbert space and U a continuous irreducible unitary linear representation of $\tilde{A}(\mathbb{R})$ acting on $\mathcal{H}$. Then the identity

$$U(v,z) \circ U(0,t) = U(0,t) \circ U(v,z)$$

holds for all elements $(v,z) \in \tilde{A}(\mathbb{R})$, $(0,t) \in \tilde{Z}$. Hence we have by the irreducibility of U and Schur's lemma

$$U(0,t) = \chi(t)\mathrm{id}_{\mathcal{H}} \qquad (t \in \mathbb{R}),$$

where χ is a unitary character of $\mathbb{R}$, i.e., a continuous homomorphism $\chi: \mathbb{R} \to \mathbb{T} = \mathbb{R}/\mathbb{Z}$. Hence

$$\chi(t) = e^{2\pi i\lambda t} \qquad (t \in \mathbb{R})$$

for some fixed $\lambda \in \mathbb{R}$. Now we have to distinguish two cases:

(i) $\lambda=0$ ("degenerate" case). Then $U(v,z) = U(v,0)$ for all $(v,z) \in \tilde{A}(\mathbb{R})$. It follows $U(v_1,z_1) \circ U(v_2,z_2) = U(v_1+v_2,z_1+z_2+\frac{1}{2}B(v_1,v_2)) = U(v_1+v_2,0)$. Hence

$$U(v_1,z_1) \circ U(v_2,z_2) = U(v_2,z_2) \circ U(v_1,z_1)$$

for all elements (v_1,z_1) and (v_2,z_2) of $\tilde{A}(\mathbb{R})$. The irreducibility of U implies, by Schur's lemma again, the existence of a unitary character η of $\mathbb{R} \oplus \mathbb{R}$ such that

$$U(v,z) = \eta(v)\mathrm{id}_{\mathcal{H}}$$

holds for all $(v,z) \in \tilde{A}(\mathbb{R})$. Therefore, there exists a linear form $\mu \in (\mathbb{R} \oplus \mathbb{R})^*$ so that

$$\eta(v) = e^{2\pi i\langle v,\mu\rangle} \qquad (v \in \mathbb{R} \oplus \mathbb{R})$$

and $\dim_{\mathbb{C}} \mathscr{H} = 1$. Let $\tilde{A}(\mathbb{R})$ act on the dual vector space $\mathfrak{n}^*$ of $\mathfrak{n}$ by the coadjoint representation Coad, the contragredient of the adjoint representation Ad of $\tilde{A}(\mathbb{R})$ in $\mathfrak{n}$. According to the Kirillov theory of unitary linear representations of connected, simply connected nilpotent Lie groups the linear forms $\mu \in (\mathbb{R} \oplus \mathbb{R})^*$ constitute the single point orbits in $\mathfrak{n}^*/\mathrm{Coad}(\tilde{A}(\mathbb{R}))$ that are in one-to-one correspondence to the isomomorphy classes of the one dimensional continuous unitary linear representations of $\tilde{A}(\mathbb{R})$.

(ii) $\lambda \neq 0$ ("generic case"). Let

$$u(v) = U(v,0) \qquad (v \in \mathbb{R} \oplus \mathbb{R})$$

denote the restriction of U to the polarized cross-section to $\tilde{Z}$ in $\tilde{A}(\mathbb{R})$ (cf. Howe [4]). Then

$$\begin{aligned} u(v_1) \circ u(v_2) &= U(v_1+v_2,\tfrac{1}{2}B(v_1,v_2)) \\ &= e^{\pi i\lambda B(v_1,v_2)} u(v_1+v_2) \end{aligned}$$

holds for all $v_1,v_2 \in \mathbb{R} \oplus \mathbb{R}$. Let $\{P,Q\}$ be a symplectic basis of $(\mathbb{R} \oplus \mathbb{R};B)$ satisfying $B(P,Q) = \frac{1}{\lambda}$. Set, for $v = \begin{bmatrix} x \\ y \end{bmatrix} \in \mathbb{R} \oplus \mathbb{R}$,

$$u(\begin{bmatrix} x \\ y \end{bmatrix}) = u(xP+yQ)$$

and

$$V(\begin{bmatrix} x \\ y \end{bmatrix}) = e^{\pi ixy}u(\begin{bmatrix} x \\ y \end{bmatrix}).$$

Since U is irreducible it is clear that V is a continuous irreducible projective linear representation of $\mathbb{R} \oplus \mathbb{R}$ acting on $\mathscr{H}$. An application of the Stone-von Neumann theorem (see Howe [4]) implies that there exists a unitary isomorphism $\mathscr{H} \to L^2(\mathbb{R})$ that maps V onto U_λ where U_λ acts on the Schwartz functions $f \in \mathscr{S}(\mathbb{R})$ according to the prescription

$$U_\lambda(\begin{bmatrix} x \\ y \end{bmatrix},z)f(t) = e^{2\pi i\lambda(z+yt+\frac{1}{2}xy)} f(t+x) \qquad (t \in \mathbb{R})$$

for all elements $(\begin{bmatrix} x \\ y \end{bmatrix},z) \in \tilde{A}(\mathbb{R})$. The affine plane $\lambda + \mathfrak{z}^0 \in \mathfrak{n}^*/\mathrm{Coad}(\tilde{A}(\mathbb{R}))$, where $\mathfrak{z}^0$ denotes the annihilator of $\mathfrak{z}$ in the real vector space dual $\mathfrak{n}^*$ of $\mathfrak{n}$, is the orbit that corresponds to the isomorphy class of U_λ under the natural bijection between $\mathfrak{n}^*/\mathrm{Coad}(\tilde{A}(\mathbb{R}))$ and the unitary dual of $\tilde{A}(\mathbb{R})$. Thus U_1, the so-called linear Schrödinger representation of $\tilde{A}(\mathbb{R})$, forms the prototype among the infinite dimensional continuous unitary linear representations of $\tilde{A}(\mathbb{R})$.

In other words, U_1 is (up to dilation and duality) the unique unitary representation of $\tilde{A}(\mathbb{R})$ of dimension > 1.

4. The Gabor Philosophy

It was emphasized by D. Gabor in his fundamental paper [2] that everyday experiences insist on a simultaneous description of signals in terms of both time and frequency. Adopting his point of view signals should be represented in two dimensions, with time and frequency as their coordinates. However, the basic idea that we want the reader to appreciate is that the three dimensional real Heisenberg nilpotent group $\tilde{A}(\mathbb{R})$ is a mathematical structure suitable to formalize the Gabor philosophy by considering the symplectic time-frequency plane $(\mathbb{R} \oplus \mathbb{R}; B)$ as the tangent plane $\mathfrak{z}^0$ to the "Schrödinger coadjoint orbit" $1+ \mathfrak{z}^0 \in \mathfrak{n}^*/\mathrm{Coad}(\tilde{A}(\mathbb{R}))$ at the point 1 in order to provide an appropriate description of signals simultaneously in both time and frequency. Thus the message which the reader should take away is that the Lie group $\tilde{A}(\mathbb{R})$ stands at the crossroads of quantum mechanics and signal theory. Indeed, the identity

$$U_1(\begin{bmatrix} x \\ y \end{bmatrix},0)f(t) = e^{2\pi i(yt+\frac{1}{2}xy)} f(t+x) \qquad (t \in \mathbb{R})$$

shows that when the linear Schrödinger representation U_1 of $\tilde{A}(\mathbb{R})$ is restricted to the polarized cross-section to $\tilde{Z}$ in $\tilde{A}(\mathbb{R})$, its action on the pulse envelope function $f \in \mathcal{S}(\mathbb{R})$ produces a time delay and a frequency shift similar to the reflection of the signal pulse s by a moving target. More precisely, our reasoning gives rise to the following geometric link between signal theory and nilpotent harmonic analysis.

THEOREM. _Let the pulse envelope_ $f \in \mathcal{S}(\mathbb{R})$ _be given. For all pairs_ (x,y) _of the time-frequency plane_ $\mathbb{R} \oplus \mathbb{R}$ _the identity_

$$H(f;x,y) = \langle U_1(\begin{bmatrix} x \\ y \end{bmatrix},0)f \mid f\rangle$$

holds, i.e., the radar auto-ambiguity function coincides with the restriction to the polarized cross-section to $\tilde{Z}$ _in_ $\tilde{A}(\mathbb{R})$ _of the matrix coefficient associated with the linear Schrödinger representation_ U_1 _of_ $\tilde{A}(\mathbb{R})$.

The preceding geometric link between harmonic analysis of the real Heisenberg nilpotent group $\tilde{A}(\mathbb{R})$ and radar theory has a series of consequences. Let us collect some of these consequences in the following list:

(1) In view of the strong Stone-von Neumann theorem, U_1 belongs to the discrete series of $\tilde{A}(\mathbb{R})$, i.e., U_1 is square integrable mod $\tilde{Z}$. Therefore the radar cross-ambiguity functions satisfy the Frobenius-Schur orthogonality relations. These relations are a consequence of the geometric fact that the "Schrödinger coadjoint orbit" $1+\mathfrak{z}^0 \in \mathfrak{n}^*/\mathrm{Coad}(\tilde{A}(\mathbb{R}))$ is flat, i.e., a two dimensional affine linear variety in $\mathfrak{n}^*$ whereas for arbitrary connected, simply connected nilpotent Lie groups the generic coadjoint orbits are affine algebraic varieties of even dimension 2m, say, in the dual vector spaces of their Lie algebras which contain affine vector spaces of dimension $\geq m \geq 1$ over $\mathbb{R}$.

(2) Since U_1 is irreducible, $H(f;.,.) \in \mathcal{S}(\mathbb{R} \oplus \mathbb{R})$ is a pure positive-definite function with respect to the convolution product of the polarized cross-section $\{(v,0) \mid v \in \mathbb{R} \oplus \mathbb{R}\}$ to $\tilde{Z}$ in $\tilde{A}(\mathbb{R})$. It follows that in every decomposition of the following form

$$H(f;.,.) = H(f_1;.,.) + H(f_2;.,.)$$

the functions f_1,f_2 in $\mathcal{S}(\mathbb{R})$ are proportional to the given pulse envelope function $f \in \mathcal{S}(\mathbb{R})$. For details, see the forthcoming paper [7].

(3) The action of the (projective) oscillator representation of $Sp(1,\mathbb{R}) = SL(2,\mathbb{R})$ yields those pulse envelope functions that generate the same radar auto-ambiguity surface over the time-frequency plane $\mathbb{R} \oplus \mathbb{R}$. This follows from I. Segal's metaplectic formula by considering those linear automorphisms of $\tilde{A}(\mathbb{R})$ that act trivially on $\tilde{Z}$. Thus the linear automorphisms of the symplectic time-frequency plane $(\mathbb{R} \oplus \mathbb{R};B)$ form the mappings which leave the radar auto-ambiguity surfaces invariant. For full details of the deep relationship between the theory of linear group representations and the symplectic geometry of the time-frequency plane the reader is referred to the papers [6], [8].

It is remarkable that not only analog signals but also digital signals can be described by means of nilpotent harmonic analysis. For instance the classical Whittaker-Shannon sampling theorem fits into this general framework. The vehicle to achieve a group-theoretic approach to the sampling theorem is the study of the so-called linear lattice representation of $\tilde{A}(\mathbb{R})$, another realization of U_1 on the compact Heisenberg nilmanifold [5]. Since the so-called nil-theta functions, a version of the classical Jacobi theta functions, are living in a natural way on the Heisenberg nilmanifold, this approach is also of interest to those people working in special functions [8].

Acknowledgements. The author is grateful to Professors Hel. Braun, Paul R. Halmos, Edwin Hewitt, and Harold S. Shapiro for their encouragements. Special thanks go to Edwin Hewitt for inviting the author to lecture on the subject at the University of Alaska. The author also acknowledges gratefully financial support by the Deutsche Forschungsgemeinschaft.

REFERENCES

1. Cartier, P., Quantum mechanical commutation relations and theta functions. Proc. Sympos. Pure Math., Vol. 9, pp. 361-383. Providence, R.I.: Amer. Math. Soc. 1966.

2. Gabor, D., Theory of communication. J. Inst. Elec. Engineers (London) 93, 429-457 (1946).

3. Hebsaker, H.-M.-Schempp, W., Graphische Darstellung räumlicher Objekte mit Anwendungen in der Radar-Ortung. Elektron. Rechenanl. 25, 32-38 (1983).

4. Howe, R., Quantum mechanics and partial differential equations. J. Funct. Anal. 38, 188-254 (1980).

5. Schempp, W., Gruppentheoretische Aspekte der Signalübertragung und der kardinalen Interpolationssplines I. Math. Meth. in the Appl. Sci. 5, 195-215 (1983).

6. Schempp, W., Radar ambiguity functions, nilpotent harmonic analysis, and holomorphic theta series. In: Special Functions: Group Theoretical Aspects and Applications. R. A. Askey-T.H. Koornwinder-W. Schempp, editors. MIA Series. Dordrecht-Boston-London: D. Reidel (to appear).

7. Schempp, W., Radar ambiguity functions of positive type. In: General Inequalities IV. L. Losonczi-W. Walter, editors. ISNM Series. Basel-Boston-Stuttgart: Birkhäuser (to appear).

8. Schempp, W., Radar ambiguity functions, the Heisenberg group, and holomorphic theta series (to appear).

9. Wigner, E.P., Quantum-mechanical distribution functions revisited. In: Perspectives in Quantum Theory, pp. 25-36. W. Yourgrau-A. van der Merwe, editors. New York: Dover Publications 1979.

10. Woodward, P.M., Probability and Information Theory with Applications to Radar. 2nd edition. New York: McGraw-Hill 1965.

VIII Differential and Difference Equations

International Series of
Numerical Mathematics, Vol. 65

FIXED POINT AND IMPLICIT FUNCTION THEOREMS AND THEIR APPLICATIONS

Joseph W. Jerome
Department of Mathematics
Northwestern University
Evanston

This paper discusses applications, to differential equations in Banach spaces, to optimization theory, to systems of partial differential equations, and to numerical functional analysis, of three fixed point theorems and a Newton iteration scheme associated with an implicit function theorem of functional analysis.

1. Introduction

This paper will discuss four applications of fixed point and implicit function theorems of analysis. The applications will include the quasi-linear Cauchy problem in Banach space, via the contraction mapping principle; quasi-variational inequalities, via the Tarski fixed point theorem; degenerate nonlinear elliptic systems, via the Schauder fixed point theorem; and convergent modified Newton iteration schemes, inherent in the proof of special cases of the Nash-Moser implicit function theorem. The emphasis throughout will be on the applications, since it is not our intent to cite variations or improvements of the underlying theorems. Unfortunately, we cannot present complete proofs of all of our results. Outlines will be presented with references. We mention that the theorem in Section 5, describing the root of an operator equation as the limit of a superlinearly convergent sequence of modified Newton iterates, is a new result. It does not require the invertibility of the Frèchet derivative. Rather, it requires only a sufficiently small initial residual, and a uniformly bounded family of approximate right inverses, for the Frèchet derivative, in a neighborhood of the root.

2. The Cauchy Problem

Consider the abstract Cauchy problem,

$$\frac{du}{dt} + A(t,u)u = 0, \quad 0 \leqslant t \leqslant T_0, \tag{2.1a}$$

(2.1b) $$u(0) = u_0,$$

in a reflexive Banach space X. Here, $A(t,w)$ is a (possibly unbounded) closed linear operator for each (t,w) in a fixed set $J \times W$; J is a given closed time interval of length $|J|$, and W is a closed ball of radius r, centered at 0 in a smooth reflexive subspace Y, densely and compactly embedded in X. It is assumed that $u_0 \in W^0$, i.e., $||u_0||_Y < r$, and that

(2.2) $$Y \subset D_{A(t,w)} \quad , \text{ for each } (t,w) \in J \times W .$$

A positive real number $T_0 \leqslant |J|$ and a function u, satisfying

(2.3) $$u \in L^\infty([0,T_0]; W) \cap H^1([0,T_0]; X),$$

are sought such that u satisfies (2.1a,b). The major result of this section, under hypotheses to be described shortly, is the existence of a real constant γ, such that a solution of (2.1) exists if u_0 and T_0 satisfy the compatibility condition

(2.4) $$||u_0||_Y \, e^{\gamma T_0} < r.$$

The following definition describes the necessary hypothesis of stability.

DEFINITION 2.1. <u>The family $\{A(t,w)\}$ is said to be stable if there are (stability) constants M and ω such that</u>

(2.5) $$|| \prod_{j=1}^{k} R(\lambda_j, -A(t_j,w_j)) ||_{X,X} \leqslant M \prod_{j=1}^{k} (\lambda_j - \omega)^{-1} \quad , \lambda_j > \omega,$$

<u>for any finite families $0 \leqslant t_1 \leqslant \cdots \leqslant t_k \leqslant |J|$ and $\{w_j\}_1^k \subset W$. Here, the resolvent operator is given by</u>

(2.6) $$R(\lambda, U) = (\lambda - U)^{-1} \quad (\text{from X to } D_U \subset X),$$

<u>and the product is time–ordered from right to left.</u>

We shall suppose that the norm in Y dominates the norm in X and is determined by an isomorphism $S: Y \to X$, i.e.,

(2.7) $$||w||_X \leqslant ||w||_Y = ||Sw||_X .$$

It is straightforward to see that $\{A_1(t,w)\}$ is stable on X if and only if $\{\tilde{A}(t,w)\}$ is stable on Y, where

(2.8a) $$A_1(t,w) = SA(t,w)S^{-1} ,$$

(2.8b) $$D_{A_1(t,w)} = \left\{ v\epsilon X : A(t,w)S^{-1} v\epsilon Y \right\} ,$$

and where $\tilde{A}(t,w)$ is the part of $A(t,w)$ in Y, i.e.,

(2.8c) $$D_{\tilde{A}(t,w)} = \{v\epsilon Y : A(t,w)v\epsilon Y\} ,$$

(2.8d) $$\tilde{A}(t,w) = A(t,w)|_{D_{\tilde{A}(t,w)}} .$$

The stability constants M_1 and ω_1 in this case are inherited by $\{\tilde{A}(t,w)\}$. We are now prepared for the theorem of this section.

THEOREM 2.1. Assume that $\{A(t,w)\}$ and $\{A_1(t,w)\}$ are stable families and define γ by

(2.9) $$\gamma := \max(\omega, \omega_1) + 1.$$

Suppose that there exist constants C and C_1 such that

(2.10a) $$||A(t,w) - A(t,w')||_{Y,X} \leqslant C||w-w'||_X,$$

(2.10b) $$||A(t,w)||_{Y,X} \leqslant C_1,$$

for $t\epsilon J$ and $w\epsilon W$ and that

$$A(\cdot,v_j)v_j \to A(\cdot,v)v \left[\text{weakly in } L^2((0,T_0); X)\right]$$

if

(2.10c) $$v_j \to v \left\{\underline{\text{weakly in}}\ L^2((0,T_0);Y)\ \underline{\text{and strongly in}}\ L^2((0,T_0);X)\right\}.$$

Then, for T_0 satisfying (2.4), there exists a solution u of (2.1) in the regularity class (2.3).

We shall merely indicate the directions of proof without actually giving the details. With T_0 selected according to (2.4), we introduce a family $P^N = \left\{t_k = k\Delta t = k\ T_0/N\right\}_{k=0}^{N}$ of uniform partitions, and we consider the fully implicit discretization, defined by

(2.11a) $$A\left(t_k, u_k^N\right) + \left(\frac{1}{\Delta t}\right)u_k^N = \left(\frac{1}{\Delta t}\right)u_{k-1}^N,\ k = 1, \dots, N,$$

(2.11b) $$u_0^N = u_0,$$

$N \geqslant N_0$. Here N_0 is an integer sufficiently large that the problems (2.11) possess solutions. In fact, the u_k^N may be shown to be fixed points of strictly contractive maps Q_k^N. We describe this as follows. First, select ζ such that

(2.12) $$M'||u_0||_Y \exp\left\{(1+\frac{1}{\zeta})\gamma T_0\right\} < r.$$

Here $M' = \max(M, M_1)$. Next, define δ and σ by

(2.13a) $$\delta := \left[r \exp\left\{-(1+\frac{1}{\zeta})\gamma T_0\right\}/(M'||u_0||_Y)\right] - 1,$$

(2.13b) $$\sigma := (1+\delta)M' \exp\left\{(1+\frac{1}{\zeta})\gamma T_0\right\}.$$

The complete metric space on which the Q_k^N act invariantly, for N suitably restricted, is given by

(2.14) $$W_0 := \left\{u \in Y : ||u||_Y \leqslant \sigma ||u_0||_Y\right\}.$$

One sees readily that $W_0 \subset W$. Now, for N fixed, define Q_k^N and u_k^N inductively as follows. Given u_{k-1}^N, set

$$Q_k^N \, \mathrm{v}: = R\left[\Delta t^{-1} - 1, -A\,(t_k, \mathrm{v})\right] \left[\Delta t^{-1}\, u_{k-1}^N - \mathrm{v}\right] \; k=1, \ldots, N,$$

and define u_k^N to be the unique fixed point of Q_k^N on W_0. The Q_k^N are strict contractions provided N satisfies each of the following two inequalities (C is given in (2.10a)):

$$N \geqslant T_0 \left\{ (1 + \delta^{-1}) M' + (1 + \zeta)\gamma \right\}, \tag{2.15}$$

$$1 > \left[\Delta t^{-1} - 1 - \omega\right]^{-1} \left\{ M + MC\sigma \| u_0 \|_X + \frac{(\omega_1 + 1) C \sigma \delta}{(1+\delta)^2} \| u_0 \|_X \right\}. \tag{2.16}$$

Moreover, the r.h.s. of (2.16) is the contraction constant of Q_k^N. It follows easily that, for such N and $k=1, \ldots, N$,

$$\| u_k^N \|_Y \leqslant r, \tag{2.17a}$$

$$\| u_k^N - u_{k-1}^N \|_X \leqslant C_1 \, r \, \Delta t \, (C_1 \text{ given by (2.10b)}). \tag{2.17b}$$

REMARK 2.1. The details of the contraction mapping discussed above may be found in the writer's work (see [6] and [7, Section 7.5]). The notion of stability which we have employed is basically due to Kato (see [10] and [11]). The conclusion of the existence theorem contained herein is somewhat stronger than that obtained by Hughes, Kato and Marsden (see [5]), where no attempt is made to correlate the sizes of $\| u_0 \|_Y$ and T_0. We have, however, assumed the compactness of the embedding $Y \to \mathrm{X}$. This implies (see Temam [17, p. 271]) that the embedding

$$L^2\left[(0, T_0); Y\right] \cap H^1\left[(0, T_0); \mathrm{X}\right] \to L^2\left[(0, T_0); \mathrm{X}\right] \tag{2.18}$$

is compact. If piecewise linear and step function sequences in time are constructed from the u_k^N, then appropriate weak-* convergent, weakly convergent, and convergent subsequences can be selected, via (2.17) and (2.18), with the same limit. Standard methods (see [6] and [7,

Chapter 5]) employing the weak formulation show that this limit is a solution of (2.1), if use is made of (2.10c). This condition is the one reasonably expected to hold in applications and motivates the compactness hypothesis. This condition, that $Y \to X$ is compact, can be weakened in the applications if local compactness holds (see [6] for the example of the convergence of the artificial viscosity method).

3. Quasi–Variational Inequalities

In 1939 Kantorovich (see [9]) proved a well-known fixed-point theorem for continuous, isotone mappings. The later version due to Tarski (see [16]) does not require continuity and, in fact, is valid for certain commuting families of mappings. We state here a special case of Tarski's result. Proofs may also be found in Jerome (see [7, pp. 88,89] and Birkhoff (see [2]).

THEOREM 3.1. Let E be a partially ordered set, and suppose $u_0 \leqslant v_0$ are elements of E, with the property that the interval $I = \{v \epsilon E : u_0 \leqslant v \leqslant v_0\}$ is a complete lattice. Suppose that S is an isotone mapping of I into E such that

$$u_0 \leqslant S(u_0), \quad S(v_0) \leqslant v_0. \tag{3.1}$$

Then the set of fixed points u of S satisfying $u_0 \leqslant u \leqslant v_0$ is nonempty and possesses a minimal element $\underline{u}$ and a maximal element $\bar{u}$.

REMARK 3.1. An isotone mapping here means an increasing mapping. A lattice is complete if every subset possesses a least upper bound and greatest lower bound in the lattice. If the continuity of S is assumed, then the convergence of the iterates of u_0 and v_0 to a fixed point can be proved if the topology is compatible, i.e., if least upper bounds (resp. greatest lower bounds) of increasing (resp. decreasing) sequences are limits of these sequences. It is desirable for the applications to avoid the hypothesis of continuity, however. We shall illustrate in the case of quasi-variational inequalities. We recall the fact that, if A is a (single-valued) pseudomonotone, bounded mapping of a (subset K of a) Hilbert space H into itself and f is specified in H, then the variational inequality

$$(A(u) - f, v - u) \geqslant 0, \quad \text{for all } v \epsilon K, \tag{3.2}$$

has a solution $u \epsilon K$, if K is closed and convex (see Browder [3]). We are interested in this section in the case when A is defined on a fixed set, but u is specified in a corresponding subset $K(u)$ depending upon u itself. This explains the term quasi-variational inequality.

DEFINITION 3.1. *We shall assume that the Hilbert space* H *is a lattice, so that* $v = v^+ + v^-$ *for each* $v \epsilon H$, *with* $v^+ = \max(0, v), v^- = \min(0, v)$, *and that* $A : H \to H$ *is a given operator satisfying*

(3.3a) $$(A(u), u - v) \leqslant \liminf_{i \to \infty} (A(u_i), u_i - v) \quad \text{(pseudomonotone)}$$

whenever

(3.3b) $$u_i \to u \text{ (weakly) and } \limsup_{i \to \infty} (A(u_i), u_i - u) \leqslant 0$$

for $u, u_i \epsilon H$; *and*

(3.4) $$A(H_b) \text{ is bounded in } H \quad \text{(bounded)}$$

for every bounded subset H_b *of* H; *and*

(3.5) $$(A(v), v)/||v|| \geqslant \lambda \, ||v|| \quad \text{(coercive)}$$

for some $\lambda > 0$, *all* $v \neq 0$. *Moreover, we assume that* A *is strictly monotone with respect to the lattice structure, i.e.,*

(3.6a) $$(A(u) - A(v), (u - v)^{\pm}) \geqslant 0,$$

(3.6b) $$(A(u) - A(v), (u - v)^+) = 0 \leftrightarrow (u - v)^+ = 0,$$

(3.6c) $$(A(u) - A(v), (u - v)^-) = 0 \leftrightarrow (u - v)^- = 0.$$

Now set

(3.7) $$H^+ = \{v \epsilon H : v \geqslant 0\},$$

and let $f \epsilon H$ *be given such that there exists* $v_0 \epsilon H^+$ *with*

$$A(\mathrm{v}_0) = f. \tag{3.8}$$

Let $M : H^+ \to H^+$ be isotone on $[0, \mathrm{v}_0]$. By the quasi-variational inequality problem is meant the *determination of* an element $u \epsilon K(u)$ satisfying

$$(A(u) - f, \mathrm{v} - u) \geqslant 0, \text{ for all } \mathrm{v} \epsilon K(u). \tag{3.9a}$$

Here, for $0 \leqslant w \leqslant \mathrm{v}_0$,

$$K(w) = \left\{ \mathrm{v} \epsilon H^+ : \mathrm{v} \leqslant M(w) \right\}. \tag{3.9b}$$

We have the following theorem.

THEOREM 3.2. Assume the hypotheses described in Definition 3.1. Then the mapping $u = S(w)$, defined by

$$(A(u) - f, \mathrm{v} - u) \geqslant 0, \text{ for all } \mathrm{v} \epsilon K(w) \quad (u \epsilon K(w)), \tag{3.10}$$

is an isotone mapping of $[0, \mathrm{v}_0]$ into H^+ satisfying

$$S(0) \geqslant 0, \quad S(\mathrm{v}_0) \leqslant \mathrm{v}_0. \tag{3.11}$$

In particular, S has a fixed point u satisfying (3.9).

REMARK 3.2. Sometimes the boundedness property is included in the definition of pseudomonotonicity as in Lions (see [12]) and the writer's book (see [7]). To see that S is increasing, let $0 \leqslant w_1 \leqslant w_2 \leqslant \mathrm{v}_0$ and let $u_1 = S(w_1)$, $u_2 = S(w_2)$. For the inequality (3.10), corresponding to $w = w_1$, set $\mathrm{v} = u_1 + (u_2 - u_1)^-$ and, for $w = w_2$, set $\mathrm{v} = u_2 - (u_2 - u_1)^-$. After addition of the two inequalities, we have

$$0 \leqslant (A(u_2) - A(u_1), \ (u_2 - u_1)^-) \leqslant 0$$

where we have used (3.6a). By (3.6c) we conclude that $(u_2 - u_1)^- = 0$, i.e., $u_2 \geqslant u_1$ and S is increasing. The verification of $S(\mathrm{v}_0) \leqslant \mathrm{v}_0$ follows, via (3.6a) and (3.6b), from setting $w = \mathrm{v}_0$,

$u = S(\mathrm{v}_0)$ and $\mathrm{v} = S(\mathrm{v}_0) - (S(\mathrm{v}_0) - \mathrm{v}_0)^+$ in (3.10), and adding the resultant inequality to

$$(A(\mathrm{v}_0) - f,\ (S(\mathrm{v}_0) - \mathrm{v}_0)^+) = 0.$$

REMARK 3.3. The application given in Theorem 3.2 is suggested by a similar application due to Lions (see [13, pp. 170-173]). A more general result in reflexive Banach spaces is stated and proved by the writer (see [7, p. 99]). Interestingly, mathematical analogies exist which identify the solutions of certain optimal stopping-time problems with Stefan problems. In these cases, the primary formulation is a quasi-variational inequality (see Friedman [4] for elaboration).

4. Nonlinear Elliptic Systems

The modeling of physical and biological systems frequently involves the interdependence of several variables of interest; typically, this might involve a potential variable and one or more concentration variables. The steady-state behavior of such systems is often modeled by reaction/diffusion/convection systems of elliptic partial differential equations. One of the most powerful tools for proving the existence of solutions of such systems is the Schauder fixed point theorem. Usually, one proceeds by defining a so-called solution map via iterative uncoupling. Thus, although such systems are not expected to be gradient systems, the individual, uncoupled equations defining the solution map are expected to be gradient equations, and make possible a well-defined map and pointwise maximum principles. The latter yield a function space domain on which the solution map acts invariantly; its fixed points are, by design, solutions. We shall illustrate by describing the system which models the movement of electrons and holes in a doped crystalline semiconductor. The system involves the three dependent variables u, n, and p, where u is the electrostatic potential, and n and p are electron and hole carrier concentrations. The quantities of primary physical interest are the current densities, expressed with unit charge modulus by

$$J_n = -\mu_n n \nabla u + D_n \nabla n, \tag{4.1a}$$

and

$$J_p = -\mu_p p \nabla u - D_p \nabla p. \tag{4.1b}$$

Here, the carrier mobilities μ_n and μ_p are dependent on the electric field, $-\nabla u$, and the products, $\mu_n \nabla u$ and $-\mu_p \nabla u$, represent the electron and hole drift velocities, respectively. The field-dependent functions D_n and D_p are diffusion coefficients. If a Brownian motion model of particle flow, with negligible fluctuations, is assumed, then the Einstein relations

$$D_n = \mu_n, \; D_p = \mu_p, \tag{4.2}$$

hold in thermodynamic equilibrium in appropriate temperature units (see Nelson [14]). If the formal change of dependent variable,

$$n = \exp(u - \mathrm{v}), \;\; p = \exp(w - u), \tag{4.3}$$

is made, then J_n and J_p assume the form

$$J_n = -\mu_n n \nabla \mathrm{v}, \tag{4.4a}$$

and

$$J_p = -\mu_p \, p \nabla w. \tag{4.4b}$$

The relations (4.3) are suggested by the so-called classical approximation, in which the carriers are assumed to satisfy Fermi-Dirac statistics, with localization in the "Boltzmann tail". The steady-state system, consisting of a Maxwell equation, relating the field and net charge density, and two mass conservation equations, is given by

$$-\nabla \cdot (\epsilon \nabla u) + n - p = N, \tag{4.5a}$$

$$\nabla \cdot J_n - U = 0, \tag{4.5b}$$

$$\nabla \cdot J_p + U = 0, \tag{4.5c}$$

where $\epsilon = \epsilon(x)$ is the dielectric function, $N = N(x)$ is the net ionized doping and $U = U(n,p)$ is the net recombination term. The uniform ellipticity of the system, when valid, is expressed by the statements that ϵ, μ_n, and μ_p are uniformly bounded above 0. The boundary conditions of physical interest are inhomogeneous Dirichlet boundary conditions on the contact portions of the semiconductor device, and homogeneous Neumann boundary conditions on the "complement". The form of U is crucial to the derivation of maximum principles, and may be expressed in normalized form as

$$U(n,p) = Q(n,p)\,(np-1), \tag{4.6}$$

where the recombination rate Q is a positive, rational function.

We are now prepared to describe the solution map T. The domain of T is defined to consist of those pairs $(v,w) \in L^2(\Omega) \times L^2(\Omega)$, whose components are constrained to have range between $\min(\inf_{\Sigma_D} \bar{v}, \inf_{\Sigma_D} \bar{w})$ and $\max(\sup_{\Sigma_D} \bar{v}, \sup_{\Sigma_D} \bar{w})$. Here Σ_D is that part of $\partial\Omega$ on which boundary values $\bar{v}$ and $\bar{w}$ are prescribed. Thus, given a pair $[\tilde{v}, \tilde{w}] \in D_T$, one carries out the fractional step involved in computing $\tilde{u}(\tilde{v},\tilde{w})$ from the gradient equation (4.5a). The functions $\tilde{u}$, $\tilde{v}$ and $\tilde{w}$ can be used in the coefficients of ∇v and ∇w, within the divergence structure of (4.5b,c), and as arguments in $U(\tilde{u}, v, \tilde{w})$ and $U(\tilde{u}, v, w)$ in (4.5b,c), respectively. The maximum principles ensure that $[v,w] \in D_T$. In addition, we have the following.

THEOREM 4.1. The mapping T is continuous and has relatively compact range, within the closed, convex subset D_T of $L^2(\Omega) \times L^2(\Omega)$, under the hypotheses: Lipschitz continuity of $\bar{u}, \bar{v}, \bar{w}, \mu_n, \mu_p$, and U; the boundedness of N; the respective decrease and increase of U in its second and third arguments; and uniform ellipticity. In particular, T has a fixed point.

REMARK 4.1. Elaboration and proofs are given by the writer (see [8]) where the case of degenerate elliptic systems is also discussed. This important example occurs in the case of an insulated gate transistor, which contains an oxide buffer introducing a reentrant corner. The field becomes unbounded in the neighborhood of such a point, and the mobilities may vanish. Modifications in the hypotheses and proofs can handle this, however.

5. Generalized Newton Iteration Schemes

A common technique in numerical analysis for the solution of nonlinear systems is to carry out computational schemes until some measurable quantity, such as the residual, is small, in conjunction with a negligible variation of the computed iterates. We shall describe in this section a generalized Newton iteration scheme, which uses as a starting value a quantity which yields a small residual, as measured in a given norm. The size of the residual depends upon the order of superlinear convergence desired. The Newton iteration is based upon the common practice of the inversion of the diagonal structure of the Frèchet derivative; in the applications this part of the linearization is commonly self-adjoint and positive definite, and yields a robust approximate right inverse. The derivative itself need not be invertible at the root, though the off-diagonal terms, when regularized by the approximate inverse, must be compatible with the residual. The scheme is suggested by the proof of the Nash-Moser implicit function theorem (see Nirenberg [15]), although here we consider the simpler case where no "loss of derivatives" occurs and hence no smoothing is necessary.

DEFINITION 5.1. *Let Y and Z be Banach spaces and let B be a closed ball in Y centered at u with radius r. Suppose*

$$F:B \to Z, \quad F(u) = 0, \quad F(v) \neq 0, v \neq u, \tag{5.1}$$

is a mapping with the following properties

(i) *F is twice differentiable with*

$$||F''(v)|| \leqslant 2M \quad (M \geqslant 1), \tag{5.2a}$$

for all $v \in B$. Here, the norms refer to standard bilinear maps.

(ii) *For every $v \in B$ there is a linear map $G(v) \in L(Z,Y)$ such that, for all $z \in Z$,*

$$||[F'(v)\, G(v) - I]z||_Z \leqslant M\, ||F(v)||_Z\, ||z||_Z, \tag{5.2b}$$

$$||G(v)z|| \leqslant M\, ||z||_Z. \tag{5.2c}$$

Define, for $u_0 \in B_{r/2}$, the formal iteration scheme

$$u_k - u_{k-1} = -G(u_{k-1})\, F(u_{k-1}), \; k = 1, 2, \cdots. \tag{5.3}$$

THEOREM 5.1. *Let t be prescribed, $1 < t < 2$. Let $\zeta > 1$ satisfy*

$$2M^3\, \zeta^{-(2-t)} \leqslant 1, \tag{5.4a}$$

$$M\zeta^{-1}/(1 - \zeta^{1-t}) \leqslant r/2, \tag{5.4b}$$

and let $u_0 \in B_{r/2}$ satisfy $||F(u_0)||_Z \leqslant \zeta^{-1}$. Then the iteration scheme defined by (5.3) satisfies

$$u_k \in B, \; k \geqslant 0, \tag{5.5a}$$

(5.5b) $$||F(u_k)||_Z \leqslant \zeta^{-t^k} \quad , \; k \geqslant 0$$

(5.5c) $$||u_k - u_{k-1}||_Y \leqslant M \, \zeta^{-t^{k-1}} \quad , \; k \geqslant 1.$$

In particular, $\{u_k\}$ satisfies the superlinear estimate

(5.6) $$||u_k - u||_Y \leqslant M\zeta^{-t^k}/(1 - \zeta^{1-t}).$$

REMARK 5.1. The verification of (5.5) proceeds by induction. Inequality (5.5c), for k, uses the definition (5.3), the hypothesis (5.2c), and the induction hypotheses (5.5a) and (5.5b) for $k-1$. This is followed by the verification of (5.5a) for k, which uses (5.5c), the inequality

(5.7) $$\sum_{k=1}^{\infty} \zeta^{-t^{k-1}} \leqslant \zeta^{-1} \sum_{k=0}^{\infty} \zeta^{-k(t-1)} = \zeta^{-1}/(1-\zeta^{1-t}),$$

and (5.4b). The verification of (5.5b) for k uses the representation

(5.8) $$F(u_k) = -(F'(u_{k-1})\, G(u_{k-1}) - I)F(u_{k-1}) + R(u_{k-1}, u_k),$$

where

(5.9) $$R(u_{k-1}, u_k) = F(u_k) - F(u_{k-1}) - F'(u_{k-1})\,(u_k - u_{k-1}).$$

The first term in (5.8) is estimated by (5.2b) and the induction hypothesis (5.5b) for $k-1$. Similarly, the second term in (5.8) is estimated by $M||u_k - u_{k-1}||_Y^2$, via (5.2a), and this quantity in turn may be estimated by (5.5c). Inequality (5.5b) follows when use is made of (5.4a) and the assumption $M \geqslant 1$. Finally, the generalized form of (5.7) implies (5.6). Implicit in this statement is the fact that u is an isolated root, and that $\{u_k\}$ is Cauchy, with a root as limit.

REMARK 5.2. Original and intricate computational procedures for reducing the residual are discussed by Bank and Rose (see [1]). Our hypotheses differ from those of these authors in that we do not assume the invertibility of the Frèchet derivative, hence do not obtain quadratic convergence.

ACKNOWLEDGMENT. This paper was written while the author was a visiting scientist at Bell Laboratories. Financial support, staff support, and availability of facilities are gratefully acknowledged.

REFERENCES

[1] Bank, R. E. - Rose, D. J., Global approximate Newton Methods, Numer. Math. 37 (1981), 279-295.

[2] Birkhoff, G., Lattice Theory (3rd. ed.), Amer. Math. Soc. Colloq. Publ. 25, Providence, Rhode Island, 1973.

[3] Browder, F. E., Nonlinear Operators and Nonlinear Equations of Evolution in Banach Spaces, Amer. Math. Soc. Symp. Pure Math. 18 (Part 2), Providence, Rhode Island, 1976.

[4] Friedman, A., Variational Principles and Free Boundary Problems, Wiley, New York, 1982.

[5] Huges, T., Kato, T., and Marsden, J., Well–posed quasi–linear, second–order hyperbolic systems with applications to nonlinear elastodynamics and general relativity, Arch. Rational Mech. Anal. 63 (1977), 273-294.

[6] Jerome, J., Quasilinear hyperbolic and parabolic systems: contractive semidiscretizations and convergence of the discrete viscosity method, J. Math. Anal. Appl. 90 (1982), 185-206.

[7] Jerome, J., Approximation of Nonlinear Evolution Systems, Academic Press, New York, 1983.

[8] Jerome, J., Consistency of semiconductor modeling: an existence/stability analysis for the stationary Van Roosbroeck system, manuscript.

[9] Kantorovich, L., The method of successive approximations for functional equations, Acta Math. 71 (1939), 63-97.

[10] Kato, T., Linear equations of "hyperbolic" type, J. Fac. Sc. Univ. Tokyo 17 (1970), 241-258.

[11] Kato, T., Linear equations of "hyperbolic" type II, J. Math. Soc. Japan 25 (1973), 648-666.

[12] LIons, J. L., Quelques Methodes de Resolution des Problemes aux Limites non Lineaires, Dunod, Paris, 1969.

[13] Lions, J. L., Sur Quelques Questions d'Analyse de Mechanique et de Controle Optimal, Univ. of Montreal Press, Montreal, 1976.

[14] Nelson, E., Connection between Brownian motion and quantum mechanics, Lecture Notes in Physics 100 (H. Nelkowski et al. eds.), Springer-Verlag, Berlin, 1979, pp. 168-179.

[15] Nirenberg, L., Topics in Nonlinear Functional Analysis, Courant Institute of Mathematical Sciences, New York Univ., New York, 1973-1974.

[16] Tarski, A., A lattice–theoretical fixpoint theorem and its applications, Pacific J. Math. 5 (1955), 285-309.

[17] Temam, R., Navier–Stokes Equations (rev. ed.), North Holland Publ., Amsterdam, 1979.

International Series of
Numerical Mathematics, Vol. 65

GRAPH THEORY IN THE APPROXIMATION THEORY OF FLUID DYNAMICS

Karl Gustafson
Department of Mathematics
University of Colorado
Boulder

Consider a viscous imcompressible flow governed by the Navier-Stokes equations. In a number of finite element schemes, one discretizes the flow equations within approximating "incompressibility subspaces." We describe the previously missing theory of dimension and bases for those subspaces.

Introduction

This work began in 1979 when we decided in a seminar to go through the excellent book by Teman [16], in which the Navier Stokes equations are treated numerically by finite element methods. As Temam pointed out, e.g., see [16, pp. 58, 138, 494, amont others], a difficulty was the lack of a suitable theory of dimension and bases for the approximating subspaces of those methods. We resolved those questions in [7,8] and have treated further interesting aspects of the resultant theory in [9,10].

My goals in the present paper are to (1) quickly expose for you the essentials of our approach with emphasis on what I shall call here "incompressibility subspaces," (2) briefly describe a number of related references, and (3) mention a few future possibilities. I shall try insofar as possible to keep the discussions and examples disjoint from those of [7,8,9,10] and to cast them within the contexts of functional analysis, approximation, and operator theory.

1. Incompressibility Subspaces and Their Analysis.

The name "incompressibility subspaces" is mine, here. They arise as follows. Assume that the vector field of velocity components of an incompressible flow is initially known and for simplicity assume also that the boundary values for the flow are to be zero. Then one has for the domain Ω in question the Navier-Stokes momentum equation

$$\vec{u}_t - \frac{1}{R}\Delta\vec{u} + (\vec{u}\cdot\nabla)\vec{u} = -\nabla p + \vec{f}$$

subject to the incompressibility constraint

$$\nabla\cdot\vec{u} = 0 . \tag{1.1}$$

In the finite element schemes under consideration, one triangulates the domain Ω and on the triangulated domain one agrees to a linear, quadratic, cubic, or the like, approximation to the unknown velocity vector field $\vec{u}$, the approximation taken in the appropriate Sobolev space and corresponding norm. A key feature to these schemes is that the resulting approximation subspace must itself satisfy the incompressibility condition (1.1). These schemes have the names APX1, APX2, APX2', APX3, APX4, APX5 in [16]. There is considerable analysis of the flow governed by the Navier-Stokes momentum equation, see [16], but here I shall keep the focus on the resulting "incompressibility subspaces" V_h, which satisfy equation (1.1).

Because versions of APX5 are quite important in the French application of these methods to airframe design, because (due to its high locality and low order) APX5 is one of the easiest of these schemes to work with, and because we did not previously highlight APX5, let me do so here. APX5 is a piecewise linear scheme in R^n, and is called nonconforming since it need not be continuous across triangulation boundaries. Because of the latter, we may initially restrict attention to a single n-simplex τ in the triangulation T of the domain Ω in R^n.

Let τ have vertices $A_0, A_1,\ldots,A_n$, μ_i be the corresponding barycentric coordinate functions, S_i the (n-1)-face on which $\mu_i = 0$, B_i the barycenter of that face, and λ_i the barycentric coordinate functions relative to B_i.

LEMMA 1.1. *The* B_i, $i = 0,\ldots,n$, *are the vertices of an* n *dimensional simplex*.

PROOF. Let $C = \sum_{j=0}^{n} A_j/n+1$. Then

(1.2)
$$\begin{aligned}\overrightarrow{CB_i} &= \sum_{\substack{j=0\\ j\neq i}}^{n} A_j/n - \sum_{j=0}^{n} A_j/n+1 \\ &= \frac{1}{n}\left(\sum_{j=0}^{n} A_j/n+1\right) - \frac{1}{n} A_i \\ &= -\frac{1}{n}\overrightarrow{CA_i}\end{aligned}$$

and

(1.3)
$$n\overrightarrow{B_iB_j} = n(\overrightarrow{CB_j} - \overrightarrow{CB_i}) = \overrightarrow{CA_i} - \overrightarrow{CA_j} = \overrightarrow{A_jA_i} \ .$$

The vectors $\overrightarrow{A_jA_i}$, $0 \le i \le n$, $i \neq j$, form a basis for R^n because the points A_i, $0 \le i \le n$ are the vertices of an n-simplex. Therefore, from (1.3), the vectors $\overrightarrow{B_iB_j}$, $0 \le i \le n$, $i \neq j$, are a basis for R^n. Hence the points B_i, $0 \le i \le n$ are vertices of an n-simplex.

THEOREM 1.2. *Let* $\vec{u}$ *be a* C^1 *divergence-free* n *dimensional vector field defined over* Ω . *On* τ *(closed) let*

(1.4)
$$\begin{aligned}\vec{u}_h(B_i) &= \int_{S_i} \vec{u}ds/m(S_i), \quad 0 \le i \le n, \\ \vec{u}_h(x) &= \sum_{i=0}^{n} \lambda_i(x)\vec{u}_h(B_i) \ .\end{aligned}$$

Then $\nabla \cdot \vec{u}_h(x) = 0$ *for* $x \in \text{int}(\tau)$.

PROOF. Because $\vec{u}_h$ is affine on $\text{int}(\tau)$, $\nabla \cdot \vec{u}_h$ is a constant c there, and the constant c is given by

(1.5)
$$\begin{aligned}cm(\tau) &= \int_{\tau} \nabla \cdot \vec{u}_h = \sum_{j=0}^{n} \int_{S_j} \vec{u}_h \cdot n_j ds \\ &= \sum_{j=0}^{n} n_j \cdot \int_{S_j} \vec{u}_h ds = \sum_{j=0}^{n} n_j \cdot m(S_j)\vec{u}_h(B_j)\end{aligned}$$

$$= \sum_{j=0}^{n} n_j \cdot \int_{S_j} \vec{u} ds = \sum_{j=0}^{n} \int_{S_j} u \cdot n_j ds$$
$$= \int_{\tau} \nabla \cdot \vec{u} = 0 .$$

Hence $c = 0$. Notations used in the above: m is Lebesgue measure, n_j is the unit outer normal to the jth face.

EXAMPLE 1.3. APX5 _in two dimensions. Let_

$$\vec{u}(x) = (x_2^2, x_1^2) , \tag{1.6}$$

let $A_0 = (0,0)$, $A_1 = (1,0)$, $A_2 = (0,1)$. _Then_

$$\vec{u}_h(x) = \frac{2}{3}(x_2,x_1) . \tag{1.7}$$

PROOF. See FIGURE 1(a). $B_0 = (1/2,1/2)$, $B_1 = (0,1/2)$, $B_2 = (1/2,0)$,

$$\vec{u}_h(B_0) = \int_{S_0} (x_2^2,x_1^2) ds/m(S_0) = \frac{1}{\sqrt{2}} \int_0^1 ((1-t^2), t^2) \sqrt{2}\, dt = (1/3,1/3) ,$$
$$\vec{u}_h(B_1) = \int_{S_1} (x_2^2,x_1^2) ds/m(S_1) = \int_0^1 (t^2,0) dt = (1/3,0) ,$$
$$\vec{u}_h(B_2) = \int_{S_2} (x_2^2,x_1^2) ds/m(S_2) = \int_0^1 (0,t^2) dt = (0,1/3) .$$

With respect to B_0,B_1,B_2, barycentric coordinates are

$$\lambda_0(x) = 2x_1 + 2x_2 - 1, \ \lambda_1(x) = 1 - 2x_1, \ \lambda_2(x) = 1 - 2x_2 ,$$

and by definition then

$$\vec{u}_h(x) = \lambda_0(x)(1/3),1/3) + \lambda_1(x)(1/3,0) + \lambda_2(x)(0,1/3)$$
$$= (2x_1 + 2x_2 - 1)(1/3,1/3) + (1-2x_1)(1/3,0) + (1-2x_2)(0,1/3)$$
$$= 2/3(x_2,x_1) .$$

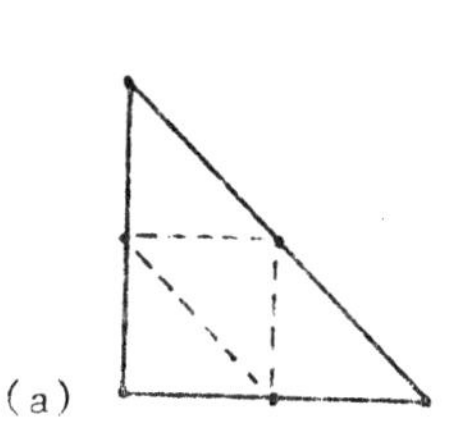

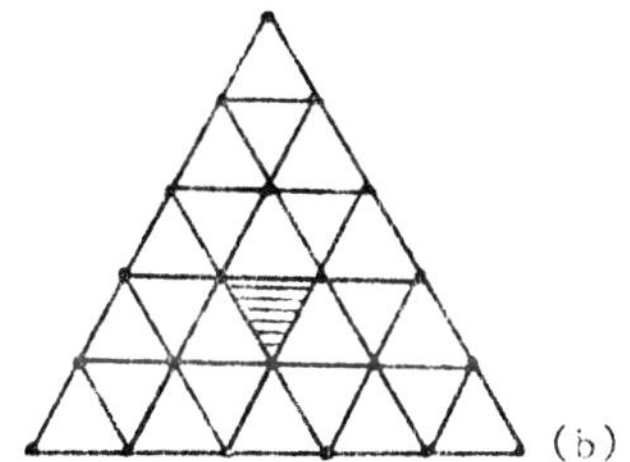

FIGURE 1. Two Examples

EXAMPLE 1.3 shows the analytic construction of th APX5 approximation on a single element. Imposition of the Dirichlet (zero on boundary vertices) boundary condition assumed in [16] and [8] would render it zero and uninteresting on a domain Ω consisting of only one element. Consider then the triangulated domain Ω consisting of 24 elements shown in FIGURE 1(b). As shown in [8] in general for acceptably triangulated Ω in R^n, one has

$$V_h(\text{APX5}) = V_2 \oplus V_3 , \tag{1.8}$$

$$\dim V_h = (n - 1)\bar{E}_1 + (\ell + m) , \tag{1.9}$$

where $\bar{E}_1$ = number of barycenters of interior $n - 1$ faces, ℓ = number of interior vertices, m = number of interior "holes". Thus for the Ω of FIGURE 1(b),

$$\dim V_h = 27 + (3 + 1) = 31 . \tag{1.10}$$

How were these general results obtained? I refer to [8] for details, but one can say that there are two main steps.

Step 1: Prove a Helmholtz decomposition theorem of the type

$$V_h(\text{APXi}) = V_1 \oplus V_2 \oplus V_3$$

where in some appropriate sense (for example, on the faces of the elements) V_1 is divergence-free and curl-free, V_2 is divergence-free and not curl-free, and V_3 is curl-free and not divergence-free.

Step 2: Employ elements of graph theory to calculate the dimensions and bases for each of V_1, V_2, V_3 .

The scheme APX5 is simplified by the absence of a V_1 component. The "Incompressibility Subspace V_h" for scheme APX5 for the triangulated domain Ω of FIGURE 1(b) has "two Helmholtz subspaces" and combined dimension 31.

2. Related References

Since 1979-80 when we developed our approach to these questions, we have become aware of two other groups who have been led (independently of one another) to use of graph-theoretic methods in the approximation theory of fluid dynamics. In summary:

School	Colorado	France	Pittsburgh
Dates (roughly)	1979 →	1979 →	1979 →
Initial Motivation	To answer Temam's questions	To resolve APX5 in 3d	To implement Russian K-L method
References	[7,8,9,10]	[12,17]	[1,14]

TABLE 1. Graph theory in fluid approximation theory.

The French school has of course a continuing interest in these methods and it is not surprising that Hecht [12] was led to the use of elements of graph theory as we were. In finding bases for the "incompressibility subspaces" APX5 he emphasized maximal trees, whereas we used cycle bases.

The Pittsburgh school was analyzing a method (1966) of Krzhivitski and Ladyzhenskaya and employed elements of network theory, notably node-link matrices and cycle bases, to reduce the operation count for the implementation of the K-L discretization. It turns out that the K-L (finite difference) scheme is essentially APX1 for the incompressibility constraint (and Scheme 5.1 of [16] for the momentum equation, in weak form).

We mention that a basic early paper finding a basis for APX5 in two dimensions was that of Crouzeix [2]. Other references for related considerations to bring you up to date are Fortin [3] and Griffiths [4]. Space limitations prevent a more complete discussion here.

3. Future Possibilities

As to future possibilities, I would like to give a few indications, connections, and interpretations.

3.1 General Theory of Finite Element Methods. We have thus far only looked at the FEM APX(i) of the French school and made connection to the Pittsburgh school's analysis of the K-L Russian Scheme. These are a "drop in the bucket" of finite element methods that have been developed for the numerical analysis of fluids, in structures, and elsewhere. One has the feeling that whenever a complicated composite of elements are used in such modeling, the graph-theoretic methods we have exposed here can play a role in finding and reducing dimension, bases, complexity. This should lead to a general graph-theoretic theory of all such methods.

3.2 The Kron Connections. In [11] I investigate, among others, a connection to the inspired work of the electrical engineer G. Kron some years ago. Kron (see [13]) tried to model and solve just about everything in terms of electrical networks. This led to the so-called tearing methods which are now becoming important in matrix theory.

Roughly, the use of duality that we employ in finding the dimension and bases for the V_3 component of any scheme, resembles tearing.

As to Kron's method of tearing, which in spite of existence and uniqueness proofs, e.g., see Roth [15], continues in its implementation to confuse both old and new alike, I would like to offer one way to look at it. Given an electrical network of, say, three impedances in parallel, suppose one wants to solve the equation

$$ZJ = V , \tag{3.1}$$

V = voltage, Z is the overall impedance, J is the current. Remember that

$$\frac{1}{Z} = \frac{1}{Z_1} + \frac{1}{Z_2} + \frac{1}{Z_3} . \tag{3.1}$$

Now consider wanting to solve a linear system

$$Ax = b \ . \tag{3.3}$$

The analogy with the network says that

$$A^{-1} = A_1^{-1} + A_2^{-1} + A_3^{-1} \ . \tag{3.4}$$

The trick, in this interpretation, would be to find the most efficient combination of component inverses A_i^{-1} .

Work in sparse matrix theory concerning the relative merits of tearing, partitioning, perturbing, and modifying, to invert A, continues. Our cycle bases lead to very sparse matrices, see [9,10].

Finally, Kron's method works [15] so long as a certain duality map L is ohmic. That L be ohmic is quivalent to the numerical range $V(L)$ not containing zero. In the context of [5] one thus has the classes

$$\underset{\text{(invertible)}}{L} \supset \underset{\text{(ohmic)}}{L} \supset \underset{\text{(partial inner product)}}{L \text{ compatible}} \supset \underset{\text{(semi-inner product)}}{L \text{ a duality map}} \tag{3.5}$$

I do not know of any systematic study of "ohmic operators".

3.3 General Vector Calculus of Graphs and Networks. In [6] we reverse the analysis: instead of the application of graph theory to fluid dynamics, what can the latter say to the former: This leads to interesting notions concerning, for example, the inherent "curl of a graph," and to the possibility of so inducing a general vector calculus of graphs and networks.

REFERENCES

[1] Amit, R. - Hall, C. A. - Porsching, T.A., An application of network theory to the solution of implicit Navier-Stokes difference equations. J. Comp. Phys. 40 (1981), 183-201.

[2] Crouzeix, M., Séminaire d'analyse numerique. Université de Paris VI (1971-72).

[3] Fortin, M., Old and new finite elements for incompressible flows. Int. J. Numer. Meth. in Fluids 1 (1981), 347-364.

[4] Griffiths, D., An approximately divergence-free 9-node velocity element (with variations) for incompressible flows. Int. J. Numer. Meth. in Fluids 1 (1981), 323-346.

[5] Gustafson, K. - Antoine, J. P., Partial inner product spaces and semi-inner product spaces. Advances in Mathematics 41 (1981), 281-300.

[6] Gustafson, K. - Harary, F., The curl of graphs and networks. (to appear).

[7] Gustafson, K. - Hartman, R., On the dimension of a finite difference approximation to divergence-free vectors. Quantum Mechanics in Mathematics, Chemistry, and Physics, Gustafson, K. - Reinhardt, W., eds. Plenum Press, New York (1981), 125-131; See also Notices Amer. Math. Soc. 1 (1980), 196.

[8] Gustafson, K. - Hartman, R., Divergence-free bases for finite element schemes in hydrodynamics. SIAM J. Numer. Anal. 20 (1983), 697-721.

[9] Gustafson, K. - Hartman, R., Graph theory and fluid dynamics, SIAM J. on Algebraic and Discrete Methods (to appear).

[10] Gustafson, K. - Hartman, R., Graph-theoretic aspects of flow calculation methods. Fifth International Symposium on Finite Element Methods in Flow Problems, Jan. 23-26, 1984, Austin, Texas (to appear).

[11] Gustafson, K., Topological considerations in finite element models, to appear.

[12] Hecht, F., Construction d'une base de fonctions P_1 nonconforme à divergence nulle dans R^3. RAIRO 15 (1981), 119-150.

[13] Kron, G., Diakoptics: The Piecewise Solution of Large-scale Systems. McDonald, London (1963).

[14] Lin, A. Cha. - Porsching, T. A., On the Krzhivitski-Ladyzhenskaya difference method for the stationary Navier-Stokes equations. 10th IMACS World Congress on System Simulation and Scientific Computation, Montreal (1982), Vol. 1, 34-36.

[15] Roth, J. Paul, An application of algebraic topology: Kron's method of tearing. Quarterly of Applied Math. 17 (1959), 1-41.

[16] Teman, R., Navier-Stokes Equations: Theory and Numerical Analysis. 2nd Edition, Elsevier-North Holland, New York (1979).

[17] Thomasset, F., Finite Element Methods for Navier-Stokes Equations. Springer, New York (1981)

International Series of
Numerical Mathematics, Vol. 65

THE SPECTRUM OF THE LAPLACIAN FOR DOMAINS IN HYPERBOLIC SPACE AND LIMIT SETS OF KLEINIAN GROUPS

Ralph Phillips
Department of Mathematics
Stanford University
Stanford

This is a report on a joint paper with Peter Sarnak [5] on the properties of the spectrum of the Laplacian, with free boundary conditions, for domains of infinite volume in hyperbolic space. We are mainly concerned with the discrete part of the spectrum, its existence or nonexistence, and lower bounds for the bottom of the spectrum. Combining these results with those of Patterson [4] and Sullivan [6] yields new estimates for the Hausdorff dimension of the limit sets of Kleinian groups.

Let H^{n+1} denote the real hyperbolic space of dimension n+1:

$$H^{n+1} = \{w = \{x,y\},\ x \in \mathbb{R}^n,\ y > 0\}$$

with line element $ds^2 = (dx^2 + dy^2)/y^2$. We denote by dw, ∇ and Δ the volume element, gradient and Laplacian, respectively, all with respect to the hyperbolic metric. In particular

$$-\Delta = y^2(\delta_y^2 + \delta_{x_1}^2 + \dots + \delta_{x_n}^2) - (n-1)y\delta_y. \tag{1}$$

Let Ω be an open connected subset of H^{n+1} and denote by $W^1(\Omega)$ the space of functions

$$W^1(\Omega) = \{f \in L_2(\Omega),\ \nabla f \in L_2(\Omega)\}. \tag{2}$$

The quadratic forms H and D on $W^1(\Omega)$ are defined as

$$H(f,g) = \int f\bar{g}dw,$$

(3)

$$D(f,g) = \int \nabla f \cdot \overline{\nabla g} dw.$$

The domain of the Laplacian with free (i.e. Neumann) boundary conditions consists of the set of all u in $W^1(\Omega)$ with square integrable Δu (defined in the weak sense) satisfying the condition

(4) $$H(\Delta u,v) = D(u,v)$$

for all v in $W^1(\Omega)$.

We shall limit ourselves to domains which are convex, geometrically finite (i.e. bounded by a finite number of geodesic hyperplanes) and are of infinite volume. The geodesic hyperplanes in H^{n+1} are either hemispheres of the form $|x-a|^2 + y^2 = r^2$, $y > 0$, or vertical Euclidian hyperplanes. Since the Laplacian is invariant under global isometries of H^{n+1}, we need not distinguish between domains which are related by such an isometry.

For such domains the proof of Theorem 4.4 in [3] can be adapted to show that the spectrum of Δ is discrete in $(0,(\frac{n}{2})^2)$ and continuous in $[(n/2)^2,\infty)$. The variational formulation for the bottom of the spectrum $\lambda_0(\Omega)$ is given by

(5) $$\lambda_0(\Omega) = \inf\{D(u);\ u \text{ in } W^1(\Omega),\ H(u) = 1\} \quad .$$

We call a domain Ω free if $\lambda_0(\Omega) = (n/2)^2$. In terms of the form

(6) $$E = D - (n/2)^2 H,$$

defined on $W^1(\Omega)$, Ω is free iff $E \geqslant 0$. In view of the above characterization of the sprectum of Δ, Ω is free iff Δ has no discrete spectrum. Free domains are the basic building blocks in this paper and the following result plays a central role:

THEOREM 1. *If* $\Omega \subset H^{n+1}$ *has* $[(n+4)/2]$ *or fewer sides then* Ω *is free; there are domains with* $[(n+6)/2]$ *sides which are not free. Here* $[c]$ *denotes the greatest integer in* c.

When Ω is free we introduce a new form:

$$G = E + K , \tag{7}$$

where

$$K(f,g) = \int_S f\bar{g}dw \tag{8}$$

and S is any compact subset of Ω. We denote the completion of $W^1(\Omega)$ with respect to G by H_G (see also [3]). It is possible for $\Delta' = \Delta - (n/2)^2$ on H_G to have 0 in its discrete spectrum. The corresponding eigenfunction v is called a null vector and satisfies the condition $E(v) = 0$; however v does not lie in $W^1(\Omega)$.

Examples of free domains with null vectors are:

(i) If P is a finite sided bounded Euclidean polyhedron in $\mathbb{R}^n$ and $\Omega = \{\{x,y\}, x \in P, y > 0\}$, then Ω is free and has a null vector $v = y^{n/2}$.

(ii) Let C_1, C_2, C_3 be three mutually tangent hemispheres, each in the exterior of the other two, and let $\Omega \subset H^3$ be the domain exterior to these hemispheres. Then Ω has a null vector.

A free domain with a null vector is very close to having an L_2 eigenfunction. More precisely, if Ω is such a domain and Ω' is obtained from Ω by excising a small hemisphere with center in $\{y = 0\}$, then Ω' is no longer free and hence has an L_2 eigenfunction.

We show for $\Omega_1 \subset \Omega$ that if $\Omega \setminus \bar{\Omega}_1$ is free then $\lambda_0(\Omega_1) \leq \lambda_0(\Omega)$, contrary to what one might expect. We also show that $\lambda_0(\Omega)$ is continuous under most deformations of the bounding sides when $n = 1$. Of more interest is the fact that the discrete spectrum is continuous under what we call simple degenerations. Essentially in such a degeneration we allow sides to degenerate in clusters of no more than $[(n+2)/2]$ sides.

Suppose now that Γ is a discrete subgroup of motions with fundamental domain Ω. The Laplacian leaves invariant the Γ-automorphic functions, i.e. functions satisfying the relation

$$u(\gamma w) = u(w) \quad \text{for all } \gamma \text{ in } \Gamma.$$

It also defines a selfadjoint operator on $L_2(\Omega)$ with boundary conditions appropriate for these automorphic functions. We denote by $\lambda_0(\Gamma) < \lambda_1(\Gamma) \leq \ldots$ the discrete spectrum of this Γ-automorphic Laplacian. It is easy to see from (5) that

$$\lambda_0(\Omega) \leq \lambda_0(\Gamma); \tag{9}$$

here Ω is the fundamental domain for Γ.

If Ω is bounded by nonoverlapping hyperplanes, we call Ω a Schottky domain. The reflections in its bounding hyperplanes generates a discrete group Γ, called a reflection group. It is easy to see that the Γ-automorphic functions satisfy Neumann boundary conditions so that in this case $\lambda_0(\Omega) = \lambda_0(\Gamma)$.

For any discrete group of motions Γ, the limit set $\Lambda(\Gamma)$ is defined as the set of limit points in $B = \partial H^{n+1}$ of any orbit

$$\Gamma w = \{\gamma w;\ \gamma \text{ in } \Gamma,\ w \text{ fixed in } H^{n+1}\} .$$

$\Lambda(\Gamma)$ is a closed subset of B invariant with repsect to Γ. Let $d(\Gamma)$ denote the Hausdorff dimension of $\Lambda(\Gamma)$. The following theorem, first proved by Patterson [4] under somewhat restrictive conditions and later proved in full generality by Sullivan [6], provides the connection between $\lambda_0(\Gamma)$ and $d(\Gamma)$.

THEOREM (Patterson-Sullivan). *If Γ is geometrically finite and $d(\Gamma) \geq n/2$, then*

$$\lambda_0(\Gamma) = d(n-d). \tag{10}$$

Returning to the concept of a null vector for Δ', we can show that

THEOREM 3. *If Ω is a Schottky domain without cusps and Γ denotes the corresponding reflection group, then Ω has a null vector iff $d(\Gamma) = n/2$.*

Beardon [2] has shown for all reflection groups Γ, corresponding to Schottky domains with m sides in H^{n+1}, that there exist constants $d(m,n) < n$ such that $d(\Gamma) \leq d(m,n)$ provided $n \geq 2$. In the same paper he raised the question of whether the $d(m,n)$ were bounded away from n. The next theorem answers this question in the affirmative for $n \geq 3$.

THEOREM 4. *For* $n \geq 3$ *there is a number* $d_n < n$ *such that for all reflection groups corresponding to Schottky domains*

$$d(\Gamma) \leq d_n. \tag{11}$$

Because of the relation (10) this theorem is an immediate consequence of the existence of a positive lower bound for $\lambda_o(\Omega)$ which holds for all Schottky domains in H^{n+1}, $n \geq 3$. When $n = 1$ (i.e. in H^2) there exist 3-sided Schottky domains for which $d(\Gamma) = 1$. When $n = 2$ we do not know if $d(\Gamma)$ has an upper bound less than 2. In the opposite direction Akaza [1] has established the existence of a Schottky domain of 4-sides in H^3 for which $d(\Gamma) > 1$. This also follows from our result on null vectors and the excision property as applied to Example (ii) stated above.

REFERENCES

[1] Akaza, T., *Singular sets of some Kleinian groups*, Nagoya Math. J. **29** (1967), 145 - 162.

[2] Beardon, A.F., *The Hausdorff dimension of singular sets of properly discontinuous groups*, Amer. J. Math. **88** (1966), 722 - 736.

[3] Lax, P.D. — Phillips, R.S., *The asymptotic distribution of lattice points in Euclidean and non-Euclidean spaces*, J. Funct. Anal. **46** (1982), 280 - 350.

[4] Patterson, S.J., *The limit set of a Fuchsian group*, Acta Math. **136** (1976), 241 - 273.

[5] Phillips, R.S. — Sarnak, P., *The Laplacian for domains in hyperbolic spaces and limit sets of Kleinian groups*, Acta Math. (To appear)

[6] Sullivan, D., *The density at infinity of a discrete group of hyperbolic isometries*, Publication I.H.E.S. **50** (1979), 171 - 209.

International Series of
Numerical Mathematics, Vol. 65

AN EXPONENTIAL FORMULA OF HILLE-YOSIDA TYPE FOR PROPAGATORS (*)

Gunter Lumer
Institut de Mathématique
Université de l'Etat
Mons, Belgique

We give an exponential formula for the representation of any propagator on a Banach space, $U : (s,t) \mapsto U(t,s)$, "pregenerated" in a certain sense by a time-dependent family of dissipative operators $\{A(t)\}$ with resolvents depending continuously on t. This formula is a direct generalization of the exponential Hille-Yosida representation $P(t) = \lim_{\lambda\to\infty} e^{tA_\lambda}$ of a semigroup $(P(t))$ with generator A. In the time-dependent case we still have an analogous limit as $\lambda \to \infty$ of an exponential expression, where that exponential expression is again computed by a power series.

1. Introduction

Let X be a complex Banach space. Consider the following evolution equation in X :

$$\frac{du}{dt} = A(t)u\ ,\ t > s\ ,\ (\text{or } t \geqslant s),$$

(1)

$u(s) = f$, f an initial value chosen in $X_0(s)$ (some given subset of X depending on s; $s,t \in J$ an interval $\subset R$,

$J \neq \emptyset$, the $A(t)$ being linear operators in X with domain $D(A(t))$. (For simplicity, henceforth "operator" will always mean "linear operator").

(*) The research in this paper was supported in part by a grant from the belgian F.N.R.S.. We also like to aknowledge a helpful conversation we had with R. Beals, at Yale U. in March 1983, concerning material in this paper.

Let us now consider first, for a moment, the case in which the A(t) are independent of t, A(t) = A for all t ∈ J, A being closed and densely defined. Then it is well know that if (and only if) the evolution equation (1) (with $t \geq s$, $X_0(s) = D(A)$) is uniformly well posed, [Z] , [Kr] ([1]), there exists a semigroup with generator A, $(P(t)) = (e^{tA})$, such that for f ∈ D(A) the solution u(t,s,f) of (1) is given by $P(t-s)f = e^{(t-s)A}f$ for s,t ∈ J. By "semigroup" we mean throughout this paper a strongly continuous one parameter semigroup of class (C_0), [H-Ph] , [Z] , and shall often, for simplicity, write P(t) or e^{tA} for either the semigroup or one operator of the semigroup when there is no risk of confusion. Going back to the specific solution semigroup $(P(t)) = (e^{tA})$, u(t,s,f) = P(t-s)f, just considered, we have the following representation formula from the Hille-Yosida theory :

$$P(t) = \lim_{\lambda\to\infty} e^{tA_\lambda} , \tag{2}$$

where $A_\lambda \in \mathcal{B}(X)$ = {all everywhere defined bounded (linear) operators on X}, λ is a real parameter $\geq \lambda_0 \in R$, lim in (2) is understood in the strong sense, in X, and where moreover the A_λ can be explicitely computed from the resolvent R(λ,A) of A, and we have in the strong sense

$$A_\lambda \to A \quad \text{in} \quad X \tag{3}$$

as $\lambda \to \infty$.

The purpose of this paper is to give, in the general time-dependent context, and exponential formula which directly extends (2) and (3), for a propagator U(t,s) associated with the general problem, general evolution equation, (1), when the latter is in some appropriate sense uniformly well posed or sufficiently close to being uniformly well posed. U(t,s) is then associated to (1) like in the previously considered simple situation (A(t) constant) via U(t,s)f = u(t,s,f) solution of (1) at t for the given initial data s and f. Notice that in the context of (2) (A(t) constant generator) the propagator boils down to U(t,s) = P(t-s). With the appropriate notations and terminology, which will be introduced and made precise later, the general exponential formula of Hille-Yosida type has the form

([1]) in the latter one uses the expression "uniformly correct" instead of "uniformly well posed".

$$(4) \qquad U(t,s) = \lim_{\lambda\to\infty} (e^{(t-s)(A_{\lambda -}+D)}1)(s), \text{ for } s \leq t \in J,$$

with $A_\lambda(t) \in \mathcal{B}(X)$, $A_\lambda(t) \to A(t)$, $\forall t \in J$.

Essentially, we do not deal with questions of existence of solutions of (1); we deal here mainly with the possibility of an exponential representation for a propagator U(t,s) associated with a family of operators A(t) under the conditions loosely described above. In so doing we shall limit ourselves essentially to the case in which the operators A(t) are dissipative.

An exponentiel representation formula such as (4) is fundamentally based on two ingredients, more specifically two levels of approximation : (i) finding an appropriate $U_\lambda(t,s)$ with $U_\lambda(t,s) \to U(t,s)$ as $\lambda \to \infty$, where U_λ corresponds to (1) with A(t) replaced by appropriate $A_\lambda(t) \in \mathcal{B}(X)$, $A_\lambda(t) \to A(t)$; (ii) finding an explicit formula of exponential type for $U_\lambda(t,s)$ (which will be essentially (4) without "$\lim_{\lambda\to\infty}$"). While (i) is based on integration techniques such as used earlier by Kato, Yosida, Kisynski, and others, (see in particular [K] section 4), together with partition of unity arguments, etc, (ii) is based on "holomorphic" and algebraic considerations (with bounded A(t) operators). We start in the next section 2 by developping all that is pertinent to (ii) while introducing at the same time notations and terminology needed later. Matters pertaining to (i) are dealt with mainly in sections 4 and 5.

2. Solutions of (1) in the "bounded holomorphic" case

Let $J =]S,T[$ be a finite interval $\subset \mathbb{R}$. Throughout this section we assume that $A(t) \in \mathcal{B}(X)$, $\forall t \in J$, and moreover that the map $J \to \mathcal{B}(X)$: $t \mapsto A(t)$, extends holomorphically to the open disc $B(S,|T-S|) = \{z \in \mathbb{C} : |z-S| < |T-S|\}$. In this situation we shall give an explicit expression for the solution of (1). We must however introduce first some notation.

Let $M \in C^\infty(J,\mathcal{B}(X))$. We shall denote by M_-, D, the operators from $C^\infty(J,\mathcal{B}(X)) \to C^\infty(J,\mathcal{B}(X))$ sending $F \in C^\infty(J,\mathcal{B}(X))$, respectively, into M_-F, DF, defined by

$$(5) \qquad \begin{aligned} (M_-F)(t) &= F(t)M(t), \ \forall\, t \in J ; \\ (DF)(t) &= F'(t), \ \forall\, t \in J. \end{aligned}$$

Of course M_-, D, could be defined on larger spaces of maps $J \to \mathcal{B}(X)$ than $C^\infty(J,\mathcal{B}(X))$, but we do not need this and would loose at the same time some algebraic properties which we want, in particular the fact that polynomials (with real or complex coefficients) in M_-, D, are well defined as operators on $C^\infty(J,\mathcal{B}(X))$.

We shall also denote by the same symbols M_-, D, the operators induced on $C^\infty(J_1,\mathcal{B}(X))$, J_1 any interval $\subset J$, by restricting t to J_1, in (5), whenever there is no risk of confusion.

We denote by 1 the corresponding number as well as the identity map on X, or the constant map $J \to \mathcal{B}(X)$ with value 1 (identity map on X), the context making it clear which is meant.

For $\lambda \in \mathbb{C}$, we write λ in lieu of $\lambda 1$ whenever there is no risk of confusion; for instance for the resolvent of an operator A in X, we write $R(\lambda,A) = (\lambda-A)^{-1}$.

Finally, let us write Δ_J for $\{(s,t) \in J \times J : s \leqslant t\}$.

DEFINITION 2.1. _A propagator on_ X, J, (²), _is a map_ $U : \Delta_J \to \mathcal{B}(X) : (s,t) \mapsto U(t,s)$, _such that_ :

$$U(s,s) = 1;\ U(t',s) = U(t',t)U(t,s) \text{ for } s \leqslant t \leqslant t',\ s,t,t' \in J\ ; \tag{6}$$
$$(s,t) \mapsto U(t,s) \text{ is strongly continuous } \Delta_J \to \mathcal{B}(X).$$

We have now the following result.

THEOREM 2.2. _Let_ A _be a map_ $J \to \mathcal{B}(X) : t \mapsto A(t)$, _which extends holomorphically to_ $z \mapsto A(z)$ _on the disc_ $B(S,|T-S|)$. _Then for every_ $f \in X$, (1) (_with_ $t \geqslant s$, $X_0(s) = X$) _has a unique solution_ $u(t,s,f)$ _which extends holomorphically to_ $z \mapsto u(z,s,f)$ _on the disc_ $B(S,|T-S|)$. _Such solutions are given by a uniquely determined propagator_ U _on_ X, J, _via_ $u(t,s,f) = U(t,s)f$, _and_ U _is given by the exponential formula_

(²) compare with the notion of "generalized semigroup" introduced in [P] p. 245. One can show that generalized semigroups correspond exactly to propagators with exponential growth, i.e. such that $\|U(t,s)\| \leqslant Me^{(t-s)\omega}$, $(s,t) \in \Delta_J$, M, ω, constants. Examples show that propagators need not have exponential growth whether J is finite or not (as an example on $J =]0,+\infty[$ consider $U(t,s) = e^{(t^2-s^2)}1$). See also footnote (³) below.

(7) $$U(t,s) = (e^{(t-s)(A_-+D)}1)(s) \text{ for } s \leqslant t,\ s,t \in J,$$

<u>where the right hand side of</u> (7) <u>is interpreted as being the strong limit on</u> X <u>of</u> $(\sum_{n=0}^{N}((t-s)^n/n!)(A_-+D)^n 1)(s)$ <u>as</u> $N \to +\infty$ (<u>this meaning in particular that the strong limit just mentioned exists on all of</u> X).

PROOF. Under our assumptions we can first apply locally, in an appropriately small disc $B(s,\delta)$, or $B(z_0,\delta)$, about a given $s \in J$, or $z_0 \in B(S,|T-S|)$, a slight modification of the usual fixed point argument, applied here to $u(z) = f + \int_s^z A(\zeta)u(\zeta)d\zeta$, or $u(z) = f_0 + \int_{z_0}^z A(\zeta)u(\zeta)d\zeta$, to obtain a unique local (holomorphic) solution of $du/dz = A(z)u$, with $u(s) = f$, or $u(z_0) = f_0$. Using this, together with analytic continuation and monodromy in the disc $B(S,|T-S|)$, we see that there exists a unique solution of (1), for $f \in X$ and $s \in J$ given, extending holomorphically to $B(S,|T-S|)$. Hence, this extension $z \mapsto u(z,s,f)$ admits a power series expansion about s converging at least in $B(s,|T-s|)$. Thus for $s,t \in J$, $s \leqslant t$, we have

(8) $$u(t,s,f) = \sum_{n=0}^{\infty} (t-s)^n a_n \ ,\ a_n = \frac{1}{n!} u^{(n)}(.,s,f) \text{ at } s \ .$$

We shall now compute explicitely $u^{(n)}(.,s,f)$ at s, which we shall denote simply $u^{(n)}(s)$. (Of course $u^{(n)}$ is as usual a notation for $(d^n/dt^n)u$; we shall also usually write ', u', u", in lieu of d/dt, $u^{(1)}$, $u^{(2)}$).

By repeatedly differentiating $u' = Au$, and replacing u' by Au, we obtain $u'' = A'u + A^2u = (A' + A^2)u$, $u^{(3)} = A''u + A'Au + (A^2)'u + A^3u = (A'' + 2A'A + AA' + A^3)u$, ..., and in general one has for all n = 1, 2, ...

(9) $$u^{(n)} = E_n(A, A', \ldots, A^{(n-1)})u \ ,$$

where E_n is the restriction to $[s,T[$ of a non commutative polynomial E_n in $A, A', \ldots, A^{(n-1)}$, which is an element of $C^\infty(J,\mathcal{B}(X))$, and where for $F \in C^\infty(J_1,\mathcal{B}(X))$, $u \in C^\infty(J_1,X)$, J_1 a subinterval of J, Fu means the element of $C^\infty(J_1,X)$: $t \mapsto F(t)u(t)$ for $t \in J_1$; (9) therefore means that $u^{(n)}(t) = (E_n(A, A', \ldots, A^{(n-1)}))(t)\, u(t)$ for all $t \in [s,T[$. We have now, writing simply E_n instead of $E_n(A, A', \ldots, A^{(n-1)})$,

$$u^{(n+1)} = (E_n)'u + E_n u' =$$
$$(E_n)'u + (E_n)Au = (DE_n + A_-E_n)u =$$
$$((D + A_-)E_n)u = ((A_- + D)E_n)u \ .$$

Hence, $\forall\, t \in [\, s,T\, [$,

$$(E_{n+1})(t)u(t) = ((A_- + D)E_n)(t)u(t),$$

and since s is arbitrary in J, while u(s) = f can take on all values in X, we conclude that we have in $C^\infty(J,\mathcal{B}(X))$ the equality

$$E_{n+1} = (A_- + D)E_n \ , \text{ for } n = 1, 2, \ldots \tag{10}$$

Moreover, $E_1 = E(A) = A = (A_- + D)1$. From (10) we have :

$$E_n = E_n(A, A', \ldots, A^{(n-1)}) = (A_- + D)^n 1 \ . \tag{11}$$

$$u^{(n)}(s) = u^{(n)}(.,s,f) \text{ at } s = ((A_- + D)^n 1)(s)f. \tag{12}$$

Going now back to (8) we have :

$$u(t,s,f) = \lim_{N\to\infty} \sum_{n=0}^{N} \frac{1}{n!}(t-s)^n((A_- + D)^n 1)(s)f =$$
$$\lim_{N\to\infty} ((\sum_{n=0}^{N} \frac{1}{n!}(t-s)^n(A_- + D)^n)1)(s)f =$$
$$(e^{(t-s)(A_- + D)}1)(s)f \ ,$$

which proves (7). ([3])

([3]) Actually, in the context of this section and under the assumptions of 2.2, one can even show by a slight variant of the argument that (7) holds also with uniform ($\mathcal{B}(X)$-norm) convergence replacing strong convergence, and this uniformly for s,t in a compact subset of J. Also U can be extended from Δ_J to $B(S,|T-S|) \times B(S,|T-S|)$. Again, 2.4 holds in the uniform sense.

REMARK 2.3. Among other things, formula (7) of 2.2 permits to compute directly the coefficients in the expansion of $u(t,s,f)$ without going through any recurrence procedure. This computation, say of (11), can be made systematically in terms of binary expansions, or for low values of n read off a graph.

REMARK 2.4. By writing in (11), $E_n = (A_-^n + \tilde{E}_n)1$, $(E_n)(s) = A^n(s) + (\tilde{E}_n 1)(s)$, one sees from 2.2, (7), and its proof, that one can use (7) to derive approximation formulas up to $O(|t-s|^k)$ for $k = 3, 4, \ldots,$ of the type

$$U(t,s) = e^{(t-s)A(s)} + \tfrac{1}{2}(t-s)^2 A'(s) + O(|t-s|^3) \tag{13}$$

(in the strong sense), (13) being the formula for k = 3.

REMARK 2.5. Of course the second member of (7) reduces to $e^{(t-s)A}$ when $A(t)$ = constant = A.

3. C^1 and $W^{1,1}$ solutions in the unbounded case

From here on , we consider the case of unbounded $A(t)$, limiting ourselves however to dissipative operators of that type. More precisely, we consider problem (1) with $J =]S,T[$ a finite open interval $\subset R$, with the alternative $t > s$ in (1). We assume that $\forall t \in J$, $A(t)$ is a dissipative semigroup generator [L-Ph] (i.e. $A(t)$ generates a contraction semigroup).

By a C^1-solution of (1), for given $s \in J$, $f \in X$, we mean a $u \in C([s,T[,X) \cap C^1(]s,T[,X)$, with $u(t) \in D(A(t))$ and $u'(t) = (du/dt)(t) = A(t)u(t)$ $\forall t \in]s,T[$, $u(s) = f$.

By a $W^{1,1}$-solution of (1), for given $s \in J$, $f \in X$, we mean a $u \in W^{1,1}(]s,T[,X)$, such that a.e. in $]s,T[$ $u(t) \in D(A(t))$, $u'(t) = A(t)u(t)$, and such that $u(s) = f$. To be more precise, for any $u \in W^{1,1}(]s,T[,X)$ there exists a unique absolutely continuous $u_1 : [s,T] \to X$, such that $\exists$ a.e. $u_1'(t)$ and $t \to u_1'(t)$ is in $L^1(]s,T[,X)$, and $u_1 = u$ a.e. (*); we consider here, when talking about u in $W^{1,1}(]s,T[,X)$,

(*) The appropriate converse also holds, i.e. any u_1 absolutely continuous, with a.e. derivative $\in L^1(]s,T[,X)$, is in $W^{1,1}(]s,T[,X)$. See [B], p. 145 proposition A.3 and corollaire A.1.

that we are refering to the continuous representative just mentioned (so, in particular, u(s) is well defined). In particular C^1-solutions are $W^{1,1}$-solutions on $[s,T'[$ for any $T' < T$ with $s < T' < T$.

It is an important fact that in our context, i.e. with dissipative operators, when a $W^{1,1}$-solution exists for a given $f \in X$ it is unique. Indeed we have the following.

LEMMA 3.1. _Suppose_ u _is a_ $W^{1,1}$_-solution, for a given_ $f \in X$, _of problem_ (1) _as considered in this section_ (_except that in this lemma we need only assume_ A(t) _dissipative but nothing else, i.e. we need not assume_ A(t) _to be a generator or even densely defined_). _Then_ $\|u(t)\| \leqslant \|f\|$ $\forall\, t \in [s,T]$, _and in particular the solution is unique, if it exists, for any given_ $f \in X$.

PROOF. Let us recall that the dissipativeness of A(t) means that $\forall\, \lambda > 0$, $g \in D(A(t))$, we have (see for instance [Y], p. 447),

$$\|(A(t) - \lambda)g\| \geqslant \lambda\|g\| \ . \tag{14}$$

Now a.e. for t in the set where u'(t) exists, and any h real > 0, we have $u(t-h) = u(t) - h\,A(t)u(t) + o(h)$,

$$\|u(t-h)\| = h\|(A(t) - \tfrac{1}{h})u(t)\| + o(h) \geqslant \|u(t)\| + o(h) \tag{15}$$

by (14); and (15) gives $(\|u(t-h)\| - \|u(t)\|)/(-h) \leqslant o(h)/h$. Since $u \in W^{1,1}(]s,T[,X)$, $t \mapsto \|u(t)\|$ is absolutely continuous from $[s,T] \to R$, and by what we have just seen we shall have a.e. $(d/dt)\|u(.)\| \leqslant 0$. Hence $t \mapsto \|u(t)\|$ is non increasing, which gives, by continuity at s, $\|u(t)\| \leqslant \|f\|$ for all $t \in [s,T[$. Hence, given two solutions u, $\tilde{u}$, corresponding to the same f, we have by linearity $\|(u-\tilde{u})(t)\| = \|u(t)-\tilde{u}(t)\| \leqslant 0$ which proves uniqueness.

In view of 3.1 we know that there is a most one unique $W^{1,1}$ (or C^1)-solution for a given $f \in X$ (and of course a given $s \in J$), and if it exists we shall denote it by

$$t \mapsto u(t,s,f) \quad \text{for} \quad t \in [s,T[\ . \tag{16}$$

We call $t \mapsto u(t,s,f)$ the $W^{1,1}$ (or C^1)-solution starting at $(s,f) \in J \times X$.

To express that a family of operators in X, $\{A(t)\} = \{A(t) : t \in J\}$, "generates" in some sense a given propagator U on X, J, is a less simple matter than for semigroups (and their time-independent generator). One can express this (like when dealing with nonlinear semigroups) by saying that U "represents" the solutions of (1) for the corresponding $\{A(t)\}$ at least for a sufficiently large set of initial values. To be able to express this, later on in section 5, in a short and precise manner, we introduce the following terminology.

Given a subset $\underline{X}_0$ of $J \times X$ such that the problem (1), corresponding to a given family $\{A(t)\}$ as considered in this section, has a $W^{1,1}$-solution starting at every $(s,f) \in \underline{X}_0$ (or alternatively a C^1-solution starting at every $(s,f) \in \underline{X}_0$), and given a propagator U on X, J, we shall say that U represents the solutions of (1) starting on $\underline{X}_0$ (and corresponding to $\{A(t)\}$), iff : $\forall\, (s,f) \in \underline{X}_0$,

$$u(t,s,f) = U(t,s)f \quad \text{for } s \leqslant t < T\,. \tag{17}$$

In the next section we collect some tools concerning approximations to $A(t)$ and $U(t,s)$, which will be needed for the proof of the exponential representation formula (4) in the general context considered in the present section.

4. Some preliminary facts concerning approximations to A(t) and U(t,s) in the general (unbounded) case

We continue to work in the context of the last section, i.e. with the $A(t)$ in (1) unbounded dissipative generators, etc., and assume we have a set $\underline{X}_0$ of initial values (s,f), and a propagator U on X, J, such that U represents the $W^{1,1}$-solutions of (1) starting on $\underline{X}_0$. Simultaneously, we consider another propagator V, on X, J, representing the C^1-solutions of (1) starting on $J \times X$ and corresponding to a family $\{\tilde{A}(t)\}$ with $\tilde{A}(t) \in \mathcal{B}(X)$ and $t \mapsto \tilde{A}(t)$ continuous from $J \to \mathcal{B}(X)$ (in this case we take of course $t \geqslant s$ and $X_0(s) = X$ in (1)). It is well known (see for instance [Kr] chapter II, section 2) that $t \to V(t,s)$, $s \to V(t,s)$ are continuously differentiable with

$(d/dt)V(t,s) = \tilde{A}(t)V(t,s)$, $(d/ds)V(t,s) = -V(t,s)\tilde{A}(s)$.

Under these circumstances, consider $(s,f) \in \underset{-}{X}_0$ and $u(.,s,f) = U(.,s)f$ solution of (1) corresponding to $\{A(t)\}$ starting at (s,f). Then it can be checked without difficulty that

$$\eta \mapsto V(t,\eta)U(\eta,s)f \quad \text{for} \quad \eta \in [s,t], \tag{18}$$

any $s < t < T$, is absolutely continuous with derivative a.e., in L^1. We have a.e.

$$\frac{d}{d\eta} V(t,\eta)U(\eta,s)f = V(t,\eta)(A(\eta) - \tilde{A}(\eta))U(\eta,s)f, \tag{19}$$

and since $\eta \mapsto V(t,\eta)U(\eta,s)f$ is in $W^{1,1}(\,]s,t[\,,X)$, we arrive using (19) at the following

LEMMA 4.1. _Under the conditions and with the notations described above in this section, we have for any_ $s,t \in J$, $s \leqslant t$, $((s,f) \in \underset{-}{X}_0)$,

$$U(t,s)f - V(t,s)f = \int_s^t V(t,\eta)(A(\eta) - \tilde{A}(\eta))U(\eta,s)f\, d\eta. \tag{20}$$

Lemma 4.1 provides a tool for the approximation to $U(t,s)$. Concerning the construction of appropriate bounded approximations to the $A(t)$, we start by recalling the well known fact that for $A(t)$ as considered here, i.e. a dissipative generator, we have the so-called Yosida approximation, $\forall t \in J$,

$$\tilde{A}_\lambda(t) = \lambda^2 R(\lambda, A(t)) - \lambda \quad \text{defined for all } \lambda > 0, \tag{21}$$

where $\tilde{A}_\lambda(t) \in \mathcal{B}(X)$ is dissipative and $\tilde{A}_\lambda(t) \to A(t)$ strongly in X (on $D(A(t))$) as $\lambda \to \infty$.

If we assume, as we shall do below, that for the given $\{A(t)\}$, and $\forall \lambda > 0$,

$$t \mapsto R(\lambda, A(t)) \quad \text{is continuous from } J \to \mathcal{B}(X), \tag{22}$$

so $t \to \tilde{A}_\lambda(t)$ is continuous from $J \to \mathcal{B}(X)$, we introduce another approximation $A_\lambda(t)$ derived from $\tilde{A}_\lambda(t)$, which we shall call the Yosida-Bernstein approximation and which is more suitable for our purpose (in fact it is an approximation to $t \mapsto A(t)$ rather than to a single operator). This approximation is defined as follows. For $\lambda > 0$ large enough so as to have $S_\lambda = S + \frac{1}{\lambda} < T - \frac{1}{\lambda} = T_\lambda$, we consider $J_\lambda =]S_\lambda, T_\lambda[$, i.e. for $\lambda > \lambda_0 = \frac{2}{T-S}$. Then on $\overline{J}_\lambda = [S_\lambda, T_\lambda]$, $A_\lambda(t)$ will be the Bernstein approximation, see [Y] p. 8 and 9, to the continuous $\widetilde{A}_\lambda(t)$, defined on $\overline{J}_\lambda$ by

$$(23) \qquad A_\lambda(t) = \sum_{k=o}^{k=n} \binom{n}{k} (T_\lambda - S_\lambda)^{-n} (t-S_\lambda)^k (T_\lambda - t)^{n-k} \widetilde{A}_\lambda(S_\lambda(1-\tfrac{k}{n}) + T_\lambda \tfrac{k}{n}),$$

where n is chosen to be the smallest positive integer for which $\|\widetilde{A}_\lambda(t) - A_\lambda(t)\| \leq 1/\lambda$ for all $t \in \overline{J}_\lambda$.

In turn, corresponding to the families $\{\widetilde{A}_\lambda(t)\}$, $\{A_\lambda(t)\}$, of operators $\subset \mathcal{B}(X)$ (with t ranging, respectively over J, J_λ) we have the propagators $\widetilde{U}_\lambda$, U_λ, respectively on X, J, X, J_λ, representing the corresponding C^1-solutions starting on $J \times X$, $J_\lambda \times X$; $\widetilde{U}_\lambda$, U_λ are thus in the same situation as the propagator V considered at the beginning of this section and in lemma 4.1. Moreover, using (21), (23), and lemma 3.1 we arrive easily at the following

PROPOSITION 4.2. <u>The operators</u> $A(t)$ <u>being dissipative</u>, $\widetilde{A}_\lambda(t)$, $A_\lambda(t)$ <u>are dissipative for all</u> t <u>involved. Hence</u> $\widetilde{U}_\lambda$, U_λ <u>are contractive, i.e. consist of contractions</u> $\widetilde{U}_\lambda(t,s)$, $U_\lambda(t,s)$.

Using 4.1, (20), and the fact that $\widetilde{U}_\lambda$, U_λ, are contractive, 4.2, we also see at once that we have the estimate

$$(24) \qquad \|\widetilde{U}_\lambda(t,s) - U_\lambda(t,s)\| \leq |T-S|/\lambda \quad \text{for } (s,t) \in J_\lambda \times J_\lambda,\ s \leq t.$$

(Actually U_λ extends continuously to $\overline{J}_\lambda \times \overline{J}_\lambda$ and (24) also holds in that case).

It also follows now from (24) and the fact that $\forall\, t \in J$, $\widetilde{A}_\lambda(t) \to A(t)$ strongly in X (in particular on $D(A(t))$) as $\lambda \to \infty$, that

$$(25) \qquad A_\lambda(t) \to A(t) \text{ as } \lambda \to \infty, \text{ in } X, \forall t \in J,$$

(which means $\{f \in X : \exists \lim_{\lambda\to\infty} A_\lambda(t)f\} = D(A(t))$ and $\lim_{\lambda\to\infty} A_\lambda(t)f = A(t)f$ for $f \in D(A(t))$, $\forall\, t \in J$).

5. The general exponential representation formula

This last section is devoted to the proof of the following result loosely anounced at the beginning of the paper, (4) section 1.

THEOREM 5.1. *Let* $\{A(t) : t \in J\}$ *be a family of dissipative semigroup generators in* X *such that* $t \mapsto R(\lambda, A(t))$ *is continuous from* $J \to \mathcal{B}(X)$ $\forall \lambda > 0$. *Assume that for a dense subset* $\underline{X}_0$ *of* $J \times X$ *there exists a* $W^{1,1}$*-solution of (1) (corresponding to the given* $\{A(t)\}$*) starting at each* $(s,f) \in \underline{X}_0$. *Assume there exists a propagator* U *on* X, J, *representing those solutions starting on* $\underline{X}_0$. *Then* U *is uniquely determined, contractive, and we have* $\forall s,t \in J$, $s \leqslant t$, *the following expression for* $U(t,s)$

$$U(t,s) = \lim_{\lambda\to\infty} (e^{(t-s)(A_{\lambda-}+D)}1)(s) \,, \tag{26}$$

the convergence as $\lambda \to \infty$ *being strong convergence on* X, *uniformly in* s,t *for* $S < s_0 \leqslant s \leqslant t \leqslant t_0 < T$. *(Notice that for* s,t *given in* J *the expression in parenthesis in the second member of (26) is indeed defined for* λ *large enough, i.e. as soon as* $[s,t] \subset J_\lambda$*).*

For the proof of 5.1 we first establish the following two lemmas (the first of which is very easy).

LEMMA 5.2. *Given any propagator* U *on* X, J, *and any compact interval* $J_1 \subset J$, $\exists$ *a constant* $K(J_1)$ *such that* $\|U(t,s)\| \leqslant K(J_1)$ *for all* $(s,t) \in J_1 \times J_1$, $s \leqslant t$. *Moreover in the context of theorem 5.1, the propagator* U *considered there is necessarily contractive.*

PROOF. $\forall f \in X$, the continuity of $(s,t) \mapsto U(t,s)f$ from the compact set $\Delta_{J_1} = \{(s,t) \in J_1 \times J_1 : s \leqslant t\} \to X$ implies that $\|U(t,s)f\| \leqslant K_0(J_1,f)$ a constant depending on J_1 and f; hence the uniform boundedness principle implies that $\exists\, K(J_1)$ with $\|U(t,s)\| \leqslant K(J_1)$ for all $(s,t) \in \Delta_{J_1}$.

In the context of 5.1, given any $f \in X$ and $s \in J$, $\exists$ by the density of $\underline{X}_0$ in $J \times X$ a sequence of $(s_n,f_n) \in \underline{X}_0$ with $s_n \to s$, $f_n \to f$. We have, for s_n in some J_1 compact whose interior contains s, t,

$$\|U(t,s_n)f_n - U(t,s)f\| \leqslant \|U(t,s_n)(f_n-f)\| + \|U(t,s_n)f - U(t,s)f\|$$

$$\leqslant M(J_1)\ \|f_n-f\| + \|U(t,s_n)f - U(t,s)f\| \to 0 \quad \text{as} \quad n \to \infty .$$

Since the $A(t)$ in 5.1 are dissipative we have from lemma 3.1 that $\|U(t,s_n)f_n\| \leqslant \|f_n\|$ for all n, hence $\|U(t,s)f\| \leqslant \|f\|$, which completes the proof of the lemma.

LEMMA 5.3. _In the context of 5.1, $\widetilde{U}_\lambda$ being associated to the given_ $A(t)$ _(or rather to the $\widetilde{A}_\lambda(t)$ derived from the_ $A(t)$_) as described in section 4, we have_

$$\widetilde{U}_\lambda(t,s) \to U(t,s) \text{ as } \lambda \to \infty, \ \forall s \leqslant t,\ s,t \in J , \tag{27}$$

uniformly for $S < s_0 \leqslant s \leqslant t \leqslant t_0 < T$.

PROOF. Choose any $(s,f) \in \underline{X}_0$, and take $\varphi \in C^\infty(\,]-\infty,T'\,],R)$ with $\operatorname{supp} \varphi \subset\,]\,s,T']$, $0 \leqslant \varphi \leqslant 1$, where $s < T' < T$. Define $F:\ ]-\infty,T'] \to X$, by

$$F(t) = \varphi(t)U(t,s)f \quad \text{for} \quad s \leqslant t \leqslant T'; \ 0 \quad \text{for} \quad t \leqslant s. \tag{28}$$

Then for any s',t with $s \leqslant s' \leqslant t \leqslant T'$ we have by 4.1, (20),

$$\begin{aligned} U(t,s')F(s') - \widetilde{U}_\lambda(t,s')F(s') &= \int_{s'}^{t} \widetilde{U}_\lambda(t,\eta)(A(\eta) - \widetilde{A}_\lambda(\eta))U(\eta,s')F(s')d\eta \\ &= \int_{s'}^{t} \varphi(s')\widetilde{U}_\lambda(t,\eta)(A(\eta) - \widetilde{A}_\lambda(\eta))U(\eta,s)f\ d\eta . \end{aligned}$$

It follows that

$$\begin{aligned} &\|U(t,s')F(s') - \widetilde{U}_\lambda(t,s')F(s')\| \leqslant \int_{s}^{T} \|A(\eta)U(\eta,s)f - \widetilde{A}_\lambda(\eta)U(\eta,s)f\|\ d\eta \\ &\qquad \text{for } s \leqslant s' \leqslant t \leqslant T' < T, \text{ and } = 0 \text{ for } s' < s . \end{aligned} \tag{29}$$

Now $\widetilde{A}_\lambda(\eta)U(\eta,s)f = \lambda R(\lambda,A(\eta))A(\eta)U(\eta,s)f$ a.e. since $U(\eta,s)f \in D(A(\eta))$ a.e., so that $\|\widetilde{A}_\lambda(\eta)U(\eta,s)f\| \leqslant \|A(\eta)U(\eta,s)f\|$ a.e.. Hence a.e.

$\|A(\eta)U(\eta,s)f - \tilde{A}_\lambda(\eta)U(\eta,s)f\| \leq 2\|A(\eta)U(\eta,s)f\|$; and $\eta \mapsto \|A(\eta)U(\eta,s)f\|$, a.e., is integrable on $[s,T]$ since $U(.,s)f$ is in $W^{1,1}$. (Of course $\tilde{A}_\lambda(\eta)U(\eta,s)f$ is measurable since it is continuous). Also $\tilde{A}_\lambda(.)U(.,s)f \to A(.)U(.,s)f$, a.e., as $\lambda \to \infty$. Hence by Lebesgue's dominated convergence theorem we see that the integral in (29) tends to 0 as $\lambda \to \infty$. Therefore

$$(30) \qquad \|U(t,s')F(s') - \tilde{U}_\lambda(t,s')F(s')\| \to 0 \text{ as } \lambda \to \infty ,$$

uniformly in $-\infty < s' \leq t \leq T' < T$.

Newt we go through a partition of unity argument to see how (30) implies the same result for a general continuous $s \mapsto G(s)$ instead of the particular F of type (28) involved in (30), (5).
Unifortunately we must do this in quite some detail to see clearly that the density of $\underline{X}_0$ suffices indeed to obtain an extension of (30) which is applicable to any G in $C([s_0,t_0],X)$. Let us therefore consider any $G \in C([s_0,t_0],X)$, s_0,t_0, as in the statement, $s_0 < t_0$. We may consider this function extended continuously to $[\tilde{s}_0,\tilde{t}_0]$ with $S < \tilde{s}_0 < s_0 < t_0 < \tilde{t}_0 < T$. Given $\varepsilon > 0$, $\exists\delta(\varepsilon) > 0$ such that $\|U(t'',s'')G(s'')-U(t',s')G(s')\| \leq \varepsilon/2$ whenever $|s''-s'|$, $|t''-t'| \leq \delta(\varepsilon)$, by the uniform continuity of $(s,t) \mapsto U(t,s)G(s)$ for (s,t) in $[\tilde{s}_0,\tilde{t}_0] \times [\tilde{s}_0,\tilde{t}_0]$, $s \leq t$. $\exists$ points $\sigma_1, \sigma_2, \ldots, \sigma_k$ in $]\tilde{s}_0,\tilde{t}_0[$, such that $\sigma_1 < s_0 < \sigma_2 < \sigma_3 < \ldots < t_0 < \sigma_k$, $0 < \sigma_{i+1} - \sigma_i = \text{constant} = \tilde{\delta} \leq \delta(\varepsilon)/2$ for $i = 1, 2, \ldots, k-1$; moreover $\exists$ real valued $\varphi_i \in C^\infty(\mathbf{R})$ with supp $\varphi_i \subset]\sigma_i,\sigma_{i+2}[$, $i = 1, 2, \ldots, k-2$, and $0 \leq \varphi_i \leq 1$, $\sum_i \varphi_i = 1$ on $[s_0,t_0]$; thus $\{\varphi_i\}$ is a C^∞ partition of unity subordinated to the covering of $[s_0,t_0]$ by the open sets $]\sigma_1,\sigma_3[$, $]\sigma_2,\sigma_4[, \ldots,]\sigma_{k-2},\sigma_k[$. Let s_i^* be a point in $]\sigma_i,\sigma_{i+2}[$ chosen to the left of supp φ_i. Then, on $[s_0,t_0]$,

$$(31) \qquad \begin{aligned} &\|(\sum_{i=1}^{k-2} \varphi_i(t)U(t,s_i^*)G(s_i^*)) - G(t)\| = \\ &\|\sum_{i=1}^{k-2} \varphi_i(t)(U(t,s_i^*)G(s_i^*) - U(t,t)G(t))\| \\ &\leq 2 \max_{|t-s_i^*|\leq\delta(\varepsilon)} \|U(t,s_i^*)G(s_i^*) - U(t,t)G(t)\| \leq \varepsilon . \end{aligned}$$

(5) The argument is related to one used in [P], p. 252-253, though here a rather different situation is treated, as mentioned below.

Write $\widetilde{F}_i$ for the function $]-\infty,t_0] \to X$, defined by $\widetilde{F}_i(t) = \varphi_i(t)U(t,s_i^*)G(s_i^*)$ for $t \geq s_i^*$, and $= 0$ for $t \leq s_i^*$, so that (31) reads $\|(\sum_{i=1}^{k-2} \widetilde{F}_i(t))-G(t)\| \leq \varepsilon$ on $[s_0,t_0]$.

Now, for each i, after s_i^* has been chosen as above, we have for any $s_i \in]\sigma_i,\sigma_{i+2}[$, $f_i \in X$, and all $t \in [\widetilde{s}_0,\widetilde{t}_0]$

$$(32)\qquad \begin{aligned} &\|U(t,s_i^*)G(s_i^*) - U(t,s_i)G(s_i^*)\| \leq \varepsilon/2k\ , \\ &\|U(t,s_i)(G(s_i^*) - f_i)\| \leq \|G(s_i^*) - f_i\| \leq \varepsilon/2k\ , \end{aligned}$$

provided $|s_i^*-s_i| \leq$ some $\delta_1(\varepsilon,k,G(s_i^*))$, and $\|G(s_i^*)-f_i\| \leq \varepsilon/2k$. By the density of $\underline{X}_0$ in $J \times X$ we can certainly find a pair $(s_i,f_i) \in \underline{X}_0$ satisfying these conditions and also such that s_i remains to the left of supp φ_i in $]\sigma_i,\sigma_{i+2}[$. Write F_i for the function $]-\infty,t_0] \to X$ defined by $F_i(t) = \varphi_i(t)U(t,s_i)f_i$ for $t \geq s_i$, $= 0$ for $t \leq s_i$; so that F_i is of the form (28), and (30) holds for each F_i in lieu of F. We have $\|\widetilde{F}_i(t)-F_i(t)\| \leq \varepsilon/k$ in view of (32) for all $t \leq t_0$. Finally, for $s_0 \leq s \leq t \leq t_0$ we have now

$$(33)\qquad \begin{aligned} &\|U(t,s)G(s) - \widetilde{U}_\lambda(t,s)G(s)\| \leq \|U(t,s)(G(s) - \sum_{i=1}^{k-2} F_i(s))\| \\ &\quad + \sum_{i=1}^{k-2} \|U(t,s)F_i(s) - \widetilde{U}_\lambda(t,s)F_i(s)\| + \|\widetilde{U}_\lambda(t-s)((\sum_{i=1}^{k-2} F_i(s))-G(s))\| \\ &\leq 2\,\|G(s) - \sum_{i=1}^{k-2} \widetilde{F}_i(s)\| + 2\sum_{i=1}^{k-2} \|\widetilde{F}_i(s) - F_i(s)\| \\ &\quad + \sum_{i=1}^{k-2} \|U(t,s)F_i(s) - \widetilde{U}_\lambda(t,s)F_i(s)\| \leq \\ &4\varepsilon + \sum_{i=1}^{k-2} \|U(t,s)F_i(s) - \widetilde{U}_\lambda(t,s)F_i(s)\| \end{aligned}$$

and the expression appearing after the last $\leq$ in (33) can be made $\leq 5\varepsilon$ uniformly in $s_0 \leq s \leq t \leq t_0$ by taking $\lambda > 0$ large enough, in view of (30). Since G(s) can be made constant, equal any given $f \in X$, on $[s_0,t_0]$, the lemma is proved.

PROOF OF THEOREM 5.1. $[s_0,t_0] \subset J_\lambda$ for $\lambda \geq$ some λ_1. We have thus, on $[s_0,t_0]$, for any $f \in X$, and $\lambda \geq \lambda_1$, in view of (24)

$$\|U(t,s)f - U_\lambda(t,s)f\| \leqslant \|U(t,s)f - \widetilde{U}_\lambda(t,s)f\| + \frac{1}{\lambda}\,|T-S|\,\|f\|. \tag{34}$$

Thus by lemma 5.2, $U_\lambda(t,s) \to U(t,s)$ strongly on X, uniformly for $s_0 \leqslant s \leqslant t \leqslant t_0$. On the other hand, by 2.2, (7), we have

$$U_\lambda(t,s) = (e^{(t-s)(A_{\lambda}-+D)}1)(s) \tag{35}$$

as soon as $J_\lambda \supset [\,s,t\,]$, thus certainly as soon as $\lambda \geqslant \lambda_1$. Hence (35), together with (34), prove (26) and theorem 5.1.

REFERENCES

[B] Brezis, H., Opérateurs maximaux monotones, North-Holland/American Elsevier, 1973.

[H-Ph] Hille, E. - Phillips, R.S., Functional analysis and semigroups A.M.S. Colloquium Publ., vol. XXXI, revised ed., 1957.

[K] Kato, T., Linear evolution equations of "hyperbolic" type, J. Fac. Sci. Univ. Tokyo, sec. I, 17 (1970), 241-258.

[Kr] Krein, S.G., Linear differential equations in Banach space, Translations of math. monographs, v. 29, American Math. Soc., 1971.

[L-Ph] Lumer, G. - Phillips, R.S., Dissipative operators in a Banach space, Pacific J. of Math., 11, (1961), 679-698.

[P] Paquet, L., Semi-groupes généralisés et équations d'évolution, Séminaire de Théorie du Potentiel, Paris n° 4, Lecture Notes in Math., vol. 713, Springer-Verlag, 243-263, 1979.

[Y] Yosida, K., Functional analysis, Springer-Verlag, sixth edition, 1980.

[Z] Zaidman, S., Abstract differential equations, Research Notes in Mathematics, vol. 36, Pitman Advanced Publishing Program, 1979.

International Series of
Numerical Mathematics, Vol. 65

GREEN'S FUNCTIONS FOR THE FINITE DIFFERENCE HEAT, LAPLACE AND WAVE EQUATIONS

Dale H. Mugler
Department of Mathematics
University of Santa Clara
Santa Clara, California

In this paper, representations are developed for the Green's functions for a partial difference formulation of an initial-value problem that includes the half-plane heat (diffusion), Laplace, and wave equations as special cases. Solutions of the partial difference equation are shown to be given by a discrete convolution that is analogous to integral representations for the continuous case. A convergence property relating each discrete Green's function to that of its associated partial differential equation is also presented. The initial research for this paper was conducted with the assistance of student Steven F. Ashby.*

1. Introduction

For the initial value problem (IVP) associated with heat propagation, assuming initial values are given by $u(x,t) = f(x)$ for $t=0$ and all real x, the Poisson representation formula or the Green's function representation of the solution is a convolution integral in x over the whole real line, $u(x,t) = f(x)*G(x,t)$, where

$$G(x,t) = \exp(-x^2/4t)/\sqrt{4\pi t}. \tag{1}$$

This function has been called the impulse-response function, source solution, influence function, or simply the Green's function for the IVP. As such, it has the characteristic as $t\to 0$ of a Dirac delta distribution, i.e. $\lim_{t\to 0} G(x,t) = \delta_x$.

We take the Green's function for the partial difference formulation of the problem as the discrete function satisfying an analogous role. A function

*Supported by National Science Foundation URP grant SPI-80-25433

will be called the Green's function for the partial difference equation if it is the kernel of a similar discrete convolution representation for solutions. As such, and writing G_m^n for this function of (m,n), it has the characteristic that $G_m^0 = \delta_m$, the Kronecker delta.

The initial value problem considered here includes a second order, linear, homogeneous partial differential equation (PDE), with initial values given on the whole real line. The PDE is initially assumed to be in one of the three canonical forms: $v_{xx} + (0)v_{tt} + \ldots = 0$, $v_{xx} + v_{tt} + \ldots = 0$, or $v_{xx} - v_{tt} + \ldots = 0$, where the dots indicate linear terms involving v and its first derivatives. For these canonical forms, a change of variables of form $u(x,t) = \exp(\alpha x + \beta t) v(x,t)$ allows for a further simplification. Choosing α and β appropriately, the IVP may be reduced to one of the three forms,

$$(2) \qquad u_{xx} - u_t = 0, \; u_{xx} + u_{tt} + \gamma u = 0, \text{ or } u_{xx} - u_{tt} + \gamma u = 0,$$

where γ is a constant. These equations are classified in the usual way as parabolic, elliptic, or hyperbolic, resp., and include the heat, Laplace, and wave equations.

A finite difference formulation of such an IVP involves placing a mesh on the (x,t) plane for fixed steplengths Δx and Δt. We write $x_m = m\Delta x$, $t_n = n\Delta t$, and $u_m^n = u(x_m, t_n)$, where the superscript is not to be interpreted as a power. In this paper, first (partial) derivatives are approximated by a forward difference, and second derivatives by a central difference. The resulting "explicit" partial difference equation (PΔE) is the type investigated in this paper, as opposed to an "implicit" method that uses a weighted average of central differences at two different time levels as an approximation.

The PΔEs corresponding to the equations in (2) have the forms

$$(3) \qquad u_m^{n+1} - (\rho u_{m+1}^n + (1-2\rho)u_m^n + \rho u_{m-1}^n) = 0,$$

for the parabolic case, and

$$(4) \qquad u_m^{n+1} \pm (\sigma^2 u_{m+1}^n - (\pm 2(1 \pm \sigma^2) - \gamma(\sigma\Delta x)^2)u_m^n + \sigma^2 u_{m-1}^n) + u_m^{n-1} = 0,$$

for the elliptic and hyperbolic cases, where the top sign refers to an elliptic and the bottom sign to a hyperbolic equation. The ratios, $\rho = \Delta t/(\Delta x)^2$ and $\sigma = \Delta t/\Delta x$, are assumed to be fixed. These equations all fit into the general form,

$$u_m^{n+1} + (b_1 u_{m+1}^n + 2b_2 u_m^n + b_1 u_{m-1}^n) + c u_m^{n-1} = 0, \tag{5}$$

where c is either 0 or 1.

A polynomial "characteristic" equation associated with a differential equation is often a useful tool, and the representation theorems for the discrete Green's functions given in section 2 involve such an idea for a PΔE. An exponential solution of (5) would have the form, $u^{*n}_m = z_1{}^m z^n$, where we allow complex base values z and z_1. Upon substitution, one finds that u* is a solution if and only if

$$F(z,z_1) = z + (b_1 z_1 + 2b_2 + b_1 z_1^{-1}) + cz^{-1} = 0. \tag{6}$$

We call the equation (6) resulting from this substitution the characteristic equation of the PΔE. The roots of this equation, here employed using z as a function of z_1, we call the characteristic roots. For the parabolic equation (3), there is only one characteristic root, given by

$$z = \rho z_1 + (1-2\rho) + \rho z_1^{-1}. \tag{7}$$

For other forms of (5), there are two characteristic roots, given by $z = -w+\sqrt{w^2-1}$, with $w = b_1(z_1 + z_1^{-1})/2 + b_2$.

For values of z_1 on the unit circle, the characteristic roots of the Laplace equation are real inverses while those of the wave equation are distinct complex conjugates. We next further classify the equations of form (2) so that this property remains typical of the elliptic and hyperbolic equations, resp. One can show that the elliptic equations will always have real characteristic roots if either $\gamma \leq 0$ or $(\Delta t)^2\gamma \geq 4(1+\sigma^2)$, and that hyperbolic equations will always have imaginary roots if both $\gamma \leq 0$ and $(\Delta t)^2\gamma \geq -4(1-\sigma^2)$.

Stability requirements for the heat and wave PΔEs are well-known, and require that $\rho \leq 1/2$ and $\sigma \leq 1$. Combining stability requirements with the conditions above, one can see that for small Δt, elliptic equations have real roots and hyperbolic equations have imaginary characteristic roots for $z_1 \varepsilon U$ if we require that $\gamma \leq 0$. In the following, we assume then, that the forms of (2) that are considered are for $\gamma \leq 0$.

The initial value problem for the wave PDE includes initial values $u(x,0)=f(x)$ and $u_t(x,0)=g(x)$, for $-\infty<x<\infty$. We use appropriate function samles for the corresponding PΔE initial values,

(8) $$u_m^0 = f_m, \quad \mu\delta_t\, u_m^0 = g_m,$$

where $\mu\delta_t\, u_m^n = (u_m^{n+1} - u_m^{n-1})/(2\Delta t)$ is the central difference approximation to the derivative. We shall show that $g_m = \mu\delta_t\, u_m^0$ in the sense that the general formula for $\mu\delta_t\, u_m^n$ evaluated at n=0 gives g_m. This differs from the usual finite difference formulas which estimate u_m^1 using values of g_m.

As is standard for elliptic problems in a half plane, we also require that u_m^n be bounded as $n\to\infty$.

2. The Integral Representations

The results presented in this section begin with the complex integral representation of the discrete Green's functions for the general equation(5), with initial values and conditions as described in section 1. The first theorem deals with the parabolic and elliptic problems (real characteristic roots), and the second with hyperbolic problems (complex characteristic roots). The corresponding trigonometric integral representations will be included as a corollary.

The main tool used in the proofs of the theorems is the Z-transform, defined for a sequence $f_m \in \ell^1$ in one variable by $Z[f_m] = F(z_1) = \sum_{m=-\infty}^{\infty} f_m z_1^{-m}$ and with inverse transform given by $f_m = \frac{1}{2\pi i}\int_U F(z_1) z_1^{m-1} dz_1$. One of the properties that will be used in the following is the shifting property, that gives $Z[f_{m+k}] = z_1^{\,k}\, Z[f_m]$.

THEOREM 2.1. _Let_ u_m^n _satisfy a discrete parabolic or elliptic_ IVP _as described above. Then_

(9) $$u_m^n = f_m * G_m^n = \sum_{j=-\infty}^{\infty} f_j G_{m-j}^n,$$

for the solution G_m^n _given by_

(10) $$G_m^n = \frac{1}{2\pi i}\int_U z^n z_1^{\,m-1} dz_1,$$

where z _is the function of_ z_1 _determined by the characteristic equation_ (_the smaller characteristic root in the elliptic case_) _and with initial values of_ u_m^n _such that_ (9) _converges_.

PROOF. Apply the Z-transform to equation (5), assuming $u_m^n \epsilon \ell^1$ as a function of m for each fixed n, and writing $U_n(z_1) = Z_m[u_m^n]$. From the shifting property, it follows that

$$U_{n+1} + (b_1 z_1 + 2b_2 + b_1 z_1^{-1})U_n + cU_{n-1} = 0, \tag{11}$$

so that U_n satisfies a difference equation in n. To solve it, the usual technique is to determine an exponential solution, $U_n = z^n$. Such a function is a solution of (11) if and only if $z + (b_1 z_1 + 2b_2 + b_1 z_1^{-1}) + cz^{-1} = 0$, which is the characteristic equation (6).

A nonzero root of this equation is also a root of the quadratic equation, $z^2 + (b_1 z_1 + 2b_2 + b_1 z_1^{-1})z + c = 0$. If this equation has roots $r_1(z_1)$ and $r_2(z_1)$, then the general solution of (11) is given by $U_n = Ar_1{}^n + Br_2{}^n$, where A and B are constants with respect to n.

In the parabolic case, the above quadratic is actually a linear equation. In the elliptic case, one of the roots has value greater than one for $z_1 \epsilon U$. The requirement that u_m^n be bounded as $n \to \infty$ leads to the conclusion that $U_n = Ar_1{}^n$ in both cases, with r_1 as the smaller root in the elliptic case. The value of A may be found by substituting n=0, giving $A = U_0 = Z[u_m^0] = \sum_j f_j z_1^{-j}$.

Inverting the transform and interchanging summation with integration, one obtains $u_m^n = \sum_j f_j \frac{1}{2\pi i} \int_U r_1{}^n z_1{}^{m-j-1} dz_1$. With G_m^n as in (10), the representation (9) follows. Also using (10), it follows that G_m^n is a solution of the PΔE, with initial values given by $G_m^0 = \delta_m$, the Kronecker delta. For any u_m^n such that (9) converges, it follows that the discrete convolution gives a solution with initial values f_m. Uniqueness of solutions then allows the removal of the initial assumption that $u_m^n \epsilon \ell^1$ for each fixed n, completing the proof.

THEOREM 2.2. *Let u_m^n satisfy a discrete hyperbolic IVP as described above. Then there are solutions G_m^n and K_m^n, related by $\mu\delta_t K_m^n = G_m^n$, such that*

$$u_m^n = f_m * G_m^n + g_m * K_m^n, \tag{12}$$

where

$$G_m^n = \frac{1}{2\pi i} \int_U T_n\left(\frac{z+\bar{z}}{2}\right) z_1{}^{m-1} dz_1 \quad \text{and} \quad K_m^n = \frac{\Delta t}{2\pi i} \int_U U_{n-1}\left(\frac{z+\bar{z}}{2}\right) z_1{}^{m-1} dz_1, \tag{13}$$

where z is a characteristic root, and T_n and U_n are the Chebyshev polynomials of the first and second kinds, resp.

PROOF. As the roots of the characteristic equation here are complex conjugates for $z_1 \varepsilon U$, this proof differs from the above in that the solution of (11) is generally written as $U_n(z_1) = A\cos(n\alpha) + B\sin(n\alpha)$, where A and B are independent of n and $r(z_1) = \cos\alpha + i\sin\alpha$ is a characteristic root. In particular, from the quadratic associated with the characteristic equation, it follows that $-\cos\alpha = b_1(z_1+z_1^{-1})/2 + b_2$.

If n=0, it follows that $A = U_0(z_1) = Z[f_m]$. To find $B(z_1)$, evaluate the expression $\mu\delta_t U_n$ at n=0 to obtain $B = \Delta t(\mu\delta_t U_0)/\sin\alpha$. In terms of initial conditions, $\mu\delta_t U_0 = \sum_j g_j z_1^{-j} = Z[g_m]$. Thus $U_n(z_1)=Z[f_m]\cos n\alpha + Z[g_m]\Delta t\sin(n\alpha)/\sin\alpha$. If z is a root of the characteristic equation in this case, it follows that $\cos\alpha = (z+\bar{z})/2$. Also, $T_n(\cos\alpha) = \cos n\alpha$ and $U_{n-1}(\cos\alpha) = \sin(n\alpha)/\sin\alpha$. Thus, using the inverse transform and interchanging summation and integration gives

$$u_m^n = \sum_j f_j G_{m-j}^n + \sum_j g_j K_{m-j}^n,$$

i.e. (12), with G_m^n and K_m^n as defined in (13). The relation that $\mu\delta_t K_m^n = G_m^n$ also follows from (13). The completion of the proof is as before.

The following results may be obtained from the theorems by a change of variable.

COROLLARY 2.3. *In the above, the functions G_m^n are of the form*

$$G_m^n = \frac{1}{\pi}\int_0^{\pi} r^n(\theta)\cos m\theta d\theta, \tag{14}$$

where $r(\theta)=(1-2\rho) + 2\rho\cos\theta$ *for parabolic,* $r(\theta)= x -\sqrt{x^2-1}$ *for elliptic, and* $r^n(\theta) = T_n(\cos\alpha)$ *for hyperbolic equations, with* x *and* α *such that*

$$x-1 = \sigma^2(1-\cos\theta- \tfrac{1}{2}\gamma(\Delta x)^2) = 1 -\cos\alpha. \tag{15}$$

3. Properties and Individual Cases

One computationally-important property, common to all of the discrete Green's functions in the previous section, is that each function is an even function of m. This follows as an easy consequence of Corollary 2.3, from which it also follows that $G_m^0 = \delta_m$. Another property that holds for the discrete heat, Laplace, and wave equations is that $\sum_m G_m^n = 1$, for $n \geq 0$. This can be established using generating functions, which may also be applied to see that the above sum does not have unit value in the case that $\gamma \neq 0$ in (4). For further properties and representations, we next concentrate on the individual equations.

3.1 The Heat (Diffusion) Equation. D.V.Widder [12,p31] listed several properties of the Green's function (1) for the heat equation. These include a description of $G(x,t)$ as a nonnegative solution having Dirac delta properties at the boundary, with $\int_{-\infty}^{\infty} G(x,t)dx = 1$ for any $t>0$. From the complex integral (10) with characteristic root given by (7), the stability condition that $\rho \leq \frac{1}{2}$ is seen to be the precise condition needed in order to likewise assure that G_m^n is nonnegative. This discrete function even models an "addition" property of Widder's, in that $G_m^{n_1} * G_m^{n_2} = G_m^{n_1+n_2}$.

An important property for computations is that the representation for solutions (9) is actually a finite sum in this case. It follows easily from the complex integral (10) that $G_m^n = 0$ if $|m|>n$. Since G_m^n is even in m, the representation $u_m^n = f_m * G_m^n$ may thus be written

$$(16) \qquad u_m^n = f_m G_0^n + \sum_{j=1}^{n} G_j^n (f_{m-j} + f_{m+j}).$$

Expanding the trigonometric integral (14), one finds a general representation for G_m^n as a finite power series in ρ,

$$(17) \qquad G_m^n = \sum_{j=|m|}^{n} (-1)^{j-m} \binom{n}{j} \binom{2j}{j-m} \rho^j .$$

There are two special cases for which G_m^n has the especially simple form of a multiple of a binomial coefficient. First, for $0 \leq m \leq n$ and $\rho=1/2$, $G_m^n = 2^{-n} \binom{n}{(n+m)/2}$ if n+m is even and $G_m^n=0$ if n+m is odd. Second, if $\rho=1/4$, $G_m^n = 4^{-n} \binom{2n}{n+m}$. These follow easily from (10).

The heat equation is intimately connected with the theory of Brownian motion, and $G(x,t)$ from (1) is the probability density function for one-dimensional Brownian motion. The formula for G_m^n at $\rho=1/2$ may likewise be seen as the probability for the position of an object undergoing a one-dimensional random walk [4,p489]. More will be said concerning this connection in section 4.

3.2 Laplace's Equation. Similar to the partial differential equation, the discrete Laplace equation requires the value at each grid point to be the average of its (four) adjacent values. It is formed from (4) with $\sigma = \Delta y/\Delta x = 1$ and $\gamma = 0$. Analogous to its PDE, it is connected to the theory of discrete analytic functions, and its solutions are called discrete harmonic functions.

Analogs of Poisson's integral formula for the upper half plane have been considered by various authors. In [2], the author developed such a formula for the even lattice of points. In [1], the authors considered an analog of a formula given in [13] for positive harmonic functions, citing [5] for the kernel employed in that formula. Although the function developed there is in a different form, it is equivalent to (10). In [10], the author found a complex integral representation, similar to (10), for the Green's function for this case. In [11,p366ff], an approach different than the one we present is used to determine Green's functions for an elliptic equation.

Of the three main equations, it is only for the Laplace equation that the Green's function is not zero outside a finite interval of m values for a fixed n, so that (9) is really an infinite series. For convergence of this representation for solutions, conditions on the initial values f_m may be derived using a result from [1]. It is shown there that $|G_m^n - n/(\pi(m^2+n^2))| < A/(n(m^2+n^2))$ for all m if n>0, where A is constant. Applying Theorem 2.1, it then follows that the convolution representation holds for discrete harmonic functions whose initial values satisfy the condition that $\sum_m |f_m|/(1+m^2)$ converges. Also shown in [1] is that G_m^n is positive for all m if n>0, a property similar to one satisfied by the Green's function for the heat equation.

For computational purposes, the values of G_m^n may be computed from (14) by Gauss-Chebyshev quadrature, and truncation error estimates for the convolution representation can be given based on the growth estimate for G_m^n above.

Note also that the Jacobi iterative method for solving Laplace's equation is precisely the same formula used in solving the heat equation in two space variables, and the method of section 3.1 extends readily to the case of more space variables.

3.3 The Wave Equation. The representation (12) of a solution to the discrete wave equation is strikingly analogous to the corresponding representation for solutions to the PDE. In [8], we showed that the solution u(x,t) was given by a convolution involving both sets of given initial values, $u(x,t)=f(x)*G(x,t)+$

$g(x)*K(x,t)$, where the kernels are distributions, with $K(x,t) = \frac{1}{2}[Y(x+t) - Y(x-t)]$, for the step function Y, and $K_t(x,t) = G(x,t)$. This corresponds precisely to the representation of Theorem 2.2, even to the point that the central difference approximation to the derivative of K_m^n is G_m^n.

As in the representation for solutions of the heat equation, the convolutions in (12) are finite sums, since it follows from the integral representations that $G_m^n=0$ if $|m|>n$ and $K_m^n=0$ if $|m|\geq n$. Each individual convolution may thus be written as in (16).

There are finite power series representations of the Green's functions in this case, similar to (17) for the heat equation. For a derivation, begin with the trigonometric integral representation for G_m^n from Corollary 2.3. Since $\gamma=0$, the term $r^n(\theta)$ is given by $T_n(1-\sigma^2(1-\cos\theta))$. Expanding $T_n(x)$ in a power series about $x=1$ and rewriting a formula from [9,p33] for the jth derivative of T_n at $x=1$ as $2^j \frac{n}{n+j} \binom{n+j}{2j} j!$, one finds $T_n(x) = \sum_{j=0}^{n} 2^j \frac{n}{n+j} \binom{n+j}{2j} (x-1)^j$. Since the values of $\frac{1}{\pi}\int_0^\pi (1-\cos\theta)^j \cos m\theta d\theta$ are $(-1)^m 2^{-j} \binom{2j}{j-m}$ if $|m|\leq j$ and zero if $|m|>j$, it follows upon substitution of the representation for r^n into the integral representation for G_m^n that

$$(18) \qquad G_m^n = \sum_{j=|m|}^{n} (-1)^{j-m} \frac{n}{n+j} \binom{n+j}{2j} \binom{2j}{j-m} \sigma^{2j}.$$

Note that this representation differs only slightly from (17).

Since $U_{n-1}(x) = T_n'(x)/n$, it follows from the power series representation for $T_n(x)$ that $U_{n-1}(x) = \sum_{j=0}^{n-1} 2^j \frac{n-j}{2j+1} \binom{n+j}{2j} (x-1)^j$. From this and the integral representation for K_m^n, a representation for K_m^n that is very similar to (18) may be derived.

In the case that $\sigma=1$, the discrete wave equation takes on an especially simple form, and the Green's functions reduce dramatically to simple expressions as well. In (15), $\cos\alpha = \cos\theta$ if $\sigma=1$, so that $G_m^n = \frac{1}{2}$ if $|m|=n$ $(n\neq 0)$, $G_m^n=1$ if $m=n=0$, and otherwise $G_m^n=0$. Similarly $K_m^n/\Delta t = 1$ if $m+n$ is odd and $|m|<n$, otherwise $K_m^n=0$. Relative to the earlier sum for G_m^n over m, we note that $\sum_m K_m^n/\Delta t = n$ for $\sigma=1$. Using the above values in (12), we obtain the simple expression for the solution, $u_m^n = \frac{1}{2}(f_{m-n}+f_{m+n}) + \Delta t(g_{m-n+1} + g_{m-n+3}+\ldots+g_{m+n-3}+ g_{m+n-1})$, as is also given in [6] for this case.

For computations, terms in the summation representation (18) for G_m^n may occasionally be so small as to cause problems of underflow. This same sum, however, may be written as a nested polynomial to reduce those difficulties.

4. Convergence

In this final section, a convergence property is shown that relates each discrete Green's function from section 2 to the appropriate Green's function of the associated PDE. The proof involves the Fourier Transform (FT), assumed to be given by $F(\omega) = \int f(x)e^{-i\omega x}dx$ for given $f(x)$. To review, the following table lists the Green's functions and their FTs.

TABLE 4.1

PDE		FT
Heat	$G(x,t) = \exp(-x^2/4t)/\sqrt{4\pi t}$	$\exp(-t\omega^2)$
Laplace	$G(x,y) = (y/\pi)/(x^2+y^2)$	$\exp(-y\|\omega\|)$
Wave	$G(x,t) = \frac{1}{2}[\delta(x+t)+\delta(x-t)]$	$\cos(t\omega)$
	$K(x,t) = \frac{1}{2}[Y(x+t)-Y(x-t)]$	$\sin(t\omega)/\omega$

In the following theorem, the discrete Green's functions G_m^n are those for the heat, Laplace, and wave equations as given in section 2, with $G(x,t)$ assumed to be for the associated PDE.

THEOREM 4.1.
$$\lim_{\Delta x\to 0} \sum_{x_1<m\Delta x<x_2} G_m^n = \int_{x_1}^{x_2} G(x,t)dx,$$
where the sum is taken over all m such that $x_1<m\Delta x<x_2$, and where $n\Delta t = t$.

The relations between Δx and Δt for the above are that $\Delta t=\rho(\Delta x)^2$ for the heat, $\Delta t= \sigma\Delta x$ for the wave, and $\Delta y= \Delta x$ for the Laplace equation.

PROOF. For the heat equation,
$$\sum_m G_m^n = \sum_m \frac{1}{2\pi}\int_{-\pi}^{\pi} (1-2\rho(1-\cos\theta))^n e^{im\theta}d\theta. \quad \text{And with } \theta= \omega\Delta x,$$
$$= \frac{1}{2\pi}\int_{-\pi/\Delta x}^{\pi/\Delta x} (1-2\rho(1-\cos(\omega\Delta x))^n \sum_m e^{i\omega m\Delta x}\Delta x \quad d\omega.$$
But $\sum_m e^{i\omega m\Delta x}\Delta x$ is a Riemann sum for $e^{i\omega x}$ over $[x_1,x_2]$. Also, $n= t/(\rho(\Delta x)^2)$, and it can be shown that $(1-2\rho(1-\cos(\omega\Delta x))$ to the power $1/(\rho(\Delta x)^2)$ approaches $\exp(-\omega^2)$ as $\Delta x\to 0$, independent of ρ. Thus, in the limit,
$$\sum_m G_m^n \to \frac{1}{2\pi}\int e^{-\omega^2 t}\int_{x_1}^{x_2} e^{i\omega x}\,dxd\omega = \int_{x_1}^{x_2}\frac{1}{2\pi}\int e^{i\omega x-\omega^2 t}\,d\omega dx,$$
as is desired for this case. Proofs for the other equations are similar. Limits that are required to complete those proofs include $\cos n\alpha\to \cos(t\omega)$ and $\Delta t\sin(n\alpha)/\sin\alpha\to \sin(t\omega)/\omega$ for the wave equation, with α as in (15), and with

the limits being independent of σ. For Laplace's equation, it is that $(x-\sqrt{x^2-1})$ to the power $1/\Delta x$ approaches $e^{-|\omega|}$, if $x = 2- \cos(\omega\Delta x)$.

There are clear links between Theorem 4.1 and theorems from probability. For the heat equation, the discrete Green's functions are thereby connected with the Gaussian distribution. For the random walk corresponding to the case that $\rho = \frac{1}{2}$, this theorem appears in [7] and in [3,p357ff]. Feller even proves a bit more. If $\rho = \frac{1}{2}$, he cites the deMoivre-Laplace approximation to conclude that for nonzero values of G_m^n, G_m^n is approximately $2\Delta x \exp(-x^2/4t)/\sqrt{4\pi t}$ as $m\Delta x \to x$ and $n\Delta t \to t$. Thus, in this case where half of the values of G_m^n are nonzero in $|m| \leq n$, $\frac{1}{2}G_m^n/\Delta x$ serves as an approximation to $G(x,t)$.

REFERENCES

[1] Allen, A.C. - Murdoch, B.H., A note on preharmonic functions. Proc. Amer. Math. Soc. 4 (1953), 842-852.

[2] Duffin, R.J., Basic properties of discrete analytic functions. Duke Math. J. 23 (1956), 335-363.

[3] Feller, W., Probability Theory and its Applications, Vol.I. J.Wiley & Sons, New York 1968.

[4] Haberman, R., Elementary Applied Partial Differential Equations. Prentice-Hall, Inc., New Jersey 1983.

[5] Heilbronn, H.A., On discrete harmonic functions. Proc. Cambridge Phil. Soc. 45 (1949), 194-206.

[6] Hildebrand, F.B., Finite-difference Equations and Simulations. Prentice Hall, Inc., New Jersey 1968.

[7] Kac, M., Random walk and the theory of Brownian motion. Amer. Math. Monthly 54 (1949), 369-391.

[8] Mugler, D.H. - Dunne, E.G., Analogies from complex analysis and heat conduction for wave propagation. SIAM J. Math. Anal. 13 (1982), 1-15.

[9] Rivlin, T.J., The Chebyshev Polynomials. J.Wiley & Sons, New York 1974.

[10] Stöhr, A., Über einige lineare partielle differenzengleichungen mit konstanten koeffizienten, III. Math. Nachr. 3 (1950), 330-357.

[11] Van der Pol, B. - Bremmer, H., Operational Calculus Based on the Two-Sided Laplace Integral. Cambridge Univ. Press, Cambridge 1964.

[12] Widder, D.V., The Heat Equation. Academic Press, New York 1975.

[13] Widder, D.V. - Loomis, L.H., The Poisson integral representation of functions which are positive and harmonic in a half-plane. Duke Math. J. 9 (1942), 643-645.

IX Prohability Theory and Miscellaneous Topics

International Series of
Numerical Mathematics, Vol. 65

ESTIMATION OF THE REGRESSION FUNCTION VIA ORTHOGONAL EXPANSION

Pál Révész
Mathematical Institute
of the Hungarian Academy of Sciences
Budapest

Our goal is to estimate a regression function $r(x) = \mathbb{E}(Y|X=x)$ $(0 \leq x \leq 1)$ from an independent sample of size n. We propose a sequence of estimators $\hat{c}_k$ of the Fourier coefficients $c_k = \int_0^1 r(x)\phi_k(x)dx$ of $r(x)$ (where $\{\phi_k(x)\}$ is a complete orthonormal sequence satisfying some regularity conditions) and prove that $\hat{c}_k \to c_k$ with probability one for any k as $n\to\infty$. We also investigate the L^2 distance between the estimate $r_n(x) = \sum_{k=1}^{N_n} \hat{c}_k \phi_k(x)$ and $r(x)$ where N_n is a suitable sequence of integers.

1. Introduction

Let $(X,Y),(X_1,Y_1),(X_2,Y_2),\ldots$ be a sequence of i.i.d.r.v.'s with

(1) $$0 \leq X \leq 1 \ , \quad Y \in R^1 \ ,$$

(2) $$\mathbb{E}(Y|X=x) = r(x) \quad (0 \leq x \leq 1) \ .$$

In the recent years a number of statisticians investigated the problem to find an estimator of the regression function $r(x)$ based on the sample (X_1,Y_1), $(X_2,Y_2),\ldots,(X_n,Y_n)$. (For a survey of the problem see [1].) Most of the authors propose and investigate

$$r_n(x) = \frac{\sum_{i=1}^{n} Y_i \lambda\left(\frac{x-X_i}{h_n}\right)}{\sum_{i=1}^{n} \lambda\left(\frac{x-X_i}{h_n}\right)}$$

as a possible estimator of $r(x)$. Here $\{h_n\}$ is a sequence of positive numbers tending to 0 and $\lambda(x)$ is a smooth enough density function.

In some cases it is natural to approximate $r(x)$ by the partial sums of its Fourier series:

$$r_m(x) = \sum_{i=1}^{m} c_i \phi_i(x)$$

where

$$c_i = \int_0^1 r(x) \phi_i(x)dx$$

and $\{\phi_i(x)\}$ is a complete orthonormal system on $[0,1]$. In such a case one might ask how can we estimate the Fourier coefficients $c_1, c_2, \ldots, c_m$ from the sample $(X_1,Y_1),(X_2,Y_2),\ldots,(X_n,Y_n)$. This paper is devoted to this problem.

2. Estimation of the Fourier Coefficients

Let $\xi_1 = X_{1:n} \leqq \xi_2 = X_{2:n} \leqq \ldots \leqq \xi_n = X_{n:n}$ be the ordered sample based on the statistic $X_1,X_2,\ldots,X_n$. Let η_i be that Y_j which corresponds to ξ_i that is to say $\eta_i = Y_j$ if $\xi_i = X_j$. Then a natural estimator of $c_i = \int_0^1 r(x)\phi_i(x)dx$ is

$$\hat{c}_i(n) = c_i = \sum_{j=1}^{n-1} \eta_j \phi_i(\xi_j)(\xi_{j+1}-\xi_j).$$

In this Section we formulate theorems stating that $\hat{c}_i(n)$ is close to c_i if n is big enough. At first we list our assumptions used in the sequel.

A_k. $r(x)$ is k-times differentiable with $|r^{(k)}(x)| \leq C_k < \infty$ $(k=1,2,\ldots)$,

B. $\mathbf{P}(X<t) = F(t) = \int_0^t f(x)dx$ $(0 \leqq t \leqq 1)$ with $f(x) \geqq \delta > 0$,

C. $\mathbf{E}((Y-r(X))^2 | X=x) = \sigma^2 > 0$ $(0 \leqq x \leqq 1)$,

D. the r.v.'s $\varepsilon_j = \eta_j - r(\xi_j)$ $(j=1,2,\ldots)$ are independent and identically distributed with $\mathbf{E}\varepsilon_j = 0$, $\mathbf{E}\varepsilon_j^2 = \sigma^2$ (cf. (2) and C) and the sequences $\{\varepsilon_j\}$ and $\{\xi_j\}$ are independent,

$E_{i,k}$. $\phi_i(x)$ is k-times differentiable with $|\phi_i^{(k)}(x)| \leqq K_{i,k} < \infty$ $(0<x<1;\ i=1,2,\ldots,\ k=0,1,2,\ldots)$,

F_k. $f(x)$ is k-times differentiable with $|f^{(k)}(x)| \leqq L_k < \infty$ $(k=1,2,\ldots)$.

Now we formulate our

THEOREM 1. Assume that conditions A_2, B, C, D, $E_{i,o}$ $(i=1,2,\ldots)$, F_2 hold true. Then for any $0<\varepsilon<\frac{1}{2}$ we have

$$\mathbf{E}\,(\hat{c}_i-c_i)^2 = \Delta_i^2\, n^{-1} + o(n^{-3/2+\varepsilon}) \quad (i=1,2,\ldots), \tag{3}$$

$$\lim_{n\to\infty} \sup_{i\leqq n^{1/2-\varepsilon}} n^{1/2-\varepsilon}|\hat{c}_i-c_i| = 0 \tag{4}$$

with probability one and

$$\lim_{n\to\infty} \mathbf{P}\{n^{1/2}\,\Delta_i^{-1}(\hat{c}_i-c_i) < x\} = \frac{1}{\sqrt{2\pi}} \int_{-\infty}^{x} e^{-u^2/2}\,du \tag{5}$$

$$(i=1,2,\ldots;\ -\infty<x<\infty)$$

where

$$\Delta_i^2 = \sigma^2 \int_0^1 \frac{\phi_i^2(x)}{f(x)}\,dx\ .$$

At first we give a few lemmas.

Let $E_1, E_2, \ldots$ be a sequence of independent, exponentially distributed r.v.'s with parameter 1 i.e.

$$\mathbf{P}\,(E_i<x) = 1-e^{-x} \quad (x\geqq 0,\ i=1,2,\ldots).$$

Then by the strong law of large numbers we have

LEMMA 1.

$$S_{n+1}^{-2} \sum_{j=1}^{n} \lambda_j\, E_j^2 = n^{-2} \sum_{j=1}^{n} \lambda_j + o(n^{-3/2+\varepsilon}) \quad \text{a.s.}$$

for any $\varepsilon>0$ where $S_n = E_1 + E_2 + \ldots + E_n$ and $\{\lambda_i\}$ is a sequence of real numbers with $|\lambda_j| \leqq K$ $(K>0)$.

LEMMA 2. Let $g(x)$ be a twice differentiable function with $|g''(x)|\leqq L<\infty$ $(0<x<1)$ and assume that conditions B and F_1 hold. Then

$$\lim_{n\to\infty} \mathbf{P}\{n^{1/2}\,D^{-1}(nT_n-m) < x\} = \frac{1}{\sqrt{2\pi}} \int_{-\infty}^{x} e^{-u^2/2}\,du \tag{6}$$

and for any $\varepsilon>0$

(7) $$T_n = \frac{m}{n} + o(n^{-3/2+\varepsilon})$$

where

$$T_n = \int_0^1 g(x)dx - \sum_{j=1}^{n} g(\xi_j)(\xi_{j+1}-\xi_j),$$

$$m = \frac{1}{2}\int_0^1 \frac{g'(x)}{f(x)}\,dx \quad \text{and} \quad D^2 = \int_0^1 \frac{(g'(x))^2}{f^3(x)}\,dx.$$

We also have

(8) $$\mathbf{E}\,T_n^2 = O(n^{-2}).$$

PROOF. It is well known that

$$\{\xi_j;\ j=1,2,\dots,n\} \overset{\mathcal{D}}{=} \left\{F^{-1}\left(\frac{S_j}{S_{n+1}}\right);\ j=1,2,\dots,n\right\}$$

(n=1,2,...). Hence by Taylor expansion

(9) $$\xi_{j+1}-\xi_j \overset{\mathcal{D}}{=} F^{-1}\left(\frac{S_{j+1}}{S_{n+1}}\right)-F^{-1}\left(\frac{S_j}{S_{n+1}}\right) = \frac{E_{j+1}}{S_{n+1}}\,\frac{1}{f\left(F^{-1}\left(\frac{S_j}{S_{n+1}}\right)\right)} + O(n^{-2}).$$

Again by a Taylor expansion

$$\int_0^1 g(x)dx - \sum_{j=1}^{n} g(\xi_j)(\xi_{j+1}-\xi_j) = \frac{1}{2}\sum_{j=1}^{n} g'(\xi_j)(\xi_{j+1}-\xi_j)^2 + O(n^{-2})$$

where by (9)

$$\sum_{j=1}^{n} g'(\xi_j)(\xi_{j+1}-\xi_j)^2 \overset{\mathcal{D}}{=} \sum_{j=1}^{n} g'\left(F^{-1}\left(\frac{S_j}{S_{n+1}}\right)\right)\frac{E_{j+1}^2}{S_{n+1}^2}\, f^{-2}\left(F^{-1}\left(\frac{S_j}{S_{n+1}}\right)\right) + O(n^{-2}).$$

Observe that

$$g'\left(F^{-1}\left(\frac{S_j}{S_{n+1}}\right)\right)\left(S_{n+1}\, f\left(F^{-1}\left(\frac{S_j}{S_{n+1}}\right)\right)\right)^{-2} - g'(F^{-1}(\tfrac{j}{n}))(nf(F^{-1}(\tfrac{j}{n})))^{-2} = O(n^{-3}).$$

Hence

$$T_n = \frac{1}{2n^2}\sum_{j=1}^{n} g'(F^{-1}(\tfrac{j}{n}))f^{-2}(F^{-1}(\tfrac{j}{n}))E_j^2 + O(n^{-2})$$

and

$$n\,T_n - m = \frac{1}{2}\sum_{j=1}^{n} g'(F^{-1}(\tfrac{j}{n})f^{-2}(F^{-1}(\tfrac{j}{n}))(E_j^2-1) + O(n^{-1}).$$

Then (6) follows from the classical central limit theorem, (7) and (8) can be obtained similarly. The details will be omitted.

Making use of the same method of proof as in Lemma 2 one has

LEMMA 3. *Let* $g(x)$ *be a differentiable function with* $|g'(x)| \le L < \infty$ $(0<x<1)$. *Then*

(10) $$\mathbb{E}\sum_{j=1}^{n-1} g(\xi_j)(\xi_{j+1}-\xi_j)^2 = n^{-1}\int_0^1 \frac{g(x)}{f(x)}\,dx + o(n^{-3/2+\varepsilon})$$

and

$$\sum_{j=1}^{n-1} g(\xi_j)(\xi_{j+1}-\xi_j)^2 = n^{-1}\int_0^1 \frac{g(x)}{f(x)}\,dx + o(n^{-3/2+\varepsilon}) \quad \text{a.s.}$$

for any $\varepsilon > 0$ *provided that conditions* B *and* F_1 *hold.*

Let

$$\bar{c}_i = \sum_{j=1}^{n-1} r(\xi_j)\phi_i(\xi_j)(\xi_{j+1}-\xi_j).$$

Then by Lemma 2 we have

LEMMA 4. *Assume the conditions of Theorem 1. Then for any* $\varepsilon > 0$ *we have*

(11) $$\bar{c}_i - c_i = o(n^{-1+\varepsilon}) \quad \text{a.s.} \quad (i=1,2,\dots)$$

and

(12) $$\mathbb{E}\,(\bar{c}_i - c_i)^2 = o(n^{-2+\varepsilon}).$$

LEMMA 5. *Assume the conditions of Theorem 1. Then*

(13) $$\mathbb{E}\,(\hat{c}_i-\bar{c}_i)^2 = \sigma^2\mathbb{E}\sum_{j=1}^{n-1}\phi_i^2(\xi_j)(\xi_{j+1}-\xi_j)^2 = \sigma^2 n^{-1}\int_0^1 \frac{\phi_i(x)}{f(x)}\,dx + o(n^{-3/2+\varepsilon})$$

for any $\varepsilon>0$ *and* $i=1,2,\dots$.

PROOF. Conditioning $\mathbb{E}(\hat{c}_i-\bar{c}_i)^2$ given $\xi_1,\xi_2,\ldots$ and by (10) one has (13).

PROOF OF THEOREM 1. Since

$$\mathbb{E}(\hat{c}_i-c_i)^2 = \mathbb{E}(\hat{c}_i-\bar{c}_i)^2 + \mathbb{E}(\bar{c}_i-c_i)^2 \tag{14}$$

applying (12) and (13) one gets (3).

By (11) in order to investigate $\hat{c}_i-c_i$ it is enough to study $\hat{c}_i-\bar{c}_i$. Conditioning this difference again given $\xi_1,\xi_2,\ldots$ and making use of the classical central limit theorem one has (5). (4) also follows by standard methods.

3. Global Theorems

Let $\{\ell_n\}$ be an increasing sequence of positive integers with $\lim_{n\to\infty}\ell_n=\infty$. Introduce the following notations.

$$\psi_n(x,y) = \sum_{i=1}^{\ell_n} \phi_i(x)\phi_i(y),$$

$$\hat{r}_n(x) = \sum_{i=1}^{\ell_n} \hat{c}_i\,\phi_i(x) = \sum_{i=1}^{\ell_n}\sum_{j=1}^{n-1} \eta_j\phi_i(\xi_j)(\xi_{j+1}-\xi_j)\phi_i(x) =$$

$$= \sum_{j=1}^{n-1} \eta_j\,\psi_n(\xi_j,X)(\xi_{j+1}-\xi_j),$$

$$\bar{r}_n(x) = \sum_{j=1}^{n-1} r(\xi_j)\psi_n(\xi_j,x)(\xi_{j+1}-\xi_j),$$

$$r_n^*(x) = \int_0^1 r(y)\psi_n(y,x)dy = \sum_{j=1}^{\ell_n} c_j\phi_j(x),$$

$$I_1=I_1(x)=\hat{r}_n-\bar{r}_n=\sum_{j=1}^{n-1}\varepsilon_j\psi_n(\xi_j,X)(\xi_{j+1}-\xi_j),$$

$$I_2=I_2(x)=\bar{r}_n-r_n^*=\sum_{j=1}^{n-1} r(\xi_j)\psi_n(\xi_j,X)(\xi_{j+1}-\xi_j)-$$

$$- \int_0^1 r(y)\psi_n(y,x)dy,$$

$$I_3=I_3(x)=r_n^*-r = \sum_{j=\ell_n+1}^{\infty} c_j \phi_j(x).$$

A trivial calculation gives

LEMMA 6.

$$\int_0^1 I_1I_3 = 0 , \quad \int_0^1 I_2I_3 = 0 , \quad \mathbf{E}\, I_1I_2 = 0.$$

Consequently

$$\mathbf{E}\int_0^1 (\hat{r}_n(x)-r(x))^2dx = \mathbf{E}\int_0^1 I_1^2 + \mathbf{E}\int_0^1 I_2^2 + \mathbf{E}\int_0^1 I_3^2 = \mathcal{I}_1 + \mathcal{I}_2 + \mathcal{I}_3.$$

LEMMA 7. Assuming the conditions of Theorem 1 we have

(15) $$\mathcal{I}_1 \sim \frac{\sigma^2}{n}\int_0^1\int_0^1 \frac{\psi_n^2(x,y)}{f(x)}\,dxdy = \frac{\sigma^2}{n}\sum_{i=1}^{\ell_n}\int_0^1 \frac{\phi_i^2(x)}{f(x)}\,dx = O\left(\frac{\ell_n}{n}\right).$$

PROOF. Applying Lemma 2 we get

$$\mathcal{I}_1 = \sigma^2\mathbf{E}\int_0^1 \sum_{j=1}^{n-1} \psi_n^2(\xi_j,X)(\xi_{j+1}-\xi_j)^2dx \sim \frac{\sigma^2}{n}\int_0^1\int_0^1 \frac{\psi_n^2(y,x)}{f(y)}\,dydx$$

which proves (15).

LEMMA 8. Assuming conditions B, E_{i2} $(i=1,2,\dots)$ and F_1 one has

(16) $$\mathcal{I}_2 = O(\ell_n^2\, n^{-2}).$$

Observe that (by $E_{i,2}$)

$$\left|\frac{\theta^2\psi_n(y,x)}{x^2}\right| = O(\ell_n)$$

by Lemma 2 follows (16).

The behaviour of I_3 depends on the properties of the function $r(x)$ and those of the system $\{\phi_j\}$. Here we only mention one well-known result.

LEMMA 9. <u>Suppose that</u> $r(x)$ <u>is differentiable on</u> $(0,1)$ <u>and its derivative has a bounded variation. Further let</u>

$$\phi_k(x) = \frac{\sqrt{2}}{\pi} \cos k\pi x.$$

<u>Then</u>

$$I_3 = O(\ell_n^{-3}). \tag{17}$$

As a consequence of the last four lemmas we obtain:

THEOREM 2. <u>Assume the conditions of Theorem 1 and the relation</u> (17). <u>Then</u>

$$\mathbf{E}\int_0^1 (\hat{r}_n(x)-r(x))^2 dx = O(\ell_n n^{-1}+\ell_n^{-3}).$$

<u>Hence the optimal choice of</u> ℓ_n <u>is</u> $\ell_n = O(n^{1/4})$ <u>and in this case</u>

$$\mathbf{E}\int_0^1 (\hat{r}_n(x)-r(x))^2 dx = O(n^{-3/4}).$$

In case $\ell_n \geqq n^{1/4+\varepsilon}$ $(\varepsilon>0)$ I_1 is much larger than I_2 and I_3 and the properties of $\mathbf{E}\int_0^1 (\hat{r}_n(x)-r(x))^2 dx$ can be studied via studying the properties of I_1 only. For example we have

LEMMA 10. <u>Assume</u>

(i) <u>the conditions of Theorem 1</u>,

(ii) $\ell_n \geqq n^{1/4+\varepsilon}$ $(\varepsilon>0)$,

(iii) $\lim_{n\to\infty} n^{-1} \sum_{i=1}^{n} \phi_i^2(x) = 1$ <u>a.s. and</u> $\lim_{n\to\infty} n^{-1} \sum_{i=1}^{n} \phi_i^2(x)\phi_i^2(y) = 1$ <u>a.s.</u>,

(iv) $\ell_n^{-1} \int_0^1\int_0^1 \psi_n^2(x,y)dxdy \to 1$ <u>as</u> $n\to\infty$

(v) $\mathbf{E}\,\varepsilon_j^4 = \rho_4$ $(j=1,2,\dots)$.

<u>Then</u>

$$\lim_{n\to\infty} \mathbf{E}\, n\ell_n^{-1} \int_0^1 (\hat{r}_n(x)-r(x))^2 dx = \sigma^2 \int_0^1 \frac{dx}{f(x)} \tag{18}$$

<u>and</u>

$$(19)\qquad \mathbf{E}\left[n\ell_n^{-1}\int_0^1(\hat{r}_n(x)-r(x))^2dx-\sigma^2\int_0^1\frac{dx}{f(x)}\right]^2\sim\frac{\sigma^4}{\ell_n^2}\int_0^1\int_0^1\frac{(\psi_n(x,y))^2}{(x)f(y)}\,dxdy.$$

PROOF. (18) follows from (15) (cf. condition (iii)). In order to prove (19) consider

$$\mathbf{E}\ {}_1^2=\mathbf{E}\,(\sum_{j,\ell}\varepsilon_j\varepsilon_\ell\psi_n(\xi_j,\xi_\ell)(\xi_{j+1}-\xi_j)(\xi_{\ell+1}-\xi_\ell))^2=$$

$$=\sigma^4\mathbf{E}\,(\sum_{j\neq\ell}\psi_n^2(\xi_j,\xi_\ell)(\xi_{j+1}-\xi_j)^2(\xi_{\ell+1}-\xi_\ell)^2)+$$

$$+\sigma^4\,\mathbf{E}\,(\sum_{j\neq\ell}\psi_n(\xi_j,\xi_j)\psi_n(\xi_\ell,\xi_\ell)(\xi_{j+1}-\xi_j)^2(\xi_{\ell+1}-\xi_\ell)^2)+$$

$$+\rho_4\,\mathbf{E}\,(\sum_j\psi_n^2(\xi_j,\xi_j)(\xi_{j+1}-\xi_j)^4)\sim$$

$$\sim\frac{\sigma^4}{n^2}\int_0^1\int_0^1\frac{(\psi_n(x,y))^2}{f(x)f(y)}\,dx\,dy+\frac{\sigma^4}{n^2}\int_0^1\int_0^1\frac{\psi_n(x,x)\psi_n(y,y)}{f(x)f(y)}\,dx\,dy\sim$$

$$\sim\frac{\sigma^4\ell_n^2}{n^2}\left(\int_0^1\frac{dx}{f(x)}\right)^2+\frac{\sigma^4}{n^2}\int_0^1\int_0^1\frac{(\psi_n(x,y))^2}{f(x)f(y)}\,dx\,dy.$$

(Here we utilized the two dimensional analogue of Lemma 2.) Hence we have (19).

Now we can formulate our main result.

THEOREM 3. _Assume the conditions of Lemma 10. Then_

$$\mathbf{P}\left\{\frac{\ell_n}{\sigma^2\theta}\left[\ell_n\,n^{-1}\int_0^1(r_n(x)-r(x))^2dx-\mu\right]<x\right\}\to\frac{1}{\sqrt{2\pi}}\int_{-\infty}^{x}e^{-u^2/2}\,du$$

_as_ $n\to\infty$ _where_

$$\mu=\sigma^2\int_0^1\frac{dx}{f(x)},$$

and

$$\theta^2=\int_0^1\int_0^1\frac{(\psi_n(x,y))^2}{f(x)f(y)}\,dx\,dy.$$

PROOF is simple using Lemma 10 and the classical method of proof of the central limit theorem.

REFERENCE

[1] Collomb, G., Estimation Non-paramétrique de la Régression: Revue Bibliographique. Int. Stat. Review 49 (1981), 75-93.

International Series of
Numerical Mathematics, Vol. 65

THE WEAK INVARIANCE PRINCIPLE WITH RATES FOR C[0,1]-VALUED RANDOM-FUNCTIONS

Paul L. Butzer and Dietmar Schulz
Lehrstuhl A für Mathematik
Aachen University of Technology
Aachen, Federal Republic of Germany

In this paper a new proof of Donsker's invariance principle for C[0,1] is presented; it is quite different to the proofs existing so far (Skorohod representation approach, tightness-concept, method of compactification). The proof is based upon the idea of interpreting the invariance principle as a particular case of the central limit theorem in C[0,1], and consists essentially of a modification of the Lindeberg-Trotter operator theoretic approach, including a suitable decomposition of the random polygonal lines as well as of the Wiener measure. The method of proof enables one to equip the convergence assertions with $\mathcal{O}$ - Rates and, under stronger assumptions, even with o - rates.

1. Introduction

This paper is concerned with the weak invariance principle, namely with the weak convergence of the distributions of random polygonal lines S_n to the Wiener measure W_R on $C = C[0,1]$, the space of all real-valued continuous functions defined on the interval [0,1]. For this principle, which was first conceived by Donsker [8] and has turned out to be one of the main results in the theory of summation of random variables (r.vs.), essentially two different methods of proof are known. One of them is the proof carried out in detail in Billingsley's text [3], which makes use of the convergence of finite-dimensional distributions of random polygonal lines as well as of the tightness concept; the other is the so - called Skorohod representation approach that can be found at length in the text by Freedman [10], for example.
A third approach, based upon the compactification of the spaces $C[0,\infty)$ and $D[0,\infty)$ (= the space of all real functions on $[0,\infty)$ that possess no discon-

tinuities of the second kind), was published by Verwaat [24] only in 1981.

In this paper another, entirely different approach to Donsker's invariance principle for C[0,1] is presented under a supplementary assumption (perhaps superfluous); it has the additional advantage that it even admits estimates for the rate of convergence. The proof is actually a suited modification of the Lindeberg - Trotter operator - theoretic approach that has effectively been used in a series of papers in studying rates of convergence for the central limit theorem (CLT), in particular for the CLT for martingale difference sequences in Banach spaces (see e.g. [7],[18],[19]).

In order to concretize some of our results, let us take particular cases of Theorems 2 and 1: If $(x_i)_{i\in\mathbb{N}}$ is a sequence of independent, identically distributed (i.i.d.) r.vs. with expectation $E[x_i]=0$ and variance $V[x_i]=1$, and if S_n stands for the partial sums of the random polygonal lines constructed from these r.vs. (cf. Lemma 1), then under condition (4.7) Corollary 1 yields

$$\lim_{n\to\infty} E[f(S_n)] = E[f(W_R)] \tag{1.1}$$

for each $f\in\overline{C}_C^2$ (see (2.6)). If in addition the third absolute moments $E[|x_i|^3]$ are finite, then Theorem 1 yields that the rate of convergence in (1.1) is of order $O(n^{-1/2})$ provided $f\in C_C^3$ (see (2.5)), namely

$$|E[f(S_n)] - E[f(W_R)]| = O(n^{-1/2}) \qquad (n\to\infty). \tag{1.2}$$

When comparing the large - O estimate with those known so far one notices right-off that none of the papers (known to us) concerned with rates of convergence for the Donsker invariance principle deal with estimates for weak convergence in the usual form; most of the time the distance between P_{S_n} and the Wiener measure W_R is expressed in terms of the so - called Lévy - Prohorov - metric L_C (see e.g. [21,1]). In this respect one should recall that weak convergence is equivalent to convergence in the Lévy - Prohorov metric provided one deals with pure convergence assertions; if these assertions are equipped with rates this equivalence fails.

Papers concerned with orders of approximation for the weak invariance principle in the frame of the Lévy - Prohorov metric just mentioned include those of Prohorov [17], Skorohod [22], Dudley [9] and Borovkov [4,5]. Furthermore, from the results of Rosenkrantz [20] and Komlós, Major and Tusnády [13] it is also possible to estimate orders for L_C (see Borovkov [5]). Sahanenko [21] and Arak [1] even showed that Borovkov's estimates cannot be improved. In case of a random polygonal line constructed from a sequence of i.i.d. r.vs. $(x_i)_{i \in \mathbb{N}}$ with expectation zero, variance one and $E[|x_i|^3] < \infty$, the latter result yields

$$L_C(P_{S_n}, W_R) = \mathcal{O}(n^{-1/8}) \qquad (n \to \infty) .$$

This estimate can however be improved to

$$L_C(P_{S_n}, W_R) = \mathcal{O}(n^{-1/2} \log n)$$

provided Cramér's condition upon the Fourier transforms of the x_i is satisfied as Borovkov [5] showed by using the above result of Komlós, Major and Tusnády. A comparison yields immediately that our speeds of convergence for the (weak) invariance principle are better than those obtained for the Lévy-Prohorov-metric; this is especially so for the little o-rates of Theorem 2. This theorem delivers in particular the estimate $o(n^{-1/2})$ which is a good deal better than that achieved by Prohorov, namely $L_C(S_n, W_R) = o(n^{-1/2} \log^2 n)$, the only paper on little o-rates known to the authors. However, our result needs the additional assumption (4.1).

A substancial survey concerning speeds of convergence for invariance principles can be taken from the papers by Dudley [9], Borovkov [5] and Major [15].

It should finally be noted that the rates of convergence to be deduced in this paper are just as good as those found by Butzer - Hahn [6] for the CLT for real r.vs. which will turn out to be a particular case of the present results on the invariance principle.

Section 2 deals with notations, definitions and some lemmas concerning Frechêt derivatives and random functions (r.f.) with values in C. Section 3 and 4 are devoted to the invariance principle with large - $\mathcal{O}$ and little - o rates, respectively. A corollary of the latter theorem gives a pure convergence assertion. Section 5 is concerned with the CLT for real r.vs. with rates as a special case of the theorems of the two preceeding paragraphs.

2. Notations and Preliminary Results

Let $C = C[0,1]$ be the class of all continuous, real-valued functions defined on [0,1] endowed with the usual supremum norm. The expectation of a C-valued r.f. $X : \Omega \to C$ on an arbitrary probabilty space (Ω,A,P) is defined by the Bochner integral $E[X] := \int_\Omega X(\omega)dP(\omega)$ and, if $E[\|X\|_C^2] < \infty$, the covariance functional $R_X : C^* \times C^* \to \mathbb{R}$ by

$$R_X(f^*,g^*) := E[f^*(X),g^*(X)] \qquad (f^*,g^* \in C^*),$$

C^* denoting the topological dual of C.

A stochastic process $(W(t))_{0\le t\le 1} = (W(\omega,t))_{0\le t\le 1}$ is called a standard Wiener process or Brownian motion process (restricted to the interval [0,1]) provided (i) $W(\omega,0) = 0$ for all $\omega \in \Omega$ (ii) for each $\omega, W(\omega,\cdot)$ is a continuous function on the interval [0,1], and (iii) for $0 \le t_1 < t_2 < \ldots < t_n \le 1$ the increments $W(t_1)$, $W(t_2) - W(t_1)$, ..., $W(t_n) - W(t_{n-1})$ are independent, Gaussian distributed r.vs. with expectation 0 and variance 1.

Note that $W(\omega,t)$ is as well a function of points $\omega \in \Omega$ as of $t \in [0,1]$ such that $W(\cdot,t)$ is A-measurable for all $t \in [0,1]$ and $W(\omega,\cdot)$ is an element of C for each $\omega \in \Omega$. Thus the Wiener process can be regarded as a r.f. with values in C. Its distribution, a measure on the class of Borel sets in C, is called the Wiener measure W_R. Properties of stochastic processes considered as random elements in function spaces are investigated in detail by R.L. Taylor in [23]. Three lemmas will be needed.

LEMMA 1. *Let* $x_0 = 0, x_1 . x_2, \ldots$ *be a sequence of i.i.d. r.vs. such that* $E[x_i] = 0$ *and* $Var[x_i] = 1$, $i \in \mathbb{N}$. *Set*

$$S_n(\omega,t) := n^{-1/2} \sum_{i=0}^{[nt]} x_i(\omega) + (nt - [nt])x_{[nt]+1}(\omega)$$

for every $n \in \mathbb{N}$, $\omega \in \Omega$ *and* $t \in [0,1]$ ([a] *denotes the greatest integer* $\le a$), *and consider the trajectories*

$$(2.1)\qquad X_{ni}(\omega,t) := \begin{cases} 0\,, & \text{for } 0 < nt \leq i-1 \\ n^{1/2}(t-\frac{i-1}{n})\, x_i(\omega), & \text{for } i-1 < nt \leq 1 \\ n^{-1/2}x_i(\omega)\,, & \text{for } i < nt \leq n\,. \end{cases}$$

Then one has

a) $(X_{ni})_{1\leq i\leq n,\ n\in\mathbb{N}}$ is an array of independent r.fs. with values in C, and $S_n(\omega,t) = \sum_{i=1}^n X_{ni}(\omega,t)$;

$$(2.2)\ \text{b)}\qquad \|X_{ni}\|_C = n^{-1/2}|x_i|\,.$$

The proof of part a), due to E. Giné [12], follows as does part b) from the definition of the r.fs. X_{ni}, $1\leq i\leq n$, $n\in\mathbb{N}$.

LEMMA 2. Let W_R be a r.f. with Wiener measure as its distribution on [0,1] having covariance function R. If $(a_{ni})_{1\leq i\leq n,\ n\in\mathbb{N}}$ is a triangular array of positive reals and

$$(2.3)\qquad A_n := (\sum_{i=1}^n a_{ni}^2)^{1/2}\,,$$

then there exist independent, Gaussian distributed r.fs. $W_{ni} := a_{ni}W_R$ such that

$$P_{A_n^{-1}\sum_{i=1}^n a_{ni}W_R} = P_{W_R}\,.$$

PROOF. First note that W_R is a Gaussian distributed r.f. (cf. [2, p. 191]). Therefore the r.f. $A_n^{-1}\sum_{i=1}^n a_{ni}W_R$, $n\in\mathbb{N}$, is also Gaussian distributed with the same convariance functional. As the convariance functional determines uniquely the distribution of a centered Gaussian r.v. (see e.g. [11, p. 314]) the proof follows.

A Feller - type condition upon a triangular array of positive reals $(a_{ni})_{1\leq i\leq n,\ n\in\mathbb{N}}$ will be assumed in Theorem 2. It states that

(2.4) $$\lim_{n\to\infty} \max_{1\leq i\leq n} \frac{a_{ni}}{A_n} = 0 ,$$

A_n being defined as in (2.3).

In the following some spaces are introduced which occur in connection with the concept of Frechêt - differentiability. Let C^r, $r \in \mathbb{N}$, denote the r - fold product space $C \times \ldots \times C$ endowed with the norm $\|X\|_{C^r} := \max_{1\leq i\leq r} \|X_i\|_C$, where $X := (X_1, \ldots, X_r) \in C^r$. Then the space $L_r = L_r((C^r, \mathbb{R}))$ of all real - valued multilinear continuous functions $g : C^r \to \mathbb{R}$ is a Banach space under the norm $\|g\|_{L_r} := \sup_{\|X\|_{C^r}=1} |g(X)|$.

The space L_r enables us to define the following function classes, needed to characterize a special case of convergence in distribution of r.fs. These are

$C_C = C_{C[0,1]} := \{g : C \to \mathbb{R} ;\ g$ continuous and bounded on $\mathbb{R}\}$,

$C_C(L_r) := \{f : C \to L_r ;\ f$ continuous and bounded on $L_r\}$,

(2.5) $$C_C^r := \{f \in C_C ;\ f^{(j)} \in C_C(L_j) ,\ 1 \leq j \leq r\} ,$$

(2.6) $$\overline{C}_C^r := \{f \in C_C^r ;\ f^{(r)} \text{ is uniformly continuous and bounded on } L_r\}.$$

On C_C^r a seminorm $|\cdot|_{C_C^r}$ is defined by $|f|_{C_C^r} := \sup_{X\in C} \|f^{(r)}(X)\|_{L^r}$. Differentiability of $f \in C_C^r$ with $\|f\|_\infty := \sup_{X\in C} |f(X)|$ (may be infinite) is always meant in the sense of Frechêt so that the jth (Frechêt) derivative $f^{(j)}$ of $f \in C_C^r$, $1 \leq j \leq r$, maps C into L_j.

For some estimates in the proofs of our theorems, a representation for $f^{(j)}(Y)$, $Y \in C$, in terms of an infinite sum is needed. Let $(e_k)_{k\in\mathbb{N}}$ be a normalized countable basis of C. Then for each $X \in C$ there exists a unique representation $X = \sum_{k=1}^\infty X^{(\nu)} e_k$ in terms of the real components $X^{(\nu)}$ with respect to the given basis. Considering the fact that $f^{(j)}(Y)$ belong to L_j, i.e., are multilinear and continuous, we obtain

$$(2.7)\quad f^{(j)}(Y)[X]^j = \sum_{\nu_1=1,\ldots,\nu_j=1}^{\infty} X^{(\nu_1)}\cdot\ldots\cdot X^{(\nu_j)} f^{(j)}(Y)(e_{\nu_1},\ldots,e_{\nu_j}),$$

where $\nu_k \in \mathbb{N}$, $1 \leq k \leq j$ and $[X]^j = (X,\ldots,X) \in C^j$. To further abbreviate (2.7), we write for every j-tuple $\nu = (\nu_1,\ldots,\nu_j) \in \mathbb{N}^j$

$$(2.8)\quad |\nu| := j\,,\ X^{\nu} := \Pi_{k=1}^{j} X^{(\nu_k)}\,,\ f^{[\nu]}(\cdot) := f^{[j]}(\cdot)(e_{\nu_1},\ldots,e_{\nu_j}) : C \to \mathbb{R}\,.$$

Then (2.7) takes on the form

$$f^{(j)}(Y)[X]^j = \sum_{|\nu|=j} X^{\nu} f^{[\nu]}(\cdot) \qquad (X,Y \in C)\,.$$

Taylor's formula for $X,Y \in C[0,1]$, $f \in C_C^r$, $r \in \mathbb{N}$ will also be employed. It reads,

$$(2.9)\quad \begin{aligned} f(X+Y) = f(Y) + \sum_{j=1}^{r} \frac{f^{(j)}(Y)}{j!}[X]^j + \frac{1}{(r-1)!}\int_0^1 (1-t)^{r-1}\cdot \\ \{f^{(r)}(Y+tX)[X]^r - f^{(r)}(Y)[X]^r\}\,dt\,. \end{aligned}$$

In order to describe smoothness properties of $f : C \to \mathbb{R}$ a special Lipschitz class of order $\beta \in (0,1]$ is introduced for the j-th (Frechêt) derivative $f^{(j)}$ of f. Indeed, $f^{(j)} \in \mathrm{Lip}(\beta, L_j)$ provided there exists a Lipschitz constant $M_{f^{(j)}} > 0$ such that

$$\sup_{0 \leq \|X\|_C \leq 1} \|\, \|f^{(j)}(X+\cdot) - f^{(j)}(\cdot)\|_{L_j} \|_{\infty} \leq M_{f^{(j)}} t^{\beta} \qquad (t > 0)\,.$$

In particular, $f^{(r-1)} \in \mathrm{Lip}(1, L_{r-1})$ with $M_{f^{(r-1)}} = |f|_{C_C^r}$ if $f \in C_C^r$.

Our third Lemma, to be found in [7] or [18], reads

LEMMA 3. *Let* X,Y *be two r.fs. with values in* C *such that* $E[\|X\|_C^j] < \infty$ *for some* $j \in \mathbb{N}$. *Then* $f \in C_C^j$ *implies*

$$E[f^{(j)}(Y)[X]^j] = \sum_{|\nu|=j} E[X^\nu f^{[\nu]}(Y)] .$$

Here $f^{[\nu]}(Y)$ and X^ν are defined according to (2.8) noting that X^ν is a real r.v.. The proof follows by noting that $X(\omega) = \sum_{k=1}^{\infty}(X(\omega))^{(k)}e_k$, using Lebesgue's dominated convergence theorem as well as the uniform boundedness of the operator norms of the projection operators $\sum_{k=1}^{n} X^{(k)}e_k : C \to C$, $n \in \mathbb{N}$. For a detailed proof see [18].

3. Invariance - Principle with $\mathcal{O}$ - Rates

It is well known that the CLT in $\mathbb{R}$ can be deduced from the weak invariance principle (see eg. [14, p. 279] and Section 5). Less known may be the fact that the invariance principle may in turn be regarded as a CLT for C[0,1] - valued r.vs. (cf. [12], [16]). In this respect several authors speak of the invariance principle as the functional central limit theorem; see e.g. [3, p. 72],[11, p. 382]. This interpretation is made possible in the present approach by use of the r.fs. X_{ni}, $1 \leq i \leq n$, $n \in \mathbb{N}$ given in Lemma 1, due to Giné [12]. They allow the random polygonal line S_n to be represented as partial sums of independent r.fs.. Since moreover the Wiener measure W_R is a particular Gauss measure (e.g. [2, p. 191]) the Lindeberg - Trotter operator method, which was effectively used to study rates of convergence for the CLT in Banach spaces, as indicated in the introduction, can also be employed in a modified form in order to prove the invariance principle (with rates). The present modification consists, apart from the use of Frechêt derivatives as a suitable smoothness concept to deduce the rates, of the introduction of appropriate functions R_{ni} (see (3.4) which permit one to avoid the Trotter operator itself which led to convergence assertions valid only in a subclass of C_C^r (and so only under more restrictive conditions)).

THEOREM 1. *Let $(x_i)_{i \in \mathbb{P}}$ be a sequence of real, i.i.d.r.vs. with mean* 0 *and variance* 1, *and let $(a_{ni})_{1 \leq i \leq n,\ n \in \mathbb{N}}$ as well as A_n be defined as in Lemma* 2. Assume that

$$\zeta_{r-1+\beta} := E[|x_1|^{r-1+\beta}] < \infty$$

for some $r \in \mathbb{N}$ and $0 < \beta \leq 1$, and for the r.fs. $(X_{ni})_{1\leq i\leq n, n\in\mathbb{N}}$ defined in (2.1) let

$$E[X_{ni}^{\nu}] = a_{ni}^{j} A_n^{j} E[W_R^{\nu}] \tag{3.1}$$

for all $|\nu| = j$, $1 \leq j \leq r-1$, $1 \leq i \leq n$, and $n \in \mathbb{N}$. Then $f \in C_C^{r-1}$ with $f^{(r-1)} \in \mathrm{Lip}(\beta, L_{r-1})$, $M_{f^{(r-1)}}$ being the Lipschitz constant, implies that P_{S_n} converges weakly to the Wiener measure W_R with rates given by

$$|E[f(S_n)] - E[f(W_R)]| \leq$$

$$\frac{M_{f^{(r-1)}}}{(r-1)!}\left\{ n^{(3-r-\beta)/2}\,\zeta_{r-1+\beta,1} + \Big(A_n^{1-r-\beta}\sum_{i=1}^{n} a_{ni}^{r-1+\beta}\Big)E[\|W_R\|_C^{r-1+\beta}]\right\}. \tag{3.2}$$

In particular, if $E[|x_1|^r] < \infty$, and if (3.1) holds, then $f \in C_C^r$ implies

$$|E[f(S_n)] - E[f(W_R)]| \leq$$

$$\frac{|f|_{C_C^r}}{(r-1)!}\left\{ n^{(2-r)/2}\,\zeta_{r,1} + \Big(A_n^{-r}\sum_{i=1}^{n} a_{ni}^{r}\Big)E[\|W_R\|_C^{r}]\right\}. \tag{3.3}$$

PROOF. First note that $f(S_n)$ and $f(W_R)$ are real integrable r.vs. for each $f \in C_C$. With help of the r.f.

$$R_{ni} := \sum_{k=1}^{i-1} X_{nk} + A_n^{-1}\sum_{k=i+1}^{n} a_{nk}W_R \qquad (1 \leq i \leq n,\ n \in \mathbb{N}) \tag{3.4}$$

one readily obtains, by summing up, the identity

$$f(S_n) - f\Big(A_n^{-1}\sum_{i=1}^{n} a_{ni}W_R\Big) = \sum_{i=1}^{n}\{f(R_{ni}+X_{ni}) - f(R_{ni}+a_{ni}A_n^{-1}W_R)\}.$$

This yields for $f \in C_C^{r-1}$ by a double application of Taylor's formula (2.9) up to the order r-1

$$f(S_n) - f(A_n^{-1} \sum_{i=1}^{n} a_{ni}W_R)$$

$$= \sum_{i=1}^{n} \sum_{j=1}^{r-1} \frac{1}{j!} \{f^{(j)}(R_{ni})[X_{ni}]^j - f^{(j)}(R_{ni})[a_{ni}A_n^{-1}W_R]^j\}$$

$$+ \sum_{i=1}^{n} \frac{1}{(r-2)!} \int_0^1 (1-t)^{r-2} \Big\{ \{f^{(r-1)}(R_{ni}+tX_{ni})[X_{ni}]^{r-1} - f^{(r-1)}(R_{ni})[X_{ni}]^{r-1}\}$$

$$- \{f^{(r-1)}(R_{ni} + ta_{ni}A_n^{-1}W_R)[a_{ni}A_n^{-1}W_R]^{r-1} - f^{(r-1)}(R_{ni})[a_{ni}A_n^{-1}W_R]^{r-1}\} \Big\} \, dt .$$

Since $f^{(r-1)} \in \mathrm{Lip}(\beta, L_{r-1})$, one has for $0 < t \leq 1$ in view of (2.2)

$$|\{f^{(r-1)}(R_{ni}+tX_{ni}) - f^{(r-1)}(R_{ni})\}[X_{ni}]^{r-1}|$$

$$\leq \| f^{(r-1)}(R_{ni} + tX_{ni}) - f^{(r-1)}(R_{ni})\|_{L_{r-1}} \|X_{ni}\|_C^{r-1}$$

$$\leq M_{f^{(r-1)}} \|X_{ni}\|_C^{r-1+\beta}$$

$$= M_{f^{(r-1)}} n^{-(r-1+\beta)/2} \, |x_i|^{r-1+\beta} .$$

Analogously one has for $a_{ni}W_R$

$$|\{f^{(r-1)}(R_{ni} + ta_{ni}A_n^{-1}W_R) - f^{(r-1)}(R_{ni})\}[a_{ni}A_n^{-1}W_R]^{r-1}|$$

$$\leq M_{f^{(r-1)}} a_{ni}^{r-1+\beta} \, A_n^{1-r-\beta} \, \|W_R\|_C^{r-1+\beta} .$$

Altogether one deduces on account of Lemma 2

$$|E[f(S_n)] - E[f(W_R)]| = |E[f(S_n)] - E[f(A_n^{-1} \sum_{i=1}^{n} a_{ni}W_R)]|$$

$$\leq |E[\sum_{i=1}^{n} \sum_{j=1}^{r-1} \frac{1}{j!} \{f^{(j)}(R_{ni})[X_{ni}]^j - f^{(j)}(R_{ni})[a_{ni}A_n^{-1}W_R]^j\}]|$$

$$+ \frac{M_{f^{(r-1)}}}{(r-1)!} \{n^{(3-\beta-r)/2} \zeta_{r-1+\beta,1} + A_n^{1-r-\beta} \sum_{i=1}^{n} a_{ni}^{r-1+\beta} \, E[\|W_R\|_C^{r-1+\beta}]\} . \tag{3.5}$$

It remains to show that

$$\sum_{i=1}^{n} \{E[f^{(j)}(R_{ni})[X_{ni}]^j - f^{(j)}(R_{ni})[a_{ni}A_n^{-1}W_R]^j]\} = 0$$

for $n \in \mathbb{N}$, $1 \leq j \leq r-1$. But according to Lemma 3 one obtains for $1 \leq i \leq n$, $n \in \mathbb{N}$ and $1 \leq j \leq r-1$

$$E[f^{(j)}(R_{ni})([X_{ni}]^j - [a_{ni}A_n^{-1}W_R]^j)] = \sum_{|\nu|=j} E[f^{[\nu]}(R_{ni})(X_{ni}^{\nu} - a_{ni}^j A_n^{-j} W_R^{\nu})]$$

$$\leq \sum_{|\nu|=j} \|f^{[\nu]}\|_{\infty} \, |E[X_{ni}^{\nu} - a_{ni}^j A_n^{-j} W_R^{\nu}]| \; .$$

Since the difference in the latter sum vanishes by (3.1), assertion (3.2) follows by (3.5). Finally (3.3) is a consequence of (3.2) and the fact that $f^{(r-1)} \in \mathrm{Lip}(1, L_{r-1})$ with constant $M_{f^{(r-1)}} = |f|_{C_C^r}$ for $f \in C_C^r$. This completes the proof.

The right sides of (3.2) or (3.3), respectively, are of practical interest only if they tend to zero for $n \to \infty$. This is the case if $r \geq 3$ and if in (3.2) $A_n^{1-r-\beta} \sum_{i=1}^n a_{ni}^{r-1+\beta} = O(n^{(3-r-\beta)/2})$, $n \to \infty$, or if in (3.3) $A_n^{-r} \sum_{i=1}^n a_{ni}^r = O(n^{(2-r)/2})$, $n \to \infty$. Because of (2.3) these conditions are fulfilled if $a_{ni} = a_{nj}$, $1 \leq i, j \leq n$.

4. Invariance Principle with o - Rates

Whereas a whole series of papers (see introduction) is concerned with O - rates of convergence for the weak invariance principle, only one paper is known to the authors that deals with o - rates, namely that of Prohorov [17]. Thus Prohorov obtained under the assumptions of Theorem 2 for $r = 3$ for the L_C-metric the estimate $L_C(P_{S_n}, W_R) = o(n^{-1/8} \log^2 n)$; this estimate is definitely not as good as the order $o(n^{-1/2})$ to be deduced in Theorem 2. However, Theorem 2 needs an additional assumption (perhaps superfluous) namely (4.1), as well as that $f \in \overline{C}_B^3$.

THEOREM 2. *Let* $(x_i)_{i \in \mathbb{P}}$ *be a sequence of real i.i.d. r.vs. with mean* 0 *and variance* 1 *such that* $\zeta_{r,1} < \infty$ *for some* $r \in \mathbb{N}$. *Let* $(a_{ni})_{1 \leq i \leq n, n \in \mathbb{N}}$ *be a triangu-*

lar array of positive reals, and let A_n, $n \in \mathbb{N}$, and $(X_{ni})_{1\leq i\leq n, n\in\mathbb{N}}$ be defined as in (2.3) or (2.1), respectively. If the $(X_{ni})_{1\leq i\leq n, n\in\mathbb{N}}$ satisfy

$$(4.1)\quad E[X_{ni}^{\nu}] = a_{ni}^{j} A_n^{-j} E[W_R^{\nu}] \qquad (|\nu| = j\,, 1\leq j\leq r\,, n\in\mathbb{N}, 1\leq i\leq n)\,,$$

and the $(a_{ni})_{1\leq i\leq n\,, n\in\mathbb{N}}$ the Feller-type condition (2.4), then one has for each $f\in\overline{C}_C^r$ and $n\to\infty$

$$(4.2)\quad |E[f(S_n)] - E[f(W_R)]| = o_f(n^{(2-r)/2}\zeta_{r,1} + (A_n^{-r}\sum_{i=1}^{n} a_{ni}^{r})E[\|W_R\|_C^r])\,.$$

PROOF. An expansion of $f(R_{ni}+X_{ni})$ and $f(R_{ni}+A_n^{-1}a_{ni}W_R)$ according to Taylor's formula, this time up to the order r, yields the estimate

$$(4.3)\quad |E[f(S_n)] - E[f(W_R)]|$$

$$\leq |[E\sum_{i=1}^{n}\sum_{j=1}^{r}\{\frac{1}{j!}f^{(j)}(R_{ni})[X_{ni}]^j - f^{(j)}(R_{ni})[a_{ni}A_n^{-1}W_R]^j\}]| +$$

$$+ |E\Big[\sum_{i=1}^{n}\frac{1}{(r-1)!}\int_0^1(1-t)^{r-1}\Big\{|f^{(r)}(R_{ni}+tX_{ni})[X_{ni}]^r - f^{(r)}(R_{ni})[X_{ni}]^r|$$

$$+ |f^{(r)}(R_{ni}+ta_{ni}A_n^{-1}W_R)[a_{ni}A_n^{-1}W_R]^r - f^{(r)}(R_{ni})[a_{ni}A_n^{-1}W_R]^r|\Big\}\,dt\Big]|\,.$$

Since $f\in\overline{C}_C^r$, $f^{(r)}$ is uniformly continuous on C[0,1], so that for each $\varepsilon>0$ there exists a $\delta=\delta(\varepsilon)>0$ such that $\|f^{(r)}(R_{ni}+tX_{ni}) - f^{(r)}(R_{ni})\|_{L_r}<\varepsilon$, provided $\|tX_{ni}\|_C<\delta$ or $\|X_{ni}\|_C<\delta$ for $1\leq i\leq n$, $n\in\mathbb{N}$, since $t\in[0,1]$. Analogously one has provided $\|a_{ni}W_R\|_C<\delta A_n$, $1\leq i\leq n$, $n\in\mathbb{N}$ that

$$\|f^{(r)}(R_{ni}+ta_{ni}A_n^{-1}W_R) - f^{(r)}(R_{ni})\|_{L_r} < \varepsilon\,.$$

Noting that for arbitrary $g\in L_r$ the inequality $|g[X]^r|\leq\|g\|_{L_r}^r\|X\|_C^r$ holds, there follows

$$(4.4)\qquad E[|\{f^{(r)}(R_{ni}+tX_{ni}) - f^{(r)}(R_{ni})\}[X_{ni}]^r|]$$

$$\leq E[\| f^{(r)}(R_{ni}+tX_{ni}) - f^{(r)}(R_{ni})\|_{L_r} \|X_{ni}\|_C^r]$$

$$\leq E[\| f^{(r)}(R_{ni}+tX_{ni}) - f^{(r)}(R_{ni})\|_{L_r} \|X_{ni}\|_C^r\{\mathbb{1}_{\|X_{ni}\|_C<\delta}+\mathbb{1}_{\|X_{ni}\|_C\geq\delta}\}]$$

$$\leq \varepsilon E[\|X_{ni}\|_C^r] + 2|f|_{C_C^r} \int_{\|X_{ni}\|_C\geq\delta} \|X_{ni}\|^r P(d\omega) ,$$

where $\mathbb{1}_A$ denotes the indicator function of the set $A\subset C$. Summing up the foregoing inequality over all i running from 1 to n and dividing by $\sum_{i=1}^n E[\|X_{ni}\|_C^r]$, one deduces on account of (2.2)

$$\text{(4.5)}\qquad \sum_{i=1}^n \{E[|\{f^{(r)}(R_{ni}+tX_{ni})-f^{(r)}(R_{ni})\}[X_{ni}]^r|]\} = o_\delta(n^{(2-r)/2}\zeta_{r,1}) ,$$

by taking into consideration that because of (2.2)

$$\left(\sum_{i=1}^n \int_{\|X_{ni}\|_C\geq\delta} \|X_{ni}\|_C^r P(d\omega)\right) \Big/ \left(\sum_{i=1}^n E[\|X_{ni}\|_C^r]\right)$$

$$= \left(\sum_{i=1}^n \int_{|x_i|\geq n^{1/2}\delta} n^{-r/2}|x_i|^r P(d\omega)\right) \Big/ \left(n^{-r/2}\sum_{i=1}^n E[|x_i|^r]\right) ,$$

and noting that the last quotient tends to zero for $n\to\infty$, because the r.vs. $(x_i)_{i\in\mathbb{N}}$ are i.i.d.. Perhaps it is worthwhile to remark that it has thus been shown that the r.fs. $(X_{ni})_{1\leq i\leq n, n\in\mathbb{N}}$ fulfill a Lindeberg-condition of order r.

As the double sum on the right side of (4.3) vanishes by (4.1), it suffices to show that the $a_{ni}W_R$, $1\leq i\leq n$, $n\in\mathbb{N}$, also satisfy a generalized Lindeberg condition of order r, namely that

$$\text{(4.6)}\qquad \sum_{i=1}^n \int_{\|a_{ni}W_R\|_C\geq\delta A_n} \|a_{ni}W_R\|_C^r P(d\omega) = o_\delta\left(\sum_{i=1}^n a_{ni}^r E[\|W_R\|_C^r]\right) .$$

For this purpose set $a_{m(n)} := \max_{1\leq i\leq n} a_{ni}$. Then

$$\{X\in C, \|X\|_C\geq\delta A_n/a_{ni}\}\subset\{X\in C; \|X\|_C\geq\delta A_n/a_{m(n)}\}\qquad (1\leq i\leq n, n\in\mathbb{N}),$$

and therefore

$$\{\sum_{i=1}^{n} a_{ni}^{r} E[\|W_R\|_C^r]\}^{-1} \sum_{i=1}^{n} \int_{\|a_{ni}W_R\|_C \geq \delta A_n} \|a_{ni}W_R\|_C^r P(d\omega)$$

$$\leq \left(\sum_{i=1}^{n} a_{ni}^{r} \int_{\|W_R\|_C \geq \delta A_n/a_{m(n)}} \|W_R\|_C^r P(d\omega)\right) \Big/ \left(E[\|W_R\|_C^r \sum_{i=1}^{n} a_{ni}^{r}\right) .$$

But the latter integral converges to zero for $n\to\infty$ on account of the Feller-type condition (2.4), and this gives (4.6).

Gathering up all the results one finally obtains the assertion (4.2) in view of (4.1), (4.5), (4.6) and (4.3).

Further, the right side of (4.2) tends to zero for $n\to\infty$ if $r\geq 2$ and $A_n^{-r}\sum_{i=1}^{n} a_{ni}^{r} = \mathcal{O}(n^{(2-r)/2})$, $n\to\infty$. Note that condition (4.1) may be superfluous in the sense that in case $r=2$ it is satisfied automatically for real r.vs. with expectation zero and variance one (cf. e.g. Theorem 3 c)).

It is of interest to note that a version of the weak invariance principle (without rates) may be deduced from Theorem 2, namely

COROLLARY 1. _Let_ $(x_i)_{i\in\mathbb{P}}$ _be a sequence of real r.vs. with mean_ 0 _and variance_ 1. _Let the r.fs._ X_{ni}, $1\leq i\leq n$, $n\in\mathbb{N}$, _defined as in_ (2.1), _fulfill the condition_

$$E[X_{ni}^{\nu}] = n^{-j/2}E[W_R^{\nu}] \tag{4.7}$$

for all $|\nu| = j$, $j=1,2$, $n\in\mathbb{N}$, $1\leq i\leq n$ (i.e. $a_{ni}=1$). _Then, for_ $f\in\overline{C}_C^2$,

$$\lim_{n\to\infty} E[f(S_n)] = E[f(W_R)] . \tag{4.8}$$

PROOF. Because of (2.3) $A_n = n^{1/2}$, and the triangular array $(a_{ni})_{1\leq i\leq n, n\in\mathbb{N}}$ satisfies condition (2.4). Considering that $\zeta_{2,1} = 1 < \infty$, we get from Theorem 2

$$|E[f(S_n)] - E[f(W_R)]| = o_f(\zeta_{2,1} + E[\|W_R\|_C]^2) , \qquad (n\to\infty)$$

and therefore the assertion.

When comparing Corollary 1 with the classical weak invariance principle it will be seen that Corollary 1 needs an additional assumption, namely (4.7).

Furthermore, the authors are not aware of the fact whether the convergence assertion (4.8) for functions $f\in\overline{C}_C^2$ is equivalent to the usual weak convergence of P_{S_n} to W_R. See the remark at end of paper.

5. The Central Limit Theorem (for real RVs)

It is our aim here to use the invariance principle to prove the CLT for real r.vs. It is basically the case $t=1$ in Theorems 1 and 2 for which the r.fs. X_{ni}, $1\leq i\leq n$, $n\in\mathbb{N}$ turn out to be the triangular array of real r.vs. $X_{ni}(1,\omega)=n^{-1/2}x_i(\omega)$, $1\leq i\leq n$, and the r.f. having Wiener measure as its distribution has just the standard normal distribution P_{X^*}. Setting $a_{ni}:=1$, $1\leq i\leq n$, $n\in\mathbb{N}$ and $A_n=n^{1/2}$, Theorems 1 and 2 yield the following CLT

THEOREM 3. *Let $(x_i)_{i\in\mathbb{N}}$ be a sequence of i.i.d. r.vs. for which*

$$E[(x_i^j)]=E[(x^{*j})] \qquad (1\leq i\leq r-1,\ i\in\mathbb{N}). \tag{5.1}$$

a) *Assume that* $\zeta_{r-1+\beta,1}:=E[|x_i|^{r-1+\beta}]<\infty$ *for some* $r\in\mathbb{N}$ *and* $0<\beta\leq 1$. *Then for all* $f:\mathbb{R}\to\mathbb{R}$, *for which* $f^{(j)}$ *is bounded and continuous on* $\mathbb{R}$ *for all* $0\leq j\leq r-1$, *and* $f^{(r-1)}$ *satisfies a classical Lipschitz condition of order* β *with Lipschitz constant* $M_{f^{(r-1)}}$, *one has*

$$\begin{aligned}&\Big|E\Big[f\Big(\frac{1}{\sqrt{n}}\sum_{i=1}^{n}x_i\Big)\Big]-E[f(x^*)]\Big| \\ &\leq \frac{M_{f^{(r-1)}}}{(r-1)!}\{n^{(3-r-\beta)/2}(\zeta_{r-1+\beta,1}+E[|x^*|^{r-1+\beta}])\}.\end{aligned} \tag{5.2}$$

b) *In particular if* $E[|x_1|^r]<\infty$, *then, for all* $f:\mathbb{R}\to\mathbb{R}$ *for which* $f^{(j)}$ *is bounded and continuous on* $\mathbb{R}$ *for all* $1\leq j\leq r$, *the estimate on the right in* (5.2) *is of order* $O(n^{(2-r)/2})$ *for* $n\to\infty$.

c) *If* (5.1) *is fulfilled for* $1\leq j\leq r$, *and if* $E[|x_1|^r]<\infty$, *one has, for all* $f:\mathbb{R}\to\mathbb{R}$ *for which* $f^{(j)}$ *is bounded and continuous on* $\mathbb{R}$ *for all* $0\leq j\leq r-1$ *and* $f^{(r)}$ *is bounded and uniformly continuous on* $\mathbb{R}$, *that*

$$|E[f(\frac{1}{\sqrt{n}}\sum_{i=1}^{n}x_i)] - E[f(x^*)]|$$

$$= o(n^{(2-r)/2}\{\zeta_{r,1} + E[|x^*|^r]\}) = o(n^{(2-r)/2}) \qquad (n\to\infty) .$$

PROOF. Parts a) and b) are particular cases of Theorem 1 under the given settings, noting that $P_{x^*} = P_{\sum_{i=1}^{n} n^{-1/2} y_i}$ where $(y_i)_{i\in\mathbb{N}}$ is a sequence of standard normally distributed r.vs.. Part c) follows from Theorem 2.

Note that parts a) and b) are large - O estimates, part c) is a little - o rate of approximation. These estimates are as sharp as those deduced directly for the CLT in [6].

The authors would like to thank an anonymous referee in connection with a paper of their collaborator [19] for raising the question whether the operator - theoretical methods developed at Aachen to study the limit theorems of probability could also give a new approach to invariance principles.

The research of the second named author was supported by DFG grant Bu 166/37 - 3.

REFERENCES

[1] Arak, T.V., On an estimate of A.A. Borovkov. Theor. Probability Appl. 20(1975), 372 - 373.

[2] Badrikian, A., Séminaire sur les Fonctions Aléatoires Linéaires et les Mesures Cylindriques. Lecture Notes in Math., Vol. 139, Berlin 1970.

[3] Billingsley, P., Convergence of Probability Measures. Wiley, New York 1968.

[4] Borovkov, A.A., On the rate of convergence for the invariance principle. Theor. Probability Appl. 18(1973), 207 - 225.

[5] Borovkov, A.A., Rate of convergence and large deviations in invariance principle. Proceedings of the International Congress of Mathematicians (Helsinki 1978, O. Lehto, Ed.), 725 -731, Acad. Sci. Fennica, Helsinki 1980.

[6] Butzer, P.L. - Hahn, L., General theorems on the rate of convergence in distribution of random variables. II: Applications to the stable limit laws and weak law of large numbers. J. Multivariate Anal. 8 (1978), 202 - 221.

[7] Butzer, P.L. - Hahn, L. - Roeckerath, M., Central limit theorem with large-O rates for martingales in Banach spaces. J. Multivariate Anal. (1983).

[8] Donsker, M., An invariance principle for certain probability limit theorems. Mem. Amer. Math. Soc. 6(1951), 1 - 12.

[9] Dudley, R.M., Speeds of metric probability convergence. Z. Wahrscheinlichkeitstheorie und Verw. Gebiete 22(1972), 323 - 332.

[10] Freedman, D., Brownian Motion and Diffusion. Holden-Day, San Francisco 1971.

[11] Gänssler, P. - Stute, W., Wahrscheinlichkeitstheorie. Springer, Berlin 1977.

[12] Giné, E., Some remarks on the central limit theorem in C(S). Probability in Banach spaces. (Proc. First Internat. Conf. Oberwolfach 1975, A. Beck, Ed.) pp. 101 - 106. Lecture Notes in Math., Vol. 526, Berlin 1976.

[13] Komlós, J. - Major, P. - Tusnády, G., An approximation of partial sums of independent RV'-s and the sample DF. Z. Wahrscheinlichkeitstheorie und Verw. Gebiete I: 32 (1975), 111 - 131, II: 34(1976), 33 - 58.

[14] Loève, M., Probability Theory II. New York 1978. 4 th edition.

[15] Major, P., On the invariance principle for sums of independent identically distributed random variables. J. Multivariate Anal. 8(1978), 487 - 517.

[16] Paulauskas, V.I., On the rate of convergence in the central limit theorem in certain Banach spaces. Theor. Probability Appl. 21(1976), 754 - 769.

[17] Prohorov, Ju.V., Convergence of random processes and limit theorems in probability theory. Theor. Probability Appl. 1(1956), 157 - 314.

[18] Roeckerath, M.Th., Der Zentrale Grenzwertsatz und das Schwache Gesetz der Großen Zahlen mit Konvergenzraten für Martingale in Banachräumen. Doctoral Dissertation, RWTH Aachen 1980.

[19] Roeckerath, M.Th., On the o-closeness of two weighted sums of Banach space valued martingales with application. (Proc. Conference Oberwolfach 1980, P.L. Butzer, B.Sz.-Nagy, E. Görlich, Eds.) pp. 395 - 405. ISNM Vol. 60, Basel 1981.

[20] Rosenkrantz, W.A., On rates of convergence for the invariance principle. Trans. Amer. Math. Soc. 129(1967), 542 - 552.

[21] Sahanenko, A.I., Estimates of the rate of convergence in the invariance principle. Soviet Math. Dokl. 15(1974), 1752 - 1755.

[22] Skorohod, A.K., A limit theorem for independent random variables. Soviet Math. Dokl. 1(1960), 810-811.

[23] Taylor, R.L., Stochastic convergence of weighted sums of random elements in linear spaces. Lecture Notes in Math., Vol. 672 (A. Dold, B. Eckmann Eds.), Berlin 1978.

[24] Verwaat, W., Une compactification des espaces fonctionnels C et D; une alternative pour la demonstration de théorémes limits fonctionnels. C.R. Acad. Sci. Paris Sér. I Math. 292(1981), 441 - 444.

REMARKS. 1) Note that it is not as yet quite clear whether the weak convergence $P_n \to_w P$ for probability measures P_n and P on C[0,1] is equivalent to $\int f dP_n \to \int f dP$ for $f \in C_C^r$, $r \in \mathbb{N}$. However, this result with C[0,1] replaced by an arbitrary Hilbert space is known to be valid; see Giné, E. - León, J.R., On the CLT in Hilbert space. In: Stochastica 4 (1980), 43 - 71. According to written communications by Professors W. Stute (Siegen) and E. Giné (Texas A & M University), it would suffice to check whether C_C^r is a "convergence determining" class.

2) (Added during publication) Professor E. Giné kindly informed the authors of a fourth proof for the invariance principle (without rates). See Giné, E. - Marcus, M., On the CLT in C(K). In: Aspects statistiques et aspects physiques des processus Gaussiens, 361 - 383, Colloques Internat. CNRS, v. 307, CNRS, Paris 1981.

International Series of
Numerical Mathematics, Vol. 65

PROOF OF A CONJECTURE BY L.A. SANTALÓ

Miguel de Guzmán
Universidad Autónoma de Madrid
Madrid, Spain

This paper presents the proof of a certain conjecture Santaló has raised in 1982 connected with some results in integral geometry. By means of this theorem one can now extend some interesting results of Bokowski |1| and of Santaló. A different proof of Santaló's conjecture has been recently obtained by M.Morales.

1. Introduction

In 1980 Bokowski |1| has proved an inequality concerning the content of the surface separating a convex body in two parts and the volume of these two parts. Santaló, in a paper to be published in the Rendiconti del Circolo Matematico de Palermo has obtained easier proofs and some extensions of the results of Bokowski. His method consists in introducing an appropriate measure and in estimating a certain integral resulting from it. In order to extend further his results in the easiest, two-dimensional case, he has raised the following conjecture, whose proof we here present.

2. The Conjecture and its Proof

THEOREM (Santaló's conjecture). Let $f:[0,\infty)\to[0,\infty)$ be a non-decreasing function. We consider the integral

$$I_f(N,B) = \iint \chi_N(x)\ \chi_B(y)\ f(|x-y|)dxdy$$

where χ_N is the characteristic function of N, a finite union of subintervals of $[0,1]$, and χ_B is the characteristic function of $B = [0,1] - N$. Then $I_f(N,B)$ reaches its maximum value for a fixed f when $N = [0,1/2]$.

PROOF. The proof will be a simple consequence of the particular case in which $f(x) = \chi_{[h,\infty)}(x)$, where $0 \leq h$. The case $h=0$ is easy. If the measure of N is n and that of B is $b=1-n$ then

$$\iint \chi_N(x)\ \chi_B(y)dxdy = bn = n(1-n)$$

and this has its maximum value, 1/4, for $n=1/2$. We therefore assume that $f(x) = \chi_{[h,\infty)}(x)$ with $0<h$. Of course we can assume also that $h \leq 1$, since if x and y are both in $[0,1]$, we always have $|x-y| \leq 1$.

The integral $I_f(N,B)$ with $f=\chi_{[h,\infty)}$, $0<h\leq 1$, admits the following twodimensional interpretation. In the square $[0,1]\times[0,1]$ we make the partition into rectangles indicated in Figure 1. The integral $I_f(N,B)$ has then the value of the shaded area situated above the line $y=x+h$ or below the line $y=x-h$.

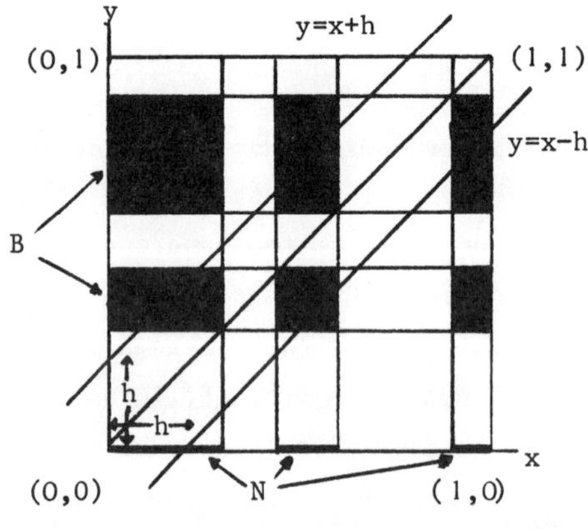

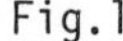

Fig.1

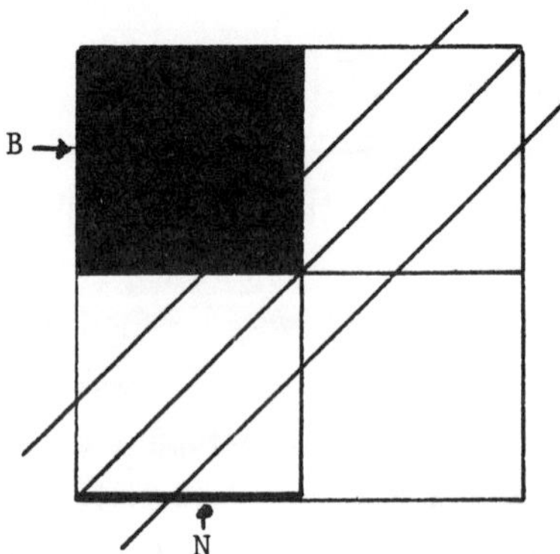

Fig. 2

We are trying to prove that this integral has its maximum value in the case of Figure 2.

For easiness we can symmetrize the geometric problem, shading the square symmetrically with respect to the diagonal $y=x$ as in Figure 3 below. We want to prove that the area $A(N,B)$, shaded portion above the line $y=x+h$ or below the line $y=x-h$, is in Figure 3a less than or equal to the corresponding one in Figure 3b.

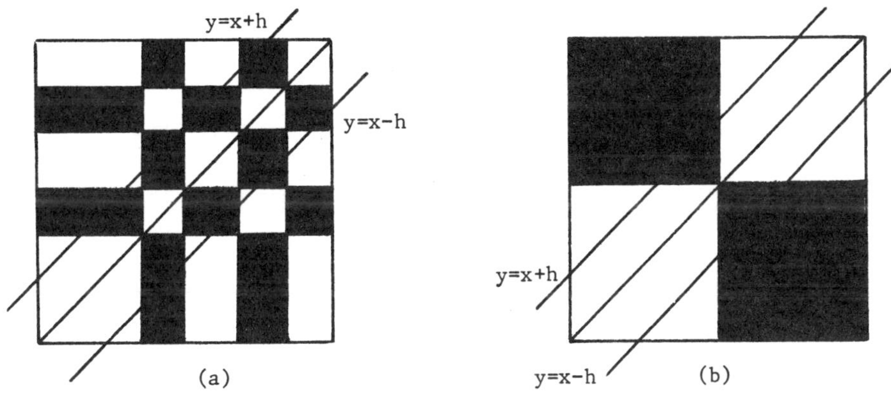

Fig. 3

Clearly we can assume $0<h<1/2$, since for $h \geq 1/2$ the statement of the theorem for this case is obvious.

Let us consider an area $A(N,B)$. We shall say that one of the vertical separating lines different from $x=0$, $x=1$, is <u>stationary</u> when any infinitesimal shift of this line produces an N',B' and a corresponding area $A(N',B')$ such that $A(N',B') \leq A(N,B)$. It is easy to see that if a vertical separating line different from $x=0, x=1$, is <u>stationary</u>, then the length of it bordering shaded area of $A(N,B)$ on its left hand side is equal to the length of it bordering shaded area of $A(N,B)$ on its right hand side.

We shall say that a <u>distribution</u> N,B <u>is stationary</u> when all their interior vertical separating lines are stationary. For example N^*,B^* with $N^*=[0,1/2]$ is stationary.

It is easy to show that if we prove that for each stationary distribution N,B we have $A(N,B) \leq A(N^*,B^*)$ with $N^*= 0,1/2$, we have proved the statement of the theorem in our case $f=\chi_{[h,\infty)}$. Therefore it will be sufficient to prove that $A(N,B) \leq A(N^*,B^*)$ for each N,B stationary. The area of $A(N^*,B^*)$ is easily computed. It is $1/2-h^2$.

In order to estimate $A(N,B)$ with N,B stationary we sweep the area by vertical lines. For each $a \in (0,1)$, such that $x=a$ is not an interior separating vertical line, $g(a)$ will be the length of $\{x=a\} \cap A(N,B)$. For $a=0$, a=1, or a such that x=a is an interior separating line we define g(a) so that y=g(x) is continuous on [0,1]. It is clear that one can do it due to the above characterization of stationary lines.

It is easy to show that the function y=g(x) is piecewise differentiable and its derivative takes the values +1 or -1. It is nonincreasing on [0,h] and nondecreasing on [1-h,1].

For an a, $0<a<1$, such that x=a is a separating line, we have

$$g(a)=\begin{cases} \dfrac{1-h-a}{2}, & \text{if } 0<a\leq h \\[2ex] \dfrac{1-2h}{2}=\dfrac{1}{2}-h, & \text{if } h\leq a\leq 1-h \\[2ex] \dfrac{1-h-(1-a)}{2}=\dfrac{a-h}{2}, & \text{if } 1-h\leq a<1 \end{cases}$$

Let us now consider separately the two possible cases:

(a) There is some separating line x=a with $a\in[h,1-h]$.

(b) There is no separating line x=a with $a\in[h,1-h]$.

Case (a). If x=a is a separating line with $a\in[h,1-h]$, then $g(a)=1/2-h$. If we show that $g(x)\leq 1/2$ for $x\in[0,1]$, then, taking into account the possible values for the derivative of g, we clearly have

$$A(N,B)=\int_0^1 g(x)dx \leq 1/2-h^2=A(N^*,B^*)$$

In order to prove that $g(x)\leq 1/2$ we first observe that in the construction of A(N,B) the squares that have y=x as diagonal are not shaded (see Figure 4).

Let the separating vertical lines between x=h and x=1-h be $x=a_1$, $x=a_2,\ldots,x=a_n$, with $h\leq a_1<a_2<\ldots<a_n\leq 1-h$. From the previous observation we deduce that if $a_{j+1}-a_j\geq 2h$, then the curve y=g(x) cannot go above the line indicated in Figure 5 between a_j and a_{j+1}.

In the same way, if $a_{j+1}-a_j<2h$, then y=g(x) cannot go above the line indicated in Figure 6.

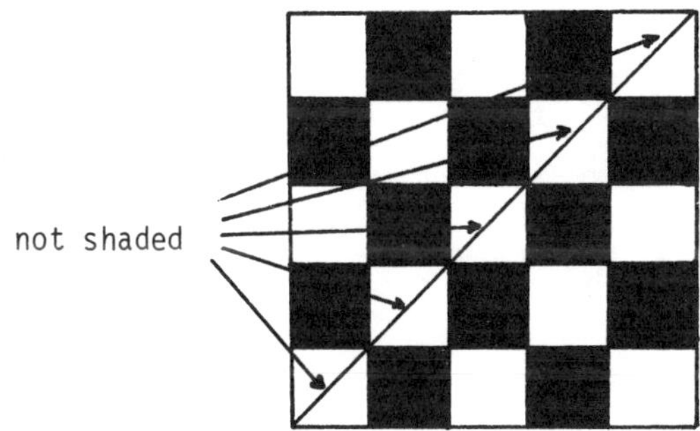

Fig.4

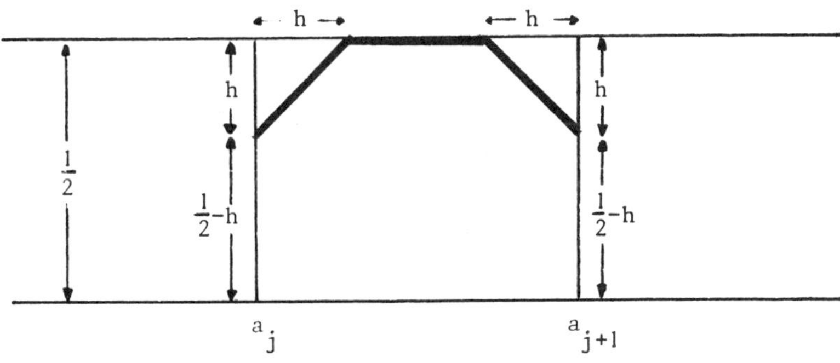

Fig.5

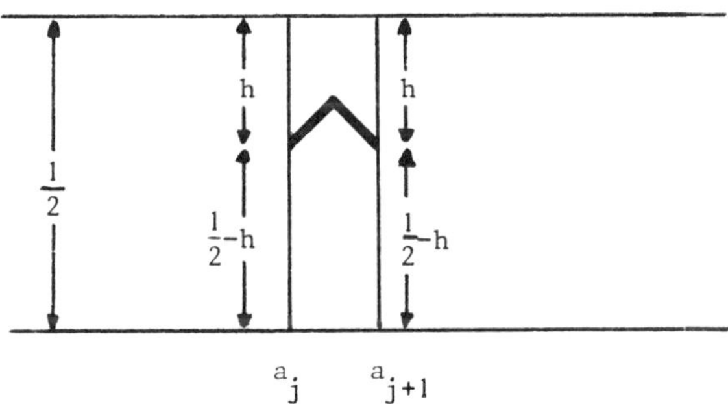

Fig.6

Therefore, for $a_1 \leq x \leq a_n$, we have $g(x) \leq 1/2$.

If to the left of a_1 there are some more interior separating lines, they will be of the form x=a, with 0<a<h. Let b be the largest of these values of a. We have g(b)=1/2-h/2-b/2. The function g is nonincreasing at the left of b and so

$$g(0) \leq \frac{1-h-b}{2} + b = \frac{1-h+b}{2} \leq 1/2.$$

Similar considerations are valid if there are no interior separating lines to the left af a_1 and also for the separating lines to the right of a_n.

Case (b). There is no separating line in [h,1-h].

The function y=g(x) can be obtained as sum of two functions, $g(x)=g_1(x)+g_2(x)$, in the following way. If x=a is not a separating line and $a \neq 0$, $a \neq 1$, $g_1(a)$ is the length of the shaded portion over x=a located above the line y=x+h and likewise $g_2(a)$ is the length of the shaded portion over x=a located below the line y=x-h. For any other a, $g_1(a)$ and $g_2(a)$ are defined so that g_1 and g_2 are continuous. The two functions so defined present the following features, which are easily established:

(i) The curve $y=g_1(x)$ is, in this case (b), piecewise differentiable with a derivative whose values are only 0 and -1. We have $g_1(x)=0$ for $1-h \leq x \leq 1$.

(ii) Likewise the curve $y+g_2(x)$ is, in this case (b), piecewise differentiable with a derivative whose values are only 0 and +1. We have $g_2(x)$ for $0 \leq x \leq h$.

(iii) We also have that

if $g_1'(x)=0$, $0 \leq x \leq 1-h$, then $g_2'(x+h)=0$, and

if $g_1'(x)=0$, $0 \leq x \leq 1-h$, then $g_2'(x+h)=+1$.

This property (iii) is readily shown by using the symmetry of the board with respect to y=x. From these three properties we easily deduce that $g_1(x)+g_2(x+h)=\text{constant}=C$ for $0 \leq x \leq 1-h$.

As in the case (a) one can show that $g(x) \leq 1/2$. For example, if x=b with $0<b \leq h$ is a separating line, then g(b)=(1-h-b)/2 and so we get as before $g(0)= (1-h+b)/2 \leq 1/2$. Hence $g(0)=g_1(0)+g_2(0)=g_1(0) \leq 1/2$, and so $C= g_1(0)+g_2(h) \leq 1/2$.

Therefore in this case we have:

$$A(N,B)=\int_0^1 g(x)dx = \int_0^1 g_1(x)dx + \int_0^1 g_2(x)dx = \int_0^{1-h} g_1(x)dx + \int_h^1 g_2(x)dx=$$
$$\int_0^{1-h}[g_1(x)+g_2(x+h)]dx \leq 1/2(1-h)$$

Since $h<1/2$ we get $A(N,B)\leq \frac{1}{2}-h^2=A(N^*,B^*)$.

This concludes the proof of the theorem for the case in which $f=\chi_{[h,\infty)}$, $0\leq h<\infty$.

Now, if f is of the form

$c_1\chi_{[h_1,\infty)}+c_2\chi_{[h_2,\infty)}+\ldots+c_k\chi_{[h_k,\infty)}$ with $c_j>0$, and h_j increasing, the theorem is just an easy consequence of the previous case.

Let finally f be a positive nondecreasing function. The function f can then be approximated from below by a nondecreasing step function as closely as we like. If there were N,B such that $I_f(N,B)>I_f(N^*,B^*)$, there would be such a step function $\bar{f}$ so that $I_{\bar{f}}(N,B)>I_{\bar{f}}(N^*,B^*)$, and this is impossible. Therefore for each f positive and nondecreasing $I_f(N,B)\leq I_f(N^*,B^*)$.

The theorem is false for a general f. Let $f(x)=\chi_{[h_1,h_2]}(x)$, where h_1,h_2 are as Figure 7 indicates. If we choose N^*,B^*, then $A(N^*,B^*)$ is the black area in Figure 7.

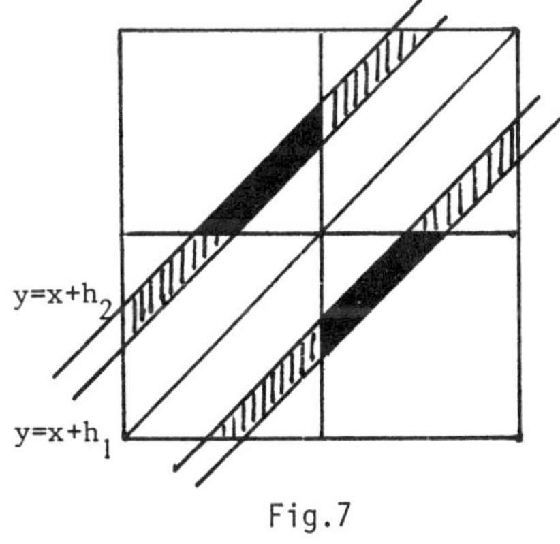

Fig.7

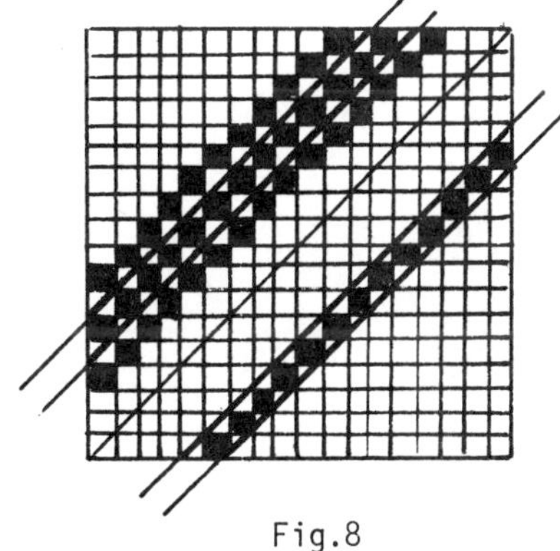
Fig.8

Instead, if we take as N,B the distribution that gives rise to the regular division of the board into very small squares as indicated in Figure 8, it is easy to see that the area of our A(N,B) can be as close as we like to a half of the area determined in the unit square by the strips of equation $h_1 \leq |y-x| \leq h_2$. By appropriately choosing h_1 and h_2 this can be made bigger than A(N*,B*).

REFERENCE

|1| Bokowski, J., Ungleichungen für den Inhalt von Trennflächen, Arch. Math. 34 (1980), 84 - 89.

X New and Unsolved Problems

International Series of
Numerical Mathematics, Vol. 65

New and Unsolved Problems

1. P.L. BUTZER - C. MARKETT: The Poisson Summation Formula for Orthogonal Systems

As the great Bochner (1899 - 1982) [2] conjectured, "the Poisson summation formula (= PSF) and Cauchy's integral and residue formulas are two different aspects of a comprehensive broad-gauged duality formula which lies athwart most of analysis". In this respect it was recently shown (cf. [4] and literature cited there) that the Kotelnikov - Whittaker - Shannon sampling theorem, one of the basic theorems of communication theory, and generalizations thereof are particular cases of the PSF. Even more is true [3] : The PSF is equivalent to the Shannon sampling as well as to the Cauchy integral formula of complex analysis in the sense that each of the three can be deduced from one of the others.

Denoting the Fourier coefficients of $f \in L^1_{2\pi}$ and Fourier transform of $f \in L^1(-\infty,\infty)$ by

$$(1)\qquad \begin{aligned} f^{\wedge}_{FC}(k) &:= \frac{1}{2\pi} \int_{-\pi}^{\pi} f(u)e^{-iku}\,du && (k \in \mathbb{Z}),\\ f^{\wedge}_{F}(v) &:= \frac{1}{\sqrt{2\pi}} \int_{-\infty}^{\infty} f(u)e^{-ivu}\,du && (v \in \mathbb{R}), \end{aligned}$$

respectively, and the classical translation operator by τ_h, i.e., $(\tau_h f)(x) := f(x+h)$, $x,h \in \mathbb{R}$, then for $f \in L^1(-\infty,\infty)$ the classical Poisson summation formula reads

$$(2) \qquad f_F^*(x) := \sqrt{2\pi} \sum_{k=-\infty}^{\infty} \tau_{2k\pi} f(x) \sim \sum_{k=-\infty}^{\infty} f_F^{\wedge}(k) e^{ikx} \qquad (x \in [-\pi,\pi]),$$

which may also be expressed as $[f_F^*]_{FC}^{\wedge}(k) = f_F^{\wedge}(k)$, $k \in \mathbb{Z}$.

The analogue of the PSF (2) for the multidimensional situation with $f \in L^1(\mathbb{R}^n)$ is also well known (cf., e.g., [18; Thm. VII 2.4]).

The restriction of PSF (2) to the case of even functions leads one to a PSF connecting the Fourier coefficients of $f \in L_w^1[-1,1]$, $w(u) = (1-u^2)^{-1/2}$, with respect to the Chebyshev polynomials of first kind, namely

$$(3) \qquad f_{CC}^{\wedge}(k) := \sqrt{\frac{2}{\pi}} \int_{-1}^{1} f(u) \frac{\cos(k \arccos u)}{(1-u^2)^{1/2}}\, du \qquad (k \in \mathbb{P}),$$

together with the cosine transform for $f \in L^1(0,\infty)$,

$$(4) \qquad f_C^{\wedge}(v) := \sqrt{\frac{2}{\pi}} \int_0^{\infty} f(u) \cos vu\, du \qquad (v \in \mathbb{R}^+).$$

In fact, under the appropriate translation operator on $\mathbb{R}^+$,

$$(5) \qquad \tau_h^C f(x) := \frac{1}{2} \{f(|x-h|) + f(x+h)\} \qquad (x,h \in \mathbb{R}^+),$$

the PSF in the Chebyshev case now states that, for $f \in L^1(0,\infty)$,

$$(6) \qquad \begin{aligned} f_C^*(x) &:= \sqrt{2\pi} \{f(\arccos x) + 2 \sum_{k=1}^{\infty} \tau_{2k\pi}^C f(\arccos x)\} \\ &\sim f_C^{\wedge}(0) + 2 \sum_{k=1}^{\infty} f_C^{\wedge}(k) \cos(k \arccos x) \qquad (x \in [-1,1]). \end{aligned}$$

This may also be expressed as $[f_C^*]_{CC}^{\wedge}(k) = f_C^{\wedge}(k)$, $k \in \mathbb{P}$. Note that $\int_{-1}^{1} |f_C^*(x)| (1-x^2)^{-1/2}\, dx \leq \int_0^{\infty} |f(\theta)|\, d\theta < \infty$.

PROBLEM. The question is to establish (one-dimensional) counterparts of the PSF which somehow connect the Fourier coefficients of an orthogonal s e r i e s on the one hand with an orthogonal t r a n s f o r m on the other. The PSF (6) suggests that such a counterpart is valid for the Fourier-Jacobi series [11] (with parameters $\alpha \geqslant \beta \geqslant -1/2$) on the left side and the Jacobi transform [10] on the right; formula (6) would then be the particular case $\alpha = \beta = -1/2$. In the Legendre case $\alpha = \beta = 0$, possible candidates for stating the PSF are the Fourier-Legendre coefficients of $f \in L^1(-1,1)$,

$$f^{\wedge}_{LC}(k) := \sqrt{k+1/2} \int_{-1}^{1} f(u)\, R_k^{0,0}(u)\, du \qquad (k \in \mathbb{P}),$$

$R_k^{0,0}(u) = {}_2F_1(-k, k+1; 1; \frac{1-u}{2})$ being the Legendre polynomials, the associated Legendre transform for $f \in L^1_w(0,\infty)$, $w(u) = \sinh 2u$, namely

$$f^{\wedge}_{L}(v) = \sqrt{\frac{2}{\pi}} \int_0^{\infty} f(u)\, \varphi_v(u) \sinh 2u\, du \qquad (v \in \mathbb{R}^+),$$

$\varphi_v^{0,0}(u) = R^{0,0}_{(iv-1)/2}(\cosh 2u)$ being the Legendre functions on $\mathbb{R}^+$, as well as the corresponding translation operator on $\mathbb{R}^+$,

$$\tau_h^L f(x) := \frac{1}{\pi} \int_0^{\pi} f(\tfrac{1}{2}\operatorname{arccosh}[\cosh 2x \cosh 2h + \sinh 2x \sinh 2h \cos\theta])\, d\theta \qquad (x,h \in \mathbb{R}^+)$$

$$= \frac{1}{\pi} \int_{|x-h|}^{x+h} f(u) \frac{\sinh 2u\, du}{([\cosh^2(x+h) - \cosh^2 u][\cosh^2 u - \cosh^2(x-h)])^{1/2}} .$$

The latter is the kernel-form representation (cf. also [15]). This translation operator, which is adapted to the Legendre transform, is not the more usual one (see e.g. [5]), namely

$$\bar{\tau}_h^L f(x) := \frac{1}{\pi} \int_0^{\pi} f(xh + \sqrt{1-x^2}\sqrt{1-h^2}\cos\theta)\, d\theta, \qquad (x,h \in [-1,1]),$$

which is adapted to the Legendre series and maps $L^1[-1,1]$ into itself. One possibility to handle this problem would be to put it into a group theoretical context. In fact, for certain discrete values of α and β the Jacobi polynomials and the Jacobi functions have interpretations as spherical functions on compact and non-compact symmetric spaces of rank one, respectively. For details see, e.g., [13],[17],[7] and the literature cited there. In her survey report A. Terras [19] shows that the PSF is a particular case of a trace formula. As a matter of fact, the Selberg trace formula (of 1956) is actually the analogue of the two-dimensional PSF (2) for the hyperbolic upper half plane H modulo SL $(2,\mathbb{Z})$. This would correspond to the Legendre case above. However, her formula (2.40) does not involve Legendre series expansions and is so complex that it may be difficult to generalize it to Jacobi functions $\varphi_v^{\alpha,\beta}$ for general α,β. Therefore it unfortunately does not give a final answer to our problem.

A further extension of the PSF would be the restriction of the n-dimensional PSF to the instance of radial functions. The n-dimensional Fourier transform is then known to reduce to the Hankel transform with parameter $(n-2)/2$ (cf., e.g., [18; Thm. IV 3.3]). PSF (6) is also the particular case $n=1$ of this instance.

In this respect let us look at the PSF in the frame of Walsh analysis. Denoting the Fourier-Walsh coefficients of $f\in L^1(0,1)$ and corresponding transform of $f\in L^1(0,\infty)$ by

$$f^{\wedge}_{wC}(k) = \int_0^1 f(u)\,\psi_k(u)\,du \quad (k\in\mathbb{P}), \quad f^{\wedge}_{w}(v) := \int_0^\infty f(u)\,\psi_v(u)\,du \quad (v\in\mathbb{R}^+),$$

respectively, $\{\psi_k\}_{k\in\mathbb{P}}$ denoting the Walsh functions (in Paley's enumeration), and $\{\psi_v\}_{v\in\mathbb{R}^+}$ Fine's generalized Walsh functions, then the present PSF reads [8]

$$(7)\qquad f^*_w(x) := \frac{1}{2}f(0) + \sum_{k=1}^{\infty}\tau_k^C f(x) \sim \sum_{k=0}^{\infty} f^{\wedge}_w(k)\,\psi_k(x) \qquad (x\in[0,1]),$$

where τ_k^C stands for the translation operator of (5). Note that x+h is not dyadic but the usual addition.

Basic in the known cases (2), (6), and (7) is a periodicity concept, enabling expansions of the f*-functions as Fourier series. Whereas these functions in (2) and (7) have period 2π and 1, respectively, regarding (6) the substitutions $x = \cos(2\pi j + \theta)$, $j \in \mathbb{P}$, $x = \cos(2\pi j - \theta)$, $j \in \mathbb{N}$, for $\theta \in [0,\pi]$ lead to the same expression in θ, independent of j. A basic problem in the general case would therefore be to find substitutes for periodicity.

It may be of interest to mention the general frame within which the unsolved problems lie. The PSF was first used to any extent (by Dirichlet) in analytic number theory (quadratic reciprocity law, Jacobi inversion formula, functional equation for zeta function, etc.) and is being gradually "identified" with it. Thus E. Hecke (1917/28) extended it to modular forms and A. Selberg to the frame of group representations in a generalized Maass theory of automorphic forms (Selberg trace formula). A very broad extension of the PSF is that in the setting of adèle groups of algebraic number theory due to K. Iwasawa and J.F. Tate (1950/67). In this respect see G.W. Mackey [14, pp. 86, 109, 173, 183, 214, 222].

Bochner [1] felt that the comprehensive duality formula embracing the above extensions he had in mind is also connected with the famous Riemann - Roch theorem on Abelian integrals. Now the latter is in turn known to be a particular case of the Atiyah - Singer index theorem (1963/65) for elliptic differential operators [16], [12]. In classical physics the Bochner formula would be the duality between an electromagnetic field on the one and electric and magnetic "discrete" charges on the other hand.

Since the (classical) sampling theorem is essentially equivalent to the classical PSF, it is to be expected that the PSF for the more general orthogonal systems posed above will somehow be connected with generalized versions of the sampling theorem. For first results in this direction for the Legendre case see e.g. [6].

The authors would like to thank Tom Koornwinder, Amsterdam, for his critical reading of the problem, and for pointing out the fine paper by Terras [19].

[1] Bochner, S., Mathematical reflections. Amer. Math. Monthly, 81 (1974), 827 - 852.

[2] Bochner, S., The emergence of analysis in the Renaissance and after. In: History of Analysis (Eds. R.J. Stanton and R.O. Wells Jr.) Rice University Studies Vol. 64 Nos. 2 and 3, 1978, pp. 11 - 56.

[3] Butzer, P.L. - Ries, S. - Stens, R.L., Shannon's sampling theorem, Cauchy's integral formula, and related results. These Proceedings.

[4] Butzer, P.L. - Stens, R.L., The Poisson summation formula, Whittaker's cardinal series and approximate integration. In: Proc. Second Edmonton Conference on Approximation Theory (Ed. Z. Ditzian et al.) CMS Conference Proc. Vol. 3, Amer. Math. Soc., Providence, R.I., 1983, pp. 19 - 36.

[5] Butzer, P.L. - Stens, R.L. - Wehrens, M., Approximation by algebraic convolution integrals. In: Approximation Theory and Functional Analysis (J.B. Prolla, ed.). Mathematical Studies 35, North-Holland, Amsterdam 1979, pp. 71 - 120.

[6] Butzer, P.L. - Stens, R.L. - Wehrens, M., The continuous Legendre transform, its inverse transform, and applications. Internat. J. Math. & Math. Sci. 3 (1980), 47 - 67.

[7] Dreseler, B., Zu Entwicklungen nach sphärischen Funktionen gehörende Approximationsverfahren auf kompakten symmetrischen Mannigfaltigkeiten. Habilitationsschrift, GHS Siegen, 1976, 101 pp.

[8] Engels, W., Written communication.

[9] Fine, N.J., The generalized Walsh functions. Trans. Amer. Math. Soc. 69 (1950), 66 - 77.

[10] Flensted - Jensen, M. - Koornwinder, T.H., The convolution structure for Jacobi function expansions. Ark. Math. 11 (1973), 245 - 262.

[11] Gasper, G., Positivity and the convolution structure for Jacobi series. Ann. of Math. 93 (1971), 112 - 118.

[12] Hirzebruch, F., Elliptische Differentialoperatoren auf Mannigfaltigkeiten. In: Festschrift zur Gedächtnisfeier für Karl Weierstrass 1815-1965 (Eds. H.Behnke&K. Kopfermann) Westdeutscher Verlag, Köln/Opladen 1966, pp. 583-608.

[13] Koornwinder, T.H., Jacobi Polynomials and their Two-variable Analogues. Doctoral Thesis, University of Amsterdam, 1974.

[14] Mackey, G.W., Harmonic analysis as the exploitation of symmetry-a historical survey. In: History of Analysis, ibidem pp. 73 - 228.

[15] Markett, C., Product formulas for Bessel, Whittaker, and Jacobi functions via the solution of an associated Cauchy problem. These Proceedings.

[16] Palais, R.S., Seminar on the Atiyah-Singer Index Theorem, with Contributions by A. Borel, F.E. Browder, and R. Soloway. Annals of Math. Studies 57, Princeton Univ. Press, Princeton, N.J., 1965.

[17] Schempp, W.- Dreseler, B., Einführung in die harmonische Analyse. Teubner, Stuttgart, 1980.

[18] Stein, E.M. - Weiss, G., Intorduction to Fourier Analysis on Euclidean Spaces. Princeton Univ. Press, Princeton, N.J. 1971.

[19] Terras, A., Noneuclidean harmonic analysis. SIAM Rev. 24 (1982), 159-193.

2. D. GAŞPAR: Translation Invariant Spaces

Let M be the space of all locally summable functions f on $(0,\infty)$ for which $\sup_{t\in(0,\infty)}\int_t^{t+1}|f(x)|\,dx = \|f\|_M$ is finite. Let T denote the associate space M with respect to L^1, i.e., $g\in T$ iff

$$(*)\qquad \sup\{|\int_0^\infty f(x)g(x)\,dx| : \|f\|_M\leq 1\}$$

is finite. It is proved in the book of Massera - Schaeffer (Linear Differnetial Equations and Function Spaces, Acad. Press. New-York, 1966) that T (endowed with the norm (*) and M are the smallest and the largest translation invariant Banach spaces, respectively. It is easy to see that $T\subset L^1\cap L^\infty$, $L^1+L^\infty\subsetneq M$, with continuous embeddings, and that each interpolation space for the Banach couple (T,M) is translation invariant. It is well known that all interpolation spaces for the couple (L^1,L^∞) are rearrangement invariant, and conversely.

QUESTION. Are all translation invariant spaces on $(0,\infty)$ interpolation spaces for the Banach couple (T,M) ? If not, how can the translation invariant spaces be characterized?

3. M. v.GOLITSCHEK: Approximation by Nomographic Functions

In Section 2.4 of my talk: "Shortest path algorithms for the approximation by nomographic functions", (These Proceedings), I have posed the following problem: Characterize the compact domains $D\subseteq S\times T\subseteq \mathbb{R}^2$ for which the infimum

$$\inf \{\sup_{(s,t)\in D} |f(s,t) - x(s) - y(t)| : x \in \ell_\infty(S),\ y \in \ell_\infty(T)\}$$

vanishes for every continuous function $f \in C(D)$.

4. K. GUSTAVSON: Remarks (1983) on Three New and Unsolved Problems (1971) in Operator Theory: (1) Principal Axis (2) Numerical Range (3) Numerical Radius

Three problems, contributed here, twelve years ago, will be briefly brought to date. One, solved, has opened wider implications. The other two problems remain, apparently, unresolved. The first goes under the names: Infinite principal axis theorem, Weyl-von Neumann, Berg-Sikonia, BDF, and the like. A general description would be the task of determining spectral invariants sufficient to classify the difference of two operators, up to unitary equivalence, as compact. The second problem may be called a Toeplitz-Hausdorff theorem for Banach space. Although convexity fails for the numerical range in passing from Hilbert space to more general contexts, topological properties remain. The third problem is fundamental to a theory of numerical range for unbounded operators in Banach space.

1. Principal Axis

1.1 Problem 1 [10]. A bounded normal operator A on a separable complex Hilbert space can be written in the form $A = D + K$, D diagonal, K compact, and in the self-adjoint case K can be taken to be Hilbert-Schmidt. What can be said about K in the normal case? Reference: Weyl's theorems, these proceedings {i.e., [9]}.

1.2 Remarks. The situation in 1971 was based on the result:

THEOREM (Berg [2], Sikonia [15]). *Every normal operator* A *can be written as the sum of a diagonal operator* D *and a compact operator* K.

Berg [2] and Sikonia [15] obtained this theorem independently, thereby extending the result known for selfadjoint operators by Weyl and Von Neumann to normal operators. Sikonia [15], who was my student and who has not received much mention, constructed a special Haar basis in terms of the spectral representation of A. Berg [2] constructed a special orthonormal basis using the resolution of the identity of A. Neither could answer whether K could be taken from the Hilbert-Schmidt class. Both we, and Berg, had intuitive arguments that it could not be so taken. But:

THEOREM (Voiculescu [16]). <u>Every normal operator</u> A <u>can be written as</u> D + K, K <u>Hilbert-Schmidt</u>.

Voiculescu's approach, is less direct than those of Berg and Sikonia, and transfers the problem to a more general question of the existence of quasicentral approximate units satisfying an almost-cummutation property in the norm of any ideal being considered. This norm-ideal approach also clarifies the suspected fact that one cannot expect to generally obtain K in the class C_p, $p < 2$.

Thus Problem 1. of [10] has been answered.

<u>1.3 Wider Implications</u>. The wider implications of Problem 1 flow, at least in the view we shall present here, in two directions: generalization and specialization.

Let us turn to generalization first.

THEOREM (Brown, Douglas, Fillmore [6]). <u>For</u> A <u>and</u> B <u>essentially normal operators with identical essential spectra and Fredholm indices, there exists a unitary operator</u> U <u>such that</u> A = U*BU + K, K <u>compact</u>.

This result was recently popularized in an excellent exposition by P. Halmos [14]. Essentially normal operators T are those for which T*T - TT* is compact. The discussion in [9] will show you how the obtaining of the representations A = U*BU + K and A = D + K are, with a little help, equivalent questions. The BDF result thus generalizes Problem 1 to essentially normal operators. To do so, it adds the condition that ind $(A-\lambda)$ - ind $(B-\lambda)$ for each λ not in the (common) essential spectrum, a condition automatically satisfied by selfadjoint and normal operators.

In [9] I gave an example showing that these results do not extend to A diagonal normal and B quasinormal: $B(B^*B) = (B^*B)B$. Sikonia [15] had given an example showing that they will not generally obtain for A and B quasinormal. Halmos [14] gives a similar example with A unitary and B quasinormal. These examples all depend on various violations of the Fredholm index theory.

Thus, in my view, a main contribution of the BDF result is the clarification of the sufficiency (beyond the necessity) of the index condition, at least for essentially normal operators.

Turning now to specialization, I shall state the:

SINGULAR SEQUENCE PROBLEM (see [12,13]). Given two selfadjoint operators A and B with identical essential spectra and singular sequences, is $A - B = K$, K compact?

A singular sequence $\{\phi_n\}$ for a given selfadjoint operator A at a λ in the essential spectrum $\sigma_e(A)$ is any sequence such that $\|\phi_n\| = 1$, $\phi_n \overset{w}{\to} 0$, $(A - \lambda)\phi_n \overset{s}{\to} 0$. Points in the essential spectrum $\sigma_e(A)$ are characterized by the existence of singular sequences there. See [9,12,13] for more specifics.

I am not sure of the degree of difficulty of this problem, but thus far it has resisted resolution. One can show [13] that it may be reduced to the case where A has pure point spectrum, $B = U^*AU$ with U unitary, and where $A-B$ is compact on the eigenspaces of A.

A version with A and B normal is immediate. Perhaps one with A and B essentially normal will follow. The singular sequence hypothesis is insufficient for quasinormal operators [12].

The idea of the BDF result is to g e n e r a l i z e the principal axis theorem to a larger class of operators. To do so, one needed to relax the index condition from that of being zero off the essential spectrum (normal operators) to that of just being equal there. One then gets a compact difference, up to unitary equivalence.

The idea of the SSP is to s p e c i a l i z e the principal axis theorem by tightening the essential spectrum condition to include also a condition (same singular sequences) akin to a strong relation between the spectral measures of A and B. The result would be a compact difference, without need of unitary transformation, i.e., so to speak, with the principal axes already lined up.

2. Numerical Range

2.1 Problem 2 [10]. A bounded operator A on a complex Banach space has numerical range $V(A) = \{x'Ax \mid x'x = \|x\| = \|x'\| = 1,\ x' \in X',\ x \in X\}$. Is $V(A)$ simply connected? Reference: F. F Bonsall and J. Duncan, Numerical Ranges of Operators on Normed Spaces and of Elements of Normed Algebras. Cambridge University Press 1971.

2.2 Remarks. The Toeplitz-Hausdorff theorem states that for any operator A in a Hilbert space, the numerical range $W(A)$ is convex, and hence (very) connected. This is easily shown: I like the proof of [8], which shows how it really is a question for real scalars, but there are many other proofs, with their merits. The situation in 1971 for the Banach space generalization $V(A)$ of the numerical range was:

THEOREM (Bonsall, Cain, Schneider [3]). *For A a bounded linear operator on a normed space, $V(A)$ is connected.*

It was known at that time that $V(A)$ need not be convex and that the individual numerical ranges $W(A)$ comprising $V(A)$ need not be connected themselves. Some results for Problem 2, notably the tear drop examples of McGregor and others, may be found in Bonsall and Duncan [4]. To my knowledge the question of whether $V(A)$ allows any holes remains open. Such an occurrence would be quite distinct topologically from the Toeplitz-Hausdorff convexity for Hilbert space.

2.3 Wider Implications. Generally speaking, there seems to be a lot more to be known about both the topology and the geometry of $V(A)$ and its closure, even for bounded operators in concrete spaces. We mention that all such questions then extend to versions for normed algebras. See [4,5].

3. Numerical Radius.

3.1 Problem 3 [10]. An unbounded operator A in a complex Banach space has numerical range $V(A) = \{x'Ax \mid x'x = \|x\| = \|x'\| = 1,\ x' \in X',\ x \in D(A)\}$. Is $V(A)$ unbounded?

3.2 Remarks. This is perhaps the most fundamental numerical range question for a numerical range theory for unbounded operators.

THEOREM (Gustafson, Zwahlen [11]). For A an unbounded linear operator with domain $D(A)$ a dense linear subspace in a Banach space X, $V(A)$ is unbounded when (i) X is a Hilbert space, (ii) $D(A) = X$, (iii) $|\sigma(A)| < \infty$, or (iv) $|V(A^*)| < \infty$.

Many other conditions, beyond those mentioned in [11], can be written down to guarantee $V(A)$ unbounded, but to my knowledge a proof of the theorem for all complex Banach spaces is not available. For real Banach spaces $V(A)$ need not be unbounded for unbounded A, see [11], and it is not fair to let $D(A)$ be a dense nonsubspace. Thus a proof would lie in the linearity and complex scalars. Preliminary investigations indicate that a counterexample, if one exists, might be found in an L^1 space.

We would like to mention that the case $D(A) = X$ is basically uninteresting. By a well-known metatheorem, almost any additional condition (e.g., $V(A)$ bounded) will then render A bounded. For example, in [11] we observed that this follows immediately by either the uniform boundedness principle or by the closed graph theorem. Somewhat later Crabb [7] and others also observed, although by other methods, that $V(A)$ not the whole complex plane means that an everywhere defined A is bounded. For unbounded operator theory the interesting case is usually that with *properly* dense domains.

3.3 Wider Implications. The wider relationships of the semi inner product structures of $V(A)$, the dual algebraic structures of partial inner products, and their common foundations in Galois connections, are investigated in [1].

[1] Antoine, J.P. - Gustafson, K., Partial inner product spaces and semi-inner product spaces. Adv. in Math. 41 (1981), 281 - 300.

[2] Berg, I., An extension of the Weyl-von Neumann theorem to normal operators. Trans. Amer. Math. Soc. 160 (1971), 365 - 371.

[3] Bonsall, F. - Cain, B. - Schneider, H., The numerical range of a continuous mapping of a normed space. Aequations Math 2 (1968), 86 - 93.

[4] Bonsall, F. - Duncan J., Numerical Ranges II. Cambridge University Press, Cambridge (1973).

[5] Bonsall, F. - Duncan, J., Numerical Ranges. Studies in Functional Analysis, MAA Studies in Math. 21, Math. Assoc. Amer. (1980), 1 - 49.

[6] Brown, L.- Douglas, R. - Fillmore, P., Unitary equivalence modulo the compact operators and extensions of C*-algebras. Proc. of a Conference on Operator Theory, Halifac, Nova Scotia, Springer Lecture Notes in Mathematics 345 (1973), 58 - 128.

[7] Crabb, M.J., The numerical range of an unbounded operator. Proc. Amer. Math. Soc. 55 (1976), 95 - 96.

[8] Gustafson, K., A simple proof of the Toeplitz-Hausdorff Theorem for linear operators. Proc. Amer. Math. Soc. 25 (1970), 203 - 204.

[9] Gustafson, K., Weyl's Theorems. Linear Operators and Approximation, Butzer, P.L. - Kahane, J.P. - Nagy, B. Sz., eds. (ISNM, vol. 20) Birkhäuser Verlag, Basel/Stuttgart (1972), 80 - 93.

[10] Gustafson, K., Problems 6. (1.,2.,3.),In:New and Unsolved Problems, (ISNM, vol. 20), Birkhäuser (1972), p. 496.

[11] Gustafson, K. - Zwahlen, B., On operator radii, Acta Sci. Math. 36, (1974), 63 - 68.

[12] Gustafson, K., The singular sequence problem. Spectral Theory Semester 1977, Banach Center Publications 8, Polish Scientific Publishers, Warsaw (1982), 289 - 294.

[13] Gustafson, K., On the singular sequence problem. (reprint).

[14] Halmos, P., BDF or the infinite principal axis theorem. Notices Amer. Math. Soc. 30 (1983), 387 - 391.

[15] Sikonia, W. Dissertation. University of Colorado (1970); Also: The von Neumann converse of Weyl's theorem. Indiana Univ. Math. J. 21 (1971), 121 - 124.

[16] Voiculescu, D., Some results on norm-ideal perturbations of Hilbert space operators. J. of Operator Theory 2 (1979), 3 - 37.

5. W.K. HAYMAN: Best Harmonic Approximants

In the paper "The best harmonic approximant to a continuous function", published in these proceedings, it was shown that under suitable conditions on the domain, such a best approximant always exists; it is unique and can be simply characterised provided that it remains continuous on the closure $\overline{D}$ of D. An example was given, where the best approximant to a function, continuous in $\overline{D}$, is no longer continuous in $\overline{D}$ but is still unique. The problem is

a) whether the best approximant is always unique, even if it is not continuous and

b) whether there is a simple characterization in the discontinuous case.

6. L. KÉRCHY: Quasi-Similarity Invariants

Is the bicommutant property $\{T\}'' = \mathrm{Alg}\,T$ a quasi-similarity invariant in the class of C_{11} contractions on Hilbert space?
Here $\{T\}''$ denotes the bicommutant of T, Alg T is the weakly closed algebra generated by I and T, and T is a C_{11}-contraction if $\lim_n \|T^n h\| \neq 0$, $\lim_n \|T^{*n} h\| \neq 0$ for every $0 \neq h \in H$.

7. T.H. KOORNWINDER: On the Completeness of the Functions $c_k^{(\alpha,\beta)}$ Related to Spherical Harmonics on the Heisenberg Group

For fixed complex α,β the generating function

$$(1 - re^{-i\phi})^{-\alpha}(1 - re^{i\phi})^{-\beta} = \sum_{k=0}^{\infty} r^k c_k^{(\alpha,\beta)}(e^{i\phi})$$

defines a system of functions $c_k^{(\alpha,\beta)}$ on the unit circle with explicit expression

$$c_k^{(\alpha,\beta)}(e^{i\phi}) = \sum_{j=0}^{k} \frac{(\alpha)_{k-j}(\beta)_j}{(k-j)!\,j!}\, e^{i(2j-k)\phi}$$

Special cases are:

$$c_k^{(\alpha,\alpha)}(e^{i\phi}) = C_k^{\alpha}(\cos\phi) \qquad \text{(Gegenbauer polynomial)},$$

$$C_k^{(\alpha,0)}(e^{i\phi}) = \frac{(\alpha)_k}{k!} e^{-ik\phi} \quad , \quad C_k^{(0,\beta)}(e^{i\phi}) = \frac{(\beta)_k}{k!} e^{ik\phi} .$$

For general α,β the functions have an interpretation as "radial" parts of spherical harmonics on the Heisenberg group. See [1] for further background. Now the problem is:

Give a "classical" analytic proof (i.e., without using analysis on the Heisenberg group) that Span $\{C_k^{(\alpha,\beta)}(e^{i\cdot})(\sin\cdot)^{|\alpha-\beta|}\}$ is dense in $(\sin\cdot)^{|\alpha-\beta|}C([0,\pi])$ with respect to the uniform norm if $\alpha-\beta \in \mathbb{Z}$ and $\min(\alpha,\beta) \in \{\frac{1}{2},1,\frac{3}{2},\ldots\}$.

In Corollary 5.3 of [1] this was proved as a consequence of the result by Gaveau and Jerison that the Dirichlet problem for the subelliptic Laplacian L_0 on the Heisenberg ball is solvable. Note that, by Weierstrass approximation theorem, the above density statement is evident if $\alpha=\beta$. If α or $\beta=0$ then Span $\{C_k^{(\alpha,\beta)}(e^{i\cdot})\}$ is dense in $C([0,\pi])$ by Mergelyan's theorem.

[1] Greiner, P.C. - Koornwinder, T.H., Variations on the Heisenberg spherical harmonics, Math. Centrum Amsterdam, Report ZW 186/83, 1983.

8. J. KOREVAAR: Square Roots of the Delta Distribution

Multiplication of distributions is a difficult subject, but for obtaining solutions of the equation $f^2 = \delta$, non-controversial local product definitions suffice. In fact, there exist lacunary Fourier series $f(x) \sim \sum c_n e^{inx}$ such that

$$\sum_k c_k c_{n-k} = 1 \qquad \forall\, n \in \mathbb{Z} .$$

[Cf. G.L. O'Brien and F.W. Steutel, Nederl. Akad. Wetensch. Indag. Math. 43 (1981), 393 - 398]. Starting with such f, B. Hanzon has studied roots of δ; there are no real roots! [Unpublished Master's thesis, Univ. of Amsterdam 1983].

QUESTIONS. Can one obtain solutions f in closed form? Could there exist integrable function solutions?

Ikehara-Wiener Theorem

For many years the standard tool in the proof of the prime number theorem has been the

I-W THEOREM. Let $a_n \geq 0$, let $\sum_1^\infty a_n/n^s$ be convergent for Re $s > 1$ with sum $f(s)$, and suppose that $f(s) - c/(s-1)$ has a continuous extension $g(s)$ to the closed half-plane Re $s \geq 1$. Then

$$\frac{1}{n} s_n = \frac{1}{n} \sum_1^n a_k \to c \qquad (n \to \infty) .$$

[To obtain the prime number theorem, one takes

$$f(s) = -\zeta'(s)/\zeta(s) = \sum_1^\infty \Lambda(n)/n^s$$

and one uses the information that $\zeta(s) \neq 0$ for Re $s \geq 1$].

The normal proof of the I-W Theorem uses some form of Wiener's Tauberian method for Fourier integrals (it could be a distributional form). A few years ago, D.J. Newman found a proof of the prime number theorem which uses no other analytic tool than Cauchy's theorem, combined with some clever estimates and standard information on $\zeta(s)$ [Amer. Math. Monthly 87(1980), 693-696]. Newman's ideas can be used to obtain a weak I-W Theorem which also suffices for the prime number theorem [J. Korevaar, Math. Intelligencer 4 (1982), 108-115]. In the weak form, one imposes the additional hypotheses that s_n/n remains bounded and that $g(s)$ be analytic or at least smooth for Re $s \geq 1$.

QUESTIONS. Can one obtain the complete I-W Theorem by complex analysis? Will relatively simple complex analysis also suffice for other Wiener-type Tauberian theorems?

9. J.J. LODDER: The Existence of Analytic Continuations of Functions Defined as Mellin Transforms or Dirichlet Series

MOTIVATION. In [1] a simple model was constructed by imposing strong sufficient conditions which are obviously not necessary. In order to see how far a symmetrical theory can be taken necessary conditions have to be found.

FORMULATION. Consider the function $g(\lambda)$ defined by

$$g(\lambda) = \int_0^a x^\lambda f(x)\, dx$$

as a local Mellin transform. Wanted are n e s e s s a r y conditions on $f(x)$ such that $g(\lambda)$ exists as an analytic function of λ in a region of the complex λ-plane, and possesses a (not necessarily single valued) analytic continuation to the entire λ-plane with the exception of at most isolated points.

INFORMALLY. Give necessary conditions on a function of a real variable such that its Mellin transform does n o t have natural boundaries.

SUBPROBLEM. The same question for functions

$$g(\lambda) = \sum_n a_n n^\lambda$$

defined by Dirichlet series.

[1] Lodder, J.J., A simple model for a symmetrical theory of generalized functions. Phys. A 116(1983), 45 - 73, 380 - 403.

This problem was communicated by T.H. Koornwinder.

10. P. NEVAI: Three Questions on Orthogonal Polynomials

1) Give asymptotics for the orthogonal polynomials p_n corresponding to the weight function $\exp(-x^4)$ which are valid around the largest zero of p_n.
2) Investigate weighted mean convergence of Fourier series and Lagrange interpolation associated with the above orthogonal polynomials.
3) Let $w = gv$ where $v(x) = \exp(-x^2)$, and let p_n denote the orthogonal polynomials corresponding to w. Assume that g satisfies some reasonable conditions, e.g., g is positive, uniformly continuous and differentiable.

Find out properties of p_n, including asymptotics and distribution of zeros.

11. J. PEETRE: Selected Problems on Invariant Function Spaces

Lagrange (or Laplace?), in later years, is said to have expressed pitty for those who came after him, because all problems were solved. For my part, I have never felt any serious lack of problems. Here follows a selection of five problems numbered -1 through 3.

-1. This is really a query, communicated to me by G. Sparr and originating from an engineering friend of his. What is known about approximation by functions taking only the values 0 and 1 (characteristic functions)? The origin of the problem in communication theory is obvious. Perhaps there is already such a theory? If we take approximation in the uniform norm the question is pretty obvious (?). (See [Ha], notably p. 52, where this is made the point of departure if an inquiry in operator theory.) And even we pass to L_2-norms, because of the essentially local nature of affairs, the issue seems equally silly. So perhaps to get anything of real mathematical interest one has to change the problem,for instance by putting some further restrictions on the approximating functions (for instance, by considering characteristic functions of intervals only).

The rest of the problems are more or less (3) related to the actual subject matter of my talk.Remember that the primary aim of my talk was to advance the thesis that the systematic study of invariant function spaces was a worthwile pursuit - and an almost inexhaustible source for new interesting problems.

0. In my talk I made allusion to the following result (for details see [AF]).

THEOREM. If X is a Möbius invariant Hilbert space of holomorphic functions in the unit disk U, the group acting in X via *uniformly bounded* operators, then X is the Dirichlet space. up to equivalence of norm.

The proof (cf. *infra*) goes via a somewhat *ad hoc* procedure so it is natural to ask whether there is not a general group theoretic principle which would allow us to conclude the desired result directly. The discussion below indicates that this is not the case but, since it involves a problem of Dixmier's [D], open for more than 30 years and still unsolved (up to my knowledge), and apparently not known in wider circles, it is perhaps pertinent to collect here some of the relevant facts.

More precisely, Dixmier's problem is the following: Let G be an arbitrary (topological) group and T a uniformly bounded (continuous) representation of G in a Hilbert space H; uniformly bounded means that $\|T_\varphi\| \leq C$ for all φ, C independent of φ. Is it true that T can be "unitarized" (i.e., does there exist an equivalent metric such that all the operators T become unitary, $\|T_\varphi x\| = \|x\|$ in this new metric)? In [D] (it is a beautiful paper!) Dixmier observed that if G is amenable (i.e., carries an invariant mean; the exact definition can be found in Greenleaf's delightful book [G]), then this is the case and he also conjectured that this property (amenability) is a necessary condition.

The case $G = Z$ (that is, the case of a single operator T with uniformly bounded powers, $\|T^n\| \leq C$) had been treated even earlier by Sz.-Nagy [SzN], and the case of finite, and more generally compact groups remotes to antiquity ("Maschke's theorem"; see any book in algebra).

The proof (one proof) of the theorem alluded to in the beginning also makes use of the fact that the Möbius group, itself not amenable, has a canonical amenable subgroup, essentially the subgroup of "dilatations" (the conformal selfmaps if U fixing a point, usally the point -1, of the boundary $T = \partial U$). On that subgroup we can therefore take the group operators tobe isometries. But then it is seen (from the explicit form of the norm) that we have indeed the Dirichlet norm.

In the same way one can see that the Möbius group has a uniformly bounded representation which is not unitarizable. Indeed, take one of the uniformly bounded representations obtained from the principal series by "analytic continuation", first discussed by Kunze and Stein [KuS]. Again it is seen that if we stipulate the group operators to be unitaries on the subgroup of dilatations it follows that the metric is uniquely determined, and this does it.

(In this connection, it is also interesting to notice that Hulanicki [Hu] (see [G], p. 61) has proved that a locally compact group G is amenable iff every irreducible unitary representation is weakly contained (in a technical sense) in the regular representation (on $L^2(G)$.)

Presumably the same argument works in the other cases too where the analytic continuation has been performed (see e.g. [KuSIV]). This suggests that a Lie group is amenable iff it in some sense does not possess a non-compact s.s. constituent.

Finally, let me mention that, in a quite different vein, that Vasilescu and Zsidó [VZ] have shown that for any group a uniformly bounded representation indeed is unitarizable, *provided*, as an extra condition, one assumes that the group operators T_φ ($\varphi \in G$) generate a *finite* von Neumann algebra, but then we have changed the problem. Besides, it seems that the representations one encounters in praxis never are of this type.

<u>1</u>. Let X be the space of holomorphic functions f in U such that $\{\hat{f}(n)\} \in l_p$, $1 \leq p \leq \infty$; this is clearly a Banach space. Is X Möbius invariant? The answer is trivially yes if $p=2$, by Parseval's formula, and no if $p=1$, by the Beurling-Helson theorem [BH] (cf. [K], p. 86). So probably the answer is no whenever $p \neq 2$. (This should be at least implicit in the literature but I know of no specific reference.) But what happens if we substitute l_p for an arbitrary r.i. sequence space Y? It is natural to conjecture that our space, still denoted by X, is Möbius invariant iff $Y = l_2$ (up to equivalence of norm). A related conjecture (Arazy): If Z is any Möbius invariant space (the isometric case) and $\{z^n\}$ is an unconditional basis for Z, then Z is a Hilbert space, thus the Dirichlet space.

<u>2</u>. This question consists of several parts.

a. What is the interpolation space $(H^\infty, \text{Bloch})_{1/2,2}$? This question is not entirely academic. Namely, the proof of Shield's inequality for the Dirichlet space D (generalizing the classical Fejér-Riesz inequality) as given in my report gives more, the same inequality with D replaced by this interpolation space,D', say, thus of the form

$$\int |f(z)|^2 \, d\nu(z) \leq C^2 \|f\|_{D'}^2 , \quad f \in D', \quad f(0) = 0.$$

b. To interpolate between H^∞ and BMO. (By BMOA I mean here of course

BMOA (analytic functions in real BMO).) Let us notice that in the real case (interpolation between L^∞ and real BMO) the issue is essentially settled by the paper by Garnett and Jones [GJ], although no explicit mention of interpolation can be found there. In particular, the K-functional can be determined (Janson and others (unpublished, as far as I know)). The analytic case probably requires a quite new approach.

c. Similarly, to interpolate between BMO and Bloch. About this I know next to nothing.

The reason for joining here these three cases is that the three spaces H^∞, BMO and Bloch are in some sense close to each other. Also they form a chain: $H^\infty \subset \mathrm{BMO} \subset \mathrm{Bloch}$. Again what comes to ones mind is the question whether not BMO is an interpolation space between H^∞ and Bloch. This in turn is a special case of a general question, apparently open already for several years. Is BMO (real or analytic case, it does not matter) an interpolation space between some natural spaces. Some results (but only negative ones) are known. For instance, Zafran (see [JNP], appendix) has shown that BMO cannot be an interpolation space between L_p $(1 \leqslant p < \infty)$ and "anything".

3. Finally, I would like to point out what I think is a most important problem, or rather an entire research program: to extend the classical theory or interpolation and approximation of analytic functions a) to harmonic vector fields (let us recall that a vector field U, in $\mathbb{R}^3$, for the sake of definiteness, is harmonic iff curl $U = 0 =$ div U; this is one generalization of the Cauchy-Riemann system, the one advocated by M. Riesz, for instance), b) to generalized analytic functions in the sense of Vekua (that is solutions f of the generalized Cauchy-Riemann equation $\partial f/\partial \bar{z} = Af + B\bar{f}$, see [Ve]). About a) I know a little (see [JP]), about b) next to nothing (perhaps such questions have been studied for generalized analytic functions, I do not know). Eventually one would like to be able to treat general elliptic p.d.e. (Cf. in this connection Shapiro's report to this meeting.) To be specific, let me mention the following analogue of the classical Carathéodory-Fejér problem (which amounts to minimizing the norm $\|f\|_\infty$ in the disk U, say, over all analytic functions f with a given Taylor expansion (up to a given order) at some point, usually the origin). Consider e.g. the model case of a strip in $\mathbb{R}^3$ (in higher dimension than two the dependence on the domain is a more sensitive issue,

because the absence of a Riemann mapping theorem, cf. [JP]). To find a harmonic vector field U in this strip which at a given point inside has a given Taylor expansion (up to some order) and which on the "sides" of the strip satisfies given bounds, in terms of the sup-norm of some other norm. In [JP] we were lead to such questions in connection with an attempt to give a several real variable extension of the classical Riesz-Thorin procedure.

ACKNOWLEDGEMENT. In preparing this list I have profited from the expert knowledge of the following mathematicians present at the meeting: -1) P. Halmos, 0) T. Koornwinder, L. Szidő, B.Sz.-Nagy, 1) & 3) H.S. Shapiro.

[AF] Arazy, J. - Fisher, S.D., The uniqueness of the Dirichlet space among Möbius invariant Hilbert-spaces. Technion Report Series No. MT-593, Haifa 1983.

[BH] Beurling, A. - Helson, H., Fourier-Stieltjes transforms with bounded powers. Math. Scand. 1 (1953), 120 - 126.

[D] Dixmier, J., Les moyennes invariantes dans les semi-groupes et leurs applications. Acta Sci. Math. (Szeged) 12 (1950), 213 - 227.

[G] Greenleaf, F., Invariant Means on Topological Groups. Van Nostrand, New York - Toronto - London - Melbourne, 1969.

[GJ] Garnett, J. - Jones, P., The distance in BMO to L^∞. Ann. of Math. (2) 108 (1978), 373 - 393.

[Ha] Halmos, P., Spectral approximants of normal operators. Proc. Edinburgh Math. Soc. 19 (1974), 51 - 58.

[JNP] Janson, S. - Nilsson, P. - Peetre, J., Notes on Wolff's note on interpolation spaces. Proc. London Math. Soc. (to appear).

[JP] Janson, S. - Peetre, J., Harmonic interpolation. Conference on Interpolation Spaces and Allied Questions of Analysis, Lund, Aug. 29 - Sept. 1, 1983 (in preparation).

[K] Kahane, J.-P., Séries de Fourier absolument convergentes. (Ergebnisse 50.) Springer Verlag, Berlin - Heidelberg - New York, 1970.

[KuS] Kunze, R.A. - Stein, E., Uniformly bounded representations and harmonic analysis of the 2×2 real unimodular group. Amer. J. Math. 82 (1960), 1 - 62.

[KuSIV] Kunze, R.A. - Stein, E., Uniformly bounded representations. IV. Adv. in Math. 11 (1973), 1 - 71.

[SzN] Sz.-Nagy, B., On uniformly bounded linear transformations in Hilbert space. Acta Sci. Math. (Szeged) 11 (1946), 152 - 157.

[VZ] Vasilescu, F.-H. - Zsidó, L., Uniformly bounded groups in finite W*-algebras. Acta Sci. Math. (Szeged) 36 (1974), 189 - 192.

[Ve] Vekua, I.N., Generalized Analytic Functions. Gosudarstv. Izdat. Fiz.-Mat. Lit., Moscow, 1959. [Russian.]

12. P. RÉVÉSZ: A Problem on Local Time

Let $f(t)$ $(0 \leq t \leq 1)$ be an arbitrary Lebesgue-measurable, square integrable function. If for any $-\infty < x < \infty$ and $0 \leq t \leq 1$ the limit

$$L(x,t,f) = L(x,t) = \lim_{\varepsilon \to 0} \frac{1}{2\varepsilon} \lambda\{s: 0 \leq s \leq t, |f(s) - x| \leq \varepsilon\}$$

exists (where λ is the Lebesgue measure), then we say that the local time of f exists and L is called the local time of f.

Let $\{\phi_k\}$ be a complete orthonormal system on $[0,1]$ and consider $S_n(x) = S_n(x,f) = \sum_{k=1}^n c_k \phi_k(x)$ where $c_k = \int_0^1 f(t)\phi_k(t)dt$. Our problem is: Having $S_n(x)$ how can we estimate the function $L(x,t)$. As an example, one can ask whether the sequence $L(x,t,S_n)$ does converge to $L(x,t)$ as $n \to \infty$ in some suitable norm. However, it is not clear at all that the best estimate of $L(x,t)$ based on S_n is $L(x,t,S_n)$.

The concept of local time is frequently used in the theory of probability. Especially the local time of the Wiener process $W(t)$ is widely studied. However in practice nobody can observe the path of a Wiener process. Instead of W we observe some estimate like $S_n(t)$ and we ask how we can estimate $L(x,t,W)$ knowing only $S_n(t.W)$ and not W itself.

The problem presented above is deterministic analogue of this stochastic problem.

Since the path functions of W are belonging to the $\frac{1}{2} - \varepsilon$ Lipschitz class for any $\varepsilon > 0$ with probability one, in the problem presented the case when f is belonging to this class seems to be especially interesting.

13. M. ROSENBLUM - J. ROVNYAK: Monotone Operator Functions of Two Intervals

Let f be a real valued Borel function on a Borel set $\Delta \subseteq (-\infty,\infty)$. We call f

a monotone operator function if whenever A,B are bounded selfadjoint operators on a Hilbert space H such that $\sigma(A)\subseteq\Delta$, $\sigma(B)\subseteq\Delta$, and $A\leqslant B$, then $f(A)\leqslant f(B)$. Monotone operator functions on an open interval are characterized by a famous theorem of Loewner [3].

Let $\Delta=(a,b)\cup(c,d)$, $-\infty<a<b<c<d<\infty$. It can be shown (Chandler [1]) that if f is a monotone operator function on Δ, then

$$f(x) = a + bx + \int_{\Delta^c}\left[\frac{1}{t-x}-\frac{t}{1+t^2}\right]d\mu(t)$$

on Δ, where a is real, $b\geqslant 0$, and μ is a nonnegative Borel measure on $\Delta^c=(-\infty,\infty)\setminus\Delta$ such that $\int_{\Delta^c}(1+t^2)^{-1}d\mu(t)<\infty$. It can further be shown that $\mu([b,c])=0$ (Chandler [1], Donoghue [2]), though the known proofs are rather indirect and the point seems obscure.

PROBLEM. Give a simple proof that $\mu([b,c])=0$.

[1] Chandler, J.D., Jr., Extensions of monotone operator functions. Proc. Amer. Math. Soc. 54 (1976), 221 - 224.

[2] Donoghue, W.F., Jr., Monotone operator functions on arbitrary sets. Proc. Amer. Math. Soc. 78 (1980), 93 - 96.

[3] Löwner, K., Über monotone Matrixfunktionen. Math. Z. 38 (1934), 177 - 216.

14. P.O. RUNCK: Beste L^1-Approximation

Die Bestimmung bester Konstanten in den Approximationssätzen von Jackson erzielt man mit Hilfe des folgenden L^1-Approximationsproblems:

Es sei $f\in L^1(0,2\pi)$ mit $f(x)=\alpha_o+\sum_{k=1}^{\infty}\alpha_k\cos(kx)$ wobei (α_k) für $k\geqslant n+1$ eine 3-fach monotone Nullfolge sei (d.h. $\alpha_k-3\alpha_{k+1}+3\alpha_{k+2}-\alpha_{k+3}\geqslant 0$). Dann ist $g(x)=\alpha_o+b_{n+1}+\sum_{k=1}^{n}(\alpha_k-b_{n+1-k}+b_{n+1+k})\cos(kx)$ mit $b_k=$ $=\sum_{m=o}^{\infty}(-1)^m\alpha_{k+(2m+1)(n+1)}$ beste L^1-Approximation mit der Abweichung $\|f-g\|_1=4\sum_{m=o}^{\infty}\frac{(-1)^m}{2m+1}\alpha_{(2m+1)(n+1)}$.

Eine analoge Aussage gilt für die Sinusentwicklung (vgl. z.B. Schönhage, Approximationstheorie, Berlin 1971, S. 175 ff).

Bei der Bestimmung von Konstanten im algebraischen Jacksonsatz wäre die Lösung des folgenden allgemeineren P r o b l e m s sehr hilfreich:

Gegeben sei $f \in L^1(0,2\pi)$ mit $f(x) = \sum_{k=-\infty}^{\infty} \alpha_k e^{ikx}$, wobei (α_k) eine reelle Folge mit geeigneten Monotonieeigneschaften sei (z.B. $(\alpha_{|k|})$ und $(\alpha_{-|k|})$ 3-fach monotone Nullfolgen). Gesucht ist die beste L^1-Approximation der Form $g(x) = \sum_{k=-n}^{n} b_k e^{ikx}$ und $\|f-g\|_1$.

Report on Problems posed at the Conferences of 1963 and 1971

Concerning the problems 1 and 2 posed by the author in "On Approximation Theory", ISNM, Vol. 5, Birkhäuser, Basel, 1964, p. 186, partial solutions are to be found in Runck, P.O., Über Konvergenzgeschwindigkeit linearer Operatoren in Banachräumen. Ibidem pp. 96 - 106.

Regarding problem 4, see Freud, G. - Knapowski, S., On linear processes of approximation (1). Studia Math. 23 (1963), pp. 105 - 112.

Concerning problem 6 posed by G.G. Lorentz in "Linear Operators and Approximation", ISNM, Vol. 20, Birkhäuser, Basel, 1972, p. 497, results for the case $k = 1$ were established in Stadler, S., Über 1-positive Operatoren. In: Linear Operators and Approximation II, ISNM, Vol. 25, Birkhäuser, Basel, 1974, pp. 391 - 403.

15. G. SCHMEISSER: A Generalization of Shannon's Sampling Theorem

Denote by H_τ the class of all functions which are holomorphic and of exponential type τ in the right half plane. A theorem of Carlson (see [1, chap. 9]) states:

If $f \in H_\tau$, $\tau < \pi$, and $f(n) = 0$ for $n = 0,1,2,\ldots$, then $f \equiv 0$.

In other words, a function $f \in H_\tau$ is uniquely determined by its values at the non-negative integers, provided $\tau < \pi$. It is therefore natural to ask for an interpolation of the form

$$(1) \qquad f(z) = \sum_{n=0}^{\infty} f(n) L_{\tau,n}(z), \qquad \operatorname{Re} z \geqslant 0.$$

However, there are reasons to conjecture:

(I) An interpolation formula of the form (1) cannot hold, even if the admissible functions f are restricted to $H_\tau \cap L^2\,[0,+\infty)$, where $\tau<\pi$.

In this connection it is interesting to mention that there exist formulae of the form

$$f(t) = \lim_{n\to\infty} \sum_{k=1}^{n} a_{kn}\, f(t-kT), \qquad T\in[0,1);$$

see [4].

Next, denote by E_τ the class of all entire functions of exponential type τ. According to the above theorem of Carlson every $f\in E_\tau$ is uniquely determined by its values at the integers, provided $\tau<\pi$. Here the sampling theorem of Whittaker-Kotel'nikov-Shannon [2] states that the formula

$$(2)\qquad f(z) = \sum_{n=-\infty}^{\infty} f(n)\,\frac{\sin\pi(z-n)}{\pi(z-n)}\,,\quad z\in\mathbb{C}$$

holds for every $f\in E_\pi \cap L^2(\mathbb{R})$. Gervais-Rahman-Schmeisser [3] developed a modified formula

$$(3)\qquad f(z) = \sum_{n=-\infty}^{\infty} f(n)\,\Lambda_{\tau,n}(z)$$

which requires $f\in E_\tau$ $(\tau<\pi)$ but even allows f to be unbounded on $\mathbb{R}$. However

$$(4)\qquad \int_{-\infty}^{\infty} \frac{\log^+|f(x)|}{1+x^2}\,dx<\infty$$

must hold as a necessary condition.

There are reasons to conjecture:

(II) A formula of the form (3) can hold at most for such $f\in E_\tau(\tau<\pi)$ which satisfy (4).

Note that if the conjecture (I) is false, then the same is the case with (II), for the following reason: Given $f\in E_\tau$ $(\tau<\pi)$, the restriction of $g(z):=e^{-\pi z}f(z)$ to the right half plane belongs to $H_\tau \cap L^2\,[0,+\infty)$. By (1) we would have

$$f(z) = \sum_{n=0}^{\infty} f(n)\, e^{\pi(z-n)} L_{\tau,n}(z)$$

for Re $z \geqslant 0$. Proceeding in a corresponding way in the left half plane we can obviously define functions $\Lambda_{\tau,n}$ on $\mathbb{C}$ such that (3) holds for all $f \in E_\tau (\tau < \pi)$.

[1] Boas, R.P., Entire Functions. Academic Press, New York 1954.

[2] Butzer, P.L., The Shannon sampling theorem and some of its generalizations; an overview. In: Constructive Funktion Theory, (Proc. Int. Conf. Varna, 1981; ed. D. Vačov) Sofia 1983, pp. 258 - 274.

[3] Gervais, R. - Rahman, Q.I. - Schmeisser, G., Sampling theorem and interpolation and approximation by entire functions of exponential type. Unpublished manuscript, Feb. 1983.

[4] Splettstößer, W., On the prediction of band-limited signals from past samples. Information Sciences 28 (1982), 115 - 130.

16. Bl. SENDOV: Polynomials with Minimal Modulus of Continuity

Let $n > k \geqslant 0$ be integers, $n + k$ even, and $\mathbb{P}_{n,k} = \{P \in H_n : E_k(P)_{C[-1,1]} = 1\}$. Find $P^* \in \mathbb{P}_{n,k}$ for which $\omega_{k+1}(P^*;[-1,1]) \leqslant \omega_{k+1}(P;[-1,1])$, $P \in \mathbb{P}_{n,k}$.

CONJECTURE. $P^*(x) = \lambda\, T_n(\alpha x) + p(x)$; $T_n(x) = \cos(n \arccos x)$, $\lambda = \text{const.}$, $\alpha > 1$, $p \in H_k$.

[1] Sendov, Bl., On the constants of H. Whitney. C.R. Acad. Bulgare Sci. 35 (1982), 431 - 434.

[2] Sendov, Bl., A new proof of the H. Whitney's Theorem. C.R. Acad. Bulgare Sci. 35 (1982), 609 - 611.

[3] Whitney, H., On functions with bounded n^{th} differences. J. Math. Pures Appl. (9) 36 (1957), 67 - 95.

17. H.S. SHAPIRO: The Gram Matrix of n Non-Negative Functions

PROBLEM. Given $n \geqslant 5$, what are (effictively verifiable) conditions on the numbers $a_{ij} (i,j = 1,2,\ldots,n)$ in order that there exist a measure space (X,μ)

and *non-negative* functions $f_i \in L^2(X;\mu)$, $i = 1,2,\ldots,n$ satisfying

$$(*) \int f_i f_j \, d\mu = a_{ij} \qquad (i,j = 1,2,\ldots,n) \ ?$$

More precisely: the conditions $a_{ij} = a_{ji}$, $a_{ij} \geq 0$, and $\| a_{ij} \|$ non-negative definite, are all obviously necessary. For $n \leq 4$ I can show they are also sufficient (in these cases we may take X to be a set with just n points) but I can show they are not sufficient for $n \geq 5$, and I ask for the further relations that $\{a_{ij}\}$ must satisfy so that $(*)$ is solvable (preferably) in the form of algebraic inequalities among the numbers a_{ij}).

REMARK. In my work upon this problem I encountered the following, which also seems of interest: given a quadratic form $\sum_{i,j=1}^{n} c_{ij} x_i x_j$ with $c_{ij} = c_{ji}$ real, what are necessary and sufficient conditions on the c_{ij} in order that this form be non-negative for all *non-negative* choices of $x_1, \ldots, x_n$? (Hopefully, one could give a finite system of algebraic inequalities among the numbers c_{ij} which would constitute a necessary and sufficient condition.)

An Inquality

PROBLEM. Let $f(z) = \sum_{j=0}^{m} a_j z^j$ and $g(z) = \sum_{j=0}^{n} b_j z^j$ be polynomials, and define sequences A_j, B_j, C_j by

$$f(z)^2 = \sum A_j z^j \ , \quad g(z)^2 = \sum B_j z^j \ , \quad f(z)g(z) = \sum C_j z^j \ .$$

If we assume both

$$\text{(i)} \quad a_j \geq 0,\ b_j \geq 0 \text{ all } j \ , \qquad \text{(ii)} \quad 1 \leq p \leq 2 \ ,$$

does there exist a finite constant K_p, depending only on p, such that

$$(*) \qquad (\sum C_j^p)^2 \leq K_p (\sum A_j^p)(\sum B_j^p) \ ?$$

REMARKS. 1) I heard of the problem from James Clunie, who attributes it to Finnbarr Holland. 2) The inequality $(*)$ is easy to prove for $p = 1$ and 2 (with $K_1 = K_2 = 1$). 3) It is easy to show that for positive p outside the range

[1,2] (*) cannot hold. 4) It is fairly easy to show that if, for some p, (*) holds it must hold for $K_p = 1$.

Thus the problem can be formulated: under assumptions (i) and (ii) does there hold

$$(**) \qquad (\sum C_j^p)^2 \leqslant (\sum A_j^p)(\sum B_j^p) \quad ?$$

5) It is easy to show that (**) holds in the special case $g \equiv 1$, even if (ii) is weakened to read $1 \leqslant p < \infty$. 6) It is fairly easy to show that for each p, $1 \leqslant p < 2$, (**) fails to hold if assumption (i) is replaced by the weaker assumption that a_j, b_j are real (then A_j^p, B_j^p, C_j^p are to be replaced in (**) by $|A_j|^p$, $|B_j|^p$, $|C_j|^p$ of course). 7) An interesting special case of (**) arises if we choose $g(z) = z^m f(1/z)$. Then f^2 and g^2 have the same ℓ^p norm and (**) asserts: the ℓ^p norm of the trigonometric polynomial $|f(e^{i\theta})|^2$ does not exceed that of the trigonometric polynomial $f(e^{i\theta})^2$.

Since ℓ^p norms are invariant with respect to translation of the coefficient sequence, it is not essential here that f be a Taylor polynomial, and we may ask:

Let $T = \sum_{-m}^{m} c_k e^{ik\theta}$ be a trigonometric polynomial with $c_k \geqslant 0$. Is it true that for $1 \leqslant p \leqslant 2$, the ℓ_p norm of $|T(\theta)|^2$, cannot exceed that of $T(\theta)^2$?

In this formulation the two norms are equal for $p = 1,2$. The answer is known to be "yes" when $n = 1$. Finally, it can be remarked that other interesting questions can be raised regarding the coefficients of $|T(\theta)|^2 = \sum \alpha_k e^{ik\theta}$ and those of $T(\theta)^2 = \sum \beta_k e^{ik\theta}$. For example, do we have $\sum \beta_k^p \leqslant \sum \alpha_k^p$ for $p > 2$, and for $0 < p < 1$?

Also, is it true, that the smallest positive coefficient of $|T(\theta)|^2$ is at least as large as the smallest positive coefficient of $T(\theta)^2$? I (and others) have amassed a fair amount of numerical evidence favoring an affirmative reply to all the above questions.

18. R. SHARPLEY: Interpolation of W_1^1 and W_∞^1

It is known [see Calderón, A.P., Studia Math. 26 (1966), 273 - 299] that the interpolation spaces of $L^1(\mathbb{R})$ and $L^\infty(\mathbb{R})$ are the rearrangement invariant function spaces X. Define the Sobolev space W(X) using the norm $\|f\|_{W(X)} = \|f\|_X + \|f'\|_X$ (in particular, $W(L^p) = L_p^1$, $1 \leq p \leq \infty$). By the work of DeVore and Scherer (Ann. of Math. (2) 109 (1979), 583 - 599) it is known that each W(X) is an interpolation space for W_1^1 and W_∞^1.

QUESTION. Can each interpolation space of W_1^1 and W_∞^1 be described in this way?

19. V. TOTIK: Approximation in the Unit Disk

Let U be the unit disk in the complex plane, and A the set of functions that are analytic in U and continuous on the closure of U. Suppose that $\alpha_{n,k} \in U$, $k = 1,2, \ldots, n; n = 1,2, \ldots$, $\alpha_{n,k'}$ if $k \neq k'$.

CONJECTURE. If $\Lambda_n = \sum_{k=1}^{n}(1-|\alpha_{n,k}|)$ is bounded, then there are no polynomials $P_{n,k}$ such that for every $f \in A$

$$U_n(f,Z) = \sum_{k=1}^{n} f(\alpha_{n,k})\, P_{n,k}(Z) \to f(Z) \qquad (n \to \infty)$$

uniformly in $Z \in U$.

REMARKS. 1) The problem makes sense for H^p functions and H^p norm approximation.

2) If $\Lambda_n \to \infty$, then there are U_n such that $\|U_n(f)-f\|_B \to 0$, $n \to \infty$, $f \in B$, where $B = A$ or $B = H^p (1 \leq p < \infty)$.

3) If $\alpha_{n,k}$ are on the unict circle, i.e. $\Lambda_n = 0$, then the answer is negative for the existence of $\{U_n\}$. This is a result of G. Somorjai [On discrete linear operators in the function space A. In: Constructive Function Theory, Proc. Int. Conf. Blagoevgrad, 1977; (Eds. Bl. Sendov and D. Vačov), Sofia 1980, pp. 489 - 496].

Alphabetical List of Papers

Mathematics Subject Classification Numbers*

* According to the 1980 Math. Subject Classification of Mathematical Reviews and Zentralblatt für Mathematik. Classification numbers were given by the authors. Numbers following subjects indicate the first page of the respective paper.

1) According to the 1970 Math. Subject Classification

Key Words and Phrases*

* Given by the authors; numbers indicate first page of respective paper.